Sustainable Tourism

ASPECTS OF TOURISM TEXTS
Series Editors: Chris Cooper *(Leeds Beckett University, UK)*,
C. Michael Hall *(University of Canterbury, New Zealand)* and
Dallen J. Timothy *(Arizona State University, USA)*

This new series of textbooks aims to provide a comprehensive set of titles for higher level undergraduate and postgraduate students. The titles will be focused on identified areas of need and reflect a contemporary approach to tourism curriculum design. The books are specially written to focus on the needs, interests and skills of students and academics. They will have an easy-to-use format with clearly defined learning objectives at the beginning of each chapter, comprehensive summary material, end of chapter review questions and further reading and websites sections. The books will be international in scope with examples and cases drawn from all over the world.

All books in this series are externally peer-reviewed.

Full details of all the books in this series and of all our other publications can be found on http://www.channelviewpublications.com, or by writing to Channel View Publications, St Nicholas House, 31–34 High Street, Bristol BS1 2AW, UK.

ASPECTS OF TOURISM TEXTS: 6

Sustainable Tourism
Principles, Contexts and Practices

David A. Fennell and Chris Cooper

CHANNEL VIEW PUBLICATIONS

Bristol • Blue Ridge Summit

DOI https://doi.org/10.21832/FENNEL7666
Library of Congress Cataloging in Publication Data
A catalog record for this book is available from the Library of Congress.
Names: Fennell, David A. – author. | Cooper, Chris – author.
Title: Sustainable Tourism: Principles, Contexts and Practices / David A. Fennell and Chris Cooper.
Description: Bristol; Blue Ridge Summit: Channel View Publications, [2020] | Series: Aspects of Tourism Texts: 6 | Includes bibliographical references and index. |Summary: "This new textbook provides a comprehensive overview of sustainable tourism framed around the UN's sustainable development goals. It examines the origins and dimensions of sustainable tourism and offers a detailed account of sustainable initiatives and management across destinations, the tourism industry, public sector and leading agencies"— Provided by publisher.
Identifiers: LCCN 2019042230 (print) | LCCN 2019042231 (ebook) | ISBN 9781845417666 (hardback) | ISBN 9781845417659 (paperback) | ISBN 9781845417673 (pdf) | ISBN 9781845417680 (epub) | ISBN 9781845417697 (kindle edition) Subjects: LCSH: Sustainable tourism. | Sustainable development.
Classification: LCC G156.5.S87 F46 2020 (print) | LCC G156.5.S87 (ebook) | DDC 910.68/4—dc23 LC record available at https://lccn.loc.gov/2019042230
LC ebook record available at https://lccn.loc.gov/2019042231

British Library Cataloguing in Publication Data
A catalogue entry for this book is available from the British Library.

ISBN-13: 978-1-84541-766-6 (hbk)
ISBN-13: 978-1-84541-765-9 (pbk)

Channel View Publications
UK: St Nicholas House, 31–34 High Street, Bristol BS1 2AW, UK.
USA: NBN, Blue Ridge Summit, PA, USA.

Website: www.channelviewpublications.com
Twitter: Channel_View
Facebook: https://www.facebook.com/channelviewpublications
Blog: www.channelviewpublications.wordpress.com

The policy of Multilingual Matters/Channel View Publications is to use papers that are natural, renewable and recyclable products, made from wood grown in sustainable forests. In the manufacturing process of our books, and to further support our policy, preference is given to printers that have FSC and PEFC Chain of Custody certification. The FSC and/or PEFC logos will appear on those books where full certification has been granted to the printer concerned.

Typeset by Nova Techset Private Limited, Bengaluru and Chennai, India.

CONTENTS

FIGURES AND TABLES

FIGURES

TABLES

ACKNOWLEDGEMENTS

David would like to thank Sam Olson, Dustin Oostendarp and Emily Bowyer for their help in tracking down some of the sources used in this book.

CHAPTER 1

INTRODUCTION TO SUSTAINABILITY

LEARNING OUTCOMES

1. To explore the history and evolution of the term 'sustainable development'.

2. To introduce the UN's 17 Sustainable Development Goals.

3. To provide an introduction to the fundamentals of the tourism industry, and illustrate why sustainable development is so important in tourism.

4. To link tourism with consumerism and consumption.

5. To explore some of the history of concern in tourism around the social, economic and environmental impacts of the tourism industry.

INTRODUCTION

There is no shortage of popular culture articles and blogs on the importance of travelling sustainably in the new age of tourism. The titles of these articles include 'Open road, clear conscience' (Bures, 2006) and 'The conscientious tourist' (Wagner, 2005). Because of the multitude of problems with innumerable people travelling and the inequitable relationship that exists between the haves and the have-nots, it is natural for some – in fact many – to speculate, as Bures (2006) does, on questions such as: How do we travel ethically? How do we know if we are doing any harm? Are we destroying the coral reefs, ancient ruins and natural wonders in a rush to see them? Are we supporting corrupt regimes by spending our money? Does the massive global warming effect of airline travel outweigh even the greenest ecotour?

Consumers are starting to demand greener products (Weeden & Boluk, 2014; Wight, 1994a); these demands have become a key challenge for the tourism industry now and into the future, with many arguing that this is our most important challenge. How can we accommodate demand for more sustainable, greener or more ethical products and experiences, while at the same time being commercially successful? At what expense do we strive for commercial success given the degree and magnitude of impacts?

These questions, and many more, rest at the heart of tourism and sustainability. We can argue that sustainable development (SD) continues to be such a global force because it takes an intermediary position on a continuum between a strict anthropocentric approach to human agency and more radical or deeper ecological views. SD appears to be the more rational and functional form of change, given the challenge that lies ahead in meeting present needs while taking into consideration the needs of future generations. As such, SD and sustainable tourism (ST) are important because they represent our best chance to move forward in manner that reconciles often competing entities and interests along economic, sociocultural and ecological lines.

This chapter discusses the origins and rationale of sustainable development along with the themes, principles and goals of SD. The focus moves into a preliminary discourse of sustainable tourism through a discussion on the nature of tourism, the nature of consumption and consumerism, and then on to the structure and conceptual framework that guides the rest of the book.

SUSTAINABLE DEVELOPMENT

ORIGINS AND RATIONALE

Sustainable development is not an idea that emerged instantaneously. It is the product of several years of questions around the impacts that development had on both planetary resources and human groups. Stabler (1996) argues that the new conservation movement of the 1950s and early 1960s was the precursor to the principles and practices that we recognize as SD today. Other important catalysts include three watershed meetings, all in 1972, which tabled issues tied to human use and damage to the natural world. These meetings included: The United Nations (UN) Conference on the Human Environment in Stockholm (UNEP, 1972); publication of the Club of Rome's report, *The Limits of Growth* (Meadows *et al.*, 1972); and UNESCO's (UN Educational, Scientific and Cultural Organization) Convention Concerning the Protection of the World Cultural and Natural Heritage (Dangi & Jamal, 2016).

These meetings translated into an expanding ecodevelopment literature which was central in terms of defining the basic structure of SD. Two early advocates that helped build this structure were Miller (1976, 1978) and Riddell (1981), both of whom captured the essence of the dissatisfaction taking place from sociocultural, economic and ecological perspectives. Ecodevelopment was premised on:

1. enlarging the capacity of individuals to fulfil the desire to be useful and wanted, thereby dignifying labour-intensive and socially directed efforts of environmentally non-degrading kinds;
2. expanding the capacity of communities to be self-sufficient, thereby leading to the replenishment of renewable resources and the careful use of non-renewable resources; and
3. enhancing the fairness and justice of society, in environmental terms, avoiding wasteful consumption.

The term 'sustainable development' was coined in 1980 by the NGO, the International Union for the Conservation of Nature and Natural Resources, through its publication *World Conservation Strategy* (Crabbé, 1998; Krueger, 2017). The Brandt Report, also of 1980, stressed the need for development to include 'care for the environment'. The World Commission on Environment and Development's (WCED, 1987) *Our Common Future* helped the term 'sustainability' gain more common usage and acceptance (Archer, 1996), by characterizing SD as a process of change in which the exploitation of resources, the direction of investments, the orientation of technological development and institutional changes are made consistent with present as well as future needs. Put succinctly, SD refers to development that meets the needs of present populations without compromising the ability of future generations to meet their own needs (WCED, 1987). It is based on the idea that economic growth should occur in a more ecologically responsible and socially equitable manner. While there were only a few definitions of sustainable development in the 1980s, by the mid-1990s there were over 300 (Dobson, 1996), attesting to how firmly academics, government and industry embraced SD during that period of time.

Some of the earliest scholarly papers on SD often used terms interchangeably. For example, Barbier's (1987) comprehensive treatment of 'sustainable economic development' (SED) also included use of the term 'sustainability' to make reference to the same set of themes. He argued that the primary aim of SED is 'to provide lasting and secure livelihoods that minimize resource depletion, environmental degradation, cultural disruption, and social stability' (Barbier, 1987: 109). An important addition to Barbier's stance is maximizing goals across biological, economic and social systems through dynamic and adaptive processes of trade-offs, and that sustainability needs to be applied across all types of economic and social activities, including forestry, agriculture and fisheries.

Other scholars argued that SD ought to be grounded in pre-established knowledge domains. For example, Nelson (1992) suggested that the theoretical basis of SD fits within heritage and human ecology. Heritage, Nelson suggests, refers to all the objects that come to us from the past, including culture, flora, fauna, language and institutions. Human ecology comprises all 'economic, technical, social and cultural ways in which human beings in different societies and places have been influenced by and have influenced the world around them' (Nelson, 1992: 7; see also du Cros, 2001). Porter (1978) characterizes human ecology as the use of ecological systems that emphasize equilibrium, balance, homeostasis and feedbacks, to explain the interconnectedness between people and environments. It is more effectively applied at scales that are small and in situations that are not overly complex. It is therefore more appropriate for small systems and cultures which are more manageable. Human ecology is a reflection of the belief that human actions are an expression of culture, and this includes human cultural actions that transform the natural world (Leighly, 1987). Berkes (1984) implemented Hardesty's (1975) theoretical argument around the notion of different human cultural groups as distinct cultural species, in the case of competition between commercial and sport fisherman on Lake Erie, Canada. Berkes argues that with any limiting resource such as food or space, competition will emerge when ecological niches overlap in a similar habitat. What Berkes found, however, was that competition was more in line with perceived conflict, with real competition not taking place at all.

Human ecology studies have also been completed in tourism contexts. De Castro and Begossi (1996) investigated local and recreational (tourist) fishermen on the Rio Grande in Brazil. While local and recreational anglers typically fish for different species in different parts of the river using different equipment at many points during the year, conflict emerges in the wet season when fish stocks are lower. In cases where there is a limited resource, conflict and competition is usually imminent. In the case of Rio Grande, it is the use of different fishing strategies between both groups that appears to be sufficient to avoid inter-group conflict. Fennell and Butler (2003) used Budowski's (1976) ideas around tourism and conservation as conflict, co-existence or symbiosis to argue that different tourism stakeholders place varying levels of pressure on each other and the natural world (see also Murphy's, 1983, ecological model of tourism development). Fennell and Butler argued that these relationships might be articulated as predatory (e.g. highly disproportionate number of tourists to locals), competitory (e.g. displacement or interference with other local activities or resources), neutral (e.g. low to negligible nature of tourism impacts), or symbiotic (e.g. locals are open and enthusiastic, xenophilic, to the arrival of and interaction with tourists).

WEAK AND STRONG CATEGORIES OF SUSTAINABILITY

Theorists have also placed considerable effort into demonstrating that there are weak and strong categories of sustainability. Fyall and Garrod (1997) illustrate that academics are generally divided into one of these two weak and strong sustainability camps. The weak sustainability camp argues that the form in which capital stocks (i.e. manmade capital, human capital, cultural capital and natural capital) are passed along to future generations is relatively unimportant. Different forms of capital may be substituted for others. By contrast, the strong sustainability camp argues that capital stocks are not open to substitution. For example, natural capital must be conserved in such a way as to pass it along in its original or authentic state. Manmade substitutes for the 'real thing' would not be acceptable in any form.

The dichotomy between weak and strong sustainability is often more formally structured according to a spectrum. This is evident in Hunter's (1997) work, which is an adaptation of the work of Turner *et al.* (1994). Table 1.1 shows that the very weak sustainability position is characterized by an anthropocentric and utilitarian approach. By contrast, the opposite end of the spectrum, very strong sustainability, is anchored by a bioethical and ecocentric approach.

THEMES, PRINCIPLES AND GOALS OF SUSTAINABLE TOURISM

Beyond a focus on definition, Mitchell (1994: 190) provides a concise set of fundamental themes that he used to define the parameters around SD. These include the need to:

Table 1.1 A simplified description of the sustainable development spectrum

Sustainability position	*Defining characteristics*
Very weak	Anthropocentric and utilitarian; growth oriented and resource exploitative; natural resources utilized at economically optimal rates through unfettered free markets operating to satisfy individual consumer choice; infinite substitution possible between natural and human-made capital; continued wellbeing assured through economic growth and technical innovation.
Weak	Anthropocentric and utilitarian; resource conservationist; growth is managed and modified; concern for distribution of development costs and benefits through intra- and intergenerational equity; rejection of infinite substitution between natural and human-made capital with recognition of some aspects of natural world as critical capital (e.g. ozone layer, some natural ecosystems); human-made plus natural capital constant or rising through time; decoupling of negative environmental impacts from economic growth.
Strong	(Eco)systems perspective; resource preservationist; recognizes primary value of maintaining the functional integrity of ecosystems over and above secondary value through human resource utilization; interests of the collective given more weight than those of the individual consumer; adherence to intra- and inter-generational equity; decoupling important but alongside a belief in a steady-state economy as a consequence of following the constant natural assets rule; zero economic and human population growth.
Very strong	Bioethical and ecocentric; resource preservationist to the point where utilization of natural resources is minimized; nature's rights or intrinsic value in nature encompassing non-human living organisms and even abiotic elements under a literal interpretation of Gaianism; anti-economic growth and reduced human population.

Source: Hunter (1997), adapted from Turner *et al.* (1994)

(1) satisfy basic human needs; (2) achieve equity and social justice; (3) provide for social self-determination and cultural diversity; (4) maintain ecological integrity and biodiversity; and (5) integrate environmental and economic considerations.

Forman (1990) illustrates that SD has four key characteristics. These include: a time period of several generations; adaptability and change in ecological and human systems; slowly changing foundation variables that are marked by irregular cycles; and mosaic stability, which permits ongoing rapid fluctuations within spatial units. While time period and adaptability in social and ecological cycles are easily interpreted, the latter two characteristics need further explanation. Slowly changing foundation variables that regulate SD include soil, biological diversity, biological productivity, freshwater and marine water from the ecological side of things, while human variables include basic human requirements of food, water, health and housing as well as fuel, and cultural cohesion and diversity. Mosaic stability refers to the fact that each landscape is a mosaic where ecosystems and human uses are fairly consistent throughout. There are different scales that may be used to characterize ecological and social conditions, e.g. New England in the USA, and these have implications for the implementation of SD parameters. Forman notes that sustainability at the biosphere scale has effects on finer scales as illustrated in hierarchy theory (Clark, 1985). The probability of attaining sustainable outcomes is very much dependent on the issue of scale.

Several organizations have developed sets of guiding principles as to how SD ought to be conceived in policy and practice. They reflect the broad mandate of SD to contribute to a better quality of life now and in the future while being mindful of economic, ecological and sociocultural priorities. Examples of these organizations include: governments such as the province of Prince Edward Island, Canada (Government of PEI Environmental Advisory Council, 2013), and the province of Quebec, Canada (Government of Quebec, 2018; see Box 1.1); universities including the University of California at San Diego (UC San Diego, 2018); academic books (Barbiroli, 2009); and international institutions. The Rio Declaration on Environment and Development (UNESCO, 1992) is built around 27 principles.

Box 1.1 The principles of Québec's Sustainable Development Act

a. *Health and quality of life*: People, human health and improved quality of life are at the centre of sustainable development concerns. People are entitled to a healthy and productive life in harmony with nature.

b. *Social equity and solidarity*: Development must be undertaken in a spirit of intra- and inter-generational equity and social ethics and solidarity.

c. *Environmental protection*: To achieve sustainable development, environmental protection must constitute an integral part of the development process.

d. *Economic efficiency*: The economy of Québec and its regions must be effective, geared towards innovation and economic prosperity that is conducive to social progress and respectful of the environment.
e. *Participation and commitment*: The participation and commitment of citizens and citizens' groups are needed to define a concerted vision of development and to ensure its environmental, social and economic sustainability.
f. *Access to knowledge*: Measures favourable to education, access to information and research must be encouraged in order to stimulate innovation, raise awareness and ensure effective participation of the public in the implementation of sustainable development.
g. *Subsidiarity*: Powers and responsibilities must be delegated to the appropriate level of authority. Decision-making centres should be adequately distributed and as close as possible to the citizens and communities concerned.
h. *Inter-governmental partnership and cooperation*: Governments must collaborate to ensure that development is sustainable from an environmental, social and economic standpoint. The external impact of actions in a given territory must be taken into consideration.
i. *Prevention*: In the presence of a known risk, preventive, mitigating and corrective actions must be taken, with priority given to actions at the source.
j. *Precaution*: When there are threats of serious or irreversible damage, lack of full scientific certainty must not be used as a reason for postponing the adoption of effective measures to prevent environmental degradation.
k. *Protection of cultural heritage*: The cultural heritage, made up of property, sites, landscapes, traditions and knowledge, reflects the identity of a society. It passes on the values of a society from generation to generation, and the preservation of this heritage fosters the sustainability of development. Cultural heritage components must be identified, protected and enhanced, taking their intrinsic rarity and fragility into account.
l. *Biodiversity preservation*: Biological diversity offers incalculable advantages and must be preserved for the benefit of present and future generations. The protection of species, ecosystems and the natural processes that maintain life is essential if quality of human life is to be maintained.
m. *Respect for ecosystem support capacity*: Human activities must be respectful of the support capacity of ecosystems and ensure the perenniality of ecosystems.
n. *Responsible production and consumption*: Production and consumption patterns must be changed in order to make production and consumption more viable

and more socially and environmentally responsible, in particular through an ecoefficient approach that avoids waste and optimizes the use of resources.

o. *Polluter pays*: Those who generate pollution or whose actions otherwise degrade the environment must bear their share of the cost of measures to prevent, reduce, control and mitigate environmental damage.

p. *Internalization of costs*: The value of goods and services must reflect all the costs they generate for society during their whole life cycle, from their design to their final consumption and their disposal.

Source: Government of Quebec (2018)

The United Nations Sustainable Development Goals

The evolution of SD is evident in the move away from principles and on to more formal and comprehensive documents that urged stakeholders to pursue specific goals and targets in achieving sustainable outcomes. The eight Millennium Development Goals (poverty and hunger; universal education; gender equality and empowerment of women; reduction of child mortality; improvements in maternal health; combating diseases like HIV; ensuring environmental sustainability; and achieving global partnership for development), developed by the UN in 2000 and in effect until 2015, was the first such approach along these lines.

The new agenda for SD at the global level for the post-2015 period is based on the UN's Sustainable Development Goals. As a replacement for the Millennium Development Goals which expired in 2015, the UN's *Transforming our World: The 2030 Agenda for Sustainable Development* (UN, 2015; https://sustainabledevelopment.un.org/post2015/transformingourworld) outlines a set of 17 goals which encapsulate the most timely social, economic and ecological issues that are most pressing in our efforts to achieve a sustainable future (Table 1.2). Nested within these 17 goals are 169 targets and a set of 330 indicators to be reached by the year 2030. According to the UN Environment Programme:

> The Sustainable Development Goals (SDGs), otherwise known as the Global Goals, are a universal call to action to end poverty, protect the planet and ensure that all people enjoy peace and prosperity.
>
> These 17 Goals build on the successes of the Millennium Development Goals, while including new areas such as climate change, economic inequality, innovation, sustainable consumption, peace and justice, among other priorities. The goals are interconnected – often the key to success on one will involve tackling issues more commonly associated with another. (UNEP, 2018)

Table 1.2 The UN Sustainable Development Goals

Goal no.	*Theme*	*Description*
1	Poverty	End poverty and all its forms everywhere
2	Zero hunger	End hunger, achieve food security and improved nutrition and promote sustainable agriculture
3	Good health and wellbeing	Ensure healthy lives and promote wellbeing for all at all ages
4	Quality education	Ensure inclusive and equitable quality education and promote lifelong learning opportunities for all
5	Gender equality	Achieve gender equality and empower all women and girls
6	Clean water and sanitation	Ensure availability and sustainable management of water and sanitation for all
7	Affordable and clean energy	Ensure access to affordable, reliable, sustainable and modern energy for all
8	Decent work and economic growth	Promote sustained, inclusive and sustainable economic growth, full and productive employment and decent work for all
9	Industry, innovation and infrastructure	Build resilient infrastructure, promote inclusive and sustainable industrialization and foster innovation
10	Reduced inequalities	Reduce inequality within and among countries
11	Sustainable cities and communities	Make cities and human settlements inclusive, safe, resilient and sustainable
12	Responsible consumption and production	Ensure sustainable consumption and production patterns
13	Climate action	Take urgent action to combat climate change and its impacts
14	Life below water	Conserve and sustainably use the oceans, seas and marine resources for sustainable development

15	Life on land	Protect, restore and promote sustainable use of terrestrial ecosystems, sustainably manage forests, combat desertification, halt and reverse land degradation and halt biodiversity loss
16	Peace, justice and strong institutions	Promote peaceful and inclusive societies for sustainable development, provide access to justice for all and build effective, accountable and inclusive institutions at all levels
17	Partnerships for the goals	Strengthen the means of implementation and revitalize the global partnership for sustainable development

Source: UN (2018a)

One of the most impressive aspects of such an ambitious process has been articulated by UN Secretary General Ban Ki-moon, who stated: 'I want this to be the most inclusive development process the world has ever known' (UNDP, 2013). Indeed, praise has emerged from several fronts on the consultation processes that unfolded in the creation of the SDGs. Global, national and regional consultations took place, and the involvement of many of the world's biggest players in business also played a part (Kharas & Zhang, 2014). It is in this realm of business involvement that the SDGs have been criticized. Several commentators have argued that even though the SDGs contend that governments, civil society and businesses have an equal say in moving forward with SD, the question remains: 'Can … profit-motivated businesses really make a meaningful contribution to achievement of the SDGs or are we likely to see "business as usual", which results in greater profits for some, and lost opportunities for many?' (Scheyvens *et al.*, 2016: 372). More specifically, these scholars are concerned over the dominant neoliberal discourse that pervades business, and the world, with a focus on short-time lines and on the bottom line, and manipulation of the terminology around SD for their own purposes, as well as partnerships. There is little assurance that social inequalities can be overcome because of power imbalances between developed and developing economies, unfair trade, debt, illicit financial flows of money, suggesting that sustainability will take place, but along the lines of the 'same old economic and social models' (Moore, 2015: 801) based on the same 'set of values and framing devices' (Moore, 2015: 806; Scheyvens *et al.*, 2016).

TOURISM AND SUSTAINABLE DEVELOPMENT

This section introduces the reader to some fundamental aspects of tourism and the tourism industry, including the definition of tourism, as well as tourism sectors and scales. The intent is to have the reader gain a sense of the complexity of tourism through recognition of the number of elements that are required to get tourists from their generating regions to their destinations, and all the services required to facilitate their experiences, and back home. There is a natural progression from this discussion onto consumption and consumerism in tourism. Tourists are one important example of the consumption ethos that is enveloping the world, which in turn has been stimulated by neoliberal forces that push people to consume more. The second part of this section moves into sustainable tourism and details more specifically its history and how it has advanced from definition through to principles and implementation challenges, and later to how ST is manifest in some forms of alternative tourism such as ecotourism.

FUNDAMENTALS OF THE TOURISM INDUSTRY

As a social science field, it is not inconceivable that tourism has been subject to a number of definitions and interpretations. Foremost among these discussions is the fundamental essence of tourism as an industry unto itself or composite of several other related subsectors (Simmons, 2013). Although dated, the following definition provides a good foundation for understanding the elements that comprise the entire tourism system, which involves

> the discretionary travel and temporary stay of persons away from their usual place of residence for one or more nights, excepting tours made for the primary purpose of earning remuneration from points en route. The elements of the system are tourists, generating regions, transit routes, destination regions and a tourist industry. These five elements are arranged in spatial and functional connections. Having the characteristics of an open system, the organization of five elements operates within broader environments: physical, cultural, social, economic, political, technological with which it interacts. (Leiper, 1979: 403–404)

This definition illustrates the complexity of the tourism system by recognizing that tourism operates in broader social and ecological systems, that it involves the movement of tourists from home through various transit routes and on to the destination, and that it involves the tourism industry. Leiper (1979) continues by identifying a number of firms, organizations and facilities that serve the needs of tourists as loosely representative of the industry. These include marketing, carriers, accommodation, attractions, miscellaneous services like duty free and specialty shops, and regulation by, for example, industry associations or government. Apart from the essential components of accommodation and carriers, different industry sectors may be emphasized more liberally because of the particular resources and geographical location of the destination. For example, in the far north of Canada the Department of Tourism and Culture of the Yukon Territory identifies the following industry sectors as priority areas for this region: accommodation; adventure tourism and recreation; attractions; events and conferences; food and beverage; tourism services (education, marketing and so on); transportation; and travel trade (e.g.

travel agents and tour operators) (Yukon Government, 2015). These services are especially important not only for tourists but also for the industry, which must compete regionally, nationally and internationally for a share of the market. It is often the case that these more peripheral destinations are characterized by the sheer number of small and medium-sized enterprises (SMEs), which poses a challenge because of the lack of skills and resources to attract tourist numbers. In fact, SMEs are critical to the fabric of the tourism industry everywhere. These companies represent 99% of the over 2.7 million businesses in European tourism, and more than 94% of these entrepreneurs are termed micro-operators because they employ fewer than 10 individuals (EU DG XXIII, 1998).

Identification and cooperation among a broad number of interrelated industry sectors in tourism is also important for success in regional development. This has prompted theorists to argue that SD ought to be, by its very nature, holistic and multi-sectoral. If we plan tourism according to a single-sector approach we are in for a load of trouble (Wall, 1996). And, as discussed by Butler (1998), getting ST right is also a function of scale. Since the impacts of tourism can only be assessed and managed in the context of a specific defined region, so too can sustainability only be comprehended and defined relative to a specific destination. Different constructs of SD and ST are likely to be necessary at local, regional, state, national and international levels, thus offering a whole new debate about the actual meanings of these terms (Hall & Lew, 1998).

The amorphous nature of tourism in view of scale, geographical space and the number of stakeholders operating in physical, cultural, social, economic, political and technological environments presents many challenges to tourism's management along sustainability lines. The tourism industry has often been measured from a demand or market perspective, which means that it is the consumer who can heavily influence the type, quality and quantity of products and services that define the breadth of tourism experiences at a destination. Demand-driven approaches have also been used in reference to tourists with a smaller ecological footprint (Dolnicar *et al.*, 2008). Destinations that develop more cautiously may reap rewards from tourists who are in search of places that are untouched and protected, and where crowding has been limited through more sustainable policies (Brau, 2008).

CONSUMPTION AND CONSUMERISM

While consumption is an ecological necessity, i.e. we all have to consume in order to satisfy physiological needs, it is the nature of our consumption that is a question of some importance (Hall, 2011a) in our efforts to secure the good life (Fennell, 2018c; Lash & Urry, 1994). The Industrial Revolution and the swift evolution of capitalist society, especially the current neoliberal form of capitalism, are at the heart of society's predilection towards mass consumption. What has aided this utilitarian thinking is the recognition that, generally speaking, tourism knowledge often revolves around getting people to consume more. Hall (2011a) argues that this is evident in the field's most important international organizations. The UN World Tourism Organization (UNWTO) and the World Travel and Tourism Council (WTTC), for example, continue to spout statistics on the increasing magnitude of the tourism industry with the assumption that more travel is better, both personally and collectively – the more we consume the happier all of us will be. Yet, as Hall (2011a) further observes, there are questions as to why travel and tourism continue to be celebrated even as costs may soar well above benefits. The system, Hall (2011a) notes, serves a narrow band of political and economic interests. It is indeed troubling that an individual act, when collectively measured, should have such a tremendous impact on so many places. Hall (2011a) argues that this mindset flies in the face of research on subjective wellbeing. While consumption is often theorized around identity, lifestyle and quality of life (indeed, we are bombarded by these messages daily through the media), more money – and therefore more things – will in the end not make us any happier. There is a wealth of research that makes this abundantly clear (Etzioni, 2003; Myers, 2003). Fennell (2018a) made this point by sourcing the work of Victor Frankl (1985), who observed that the pursuit of pleasure is destroyed if made the primary goal of one's life. Pleasure and happiness ensue – they are by-products of our daily endeavours – and the more one directly pursues pleasure, the more one is bound to miss bits because of the inherent self-interest that is built into this mindset.

Still, there are pockets of examples where this may not hold to be true. Chi *et al.* (2017) found that residents in China with higher economic status were found to have higher subjective wellbeing in consideration of both cognitive and affective domains. They conclude that 'Residents with higher household income, better housing conditions, and a higher standard of living are more satisfied with their life and are happier' (Chi *et al.*, 2017: 217). These authors echo the concerns of Woo *et al.* (2015), who note that ST might only be achieved if residents have a higher degree of subjective wellbeing, which means, in turn, that there is some incentive or willingness to support tourism development from this subjective wellbeing perspective. This latter series of studies indicates that there is a level, basic as it may be, of secured needs that is required to make us happy.

Carrying on from Hall's (2011a) work, Higgins-Desboilles (2010) illustrates that we live in a world of two contrasting value systems in tourism. The first, termed corporatized tourism, is based on a culture ideology of consumerism and is inherently unsustainable, while the second is a direct alternative to the first (see Figure 1.1). One of the paths to transformation from the left-hand side to the right-hand side of Figure 1.1 is a change in consumer behaviour, according to Higgins-Desboilles (2010). Consumers must make meaningful changes to their consumption values in order to secure experiences that place cultures, societies and environments on a higher plane of moral consideration. This is reinforced by Leslie (1994), who argues that the lavish and wasteful consumerist lifestyle of the West needs a significant overhaul: norms, values and attitudes need to change away from the 'must have it' and 'consume today – discard tomorrow' ethos to an approach that places the interest of the natural world on a higher level.

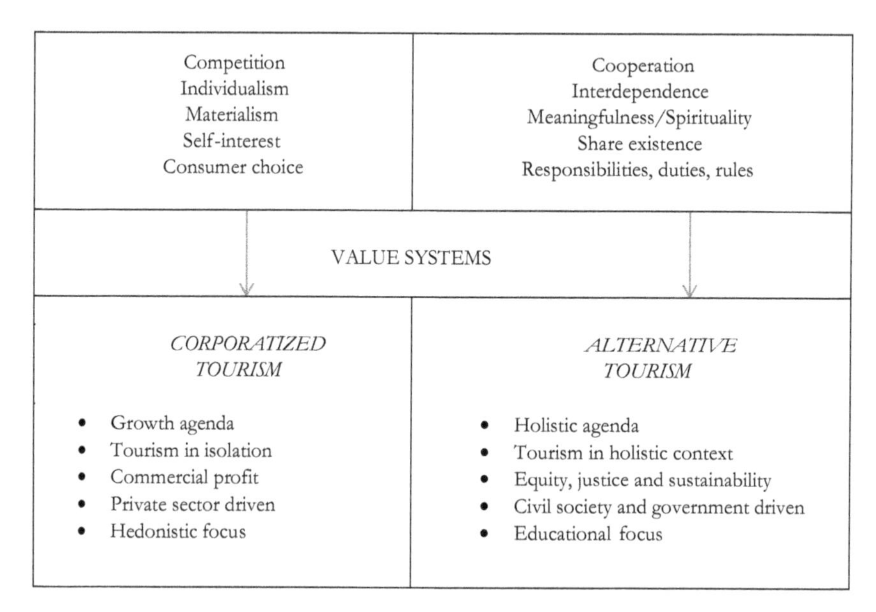

Figure 1.1 Contrasting value systems for the corporatized and alternative tourism paradigms
Source: After Higgins-Desboilles (2010)

Budeanu's (2007) work on this topic is important because it places into context many challenges and opportunities for the development of a more sustainable tourism industry. Budeanu found that, even though tourists expressed positive attitudes towards sustainability in tourism, very few matched their attitudes and intentions with actual behaviour – by purchasing responsible tourism products, choosing more environmentally friendly transportation options and adopting responsible travel behaviours at the destination. Budeanu argues that there are several tools that should be implemented in moving towards more sustainable travel behaviour. These tools include increasing the costs of environmentally destructive behaviour (and, one would expect, socially and culturally destructive behaviour), decreasing the costs of choosing good, environmentally friendly options, increasing education to tourists, educating tourists about environmentally destructive behaviours, and using resources in a more efficient manner.

Budeanu's (2007) work is substantiated in other studies. Using a willingness-to-pay methodology involving 1118 respondents visiting the western Costa del Sol (Spain), Pulido-Fernández and López-Sánchez (2016) found that only 26.6% of respondents were willing to pay more for this destination to become more sustainable. The authors also split the respondent pool into three different segments based on their 'sustainability intelligence': reflective tourists, unconcerned tourists and pro-sustainable tourists. Of the pro-sustainable segment, 97.8% indicated that they would pay more for the destination to become more sustainable than at present, but the vast majority (85.4%) said they would only pay 10% more for these changes: 4.4% said they would pay 20% more; 0.7% said they would pay 30% more; 0.7% said they would pay 40% more; and 2.2% said they would pay 50% more than the cost of their vacation.

Much of this behaviour is a function of the complexity behind self-interest and altruism, a topic discussed in Chapter 3. Miller (2003) suggests that, although tourists are making decisions based on social, cultural and environmental considerations in the purchase of tourism products, there is an element of selfishness built into this altruism. Following Ottman's (1993: 3) definition of green consumerism as 'individuals looking to protect themselves and their world through the power of their purchasing decisions', when tourism products are purchased the consequences of this decision are different from the purchase of other products. This is because tourism is produced and consumed at the destination, so ignoring reports about the impacts of tourism at a specific destination means that the tourist visiting such a place must be faced with this dilemma first hand. Consumers of other products, Miller claims, can 'switch-off' concern because these products are produced physically in other places, so the consumer will not see the environmental and social costs associated with production. And even though ST is 'the most prominent feature of contemporary tourism discourse' (Higgins-Desboilles, 2010: 116), it often does not figure prominently in marketing (McKercher, 2016). There continues to be a void between scholarly works on ST and how it is conceived and used in practice – but this is changing, as we shall see.

Other studies on tourism and consumption have emphasized the transition taking place in tourism away from resort-type experiences to more of a focus on individual experiences (Voase, 2007). Flexibility is now the norm, with tourists travelling differently in space and time. Voase's thesis is that, far from being an expression of pure individualism, tourism provides the opportunity for people to discover a new form of the collective. The key for this idea is found in Veblen's (1899) *Theory of the Leisure Class*, in which Veblen argues that consumers are not *imitating* others along the lines of 'keeping up with the Joneses', but rather *emulating* them, which lies at the heart of ownership and the expression of leisure. People are motivated to compete with or equal those that reside within their social domains, which leads to a type of uniformity within an emerging class structure. Tourism is a form of symbolic consumption in the same way as luxury goods like wine, nice cars, and jewellery have a social dimension attached to them (Hjalager, 1999).

In the same vein as Voase's (2007) work, research also indicates that the ways we consume in the present day are becoming more complex and fascinating (Sonnenburg & Wee, 2016). Tourism is an apt example of this complexity and fascination as newer forms of travel move away from the 'break-away from the everyday' form of consumption to styles that do not differentiate between travel and everyday life – tourism and life collapsing into one. These new leisure- or tourism-scapes create new lines of consumption in conscious and reflective ways in the production of meaningful experiences (Sonnenburg & Wee, 2016; see also Hughes, 1995; Humphreys & Grayson, 2008; Wilson, 1992). Accordingly, there are a growing number of consumers who are consciously moralizing their consumption practices (Nava, 1991) because of discontent around the pursuit of pleasure at all costs. Soper (2008) introduced the concept of alternative hedonism – or the need for consumers to gain pleasure in a different way, by purchasing products that are good for the environment and other people. These tourists willingly forego the many comforts of tourism to ensure that their money is contributing to a greater good.

Slow tourism and staycations are two examples that resonate with alternative hedonism. Slow tourism is characterized as a middle-class phenomenon, which emphasizes lower carbon forms of travel instead of air and car travel, shorter distances travelled, longer stays at the destination and fewer trips (Dickinson & Peeters, 2014). Staycations, or the practice of taking holidays at home, emerged during the summer of 2008 because of escalating petrol prices and a poor economy that scuttled the vacation plans of the suburban middle class in the USA (Molz, 2009). Pyke *et al.* (2016) found in a qualitative study (focus groups) of tourism investors that staycations and shorter holidays were increasing in the UK as a market trend, acting as a barrier to the promotion of wellness tourism, i.e. the short-stay visitation pattern of tourist arrivals acts as a constraint to the development of wellness tourism products. Others have found that the differences between participation in leisure activities while staying at home versus travelling away from one's home are marginal at best in the assessment of subjective wellbeing (de Bloom *et al.*, 2017).

HISTORY OF CONCERN IN TOURISM

Initial alarm over the impacts of tourism from a broad-based perspective intensified during the 1950s, when the International Union of Official Travel Organizations' (the precursor to the UNWTO) Commission for Travel Development started the critical discourse on how best to minimize destination impacts (Shackleford, 1985). Popular culture articles began to emerge in the 1950s and 1960s on the darker side of tourism, exposing a vulnerable underbelly of tourism as the so-called 'smokeless' industry – a moniker extended to tourism because of its seemingly benign effects. Examples include Acapulco (Cerruti, 1964), the Balearic Islands (Naylon, 1967) and Gozo (Jones, 1972). While these examples make reference to impacts taking place in destination regions of the 'South' or developing countries, similar questions were being asked in developed country contexts like the UK, where Harrington (1971) documented how unregulated hotel development led to a lower quality of life in London.

Scholars followed suit during the 1970s, at the dawn of the tourism studies field, through seminal pieces that critically examined the effects of tourism, especially on the environment. Budowski (1976), for example, described the relationship between tourism and conservation as coexistence moving towards conflict, while Krippendorf (1977) and Cohen (1978) discussed how poorly planned tourism and development led to environmental impacts. An important general theme moving through the 1980s into the 1990s and beyond has been captured by D'Sa (1999), who argued that 'today's form of tourism is highly exploitative and socially damaging. It is manifestly unethical and unjust to foist it on traditional communities' (D'Sa, 1999: 65–66).

The main culprit responsible for these impacts was said to be mass tourism. In the past, and just as relevant in the present day, there was a belief that mass tourism is almost entirely unsustainable because of its volume and magnitude, the seasonal nature of its employment, its non-local orientation of management and administration, and because of the unsavoury leakage of money out of the local economy. (See Butler, 1996, who argued that there is no overwhelming proof to suggest that mass tourism is always unsustainable, and that the biggest challenge in world tourism is in fact to make mass tourism more sustainable.) The dissatisfaction with mass tourism germinated new ways of thinking about tourism and new practices. Alternative tourism (Table 1.3) was designed simply as an alternative to mass tourism, creating two polar paradigms through new forms of tourism such as ecotourism, pro-poor tourism, fair trade tourism, green tourism, responsible tourism and justice tourism.

Others, however, called for more caution in accepting alternative tourism outright as the way forward. Cohen (1987) argued that alternative tourism still involves commercialization and changes to the customs of local people, and those seeking unspoiled nature are something of a paradox because they end up being a threat to nature as they penetrate deeper into the backcountry. Butler (1990) argued that it is hard to disagree with it in principle, and that every form of tourism, even alternative tourism, can be damaging to the environment. Wheeller (1992) followed suit by suggesting that many of the practical issues of the implementation of alternative tourism were not being considered, especially in policy documents, suggesting that caution and scepticism should go before outright acceptance.

An alternative form of tourism that continues to garner great support in industry and scholarship is responsible tourism (RT). Dictionary definitions of 'responsible' make reference to the need to respond, to be answerable and accountable, to be trustworthy and reliable or to have an obligation or duty towards these others. RT ought to be built around this foundation of responding to the needs of others as a duty (Fennell, 2008a) and so it should emphasize, for example, a more participatory and inclusive approach to community involvement in planning (Haywood, 1988). In their Development of Responsible Tourism Guidelines for South Africa, Goodwin *et al.* (n.d.) argue that RT requires a proactive approach to tourism planning and development, and it is underpinned by ST with respect to: assessing environmental impacts; using local resources sustainably; maintaining natural diversity; involving local communities in planning; ensuring that locals benefit; assessing social impacts; and maintaining and encouraging social and cultural diversity.

McLaren (2006) offers a much more detailed definition, stating that responsible tourism:

- generates greater economic benefits for local people, enhances the wellbeing of host communities, improves working conditions and access to the industry;
- involves local people in decisions that affect their lives and life chances;

Table 1.3 Alternative tourism

Accommodation • Does not overwhelm the community • Benefits (jobs, expenditures) are more evenly distributed • Less competition with homes and businesses for the use of infrastructure • A larger percentage of revenues accrues to local areas • Greater opportunity for local entrepreneurs to participate in the tourism sector
Attractions • Authenticity and uniqueness of community is promoted and enhanced • Attractions are educational and promote self-fulfilment • Locals can benefit from existence of the attractions even if tourists are not present.
Market • Tourists do not overwhelm locals in numbers; stress is avoided • 'Drought/deluge' cycles are avoided, and equilibrium is fostered • A more desirable visitor type • Less vulnerability to disruption within a single major market
Economic impact • Economic diversity is promoted to avoid single-sector dependence • Sectors interact and reinforce each other • Net revenues are proportionally higher; money circulates within the community • More jobs and economic activity are generated
Regulation • Community makes the critical development/strategy decisions • Planning to meet ecological, social and economic carrying capacities • Holistic approach stresses integration and wellbeing of community interests • Long-term approach takes into account the welfare of future generations • Integrity of foundation assets is protected • Possibility of irreversibilities is reduced

Source: Weaver (1993)

- makes positive contributions to the conservation of natural and cultural heritage, to the maintenance of the world's diversity;
- provides more enjoyable experiences for tourists through more meaningful connections with local people, and a greater understanding of local cultural, social and environmental issues;
- minimizes negative economic, environmental and social impacts; and
- is culturally sensitive, engenders respect between tourists and hosts, and builds local pride and confidence.

This way of *doing* tourism, or the need for an applied focus in RT, emphasizes the importance of striking a balance between the social (community) and ecological components of tourism planning (Husbands & Harrison, 1996) and the day-to-day practices of tourism operators. The principles of RT can be seen in the web pages and promotional materials of many tourism operators. For example, the Australian tourism operator *Adventure World*, based out of Perth, has as their primary commitment the aim of:

> [Promoting] responsible travel to areas that conserve the natural environment and improve the well-being of those indigenous people. We believe that responsible tourism promotes positive cultural and environmental ethics and practices. Understanding the potential harm that can come from promoting tourism through many of the delicate environments in which we operate is reflected on our ideology of 'leave no trace' tourism. Working with our ground operators, several who have won the prestigious AITO (Association of Independent Tour Operators) Responsible Tourism Award, we promote responsible, sustainable tourism in all areas we visit. (Adventure World, 2018)

SUSTAINABLE TOURISM

Earlier in this chapter, the emergence of SD was discussed through three watershed conferences in 1972 that opened the door to a deeper discourse on ecodevelopment, which later translated more formally into sustainable development in 1980. D'Sa (1999) documents the evolution of ST through a series of international meetings, the first of which was the Manila Declaration of 1980. This meeting emphasized how disadvantaged Third World countries were in the planning, development and management of tourism. Following Manila, the Bad Boll Conference in 1986 was one of the first to envision a 'new tourism order', with a basis in justice, participation, sensitivities to cultures, economic benefits to both sides, and the need for support for the victims of tourism, such as those involved in prostitution. Globe '90 (1991) was an international conference and trade fair held in Vancouver, British Columbia, Canada in March 1990 on SD technologies. The conference focused on the environmental challenges faced by 12 industrial sectors, including tourism. Tourism delegates quickly realized that the successful development of tourism might only take place through 'sustainable development based on integrated and strategic resource use planning' (Cronin, 1990: 13). Globe '90 defined ST as development for the purpose of 'meeting the needs of present tourists and host regions while protecting and enhancing opportunity for the future', and further that 'sustainable tourist development is envisaged as leading to management of all resources in such a way that we can fulfil economic, social, and aesthetic needs while maintaining cultural integrity, essential ecological processes, biological diversity and life support systems' (Globe '90, 1991: 3). The document emphasized implications for policy, and the role of governments, non-governmental organizations, the tourism industry, individual tourists and international organizations in promoting ST (Globe '90, 1991).

The SD approach is essential to the tourism industry because the sector relies heavily on the natural environment and the cultural and historic heritage of destinations. ST development can be understood as the proficient 'management of all resources in such a way that economic, social, and aesthetic needs can be fulfilled while maintaining cultural integrity, essential ecological processes, biological diversity, and life support systems' (UNWTO, 1998: 21). The degradation or destruction of these resources will hinder the long-term viability of the tourism industry, as it is dependent on the conservation of local resources for continuous and responsible use in the future. Another of the principles of ST development, as identified by Cronin (1990), requires that planning should involve the local population and proceed only with their approval with an eye to local control. For ST to become a reality, local people and government authorities must be allowed to actively participate in the shaping of the local tourism industry. ST development must acknowledge the contributions that local communities and cultures make to the tourist experience; hence local people should receive benefits resulting from tourism developments (Cronin, 1990). ST development posits that socioeconomic benefits should be spread widely across the community because if benefits are maximized for local residents they are more likely to support tourism development (UNWTO, 1998). Thus, the concept emphasizes recognition of the local sense of stewardship and asserts that the tourism industry has a responsibility to ensure that the conditions in which the host community lives do not deteriorate as a result of tourism development (May, 1991). Hawkes and Williams (1993) state that the concept of ST embodies a challenge to develop the world's tourism capacity and the quality of its products without adversely affecting the environment that maintains them and while not impeding its ability to sustain them.

Tourism scholars quickly recognized the importance of taking an active position on SD soon after its inception in 1987. Butler (1993) argued that whereas it is one thing to adopt SD in tourism, it is quite another thing to adopt SD without properly defining it for tourism purposes. The word 'development' has tended to be the more offensive term in 'sustainable development' because of too much emphasis on improper planning, development and management in the tourism industry. However, as Butler (1996) suggests, 'sustainable' (to sustain something over long periods of time) can be just as misleading: tourism in such places as Niagara Falls, Paris, London or Rome is eminently sustainable because these destinations have been in existence for centuries, with no signs of waning interest. The question remains whether these places are using resources in the most efficient manner. Butler (1996) argues that ST implies something very different.

Furthermore, there is little understanding of the true preferences of current generations for tourism and, moreover, there is no reliable notion of the needs of future generations on which the tenets of ST are based. Tourism forecasting is normally based on current conditions being projected into the future. Even if there were some way of producing accurate numbers to describe the needs and desires of future tourists, how could one prioritize tourism versus other sectors at global, down to local, levels? Will it become a more popular leisure activity in the future, producing greater benefits and thus receiving greater importance? What forms of tourism will be most popular? Butler (1998) questions how the needs of current and future tourists can be balanced against the needs of future local populations for resources and space, who could formulate such an equation, and how the inevitable disagreement over the results could be mitigated. This area of concern is not simply a question of how the needs of future generations are to be met, but also involves the more rational dilemma of trying to determine what a future generation might require, let alone how to achieve such criteria. Indeed, any attempt to create ST which meets the needs of people who do not yet exist is a major challenge, while somehow measuring the long-term implications for future generations is impossible (McMinn, 1997).

Additionally, the management and commercial sides of tourism represent two very different systems operating with vastly different assumptions (Kaltenborn, 1996). The challenge inherent in these differences is that both lay claim to the same resource base. Kaltenborn suggests that an additional challenge is getting both sides to work together to shared ends – making sure businesses are successful but not at the expense of the resources upon which this success is derived. Technologies continue to evolve, and there are some noteworthy examples of companies setting the pace, especially among some of the biggest users of resources in the tourism industry (see Case Study 1.1).

Given some of the questions and challenges noted above, sustainable tourism is defined as:

> tourism which is developed and maintained in an area (community, environment) in such a manner and at such a scale that it remains viable over an infinite period and does not degrade or alter the environment (human and physical) in which it exists to such a degree that it prohibits the successful development and well being of other activities and processes. (Butler, 1993: 29)

This definition emphasizes key elements including the notion that ST is likely more achievable in site-specific situations. Scale is also important, as noted above, because it implies that on a large scale, the tourism industry might not be viable over a prolonged period of time, and also that it may not be capable of taking care of the needs and interests of elements of the social and natural realms of both social and natural worlds. There is also the implicit recognition that tourism may have a negative impact on other types of activities in the area. These means that, for example, resources are competed for by the tourism industry in such a way as to compromise the success and wellbeing of other industries (e.g. farming and agriculture) or groups of people (marginalized groups such as women, children and animals).

Furthermore, ST has come to represent a set of principles, policies and management schemes to protect a destination's ecological, social and cultural resource base. Even though ST is a relatively young concept, it has had several different perceptions as to how it should be conceived. Clarke (1997) has proposed a framework of approaches to ST that is chronologically sequenced according to the knowledge about ST at different points in time. The first, 'polar opposites', illustrates that mass tourism and ST (referenced as

Chapter Case Study 1.1 Norwegian Airlines and sustainable policy

The International Council on Clean Transportation (ICCT)[1] white paper analysed the fuel efficiency of the 20 leading airlines on routes between Europe and the United States in 2017. Norwegian once again rose to the top as the most fuel efficient airline on transatlantic routes. This achievement marks the second time that the airline has been recognized as the clear leader by the ICCT following their initial report released in 2015.

Findings showed Norwegian, on average, achieved 44 passenger kilometres per litre (pax-km/L), which is 33% higher than the industry average – soaring past 19 of its competitors. Norwegian flies one of the youngest fleets in the world, comprised of Boeing 787 Dreamliners, 737-800s and 737-MAXs.

According to the study, Norwegian eclipsed its competition as the most fuel efficient on sample routes from New York to London, Los Angeles to London, and New York to Paris. In fact, on the New York to London route, Norwegian's competition – including Virgin Atlantic, American, Delta, United and British Airways – burned 33–78% more fuel per passenger-km. The results show that British Airways, which ranked last in the results, burned a staggering 63% more fuel per passenger-km than Norwegian, a marked increase since the last assessment in 2014. Additionally, Norwegian has reduced its per-passenger emissions by 30% since 2008, thanks to its investment in new aircraft.

> The most important thing that an airline can do for the environment is to invest in newer aircraft which use the latest technology to be as fuel efficient as possible. Our strategy to have a modern fleet is paying dividends not only for our business and customers but also our planet. This recognition from ICCT is truly the highest form of industry praise and is validation that we're moving in the right direction with more environmentally friendly planes. For customers, this offers yet another reason to fly with us, to help reduce their carbon footprint. (Bjørn Kjos, CEO of Norwegian)

> One of the biggest changes in the transatlantic market between 2014 and 2017 was an increase in operations from European low-cost carriers and the further utilization of newer, fuel-efficient aircraft. (ICCT's Brandon Graver, lead author of the study)

Norwegian's inflight magazine (Norwegian, 2018) reports that the average age of its fleet of 150 aircraft is 3.7 years, making it one of the world's greenest airlines. The airline has 200 aircraft on order with plans to replace seven- to eight-year-old aircraft on a regular basis. The Dreamliner alone uses up to 25% less fuel than its counterparts. The MAX consumes 14% less fuel than the 737-800.

Source: Orban (2018)

Notes: [1]The International Council on Clean Transportation is an independent non-profit organization founded to provide first-rate, unbiased research and technical and scientific analysis to environmental regulators. Its mission is to improve the environmental performance and energy efficiency of road, marine, and air transportation, to benefit public health and mitigate climate change.

alternative tourism) were positioned as 'bad' and 'good', with the often extreme call for the total replacement of mass tourism by sustainable tourism. The second position, termed 'continuum', recognizes that sustainable tourism and mass tourism were not absolutely discrete entities, but that both forms of tourism often used similar infrastructure. It would be impossible to replace mass tourism with sustainable tourism, and so aspects of scale were important in delineating various forms of tourism along this continuum. The third position, 'movement', was marked by the belief that mass tourism itself became the subject of improvement and that it could and should be changed to more sustainable forms through the operationalization of expanding knowledge on ST. As illustrated in Figure 1.2, large-scale tourism operators had the power to implement changes in fostering a higher appreciation of the natural world, between not only suppliers but also customers. The final position, 'convergence', is premised on the notion that ST is a goal that all forms of tourism, regardless of their size, ought to pursue. Environmental and social conditions are recognized as essential elements in the formulation of successful business plans and practices. So, for example, recent changes in policy for some large tourism operators now include animal welfare standards, whereas only a few years earlier these were completely absent (see Chapter 11).

As such, ST seems to be an exceedingly complex concept with varied definitions due to different interpretations of the meaning and use of the concept (McKercher, 1993a). Welford *et al.* (1999) state that, quite expectedly, the concept of ST had not been precisely defined and, like its roots in SD, the term is open to wide interpretation. Unfortunately, the lack of widely known and accepted definitions can lead to some confusion as to what sustainable tourism means.

Tourism continues to play a role at the world's largest international meetings on the environment. In the past this role has been marginal, but more recently it has expanded through recognition of tourism's potential to address global issues like employment, poverty, justice and global relations. This evolution can be seen through the Rio de Janeiro

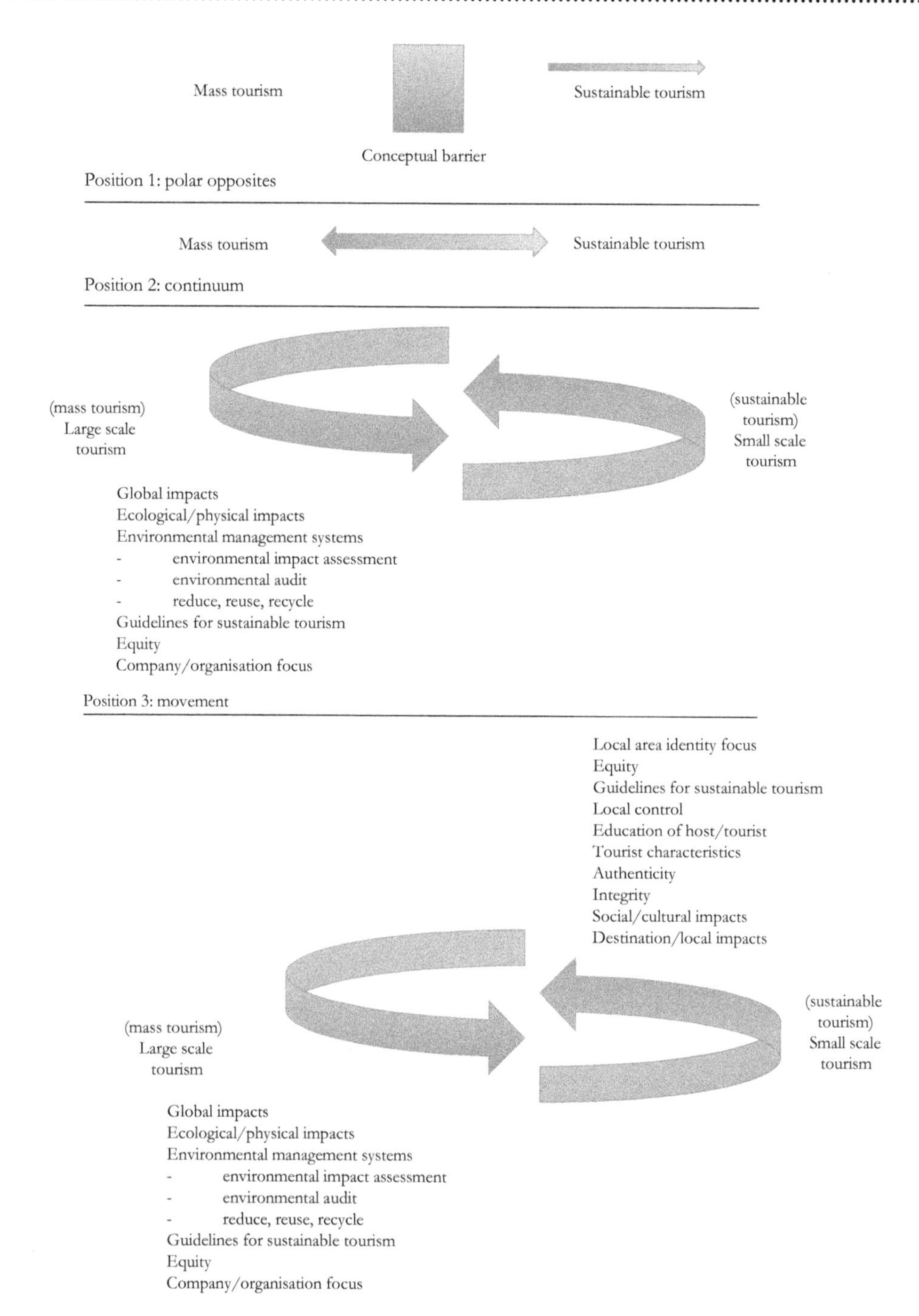

Figure 1.2 Framework of approaches to ST
Source: After Clarke (1997)

Earth Summit of 1992 and the Johannesburg World Summit on Sustainable Development in 2012 (Rio+20), the latter of which emphasized the implementation and achievements of SD over the course of 20 years (Jovicic, 2014). The UN Millennium Development Goals (UN-MDGs) emerging from the UN Millennium Summit in New York in 2000 cast tourism in a slightly stronger light, given that 80% of 56 nations that have policies and strategies to reduce poverty cite tourism as a vehicle by which to reduce poverty and bolster employment (Hawkins & Mann, 2007). But as Novelli and Hellwig (2011) observe, there is still very little research connecting tourism with the UN-MDGs. The creation of the SDGs, an important cornerstone of this book, continues the trend of being comprehensive of a number of key hot buttons of sustainability and at the same time having general applicability across industry sectors. The task, as we emphasize here, is weaving industry sectors (like tourism) into the fabric of the SDGs.

Tourism, via the UNWTO, has embraced the aforementioned SDGs as an overarching template through which to implement sustainability now and in the future (Ferraz *et al.*, 2018). Furthermore, the UN General Assembly declared 2017 to be the International Year of Sustainable Tourism for Development, built around five key themes (UNWTO, 2016): (1) inclusive and sustainable economic growth; (2) social inclusiveness, employment and poverty reduction; (3) resource efficiency, environmental protection and climate change; (4) cultural values, diversity and heritage; and (5) mutual understanding, peace and security. These priorities are an offshoot of the aforementioned UN *2030 Agenda for Sustainable Development*, where sustainable tourism was targeted specifically in SDGs 8, 12 and 14 (Bricker, 2018). This recognition for tourism as an important international sector is justified because tourism is a global economic powerhouse with trillions earned from international and domestic tourism, representing just over 10% of global GDP (Bricker, 2018). At the same time, however, it does mean that the agenda for tourism has grown tremendously because it must keep pace with changes that are taking place both inside and outside the tourism industry, with the necessity of working within the boundaries of the SDGs listed in Table 1.2.

Examples of recent tourism studies that reference the SDGs include work by Peeters *et al.* (2019) on configuring desirable transport futures in the pursuit of SDGs. Robinson *et al.* (2019) argue that there is a dearth of tourism studies research that focuses on workforce themes even though this is emphasized as an SDG (Goal 8). They observe that even though sustainability argues for decent work and economic growth, there are still deep social cleavages leading to precariousness of work, precariousness at work, and precariousness of life in general, because sustainability is still nested within a neoliberal set of assumptions. There may be problems in satisfying the targets and criteria of the SDGs in Southeast Asia, according to Bushell (2018), because enormous growth in tourism comes at a cost of economic uncertainty, a widening gap between rich and poor, and environmental degradation. And finally in Japan, global hotel chains are starting to align their targets with the SDGs for 2030, with the aim of securing return on investment for shareholders. The problem, according to Jarman-Walsh (2018), is that a lack of corporate policy in the area of sustainability is lacking as a platform to meet SDG targets. Thompson *et al.* (2017) argue that the hospitality industry has a responsibility to better accommodate volunteers who travel specifically to fundraise and contribute to the welfare of others. Xiao *et al.* (2018) used geoinformatics to document the challenges and opportunities of enhancing cultural heritage for protection and the promotion of SD. For example, SDG 8.9 places emphasis on promoting the creation of jobs that emphasize local culture and local products in the pursuit of SD.

These 17 SDGs will be used liberally in the coming pages to help fortify the direction taken in this book, as well as to provide stimulus for further research in the area of ST. The UN's Sustainable Development Goals Report in full for 2018 can be found at https://www.un.org/development/desa/publications/the-sustainable-development-goals-report-2018.html.

STRUCTURE AND CONCEPTUAL FRAMEWORK OF THE BOOK

CONCEPTUAL FRAMEWORK (FIGURE 1.3)

Following the lead of Griggs *et al.* (2013), this book reconfigures the Bruntland Report's (1987) definition of SD, and the UN's three pillars of SD paradigm (economic, social and environmental) based on the recognition that it is the stable functioning of the Earth's systems that is most important in all of our endeavours. Materials use, clean air, nutrient cycles, water cycles, ecosystem services, biodiversity and climate are all planetary must-haves, according to Griggs *et al.* (2013), as all are required for a global society, in all of its myriad endeavours, to flourish. The following definition based on different nested priorities is used in this book:

> Sustainable development in the Anthropocene is development that meets the needs of the present while safeguarding the Earth's life-support system, on which the welfare of current and future generations depends. (Griggs *et al.*, 2013: 306)

Combining the virtues of Butler's (1993) definition of ST with the emphasis placed on life support systems, by Griggs *et al.* (2013), provides us with a definition of ST that fits within the context of this book. ST is defined here as:

> Tourism which is developed and maintained in an area in such a manner and at such a scale that it remains viable over an infinite period while safeguarding the Earth's life-support system on which the welfare of current and future generations depends.

The 17 SDGs take a central position in the framework and emphasize how essential SD is to the manner in which ST ought to be configured. In essence, ST should follow the lead of the SDGs in mapping out a more inclusive and dynamic tourism future. The breadth of tourism research that currently exists on ST is located in the five accompanying blocks

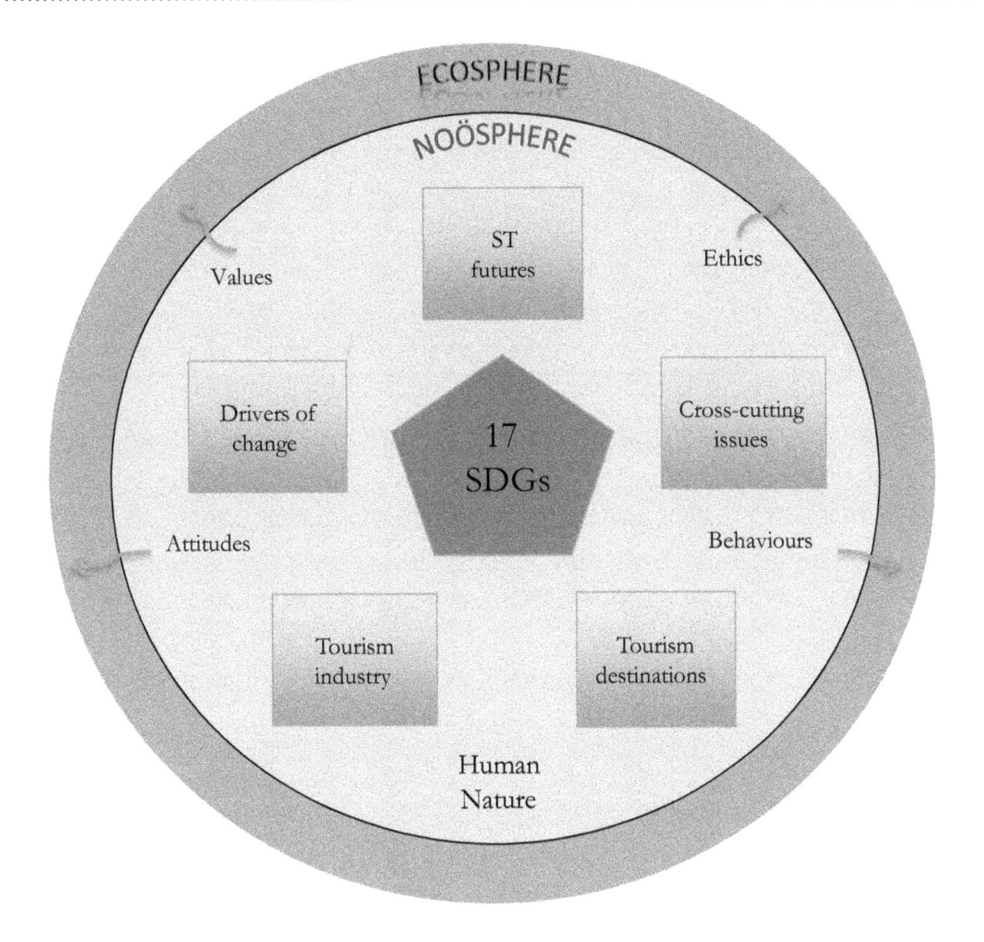

Figure 1.3 Conceptual framework

that represent key aspects of the book. Tourism destinations and the tourism industry (Chapters 5–7) include the many components of the tourism industry such as attractions, accommodation, food and so on, while drivers of change (Chapter 8), cross-cutting issues (Chapter 10) and ST futures (Chapter 11) underscore the manner in which tourism and ST have progressed, but also the many, and necessary, innovative strategies, tools and processes that are required to push the ST agenda forward. All of these key elements are nested within the noösphere, which in turn is nested within the ecosphere as the natural resource base upon which the whole of human enterprise exists. Straddling the ecosphere and noösphere circles are five fundamentals that are deemed essential in the relationships that exist between people and the natural world now and into the future, including values, ethics, attitudes, behaviours and human nature. Unfortunately, not enough research is taking place in these five areas, and we may subsequently discover that with increased focus on these fundamentals we may come closer to better understanding how ST can make a better future for all.

STRUCTURE

Chapter 2, Introduction to the Ecosphere, provides an in-depth overview of the concepts around natural capital and ecosystem services, with the latter now essential as an agent of conservation owing to its focus on quantifying the services that nature provides for us free of charge. Policy makers require this information because it is often the economic bottom line that induces these groups to act on behalf of the natural world. The chapter then embarks on a discussion of six of the UN SDGs. These relate to natural capital and ecosystem services, biodiversity, climate change, water, energy, and the biophilia hypothesis, or the innate connection that people have with nature.

Chapter 3, follows a similar trajectory on the use of SDGs after a brief introduction to the noösphere. Poverty and hunger, health and wellbeing, education, gender equality, decent work and economic growth, industry and technology, reduced inequalities, and peace and justice, partnerships and community development form the foundation of a radical change in tourism research that has been taking shape over the course of the last few years. The chapter concludes with a focus on aspects of human nature that are thought to provide a deeper explanation as to why tourists and tourism service providers often have their own best interests in mind, based on theories of reciprocal altruism and inclusive fitness, akrasia or weakness of will, and the *summum bonum*.

Chapter 4 switches to a more comprehensive overview of ST. After emphasizing the importance of resources and the commons, including Hardin's tragedy of the commons, the discussion looks at the difference between mass tourism and alternative tourism paradigms. While some argue that mass tourism will always struggle to achieve sustainability status because of its sheer magnitude, there are also questions as to whether the so-called green or softer forms of tourism like ecotourism can be sustainable. A section on impact-mitigating models and strategies allows the reader to gain an appreciation of some of the older approaches to fixing some of tourism's ecological and social problems. Examples include carrying capacity, ecological footprint and the precautionary principle, followed by a discussion of principles and indicators of ST, as well as a brief mention of the scale and sectors involved in ST. The chapter concludes with an introduction to ST in action, including implementation of ST, regulation, certification and the use of ecologos, auditing ST, and codes of ethics.

While Chapters 1–4 looked at the general literature on SD, with a few examples of tourism sustainability included, Chapter 5 deals with the spectrum of different contexts in which tourism takes place. These include marine and freshwater environments, urban and rural areas, islands, polar regions, mountains and tropical rainforests. Select case studies are included in this chapter (and others) to exemplify specific scenarios in which ST is used.

Chapters 6 and 7 investigate ST as it relates to the many different sectors of the tourism industry. The first of these involves the principal sectors of the industry, including accommodation, transportation, attractions and facilities. An effort is made to couple these sectors with the tools, corporate social responsibility, monitoring and reporting, and examples of state-of-the-art research and practice that are driving each sector. Chapter 7 follows suit but switches to other supporting sectors including food, water, waste management and energy and technology.

Given the significant amount of change that is taking place in sustainability, either as it relates to tourism or outside of tourism, it is essential to document these recent changes in a stand-alone chapter. Chapter 8 is organized according to the three different sectors that play an active role in the planning, development and management of tourism, and within sociocultural, economic and ecological dimensions. NGO, public sector and private sector innovations are discussed along with how these innovations are altering the ST landscape. For example, benefit corporations are mandated through policy and law to implement CSR measures in an effort to create positive social change through their commercial activities. Tourism companies have a role to play in furthering this agenda. The final section of the chapter deals with how education for ST has advanced over the years, with a number of programmes taking the lead in this area, including the Tourism Education Futures Initiative.

A chapter on parks and protected areas is included in the book (Chapter 9) because of the sheer volume of research that has been completed on the relationship between tourism visitation and conservation – not always an easy task to balance these two main pillars of protected areas management. The chapter opens up an opportunity to shed more light on forms of tourism like ecotourism, wildlife tourism and nature-based tourism that use protected areas as supply for the many forms of outdoor recreation such as fishing, hunting, bird watching, hiking and nature photography. Examples of protected areas at a variety of scales and jurisdictions are included in the chapter, from both developed and developing country contexts, for the purpose of exposing the problems and challenges of being sustainable in some of the world's most sensitive and spectacular areas.

Chapter 10 is an amalgam of many of tourism's most vexing issues, many of which correspond to the SDGs that provide the foundation for this work. Sections are included on climate change, and energy and technology. The largest section is devoted to marginalized groups and empowerment, including human resources, poverty and work, gender inequalities, indigenous rights, people with disabilities, and animals. While all of these are hot topics in tourism studies research, the recent focus on animal ethics in tourism serves notice that *all* stakeholders involved in the tourism industry need to have their interests taken into consideration. Several theories will be incorporated into the analysis in these broad areas to emphasize the value of theory in pushing the ST agenda forward.

The final thematic chapter (Chapter 11) is well situated at the end of the book for the purpose of discussing where ST tourism will stand in the near and long-term future. The chapter acts as a catalyst for opening up new avenues of theory and practice based on new knowledge in areas such as virtual technologies, governance, resilience and destination adaptation.

CONCLUSION

Sustainable tourism is a popular subject area in tourism studies research, with countless sources that have emerged in the form of books (see, for example, Edgell, 2006; Epler Wood, 2017; Gössling *et al.*, 2008; Graci & Dodds, 2010; Harris *et al.*, 2012; McCool & Bosak, 2015; Swarbrooke, 1999; Weaver, 2015). The *Journal of Sustainable Tourism* (*JOST*) has been active since 1993, and stands as one of the most successful and popular journals in the tourism field. Key themes on ST that have emerged in *JOST* and in other journals include paradigm, ST development, market research and economics, policymaking and infrastructure (e.g. aviation) and, to a lesser extent, modelling and planning, rural tourism, environment and crisis management, ecosystems and ecotourism, climate change, ecology, culture and heritage, human resource management and energy and material saving (Zolfani *et al.*, 2015). Other important topics in the evolution of ST through the first decade of the new millennium include climate change and global environmental change, inequality and social justice and environmental risks (Bramwell & Lane, 2008).

In acknowledgement of all this research at hand, the intent of this book is to be different from previous volumes on the topic of ST by being more comprehensive and representative of the theory and practice of ST, adhering to the SDGs as the leading values and priorities in SD through a better understanding of values, ethics and human nature, and by being interdisciplinary. Readers will hopefully recognize the benefits of this approach, and this becomes immediately apparent in view of the range of studies covered from a broad array of disciplinary perspectives. Even with the sheer weight of scholarship on the topic of ST, some authors argue that:

- 'there is little evidence of any significant change in tourism practice' (Moscardo & Murphy, 2014: 2538);
- ever since its inception, scientists have claimed that theories of sustainable development rarely work in practice (Linden, 1993);
- there are few efforts stemming from ST research that are truly innovative in the tourism industry (Bramwell & Lane, 2012);
- most innovations come from outside the field and are applied in a tourism or sustainable tourism context (Fennell, 2018a); and
- most innovations are only incremental instead of 'breakthrough' moments (Carlsen *et al.*, 2008).

These are interesting and valid observations. We shall see how theory intersects with the practice in sustainable tourism, and discuss ways of improving this unstable marriage.

CHAPTER 1 DISCUSSION QUESTIONS

1. How did the term sustainable development originate, and which organizations were central to its development?

2. Identify the weak and strong categories of sustainability.

3. What is alternative tourism, and how does it link to sustainability?

4. What is Norwegian Airlines doing with regard to sustainability that is making such a positive impact on the transportation sector?

CHAPTER 2

INTRODUCTION TO THE ECOSPHERE

LEARNING OUTCOMES

This chapter, as the title suggests, includes several of the UN's Sustainable Development Goals (SDGs) that have more of a focus on terrestrial and marine habitats – the ecosphere. While this creates a great deal of imbalance between this chapter and the next regarding these goals, separating ecological components (here) and human elements (Chapter 3) provides a more cohesive discussion for what is to follow in the book. We recognize that these goals and issues are largely inseparable in view of the connection between ecological and human components. This chapter, therefore:

1. Provides an in-depth discussion on natural capital and ecosystem services.

2. Explores the intricacies of biodiversity and biodiversity conservation, including the primary causes of biodiversity loss.

3. Discusses climate change at an introductory level with a connection to tourism.

4. Looks at water and the conditions that contribute to water scarcity.

5. Shows how energy is used and the implications for energy consumption in tourism, while in the process underscoring the importance of gaining a handle on waste management.

6. Discusses the biophilia hypothesis as an explanation as to why humans have an innate (evolutionary) reverence for the natural world, and thus opens the door as to why it is in our best interests to conserve nature.

INTRODUCTION

The purpose of this chapter is to introduce the reader to the ecosphere, which will provide the necessary background from which to examine sustainable tourism (ST) from a more ecologically informed position. The chapter begins with a discussion of natural capital, which includes natural assets like forests and rivers that provide a flow of goods and services now and into the future. The discussion moves on to a dialogue on ecosystem services, which is essentially the quantification of the value of all ecological services that the natural world provides for the planet free of charge. There is a return to the discussion on SDGs and how these relate to biodiversity, climate, water, energy, waste management and food. The chapter concludes with an overview of Wilson's (1984) biophilia hypothesis, which we view as a bridge between this chapter and the following chapter on the noösphere because of the innate tie that humans have with the natural world.

The title of this chapter, 'Introduction to the Ecosphere', was selected because there is a longstanding debate in ecology about the term that best describes the Earth and all of its inhabitants. Following Huggett (1999), three terms have been at the centre of this debate: Gaia, which refers to the planet as a living entity; biosphere, as the totality of *living* things that are found on Earth; and ecosphere, which refers more broadly to the totality of *living* organisms on Earth as well as the *non-living*/inorganic elements of the environment that sustain them. 'Biosphere' has a long-established history which traces back to the work of the Russian scientist Vernadsky (1926, 1929) and his pioneering work on how our living planet functions (Ghilarov, 1995). 'Ecosphere' was the term chosen by Commoner (1972) in framing the intricacies and challenges of the environmental crisis in the 1960s and early 1970s, which follows from Aldo Leopold (1949/1989) who developed a land ethic based not just on living entities, but also on the non-living components that comprised the whole. Leopold's land ethic is one of the earliest direct references to an emerging science of ecocentrism that emphasizes the connectedness between living and non-living entities. 'Ecosphere', Huggett (1999: 430) concludes, is the most appropriate term 'for all situations where living things and their supporting environment are taken as a whole'.

NATURAL CAPITAL AND ECOSYSTEM SERVICES

NATURAL CAPITAL

We owe a debt of gratitude to the economist Erich Zimmermann (1951), who provided one of the first comprehensive studies on the natural, social and cultural mechanisms that define the nature of resources. The central and most enduring principle of Zimmermann's work is the notion that resources 'are not, they become; they are not static but expand and contract in response to human actions' (Zimmermann, 1951: 15). It is a simple but powerful way of suggesting that elements of the natural world, what he termed 'neutral stuff', are put into service at a given time for a given function to satisfy human wants and needs. Important in his configuration of resources is that these entities were not of a knowable and fixed quantity – a matter Zimmermann argues is for natural scientists rather than economists. For example, mountains are considered to be neutral stuff until we use them to satisfy human interests in activities like mining and forestry. Furthermore, resources have a cultural element tied to them, suggesting that different cultures value resources in different ways. Traditionally, the Maasai of Africa use cattle as currency, as well as sources of meat, milk and blood for consumption. A Maasai's wealth, therefore, is measured by the number of cattle he has (as well as children). In sum, resources are not actually resources in the absence of humans, which makes the interaction between humans and the natural world pivotal. This was in contrast to other economists like Jevons who worked from the perspective of known quantities of resources, the value of which would increase as they became scarcer (Bradley, 2009).

It is not such a difficult stretch to see how natural resources, in the way Zimmermann outlined them above, play a pivotal role in the tourism industry. Mountains become resources for hiking, mountain biking and/or skiing depending on the needs of a human group in time. Or mountains once mined for the aggregate industry can later be 'reclaimed' and used for other purposes such as golf courses. In fact, the tourism industry is founded upon and sustained by a number of different categories of resources that are cultural, social, physical and ecological in nature, that once combined result in a dizzying array of attractions, transportation, facilities and types of accommodation that facilitate the supply and demand needs of the tourism industry.

The science behind the use of the natural resource base, and the management of these resources, has spawned a number of different terms that are often used interchangeably when explaining the systems, components and processes of the natural world. Some of these include natural capital, ecosystem assets and natural resources, and it worth differentiating these terms. Natural capital is defined as 'the stock of properly functioning natural assets (such as forests, wetlands, rivers, coasts)' (van den Belt & Blake, 2015: 668), that 'yield a flow of valuable goods and services into the future' (Costanza & Daly, 1992: 38). Natural capital is made up of ecosystem assets and natural resources (Dickson *et al.*, 2014), and so can also be defined as the world's stock of natural resources. For example, soil is an example of an ecosystem asset, while land used for agricultural purposes is an example of a natural resource. Natural assets are also referred to as 'stock' resources, which refer to non-renewable resources like coal and oil, and 'flow' resources, which include renewable resources like crops, forests, rivers and fisheries. Flow resources are especially important, as is their management. If they are properly managed, there is no limit to the flow of goods and services that humans might use to our advantage (Stabler, 1996).

ECOSYSTEM SERVICES

Van den Belt and Blake (2015: 668) continue by observing that 'The stock of natural capital, specifically ecosystem assets, provides an ongoing flow of goods and services, which are referred to as "ecosystem services"'. In Costanza *et al.*'s (1997) account, ecosystem services may be defined as 'flows of materials, energy and information from natural capital stocks which combine with manufactured and human capital services to produce human welfare' (Costanza *et al.*, 1997: 254). Ecosystem services are the benefits that humans obtain from ecosystems (Daw *et al.*, 2011; MEA, 2005) to produce or increase, either actively or passively, human wellbeing (Fisher *et al.*, 2009). This latter definition underscores the notion that ecological functions and processes become services when people benefit from them – if there were no human beneficiaries they would not be services – and that these benefits could be direct (clean water for consumption) or indirect (nutrient cycling). Figure 2.1 differentiates between natural capital, ecosystem assets and natural resources (see Box 2.1 for related definitions). Natural capital such as trees or fish is combined with human capital (human bodies) and manufactured capital (buildings and machines) to enrich our lives in any number of different ways.

Van den Belt and Blake (2015) argue that despite several years of theorizing on the importance of natural capital and ecosystem services, they are still undervalued because their value is rarely quantified. The authors propose a model whereby return on investment (ROI) is used to invest in natural capital, which may be aided by use of the following tools (Crossman & Bryan, 2009):

- Forecast and quantify the ROI (for ecosystem services and their benefits) in natural capital.
- Carry out participatory trade-off assessments and model these in a spatially explicit and dynamic way.
- Model the spatial and temporal dynamics in landscape service supply and societal demand.
- Show the feedbacks and interactions between society, land-use decisions and the environment over time.

NATURAL CAPITAL	
Ecosystem Assets	***Natural Resources***
Biodiversity – the stock of plants (including trees) & animals (including fish), fungi & bacteria (e.g. for food, fuels, fibre & medicine, genetic resources for developing new crops or medicines, or as a tourism asset, etc.)	The recoverable stock of fossil fuels (i.e. coal, oil & gas)
	The recoverable stock of minerals (including metals, uranium etc.)
Soils for producing crops (note that the crops themselves, i.e. the commercial seeds & livestock, are better considered as a produced asset in this instance)	Aggregates (including sand)
	Fossil water stores (i.e. deep underground aquifers replenished over centuries)
Surface fresh waters (e.g. for drinking water, hydropower, watering crops, washing, etc.)	Deep ocean stores of carbon
	Land (i.e. space for activity to take place)
The store of organic carbon (held in terrestrial plants & soils, as well as in marine organisms)	Ozone layer (protective value)
Landscapes (in terms of aesthetic values for enjoyment, including tourism use)	Solar energy (i.e. as a source of energy, including plant growth)

Figure 2.1 Natural capital and related terms
Source: Adapted from UNEP (2014)

Box 2.1 Ecosystem service definitions

- *Ecosystem services*: Contributions of ecosystem structure and function – in combination with other inputs – to human wellbeing (Burkhard *et al.*, 2012a: 2).
- *Ecosystem processes*: Changes or reactions occurring in ecosystems; either physical, chemical or biological; including decomposition, production, nutrient cycling and fluxes of nutrients and energy (MEA, 2005: 33).
- *Ecosystem functions*: Intermediate between ecosystem processes and services and can be defined as the capacity of ecosystems to provide goods and services that satisfy human needs, directly and indirectly (de Groot *et al.*, 2010: 262).
- *Supply of ecosystem services*: Refers to the capacity of a particular area to provide a specific bundle of ecosystem goods and services within a given time period (Burkhard *et al.*, 2012b: 18).
- *Demand for ecosystem services*: The sum of all ecosystem goods and services *currently* consumed or used in a particular area over a given time period (Burkhard *et al.*, 2012b: 18).
- *Ecosystem service footprint*: Calculates the area needed to generate particular ecosystem goods and services demanded by humans in a certain area at a certain time (Burkhard *et al.*, 2012b: 18).

- Demonstrate and communicate the value of investments and resource management decisions to stakeholders.
- Model the achievement of multiple benefits for multiple stakeholders.
- Take into account the particular issues related to time and geographic scale and values outlined in this article by Crossman and Bryan (2009).
- Find complementarity between multiple interests in the same landscape through the use of concepts such as joint production and hotspots.

Costanza *et al.* (1997) made an attempt to quantify the global value of ecosystems services. They found that the total value of 17 ecosystem services combined for the entire ecosphere is estimated to range from $16 trillion to $54 trillion. This value is also said to be infinite by Costanza *et al.* because the economies of the world would grind to a halt without these ecosystem services in place for our use. Estimating the value of ecosystem services is essential, therefore, because it places a monetary value on the role that these services play in allowing human beings and all other species to survive. It forces policymakers to focus on the value of these services with the hope that such will induce actions that are geared towards their conservation. De Groot *et al.* (2002) extend the number of ecosystem services to 23 in a classification of ecosystem functions, processes and goods and services (Table 2.1).

Table 2.1 Ecosystem functions, processes and components, and goods and services

Functions	*Ecosystem processes and components*	*Goods and services (examples)*
Regulation functions: maintenance of essential ecological processes and life support systems		
1. Gas regulation	Role of ecosystems in bio-geochemical cycles (e.g. CO_2/O_2 balance, ozone layer, etc.)	1.1 UVB protection by O_3 (preventing disease) 1.2 Maintenance of (good) air quality 1.3 Influence on climate (see also Function 2)
2. Climate regulation	Influence of land cover and biologically mediated processes (e.g. DMS production) on climate	Maintenance of a favourable climate (temperature, precipitation, etc.) e.g. for human habitation, health, cultivation
3. Disturbance prevention	Influence of ecosystem structure on dampening environmental disturbances	3.1 Storm protection (e.g. by coral reefs) 3.2 Flood prevention (e.g. by wetlands and forests)
4. Water regulation	Role of land cover in regulating runoff and river discharge	4.1 Drainage and natural irrigation 4.2 Medium for transport
5. Water supply	Filtering, retention and storage of fresh water (e.g. in aquifers)	Provision of water for consumptive use (e.g. drinking, irrigation and industrial use)
6. Soil retention	Role of vegetation root matrix and soil biota in soil retention	6.1 Maintenance of arable land 6.2 Prevention of damage from erosion/siltation
7. Soil formation	Weathering of rock, accumulation of organic matter	7.1 Maintenance of productivity on arable land 7.2 Maintenance of natural productive soils

8. Nutrient regulation	Role of biota in storage and recycling of nutrients (e.g. N, P & S)	Maintenance of healthy soils and productive ecosystems
9. Waste treatment	Role of vegetation & biota in removal or breakdown of xenic nutrients and compounds	9.1 Pollution control/ detoxification 9.2 Filtering of dust particles 9.3 Abatement of noise pollution
10. Pollination	Role of biota in movement of floral gametes	10.1 Pollination of wild plant species 10.2 Pollination of crops
11. Biological control	Population control through trophic–dynamic relations	11.1 Control of pests and diseases 11.2 Reduction of herbivory (crop damage); maintenance of biological & genetic diversity (and thus the basis for most other functions)
Habitat functions: Providing habitat (suitable living space) for wild plant and animal species		
12. Refugium function	Suitable living space for wild plants and animals	Maintenance of commercially harvested species
13. Nursery function	Suitable reproduction habitat	13.1 Hunting, gathering of fish, game, fruits, etc. 13.2 Small-scale subsistence farming & aquaculture
Production functions: Provision of natural resources		
14. Food	Conversion of solar energy into edible plants and animals	14.1 Building & manufacturing (e.g. lumber, skins) 14.2 Fuel and energy (e.g. fuel wood, organic matter) 14.3 Fodder and fertilizer (e.g. krill, leaves, litter)

15. Raw materials	Conversion of solar energy into biomass for human construction and other uses	15.1 Improve crop resistance to pathogens & pests 15.2 Other applications (e.g. health care)
16. Genetic resources	Genetic material and evolution in wild plants and animals	16.1 Drugs and pharmaceuticals 16.2 Chemical models & tools 16.3 Test and essay organisms
17. Medicinal resources	Variety in (bio)chemical substances in, and other medicinal uses of, natural biota	
18. Ornamental resources	Variety of biota in natural ecosystems with (potential) ornamental use	Resources for fashion, handicraft, jewellery, pets, worship, decoration & souvenirs (e.g. furs, feathers, ivory, orchids, butterflies, aquarium fish, shells, etc.)
Information functions: Providing opportunities for cognitive development		
19. Aesthetic information	Attractive landscape features	Enjoyment of scenery (scenic roads, housing, etc.)
20. Recreation	Variety in landscapes with (potential) recreational uses	Travel to natural ecosystems for eco-tourism, outdoor sports, etc.
21. Cultural and artistic information	Variety in natural features with cultural and artistic value	Use of nature as motive in books, film, painting, folklore, national symbols, architecture, advertising, etc.
22. Spiritual and historic information	Variety in natural features with spiritual and historic value	Use of nature for religious or historic purposes (i.e. heritage value of natural ecosystems and features)
23. Science and education	Variety in nature with scientific and educational value	Use of natural systems for school excursions, etc.; use of nature for scientific research

Source: de Groot *et al.* (2002)

Even more essential is the need to document and act on the threats that these ecosystem services are under in managing our use of natural resources now and in the future. The primary threats to ecosystem services have been documented by Daily *et al.* (1997), and include land use changes that contribute to losses in biodiversity along with disruption of biogeochemical cycles, exotic species introductions, toxic substances, climate change and ozone depletion – many of the topics explored below. The worry in the case of biodiversity is that as species numbers decline, the recovery potential from perturbations, stability, the capacity to provide food, and water quality decrease exponentially (Worm *et al.*, 2006). Based on available scientific evidence, the dramatic alteration of all remaining natural ecosystems, which will ultimately cause deterioration in ecosystem services, is well under way (Daily *et al.*, 1997).

It is also important to recognize that the rate of change taking place within global ecosystems is not uniform, as a result of political, social and economic conditions. For example, some developing countries such as Brazil and Indonesia are responsible for a larger percentage of the loss of some global ecosystem services than developed countries, both on national scale (absolute terms) and per capita bases. This is because forest biomass is decreasing in these two counties, whereas in some developed countries such as Germany forest biomass is increasing in size (Davidson, 2017). As such, ecological restoration projects have been instrumental in increasing ecosystem services in many places around the world, sometimes on the basis of synergistic relationships. In the three-rivers headwater region of China, Han *et al.* (2017) found that increases in net primary productivity simultaneously increased levels of water yield and soil conservation.

Furthermore, there is mounting recognition of the fact that if we are to maintain healthy, diverse and resilient ecosystems, management needs to be based on maximizing not only one ecosystem service but as many as possible (Smith *et al.*, 2017; see also Maseyk *et al.*, 2017, who have developed a stepwise decision-making process for managing ecosystems in an effort to gain sustainable yield across multiple ecosystem services). Areas of desirable and undesirable sets of ecosystem services may be identified through the ecosystem bundles framework – multiple ecosystem services that occur across a specific landscape – and these bundles may determine which trade-offs are needed according to the social-ecological demands of the region.

Scholars have also identified generic plans to identify ways in which to build resilience in ecosystems, and their services, for human survival. Mooney *et al.* (2009: 52) identify the following four points in working towards this end. We need to:

a. develop an integrated system for mapping the stocks and flows of ecosystem services, and their values, at multiple scales;
b. strengthen our basic science programme to bolster our understanding of the linkages among biological diversity, ecosystem functioning, ecosystem services and societal needs and adaptability;

c. carry out bold new experiments and model developments that incorporate the full suite of global change drivers in order to prepare adaptation strategies for a variety of ecosystems, including our crops;
d. develop conservation, restoration and natural resource management plans that are proactive and based on maximizing ecosystem service delivery, considering trade-offs among services, and that are resilient to projected global changes. These plans must take into account that we may not be able to manage to return natural ecosystems to previous states or conditions.
e. focus more scientific attention towards adaptation in light of inevitable changes in ecosystem functioning and services in the coming years; at the same time we need to identify practices that, if modified, will mitigate the drivers of climate change.

The ecosystem services typology has value to tourism studies through recognition of the important functions that recreation (Function 20), culture (Function 21), spiritual and historic information (Function 22) and science and education (Function 23) play (Table 2.1; de Groot *et al.*, 2002). This would implicitly include psychological benefits that people experience through nature in many different ways which lead to more happiness and positive mood (Barton & Pretty, 2010) as well as the reduction of stress (Ulrich, 1979).

Tourism scholars are now recognizing the value of implementing the ecosystem services perspective into their work. For example, Willis (2015) has shown that cultural ecosystem services can have a positive effect on the interconnection of tourism, nature and wellbeing, through her work on the Jurassic Coast in the UK. Cultural services include those non-material benefits that people derive from ecosystems (MEA, 2003). So-called 'blue' spaces, which include coastal and marine environments as well as lakes and rivers, are particularly effective in generating non-material benefits including calmness, relaxation and revitalization. Willis (2015) concludes by suggesting that tourism management may be improved by arranging more opportunities for sustainable experiences in nature and recognition of the importance of non-material benefits. Raudsepp-Hearne *et al.* (2010) identified a number of ecosystem service bundles in a mixed-use landscape involving 12 ecosystem services and 137 municipalities in Quebec, Canada. Examples include provisioning services such as crops, pork, drinking water and maple syrup. The cultural bundle identified in this study includes deer hunting, tourism, nature appreciation, summer cottages and forest recreation. The authors note that it is often difficult to manage multiple ecosystem services and as such there are often trade-offs between certain types of ecosystem service bundles. The authors conclude:

> It has been suggested that the loss of regulating [e.g. carbon sequestration] and cultural services in areas of high provisioning service production may undermine the sustainability of this production, diminish the possibility of diversifying economic activities, and impact local human wellbeing directly. (Raudsepp-Hearne *et al.*, 2010: 5246)

Tourism operators may need to suspend water recreation in view of other resource demands on the region; knowing where these trade-offs are occurring makes the management of diverse landscapes possible (Raudsepp-Hearne *et al.*, 2010).

The ability to access and benefit from ecosystem services is subject to variability because of individual capacities and broader access mechanisms. McClanahan and Kaunda-Arara (1996) found that the creation of a marine park created winners and losers among fishers because, while some fishers lost their livelihoods, others benefited because of the utilization of skills and opportunities in the tourism industry. These differences in opportunity are contingent upon a number of village, gender and livelihood dimensions (Rönnbäck *et al.*, 2007).

Ecosystem service research has also been instrumental in addressing issues of poverty. Daw *et al.* (2011) argue that different global groups derive different benefits and costs from ecosystem services, creating a climate of winners and losers. They use the example of tourism to suggest that the poor in developing countries are said to benefit by selling coastal tourism to wealthy consumers, but the value of these resources is typically measured by willingness-to-pay (WTP) metrics. The high costs paid to consume natural and cultural resources of the destination typically lead to an inflation of tourism values that overshadows locally held values, which in turn do not accurately reflect the poverty alleviation potential of the tourism industry. The tourism literature is abundant with examples of how leakages and import substitution have led to winners and losers between developed economies and lesser developed ones.

GENERAL ISSUES

SDG 14: LIFE BELOW WATER AND SDG 15: LIFE ABOVE LAND

The ecosphere maintains an incredible diversity of biological life (biodiversity) which is critical to the ecological functioning of life on the planet and for the survival and prosperity of humans. The discussion above on natural resources and ecosystem services serves to emphasize this point. However, the magnitude of human activities has accelerated to such a point that we are now in the midst of another mass extinction event which threatens biodiversity on all levels. This event is referred to as the Anthropocene extinction, defined as a human-dominated geological epoch characterized by a fundamental change in the relationship between humans and the Earth (Lewis & Maslin, 2015). The chief threats to biodiversity include habitat loss, invasive species, pollution, population growth and overexploitation (Torrence, 2010). These five 'HIPPO' factors will be discussed in more detail below, after a more thorough discussion of biodiversity.

Biodiversity is defined as 'the variability among living organisms from all sources including, inter alia, terrestrial, marine and other aquatic ecosystems and the ecological complexes of which they are part, such as diversity within species, between species and of ecosystems' (Convention on Biological Diversity, 2019). Biodiversity, therefore, includes all of the ecosystems on Earth, all of the flora and fauna that reside within these habitats and all of the genes that make up the organisms. Considering the example of the Great Barrier Reef, an ecosystem off the coast of Australia, the species biodiversity includes all of the different types of organisms living in it such as the humpback whale, sea turtle, clownfish, marine algae and coral. Genetic diversity can occur within or between species and includes all of the traits that make up an individual or a group of individuals. This would include the traits that differentiate one species from another, such as the behaviour and body shape difference between a yellow tang fish and a butterfly fish. Within the damselfish species there is a wide range of sizes and colours that differentiate an individual from others. More recent studies have expanded upon this 'biodiversity trilogy' (Kaennel, 1998) to include species evenness (how close in number species are in a given environment), species composition (the identity of all organisms that make up a community), functional composition (specific functional traits) and landscape units (Bermudez & Lindemann-Matthies, 2018; Diaz *et al.*, 2015).

Valuing biodiversity conservation in economic terms, as outlined briefly above in the discussion on ecosystem services, is important because it means that such will need to be built into our economic accounting systems (Edwards & Abivardi, 1998). This is the principle reason why the concept of ecosystem services was formulated as 'a powerful new argument for protecting biodiversity' (Edwards & Abivardi, 1998: 240). Munasinghe (1994) has categorized economic values that are linked to environmental assets into personal use values and non-use values, as follows:

Personal use values:

- *Direct use value*: Outputs that can be consumed directly (food, biomass, recreation, health);
- *Indirect use values*: Functional benefits (ecological functions, flood control, storm protection);
- *Option values*: Including future direct and indirect use values (biodiversity, habitat conservation).

Non-use values:

- *Bequest values*: The value of leaving use and non-use values for offspring (habitats, irreversible change);
- *Existence values*: Value from knowledge of continued existence based on moral conviction (habitats, endangered species).

Munasinghe argues that as one moves down the list, i.e. from the top of personal values to the bottom of non-use values, there is decreasing tangibility of value to individuals. We value the ability to use resources directly and immediately for our own purposes with decidedly less emphasis on the needs of others now and in the future. A classic example of this type of thinking comes from Clark (1973), who explored the concept of economic value from the perspective of extinction of animal species. He wrote that a dead blue whale would only be valued for its oil and blubber. It is a one-time use of an animal that has no value to future generations. These sorts of examples of use are rather like barbarism, Clark adds, because the value of the animal/species to science, aesthetics, recreation and so on cannot be fully estimated, which is a sad commentary on the general worth that society places on the preservation of environmental attributes (Pigram, 1990).

Even though non-use values are said to outweigh the values of more conventional uses like clear-cutting, direct use values are central in cost-benefit analyses (Gössling, 1999b). As a direct use, tourism can play a vital role because the sustainable use of tropical forests, for example, through ecotourism may in fact outweigh the costs of conservation. Species and ecosystems would no longer persist if not for the benefits of tourism, and direct money transfers and debt-for-nature swaps are mechanisms that will provide the right incentives for conservation (Gössling, 1999b). This is what is so attractive about non-consumptive forms of tourism like ecotourism: the animal may be 'used' again and again.

An innovative scheme that is commanding interest in the area of poverty alleviation is payments for ecosystem services. An example of such a scheme is in the Maasai Mara ecosystem in Southwestern Kenya, where pastoral landowners receive direct payments from five tourism operators in return for voluntary resettlement and the prohibition of livestock grazing on their subdivided territories (Osano *et al.*, 2013). The lands, in the absence of human occupants, are dedicated to wildlife tourism and ecotourism, and the payments received by the community are enough to eliminate poverty – which is a testimony to the success of the wildlife tourism industry in the region. Co-benefits include employment opportunities in conservancy and in a range of social services.

A constraint to moving in the direction of ecological accounting methods is the lack of emphasis that economic and political institutions have placed on providing the right incentives to conserve biodiversity. Osano *et al.* offer five reasons for such failure:

1. *Short time horizons*: Short time horizons for politicians in power lead to discounting the future in favour of immediate needs.
2. *Failures in property rights*: When resources are not individually owned there is no individual interest in maintaining or improving the resource.
3. *Concentration of economic power*: Extreme inequalities exist in the distribution of income and assets, and politically connected individuals exploit the resource at the expense of traditional occupants.
4. *Immeasurability*: Then value of biodiversity is not appreciated because the reliability of valuation techniques is disputed. Simplicity is favoured by policymakers, who prefer the direct measurables of biodiversity such as harvesting and tourism use. Indirect uses are not taken seriously.
5. *Institutional and scientific uncertainty*: Lack of knowledge about the role that species and habitats play in life-support functions, which adds up to an imperfect manner in which to assign precise values to these entities and systems.

The problem frequently becomes exacerbated in the developing world where most of the world's biodiversity exists, because donor funds often get misappropriated through political corruption and distorted priorities. This is because: external funds are easier to misappropriate than internal funding; government officials are poorly paid so the acceptance of bribes is common; conservation departments lack political weight so other more powerful departments like armed forces may skim funds; and measuring the success of conservation projects is difficult (Smith *et al.*, 2003).

Scholars argue that one of the most important drivers of getting conservation of biodiversity and local people in developing counties onto the same page is getting the right incentives in place. The conservation of biodiversity in developing countries involves the investment of billions of dollars, with a range of both direct and indirect incentives to induce these countries to embrace conservation. Ferraro and Kiss (2002) illustrate the breadth of these approaches, with countries of the more developed world embracing direct incentives (land purchases, leases, easements, performance payments and tax relief), and with conservation initiatives in the developing countries siding more with indirect incentives (integrated conservation and development projects, community-based natural resource management). Figure 2.2 outlines the various types of conservation investment (from least direct to most direct) and a series of examples of each. Ferraro and Kiss argue that even after decades of use, the more indirect incentives have been shown not to work. The argument follows that direct payments are a more fruitful avenue for those living in the developed world, with residents receiving income that is worth double that which an indirect investment project would furnish. Ferraro and Kiss contend that the cheapest way

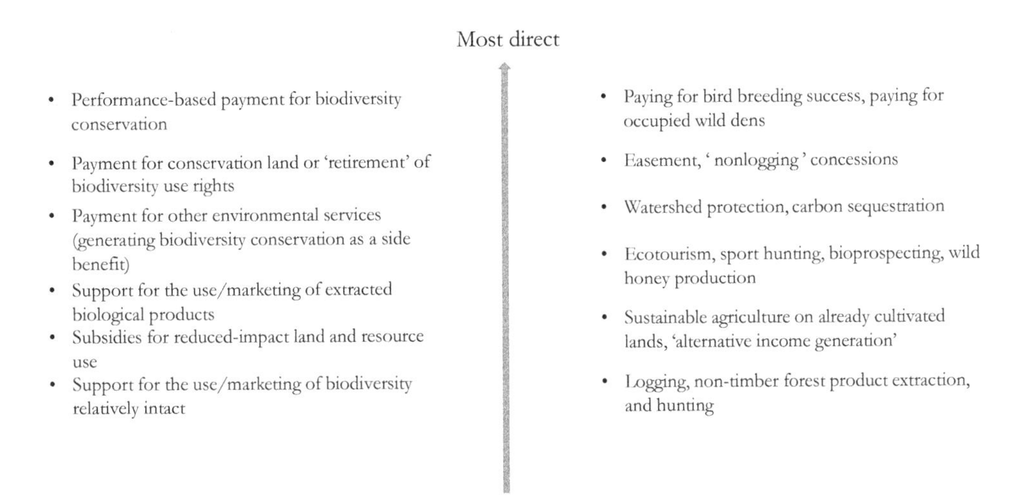

Figure 2.2 Potential investments for biodiversity conservation
Source: After Ferraro and Kiss (2002)

to get what you want (e.g. an intact tropical rainforest) is to pay directly for it as opposed to paying for it indirectly (e.g. money to improve the ecotourism industry).

The UN International Year of Biodiversity (2010) was supposed to be a turning point in efforts to control biodiversity loss, but biodiversity continues to decline and pressures continue to escalate. Hanski (2011) writes that we are now armed with better indicators and better science to take on this monumental task through the 2020 Aichi Biodiversity Targets organized by the Convention on Biological Diversity (CBD, 2018: Appendix 9.1). The main strategic goals of the Aichi Targets include: (1) address the underlying causes of biodiversity loss by mainstreaming biodiversity across government and society; (2) reduce the direct pressures on biodiversity and promote sustainable use; (3) improve the status of biodiversity by safeguarding ecosystems, species and genetic diversity; (4) enhance the benefits to all from biodiversity and ecosystem services; and (5) enhance implementation through participatory planning, knowledge management and capacity building. The importance of turning the tide on biodiversity loss is so important that we are now in the midst of a UN *Decade* on Biodiversity, spanning 2011 to 2020.

Two of the UN SDGs focus specifically on the conservation of biodiversity. The first is SDG 14, the conservation and sustainable use of the oceans, seas and marine resources for sustainable development, while the second is SDG 15, to protect, restore, and promote sustainable use of terrestrial ecosystem, sustainably manage forests, combat desertification, and halt and reverse land degradation and halt biodiversity loss. The progress of both of these goals is identified in Box 2.2

Box 2.2 Progress of SDGs 14 and 15

Advancing the sustainable use and conservation of the oceans continues to require effective strategies and management to combat the adverse effects of overfishing, growing ocean acidification and worsening coastal eutrophication. The expansion of protected areas for marine biodiversity, intensification of research capacity and increases in ocean science funding remain critically important to preserve marine resources.

The global share of marine fish stocks that are within biologically sustainable levels declined from 90% in 1974 to 69% in 2013.

Studies at open ocean and coastal sites around the world show that current levels of marine acidity have increased by about 26% on average since the start of the Industrial Revolution. Moreover, marine life is being exposed to conditions outside previously experienced natural variability.

Global trends point to continued deterioration of coastal waters due to pollution and eutrophication. Without concerted efforts, coastal eutrophication is expected to increase in 20% of large marine ecosystems by 2050.

As of January 2018, 16% (or over 22 million km^2) of marine waters under national jurisdiction – that is, 0–200 nautical miles from shore – were covered by protected areas. This is more than double the 2010 coverage level. The mean coverage of marine key biodiversity areas (KBAs) that are protected has also increased – from 30% in 2000 to 44% in 2018.

Progress of SDG 15 in 2018

Protection of forest and terrestrial ecosystems is on the rise, and forest loss has slowed. That said, other facets of terrestrial conservation continue to demand accelerated action to protect biodiversity, land productivity and genetic resources and to curtail the loss of species.

The Earth's forest areas continue to shrink, down from 4.1 billion ha in 2000 (or 31.2% of total land area) to about 4 billion ha (30.7% of total land area) in 2015. However, the rate of forest loss has been cut by 25% since 2000–2005.

About one-fifth of the Earth's land surface covered by vegetation showed persistent and declining trends in productivity from 1999 to 2013, threatening the livelihoods of over one billion people. Up to 24 million km^2 of land were affected, including 19% of cropland, 16% of forest land, 19% of grassland and 28% of rangeland.

Since 1993, the global Red List Index of threatened species has fallen from 0.82 to 0.74, indicating an alarming trend in the decline of mammals, birds, amphibians, corals and cycads. The primary drivers of this assault on biodiversity are habitat loss from unsustainable agriculture, deforestation, unsustainable harvest and trade, and invasive alien species.

Illicit poaching and trafficking of wildlife continues to thwart conservation efforts, with nearly 7000 species of animals and plants reported in illegal trade involving 120 countries.

In 2016, bilateral ODA in support of biodiversity totalled $7 billion, a decrease of 21% in real terms from 2015.

Source: UN Report of the Secretary-General (2018g, 2018h)

While the SDGs have targets that are geared towards ameliorating problems like biodiversity loss by 2030, the realization of these goals will be challenging, especially with respect to biodiversity. Biggs *et al.* (2008) focused on scenarios for biodiversity loss in Southern Africa for the present century and found that there will be large declines in the average population sizes of terrestrial plants and vertebrates, which are reported to be two to three times greater than reductions recorded since 1700. Sala *et al.* (2000) argue that the drivers of change for terrestrial biodiversity loss by the year 2100 will, in rank order, be land-use change, climate change, nitrogen deposition, biotic exchange and elevated carbon dioxide concentrations. In freshwater systems, biotic exchange is said to be much more important. The ecosystem type that will experience the greatest amount of change will be Mediterranean climate and grassland ecosystems because these regions must endure all the drivers of change. Northern regions will experience the least amount of change mainly because these areas have already experienced a great deal of land-use change.

Biodiversity hotspots

A biodiversity hotspot is an area 'featuring exceptional concentrations of endemic species and experiencing exceptional loss of habitat' (Myers *et al.*, 2000: 853). Almost half of the world's vascular plants and about one-third of terrestrial vertebrates are endemic to 25 global hotspots of biodiversity (Brooks *et al.*, 2002). None of these hotspots has more than one-third of its unspoiled habitat remaining, and where once they covered approximately 12% of the Earth's terrestrial surface, in 2002 they covered a mere 1.4% of the land (Brooks *et al.*, 2002). The implication of this tremendous reduction in habitat is the feeling that many hotspot endemics would either be extinct or threatened without conservation.

Scientists argue that priorities need to be set for the preservation of certain species and habitats given that there is a critical lack of conservation funding. Myers *et al.* (2000) argue that these limited funds ought to be dedicated first to the hottest of the hotspots. These are hotspots that, when measured against all other hotspots according to endemic plants, endemic vertebrates, endemic plants/area ratio (species per 100 km^2), endemic vertebrates/area ratio (species per 100 km^2), and remaining primary vegetation as percentage of original extent, appear at least three times in the top ten listing for each factor (Myers *et al.*, 2000). The world's hottest hotspots in rank order include Madagascar, Philippines, Sundaland, Brazil's Atlantic forest, Caribbean, Indo-Burma, Western Ghats/Sri Lanka, Eastern Arc and the coastal forests of Tanzania and Kenya.

Biodiversity hotspots are attractive from an ecological standpoint, but what makes them attractive ecologically also makes them attractive from a development perspective. Tourism is simply another industry, in a long list, that contributes to land clearance, pollution and climate change and the loss of biodiversity (Hall, 2010a; Huang *et al.*, 2011). Coastal areas can be particularly hard hit as ecosystems succumb to land clearance, urbanization and the draining of wetlands (Gössling & Hall, 2006).

Tourism's positive contributions to biodiversity are rare and typically isolated to individual species and small areas of habitat (Hall, 2010a; Table 2.2). Parks and protected areas are excellent alternatives to other forms of land use such as agriculture and animal husbandry, which are the main causes of deforestation, and more emphasis needs to be placed on weighing negative effects against positive effects on biodiversity conservation (van der

Table 2.2 Positive and negative contributions of tourism to biodiversity conservation

Positive	*Negative*
An economic justification for biodiversity conservation practices, including the establishment of national parks and reserves (public and private)	Outside of national parks and reserves contributes to the fragmentation of natural areas and a reduction in their size
A source of financial and political support for biodiversity maintenance and conservation	Contributes to changes in ecosystem conditions, particularly ecosystems in high-value amenity areas such as the coast and alpine areas, as well as more localized effects such as trampling
An economic alternative to other forms of development that negatively impact biodiversity and to inappropriate exploitation or harvesting of wildlife, such as poaching	Is a major vector for the introduction of exotic species and disease; in some cases introductions may be deliberate, e.g. for hunting, fishing or aesthetic reasons
A mechanism for educating people about the benefits of biodiversity conservation	Is a significant contributor to climate change
Potentially involves local people in the maintenance of biodiversity and incorporating local ecological knowledge into biodiversity management practices	The presence of tourists can lead to changes in animal behaviour, while consumptive tourism such as hunting and fishing, if poorly managed, can lead to species loss

Source: Hall (2010a)

Duim & Caalders, 2002). Since tourism businesses in areas such as transportation, accommodation, food and restaurants and other leisure and entertainment services depend directly on biodiversity, then biodiversity conservation should be a central component of business sustainability (Habibullah *et al.*, 2016).

HABITAT LOSS

Of all the causes of biodiversity loss, habitat destruction is viewed as the greatest contributing factor, and it is studied on a number of different scales including habitat patches, landscapes, biomes and ecoregions, and in all regions of the world. The Millennium Ecosystem Assessment (MEA, 2005) reported that over half of the Earth's biomes had been converted by 1990 into agricultural lands, pastures, plantations, built areas and infrastructure (Hanski, 2011). The problem is so significant that we now have a 'biome crisis', where habitat conversion exceeds habitat protection in temperate grasslands and Mediterranean biomes by a ratio of 8:1, and by 10:1 in over 140 ecoregions in all continents of the world except Antarctica (Hoekstra *et al.*, 2005).

A study by Reyers *et al.* (2009) combines habitat loss, ecosystem services, biodiversity hotspots and the concept of trade-offs in the semiarid Little Karoo region of South Africa, which is dominated by high-density livestock enterprises (mainly ostriches). The authors found that five ecosystem services (production of forage, carbon storage, erosion control, water flow regulation and tourism) experienced 20–50% in declines because of land-cover changes in the region and poor management. These declines in ecosystem services have led to unemployment, vulnerability and the narrowing of future options for people living in Little Karoo. Restoration, new streams of funding and the development of new future economic activities (tourism) have been essential in building a more resilient social-ecological system in the Little Karoo region through diversification and adaptation. Furthermore, the massive urban expansion dynamic in certain parts of China is having a significant impact on habitat loss. In the Pearl Delta, for example, 25.8% of the natural habitat and 42% of local wetlands were lost over a period of 20 years, from 1992 to 2012 (He *et al.*, 2014).

INVASIVE ALIEN SPECIES

Studies over decades have documented the catastrophic consequences that the introduction of invasive alien species (IAS) has on native species and environments. This is because ecosystems are not resilient enough to withstand the pressures that these new species have on the balance of life – introduced species simply outcompete native species. Many of the introductions in the past were deliberate. For example, the introduction of Nile perch into Lake Victoria in 1954 for the purpose of increasing fish stocks for fishing (the lake had been over-fished) resulted in the extinction of over 200 native fish species (ISSG, 2004). Most introductions, however, are not intentional and are the result of careless behaviour or accidents. In marine environments, a study of the global threat of invasive species found that on the basis of over 350 databases and 329 marine invasive species, just 16% of marine ecosystems are devoid of invasions. International shipping and aquaculture operations are the major reasons behind these introductions (Molnar *et al.*, 2008).

Tourism is also an important conduit for the introduction of IAS. Due to warming trends in Antarctica, new species, including the common housefly, have been introduced from ships and are thriving at Antarctic bases. Species are also introduced from backpacks and camera bags which tourists take from one continent to another (The Guardian, 2017). Furthermore, countries like France, Spain and Italy that host more tourists have a higher density of introduced plant species than countries with less tourist visitation (Vilà & Pujadas, 2001).

Two global programmes designed to control IAS are of note. The Global Invasive Species Programme is an independent, not-for-profit association whose mission is to conserve biodiversity and sustain human livelihoods by minimizing the spread and impact of IAS (World Bank, 2018). The Global Invasive Species Database is a free, online database that documents alien species that have an impact on global biodiversity, with knowledge transfer on awareness and management activities. This organization is administered by the Invasive Species Specialist Group of the IUCN Species Survival Commission (ISSG, 2018). Their work is both timely and seems unending, as studies reveal that the targets set by the CBD are not being met, with IAS pressure being responsible for declines in native species diversity globally (McGeoch *et al.*, 2010). Only about half of the countries that are signatory to the CBD have invasive species relevant legislation, McGeoch *et al.* add, and even those that have such legislation are inconsistent in their implementation strategies to mitigate biodiversity impacts.

POPULATION GROWTH

The consequences of population growth were popularized by Paul Ehrlich (1968) in his major work, *The Population Bomb*, in which the world during the 1970s and 1980s was marked by mass starvation due to population growth outpacing food production. Ehrlich's dire predictions did not all materialize, thankfully, but his message was abundantly clear: we are growing much too fast as a species, and the ability of the world's natural resources to keep pace is being seriously compromised.

The UN forecasts population growth to continue to rise to 8.6 billion by 2030, 9.8 billion by 2050 and 11.2 billion by 2100 (UNDESA, 2017). The report indicates that most of the *future* growth will take place in just nine countries: India, Nigeria, the Democratic Republic of the Congo, Pakistan, Ethiopia, the United Republic of Tanzania, the USA, Uganda and Indonesia, ordered by expected contribution to growth. The report also indicates that there will continue to be large movements of refugees and other migrants between regions, typically from low- and middle-income countries to high-income nations. The net inflow of migrants is in the magnitude of approximately 3.2 million per year.

Population growth is widely regarded as one of the principal causes of biodiversity loss (Hinrichson, 1994). Modelling the relationship between the number of threatened mammal and bird species and human population density shows how imminent the problem is. The number of threatened species in the average nation will increase to 7% by 2020 and to 14% by 2050, based on human population growth alone (McKee *et al.*, 2003). Escalating population numbers, especially in lesser developed countries, can have serious impacts on the management of sensitive regions like protected areas because of the demands on natural resources (Soule & Sanjayan, 1998). In the Danau Sentarum National Park, Indonesia, population in the park has grown by 50% over a 10-year period, with heavy impacts both on fishing as well as on trees used for the production of rubber (Indriatmoko, 2010).

Tourism continues to experience strong, sustained growth internationally on a year-by-year basis. In 2017, 1.326 billion tourists travelled internationally, generating US$1.340 trillion in receipts (UNWTO, 2018b). Tourism arrivals increased by +7% over the course of 2017, surpassing the UNWTO's forecasts for the year. The problem with all this growth is that there are destinations that are being 'invaded' by tourists. This phenomenon has been termed 'overtourism', and includes cities like Barcelona and Venice, for example, where local people must share their cities with numbers of visitors that far exceed city population sizes. The number of tourists staying in hotels in Barcelona in 2017 was 8.9 million, which represents an increase of over 7 million since 1990 when only 1.7 million stayed in the city (Statistica, 2018). The permanent population of Venice is 55,000 but over 20 million tourists visit the city per year (DW, 2018). Demarketing, or actions that go into decreasing demand for a product, is a topic that will be discussed in Chapter 9.

POLLUTION

Pollution is defined as 'an alteration and contamination of the natural conditions in the environment due to physical, chemical, or biological factors' (Falchi, 2019: 147). An example of physical pollution is artificial light at night, which operates against the natural light and dark cycles of the Earth on its axis (Falchi, 2019). Chemical pollution is recognisable as a form of pollution because of its effects on the natural world (Travis & Hester, 1991), with mercury poisoning being a case in point. Bruser and Poisson (2016) report on how significant the effects of mercury poisoning from the pulp and paper industry were, and are, on fish and on the indigenous populations of Canada's north. Freshwater systems were being contaminated by industry discharges of mercury, and because human populations consume fish as a staple of their diet, and because mercury bioaccumulates in aquatic ecosystems, the long-term effects on the indigenous populations were significant. Heavy metals and pesticides (fungicide, insecticide and herbicides) have been found to have the most deleterious impact on aquatic habitats. In sufficient levels, these compounds affect important tissues such as gills, liver and kidney to the point where failures in stock recruitment are imminent and population declines (Khoshnood, 2017). An example of biological pollution is a virus or bacterium that has an impact on the fitness of another organism (Elliott, 2003). Elliott provides an overview of terms and their relationship between chemical and biological pollution (Table 2.3).

The extent to which pollution can have an environmental and economic effect on the tourism industry is demonstrated in the case of Geoje Island, South Korea. In July 2011 a huge amount of marine debris (mainly wood in the form of sticks, branches and garbage) washed up on the shores of Geoje as a result of heavy rainfall taking place that month. Although it took 15 days to clean up the mess, it coincided with the summer vacation season. This event resulted in a 63% reduction of visitors (890,435 in 2010 to 330,207 in 2011), with an estimated loss of revenue of US$29–37 million (Jang *et al.*, 2014). Marine

Table 2.3 Definitions of terms used in chemical pollution and their translation for biological pollution

Term	*Chemical-based definition*	*Translation to biological pollution*
Pollutant	A substance introduced into the natural environment as a result of man's activities and in quantities sufficient to produce undesirable effects	The input and effects of micro- and macro-organisms on the condition that adverse effects can be demonstrated
Pollution	(i) is a change in the natural system as a result of man' activities (ii) has occurred if it reduces the individual's/population's/ species'/ community's fitness to survive; the introduction by man, directly or indirectly, of substances or energy into the marine environment (including estuaries) resulting in such deleterious effects as to harm living resources, hazards to human health, hindrance to marine activities including fishing, impairment of quality for use of seawater, and reduction of amenities (GESAMP)	The effects of introduced, invasive species sufficient to disturb an individual (internal biological pollution by parasites or pathogens), a population (by genetic change) or a community (by increasing or decreasing the species complement); including the production of adverse economic consequences
Contamination	An increase in the level of a compound/element ('pollutant') (as the result of man's activities) in an organism or system which not necessarily results in a change to the functioning of that system or organism	The introduction of species without noticeable effects (e.g. microbes which are killed immediately by natural conditions, possibly to be extended to species occupying available and vacant niches)

Source: Elliott (2003: 276)

debris is a conspicuous pollutant which has the effect of making beaches aesthetically unappealing. Studies have found that 85% of beach users would avoid a beach if there was excessive litter to be found per unit area (Krelling *et al.*, 2017). Other studies on tourism with pollution show similar results. Ahmed *et al.* (2018) investigated the extent to which the unique ancient culture and spectacular natural scenic spots of the One Belt One Road provinces of Western China have been impacted by tourism and economic development and associated levels of environmental pollution. They conclude as follows:

> At present, the negative impact from tourism and energy use in several sectors outweighs the positive effect in majority of our sample. The main factors of environmental degradation in the Northwestern part of OBOR are the coal based industrial development, coal mining, arid climate, and the overall underdevelopment of area and masses. To improve the environmental situation and to develop the tourism sector, the government should engage the local community to promote tourism and environmental protection. (Ahmed *et al.*, 2018: 19)

OVERHARVESTING

The conservation through sustainable use of wildlife (CSU) debate is one that has specific reference to tourism and is perhaps one of the most contentious issues in conservation research and practice. The harvesting of wildlife, although sensitive, has been a conservation strategy for many years. On one side of the debate, proponents argue that the harvesting of a wildlife species is instrumental in guaranteeing its conservation. Some studies indicate that wildlife is able to adapt to harvest reductions, suggesting a level of dynamism in populations that has often been left out of the debate on the moral legitimacy of the activity (Webb, 2002). And it is not so much an issue of stopping harvesting, Webb continues, but rather how best to sustain these activities.

On the other hand, opponents suggest that these sorts of practices are simply an excuse to exploit wildlife (Allen & Edwards, 1995). For example, scientists have documented the devastating decline in forest elephants in Central Africa (Maisels *et al.*, 2013). They found that the population size declined by 62% between 2002 and 2011, with the species losing 30% of its range. Reasons for such a significant decline include hunting and increasing human population density, among other factors.

The legitimacy of hunting as a conservation practice has recently been called into question through the killing of Cecil the lion in July 2015, and the legal hunt of a critically endangered male black rhino, also in 2015, at a cost of US$350,000 via an auctioned permit (Herskovitz, 2014). While hunting can and does increase funding for conservation, there are questions about how much of this money gets into the hands of local people. Furthermore, from an ecological standpoint there are issues tied to the introduction of new species into game ranches with the risk of these animals becoming invasive, and competing and even hybridizing with indigenous species. The practice of erecting fences for canned hunting (keeping and breeding lions and other large prey within fences so that hunters can easily dispatch these animals) disrupts the dispersal of these animals and also promotes inbreeding and the loss of heterozygosity (Ripple *et al.*, 2016).

Baker (1997) argues that the ongoing preservation of species on a large scale is no longer possible because of HIPPO, as discussed earlier in this section. Instead, innovative management schemes are required in an effort to use wildlife on a sustainable basis. Trophy hunting is one such use and, according to Baker, it must be based on incentives (benefits) for local acceptance and control, and decision makers need evidence of the economic value of wildlife resources. In Africa, Baker argues that the following conditions must be met in order for trophy hunting to be a beneficial and sustainable use of wildlife (Baker, 1997: 313–317):

- quotas must be based on scientific population estimates;
- quotas can only result in sustainable off-takes if they are comprehensive and enforced;
- hunting concession awards must be limited to reputable, experienced outfitters and allocation must be perceived as fair, transparent, consistently administered and incorruptible;
- the system governing the receipt of fees and return of revenues to local communities must be transparent and accountable;
- the hunting industry must be completely regulated and internally and/or externally policed to prevent abuses; and
- returning economic benefits to local communities will not guarantee equitable sharing of the resources or protection of wildlife.

These conditions recognize that indigenous people need to play a vital role in these enterprises and that the proper institutional arrangements are implemented to ensure that individual and organizational self-interest (the tourism industry) does not overrun the system. Other confounding factors in the sustainable use of wildlife include the acceptability of traditional practices and time (Frazier, 2007). In the former case, there are questions over the acceptability of allowing Aboriginal people to harvest wildlife even though traditional (i.e. non-Western) methods are used. Short Hills Provincial Park in Ontario, Canada is a case in point. Every year Aboriginal people are allowed to harvest deer in this small park because of high deer numbers (Ministry of Natural Resources and Forestry, 2017). There are no natural predators (wolf) left to balance the deer population, and the abundance of deer has an adverse effect not only on park vegetation but also on the produce of local farmers. Pro-wildlife groups protest the cull each and every year (Gignac, 2017). In the latter case, time to assess our efforts at being sustainable is simply too short. Trends over a few years or even a few decades to gauge sustainability are said to be too short and therefore of limited value. Instead, Frazier contends that archaeozoology is of greater value through the use of unique tools that can gauge trends on human–environment uses over centuries. While an assessment of longer trends is of obvious value, decision makers often require shorter periods of time to anticipate and respond to the needs of different stakeholder groups almost immediately.

SDG 13: CLIMATE ACTION

Climate change is a change in the statistical properties of the climate system that persists for several decades or longer – usually at least 30 years. These statistical properties include averages, variability and extremes. Climate change may be due to

> natural processes, such as changes in the Sun's radiation, volcanoes or internal variability in the climate system, or due to human influences such as changes in the composition of the atmosphere or land use. (Australian Academy of Science, 2018)

There is overwhelming evidence to suggest that the changes identified in the quotation above have been accelerated by human actions (Mooney *et al.*, 2009). These actions have increased atmospheric concentrations of methane, water vapour and carbon dioxide (CO_2) – greenhouse gases (GHGs) – and prevent the outward flow of infrared energy from the Earth to outer space. Radiation is absorbed as it leaves the Earth's atmosphere and becomes re-radiating back to Earth. Evidence implicating human action as the main cause comes from many corners of the scientific community. Cook *et al.* (2013) analysed almost 12,000 scientific papers by over 29,000 authors, published in 1980 journals, with over 97% consensus on global warming being caused by human actions. These results are corroborated by the Intergovernmental Panel on Climate Change, the National Academy of Sciences, the American Meteorological Society, the American Geophysical Union and the American Association for the Advancement of Science (Oreskes, 2004).

The range of papers on the topic of climate change is enormous, connecting it to a diverse range of impacts. Some of these include threats to: general widespread biodiversity loss (MEA, 2005; Thomas *et al.*, 2004); coral reefs (Hughes *et al.*, 2003); extreme floods (Parsons, 2018); heightened extinction risks (Thomas *et al.*, 2004); risk of wildfires (Hamilton *et al.*, 2018); decline in floral diversity (Dirnböck *et al.*, 2017); reproduction problems in dairy animals (Yadav *et al.*, 2018); forestry biodiversity (Pawson *et al.*, 2013); and poverty (Beckford, 2019). In this latter case, Beckford argues that poverty must be broadened to include not only the absence of money, but also poor access to non-polluting energy sources, as well as to clean air and local conditions that are not subject to climate change instabilities. Beckford calls upon the main global power brokers to develop a new ecological approach to development that rewards ethical behaviour through the establishment of new standards based on ecological inclusivity rather than exclusivity – which Beckford refers to as transformative anthropocentrism.

The importance of climate change research and its applications in practice is represented in the UN SDGs, specifically SDG 13, where urgent action is needed to combat climate change and its impacts. The UN Report of the Secretary-General (2018f) states that that 2017 was recorded to be one of the three warmest years on record at 1.1°C above the pre-industrial period. Facts and figures from UNEP (2018) on the climate change SDG are:

- From 1880 to 2012, average global temperature increased by 0.85%.
- The Arctic's sea ice extent has shrunk every decade since 1979, with 1.07 million km^2 of ice lost every decade.
- From 1901 to 2010, the global average sea level rose by 19 cm as oceans expanded due to warming temperatures and melting ice.

- GHG emissions continue to rise and are now more than 50% higher than their 1990 level.
- For each 1°C of temperature increase, grain yields decline by about 5%.
- Since 1970, the number of natural disasters worldwide has more than quadrupled to around 400 a year.

Furthermore, World Meteorological Organization (2017) records show that the five-year average global temperature from 2013 to 2017 was the highest on record. Rising sea levels, extreme weather conditions (the North Atlantic hurricane season was the costliest ever recorded) and increasing concentrations of GHGs call for urgent action by countries as they implement their commitments to the Paris Agreement on Climate Change. Progress of the climate change SDG in 2018 is found in Box 2.3

Several tourism scholars have documented the significant impact that climate change is having on the tourism industry. Scott (2011) argues that it is imperative that tourism industries urgently respond to climate change as a prerequisite to achieving sustainable development. He further argues that global warming will be of the magnitude of +4°C and that society will need to adapt to a carbon-constrained economy. The implications for the tourism industry are dire and numerous. For example, Scott *et al.* (2012b) observe that a 1-metre rise in sea level would partially or fully inundate 29% of 906 coastal resorts in the Caribbean with water, while between 49% and 60% of these properties would be at risk of beach erosion.

What is problematic about these sorts of scenarios is the degree to which stakeholders in the tourism industry are willing to react to the consequences of climate change. Scott *et al.* (2012a) write that stakeholders in 11 countries report being relatively well informed

Box 2.3 Progress of SDG 13 in 2018

- As of 9 April 2018, 175 parties had ratified the Paris Agreement and 168 parties (167 countries plus the European Commission) had communicated their first nationally determined contributions to the UN Framework Convention on Climate Change Secretariat.
- In addition, as of 9 April 2018, 10 developing countries had successfully completed and submitted the first iteration of their national adaptation plans for responding to climate change.
- Developed country parties continue to make progress towards the goal of jointly mobilizing $100 billion annually by 2020 to address the needs of developing countries in the context of meaningful mitigation actions. (UN Report of the Secretary-General, 2018a)

Source: UN Report of the Secretary-General (2018f)

about the dangers of climate change, but are not overly concerned about climate change impacts to tourism businesses. There seems to be an implicit sense of optimism on the part of tourism operators about the chances of overcoming the challenges of climate change. At odds in the relationship between climate change and tourism businesses is the long-term temporal nature of climate change and climate change scenarios versus the relatively short-term nature of industry planning (two to five years), and the resultant downgrading of climate change realities to a lower priority status, i.e. 'Climate change will not be a problem during our business cycle'. This appears to be the case in South Africa on the basis of the investigation of 31 key stakeholders in tourism. Stakeholders consistently negate the possibility of impacts from climate change even in the face of the conclusions of the international climate change academic community (Pandy & Rogerson, 2018), as noted above.

SDG 6: CLEAN WATER AND SANITATION

The abundant supply of freshwater that many counties of the developed world enjoy, like Canada and Norway, contrasts with the shortage of freshwater that so many inhabitants of the world struggle with. Approximately 80% of the world's population is facing either water-related biodiversity issues or water security risk (Vörösmarty *et al.*, 2010). Research continues to document water-related problems that exist in these water-impoverished regions in the form of water security, water quality, water governance and the perceptions that people have about the quality of their water.

Water security is defined as 'the availability of an acceptable quantity and quality of water for health, livelihoods, ecosystems and production, coupled with an acceptable level of water-related risks to people, environments and economies' (Vörösmarty *et al.*, 2010: 545). Bakker (2012) argues that water security has attracted considerable attention from the social, natural and medical sciences and policy makers for the following main reasons:

1. threats to drinking water supply systems from contamination, human impacts and poor access;
2. threats to economic growth and human livelihoods from water-related hazards such as floods, droughts, water scarcity, water stress and energy security;
3. threats to water-related ecosystem services from point- and non-point source pollution and increased water consumption;
4. increased hydrological variability from climate change, especially the increase in the amplitude and frequency of droughts and floods. (Bakker, 2012: 914)

In Africa's semiarid countries there is an immediate need to improve the quality of life and food security, but this might only take place through a radical improvement in water availability. This is not a new phenomenon. Falkenmark (1989) argued for such changes in the 1980s especially among African leaders because of the growth in population numbers and the prospects for repeated water security collapses from recurrent droughts. More recent analyses suggest that 51 of 53 African large basins will become more arid, with consequences such as unsustainable food production in the northern part of the continent, social tensions over water in the Sahel, flood risk in Central Africa and considerable aridification in the southern part of the continent (Piemontese *et al.*, 2018). Water scarcity issues are also prevalent in more developed countries like the USA where specific regions have high agricultural demands because of an expanding population base (Yin *et al.*, 2016).

Water quality issues continue to be a problem for sustainable development, with persistent organic pollutants having an impact globally for over five decades, according to Schwarzenbach *et al.* (2010). They argue that geogenic pollutants (naturally occurring pollutants from geological processes, such the release of arsenic in water bodies), mining operation pollution and hazardous waste sites have long-term regional and local effects, while agricultural chemicals and wastewater sources have shorter term effects at these scales.

As sinks that capture the full impact of anthropogenic activities, marine biodiversity is threatened in all of the oceans, with acceleration in the extinction of marine organisms. Zhao *et al.* (2016) attributed the causes of fisheries exhaustion in the East China Sea from the stoichiometric (calculation of reactants and products in chemical reactions) composition of seawater (concentrations of N and P), toxic effects of marine pollution, marine habitat destruction, increased seawater temperatures, ocean acidification, pressure from overfishing and the spread of marine pathogenic bacteria. While these factors are significant on their own, it is the synergistic effect of these factors and their interactions that may pose the greatest threat to fisheries resilience.

A lack of information about the threats to marine biodiversity has led to a failure to enact the proper governance to mitigate these impacts (Craig, 2012b). Water governance is expressed as the combination of political, social, economic and administrative influences that manage water resources and provide services at different levels of society (Parven & Hasan, 2018). Governance is itself a climate change adaptation measure, and some communities have been using climate change impacts as a means to reduce their climate footprint. Australia's Great Barrier Reef is a case in point (Craig, 2012a). The effects of climate change on water quality due to warming and the consequences of extreme events continue to be a topic of paramount interest among scholars (Delpla *et al.*, 2009).

Progress in SDG 6, to ensure the availability and sustainable management of water and sanitation for all, is outlined in the UN Report of the Secretary-General (2018b). Some of the main issues tied to water include (http://www.undp.org/content/undp/en/home/sustainable-development-goals/goal-6-clean-water-and-sanitation.html):

- Water scarcity affects more than 40% of the global population, and this figure is projected to rise.
- 663 million people are still without improved drinking water.

- Each day, nearly 1000 children die due to preventable water and sanitation-related diseases.
- Women in sub-Saharan Africa collectively spend about 40 billion hours a year collecting water.
- 2.4 billion people worldwide do not have access to basic sanitation services like toilets or latrines.
- 80% of wastewater from human activities is discharged into waterways without any pollution removal.

The report notes the following changes taking place globally for the year 2018 (Box 2.4).

Apart from the actual environmental impacts on water, there are important considerations over the *perceptions* that people have regarding the quality of water available to them for consumption. Doria (2010) suggests that these perceptions are subject to a broad range of factors including risk perception, attitudes towards chemicals that may or may not be in the water, issues with the supply system, lack of familiarity with the properties of water, trust in suppliers, past problems that may have taken place with the supply of water, and the credibility of information provided by interpersonal sources and the media.

Box 2.4 Progress of SDG 6 in 2018

In 2015, 29% of the global population lacked safely managed drinking water supplies, and 61% were without safely managed sanitation services. In 2015, 892 million people continued to practise open defecation.

- In 2015, only 27% of the population in least developed countries (LDCs) had basic handwashing facilities.
- Preliminary estimates from household data of 79 mostly high- and middle-income countries (excluding much of Africa and Asia) suggest that 59% of all domestic wastewater is safely treated.
- In 22 countries, mostly in the Northern Africa and Western Asia region and in the Central and Southern Asia region, the water stress level is above 70%, indicating the strong probability of future water scarcity.
- In 2017–2018, 157 countries reported average implementation of integrated water resources management of 48%.
- Based on data from 62 out of 153 countries sharing transboundary waters, the average percentage of national transboundary basins covered by an operational arrangement was only 59% in 2017.

Source: UN Report of the Secretary-General (2018b)

On Holbox Island, Quintana Roo, Mexico, Tran *et al.* (2002) studied the public's (various stakeholder groups including tourists and local people) perceptions of tourism development with regard to water quality pollution. They found that local people did not perceive water quality to be an issue in the coastal community, despite the fact that there are water quality problems there. Almost 50% of local people interviewed felt that visitors and tourism had no effect on the negative changes taking place on the island. By contrast, a study by Wu *et al.* (2008) found that there was recognition of the direct contamination of Lugu Lake, Yunnan Province, China from tourism, specifically due to overdevelopment of the region.

SDG 7: AFFORDABLE AND CLEAN ENERGY

The focus on energy in the SDGs is of critical interest because of the enormous demand that humanity places on the environment. SDG 7, to ensure access to affordable, reliable, sustainable and modern energy for all, emphasizes the importance of providing energy for the world's growing population, but also the need to ensure that there is a move towards sustainable forms of energy production and consumption. Critical facts and figures offered by UNEP (2018c) on global energy issues include:

- Energy is the dominant contributor to climate change, accounting for around 60% of global GHG emissions.
- More than 40% of the world's population, 3 billion people, rely on polluting and unhealthy fuels for cooking.
- Globally, as of 2011, more than 20% of power is generated through renewable sources.

Progress in energy is documented in the Sustainable Development Report of 2018 (UN Report of the Secretary-General, 2018c) shown in Box 2.5.

The International Energy Agency's (IEA, 2017a) World Energy Outlook reports that there will be four main shifts on the global energy scene in the coming years. These include the falling costs of green energy, growing demand for electric energy, a shift towards more of a service-oriented economy in China (where there is an energy revolution taking place), and the resilience of gas and oil in the USA. The report illustrates that although energy demand is rising more slowly than in past years and decades, energy needs will expand by 30% between now and 2040. This is akin to adding another India and China to the world demand for energy – India's share of the global energy use will rise to 11% by 2040.

Furthermore, the types of energy used in the future are of critical importance too. Non-renewable sources of energy, which we rely on to such a significant degree, include oil, coal, natural gas and nuclear power, while renewable sources of energy have diversified significantly over the last few decades. Some scholars believe that countries like Denmark are capable of achieving a 100% renewable energy system through further technological improvements of the energy system. Technological change in the area of sustainable energy development strategies usually follows three pathways: demand-side

Box 2.5 Progress of SDG 7 in 2018

Ensuring access to affordable, reliable and modern energy for all has come one step closer due to recent progress in electrification, particularly in LDCs, and improvements in industrial energy efficiency. However, national priorities and policy ambitions still need to be strengthened in order to put the world on track to meet the energy targets for 2030.

- From 2000 to 2016, the proportion of the global population with access to electricity increased from 78% to 87%, with the absolute number of people living without electricity dipping to just below 1 billion.
- In the LDCs, the proportion of people with access to electricity more than doubled between 2000 and 2016.
- In 2016, 3 billion people (41% of the world's population) were still cooking with polluting fuel and stove combinations.
- The share of renewables in final energy consumption increased modestly, from 17.3% in 2014 to 17.5% in 2015. Yet only 55% of the renewable share was derived from modern forms of renewable energy.
- Global energy intensity decreased by 2.8% from 2014 to 2015, double the rate of improvement seen between 1990 and 2010.

Source: UN Report of the Secretary-General (2018c)

energy savings, efficiency in energy production and replacement of fossil fuels by renewable sources (Lund, 2007). Ellabban *et al.* (2014) provide an overview of the breadth of renewable energy sources that should be used in the development of a sustainable future (Figure 2.3). These authors argue that renewable sources of energy are attractive because in principle they could exceed the world's demand for energy through the changing and adaptation of technologies.

Despite their importance now and in the future, and the fact that they are used now more than ever, renewable energy technologies still represent a small percentage of the global renewable–non-renewable energy mix (Stigka *et al.*, 2014). Statistics from the IEA (2017b) illustrate that in 2015 oil represented 31.7% of the total primary energy supply (TPES), followed by coal at 28.1%, natural gas at 21.6% and nuclear power at 4.9%. The total TPES for these non-renewable sources is 86.3%. By contrast, renewable sources such as biofuels and waste (9.7%), hydro (2.5%), and other sources including geothermal, solar, wind, tide/wave/ocean and other minor sources, amounted to a mere 1.5%, for a total of 13.7% of renewable TPES. US energy consumption in 2017 is outlined in Table 2.4.

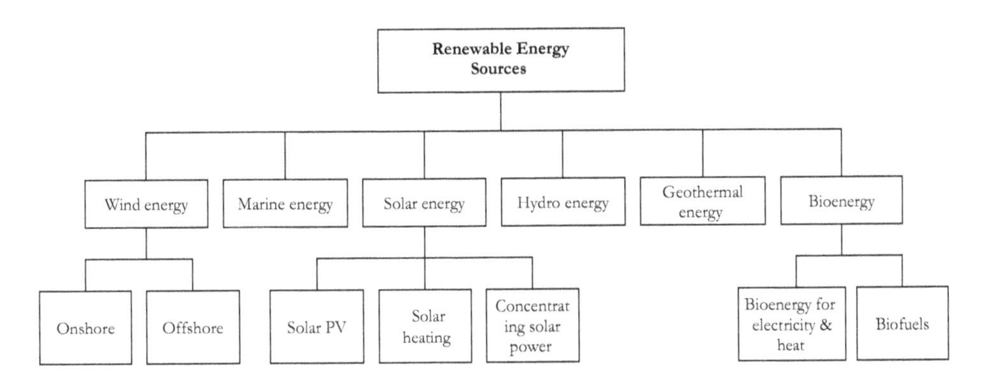

Figure 2.3 Overview of renewable energy sources
Source: After Ellabban *et al.* (2014)

Table 2.4 US energy consumption by energy source, 2017

Non-renewable sources	%	*Renewable sources*	%
Petroleum	37	Hydroelectric	2.7
Natural gas	29	Wind	2.3
Coal	14	Biomass – biofuels	2.3
Nuclear electric power	11	Biomass – wood	2.1
TOTAL	91	Solar	0.7
		Biomass – biomass waste	0.4
		Geothermal	0.2
		TOTAL	10.7

Source: US EIA (2018)
Note: Sum of components does not add up to 100% because of independent rounding.

In Finland, a study on the social acceptability of renewable energy technologies found that 63% of respondents were willing to pay more for green energy and that the public sector should take more responsibility for stimulating incentives and new business models in the implementation of renewable energy technologies (Moula *et al.*, 2013). In South Korea, consumers were willing to pay an extra US$3.21 per month for renewable energy – a low value according to Lee and Heo (2016) relative to other advanced countries, with the suggestion that more knowledge creation needs to take place among local populations. WTP for green electricity is often a function of education and environmental awareness, with household income as the most important factor (Zorić & Hrovatin, 2012).

A detailed account of the use of energy in tourism comes from Kelly and Williams' (2007) investigation of the resort town of Whistler, BC, Canada. While some energy is typically derived from local systems such as wind turbines and photovoltaic cells, most energy comes from outside the destination through the transformation of fossil fuels such as natural gas, coal, oil and electricity supplied by hydro generation into energy. In the process of transformation there is the production of harmful chemical substances that are released into the atmosphere, including CO_2, carbon monoxide (CO), nitrogen oxide (NO_x), sulphur oxide (SO_x) and particulate matter. Those fuels that are not completely burned are termed volatile organic compounds and may also enter the atmosphere. Kelly and Williams provide a breakdown of the direct effects (e.g. energy consumed by hotels in providing accommodation for tourists) and indirect effects (e.g. energy consumed by a business in providing office services to tourism operators). These effects can be seen for the following uses: residential; passenger transportation; commercial, industrial and institutional; municipal buildings and infrastructure; and public transportation (Table 2.5).

Energy use in tourism is typically greater than for other non-tourism related uses (based on a comparison of similar-sized uses) because of the need to deliver superior services to tourists. In Hawaii, for example, tourists were found to account for about 60% of all energy and fuel used in the state (Tabatchnaia-Tamirisa *et al.*, 1997). As tourist numbers continue to escalate, there will be associated increases in the demand for energy to accommodate this need.

WASTE MANAGEMENT

The Global Partnership on Waste Management (GPWM), launched in 2010, is a collection of international organizations, governments, academics, local authorities and NGOs, interacting for the purpose of building knowledge, cooperation, information sharing, identification of gaps and the promotion of resource efficiency and resource conservation (UNGPWM, 2018). The main partners include the UN Environment Programme, the International Solid Waste Association and the International Telecommunications Union. The major themes of the partnership include: (1) waste and climate change; (2) waste agricultural biomass; (3) integrated solid waste management; (4) e-waste management; (5) marine litter; and (6) waste minimization as well as hazardous waste management and metal recycling. This last theme is particularly problematic for human health as the effects of lead, for example, include damage to the brain and the renal, reproductive, blood and nervous systems, with lead accumulations in the environment contributing to acute and chronic effects in humans (Kiddee *et al.*, 2013).

Although not a specific UN SDG, waste does factor prominently in other goals. Goal 11 (Make cities and human settlements inclusive, safe, resilient and sustainable) illustrates that, as of 2018 and based on a dataset of 214 cities and municipalities, three-quarters of municipal solid waste generated is being collected (UN Report of the Secretary-General,

Table 2.5 The effect of tourism on destination energy and GHG emissions

Sector	*Direct effects (included)*	*Indirect effects (included)*	*Induced effects (not included)*
Residential	Energy consumed domestically by tourists staying in residential dwellings		Energy consumed domestically by resort workforce
Passenger transportation	Energy consumed by tourists using private transportation		Energy consumed by resort workforce using private transportation
Commercial, industrial and institutional	Energy consumed by businesses in providing tourists with products and services (e.g. accommodation or recreation services)	Energy consumed by businesses in providing other tourism businesses with products and services (e.g. office or construction services)	Energy consumed by businesses in providing resort workforce with products and services (e.g. restaurant or retail services)
Municipal buildings and infrastructure	Energy consumed in providing tourists with municipal services	Energy consumed in providing tourism businesses with municipal services	Energy consumed in providing resort workforce with municipal services
Public transportation	Energy consumed in providing tourists with public transportation		Energy consumed in providing resort workforce with public transportation

Source: Kelly and Williams (2007)

2018d). However, even though recycling is on the rise, and people recognize their ethical obligation to recycle, 85% of landfill sites are filled with unwanted waste, including e-waste, glass, metal, plastic, paper and textiles (Nodoushani *et al.*, 2016). Furthermore, Goal 12 (Ensure sustainable consumption and production patterns) illustrates that, according to a recent report from KPMG, 93% of the world's 250 largest companies (in terms of revenue) are now reporting on sustainability, as are three-quarters of the top

100 companies in 49 countries (this would include reporting on waste) (UN Report of the Secretary-General, 2018e).

The problem of waste is particularly acute in developing country cities where escalating human populations in association with poor waste management infrastructure and poor education and awareness are just a few of the factors that limit success in waste management (Guerrero *et al.*, 2013). Studies on tourism, waste and developing countries often show similar results. In the hotel sector in Vietnam, for example, Trung and Kumar (2005: 113) report that:

- Most of the wastewater is discharged into the public sewage system, river or sea. The wastewater treatment system does not handle all the wastewater. Some hotels and resorts do not have the capital, technical support or access to suitable small-scale water treatment systems.
- Recycling wastewater for other purposes (watering, gardening, etc.) is not common.
- In some hotels, solid waste is separated for recycling, but mainly for paper, plastic and aluminium cans.
- In some resorts, efforts are not made to compost organic waste and produce fertilizer.
- Official letters are still used for internal communication in 4- and 3-star hotels.

A study of 120 hotels in Hoi An, Vietnam, found that the mean amount of waste generated in the hotels was 2.28 kg/guest/day, and that higher scale hotels have a higher waste generation rate (Pham Phu *et al.*, 2018). Like energy consumption, above, where tourists use more energy than local people, it is also the case that tourists generate more waste than locals, which places even more strain on destinations that have waste management challenges to begin with (Wang *et al.*, 2018).

BIOPHILIA

During the 1960s there were important signals emerging from the academic community regarding the shift taking place in ecological values. Rachel Carson's *Silent Spring* (1962) documented how the indiscriminate use of chemicals (pesticides) was having an adverse affect on many aspects of the natural world. During the same period of time, Fromm (1964, 1968) observed that humans were experiencing an existential dichotomy based on an inability to connect with nature as a consequence of the affinity towards human-made structures and things – elements that are not alive but rather mechanical. This resulted in an array of psychological problems such as lostness, meaninglessness, insignificance and fear, among other factors (see also Gunderson, 2014).

Two decades later, E.O. Wilson (1984) authored a book illustrating that this movement away from a connection to the natural world was itself rather unnatural. His biophilia hypothesis – biophilia refers to the 'innate love of nature' – includes reverence for natural elements and processes such as animals and landscapes, along with the meaning and pleasure derived from these relationships (Joye & De Block, 2011). Wilson (1993) argued that biophilia evolved in humans via natural selection. Indeed, there is evidence to suggest that *biophobia*, or the intense fear of living things, evolved in humans as an evolutionary survival strategy against snakes, spiders and other taxa in nature that posed a threat to our wellbeing (Gullone, 2000). Others have argued that biophilia is learned and experiential (Simaika & Samways, 2010) and that the evolutionary or gene-centred view of biophilia is tenuous at best (Joye & De Block, 2011).

Researchers have tried to shed light on the intricacies of biophilia as it relates to different populations and in different regions. Zhang *et al.* (2014) surveyed children in both rural and urban schools in China and found that children who had higher levels of contact with nature, typically in rural schools, were more likely to conserve animals and support conservation. Other studies in China point to the fact that the young have more positive attitudes towards animals in comparison to middle-aged and older citizens (Su & Martens, 2017). In a study of the sensitivities to a biocentric view of pre-schoolers (aged five) in Northern Italy, Margoni and Surian (2017) found that children were more willing to support a reward for an agent who had biocentric intentions over those who had anthropocentric intentions. The authors suggest that this shows that children can have biocentric sensitivities at an early age, and this has relevance to Wilson's views about the innate love of nature, which might nevertheless erode over time and with exposure to the non-natural elements of civilization.

Cross-cultural studies have also been conducted on biophilia, and some of these have a direct link to tourism. Fox and Xu (2017) found that biophilia is influenced by the sociocultural environment (national culture) as well as by current living conditions. Based on a survey of visitors to a national park in England and one in China, the authors found that the biophilic construct used in the study (i.e. the assumption that biophilia is an innate, evolutionary characteristic of humanity and would not vary between samples) was in fact variable. The British held stronger feelings towards biophilic statements than the Chinese sample. In a related study, Xu and Fox (2014) discovered that people's attitudes to nature predict their attitudes about the relationship between the environment and tourism, and further about the relationship between ST and national parks in reference to resource management.

Stedman and Ingalls (2014) have drawn a connection between Wilson's biophilia and Tuan's (1980) concept of topophilia, the latter defined as 'the love of place', which may include neighbours and neighbourhoods, social relationships, memories, particular landmarks and other elements of the built environment tied to place. These two concepts may be used in a strategy to help red zone areas recover from social, economic and environmental catastrophes. Red zone areas, following Tidball and Krasny, are settings, in spatial and temporal terms, that the authors characterize as 'intense, potentially or recently hostile or dangerous' areas or periods of time. These areas may be urban environments that have suffered erosion or decline because of economic stagnation, or they may be areas that have suffered because of natural disasters, warfare, environmental accidents and/or other highly visible crises. Stedman and Ingalls (2014) argue that both biophilia and topophilia may be operationalized as greening strategies for the rejuvenation of red zone areas through the coming together and engagement of people from different walks of life in building social networks, trust and cooperation in the creation of resilient communities. Biophilia has also been applied practically through the concept of biophilic design (Kellert, 2008), in the design of cityscapes (Beatley, 2011). An array of these strategies has been reported by Downton *et al.* (2016), and includes, for example, green gardens, sky gardens, green walls, green courtyards, green streets, sidewalk gardens, edible landscaping, ecology parks and gardens, community gardens, greening greyfields and brownfields, urban creeks and riparian areas, green schools and the greening of utility corridors.

Finally, several authors have emphasized the value of biophilia as a foundation for a deeper conservation ethos (Simaika & Samways, 2010). Santas (2014) argues that if such is to materialize it must be taken more broadly and incorporated as an interconnected feature of biotic systems. Drawing on Aristotelian ethics in an effort to make this connection, particularly Aristotle's concept of *philia* (friendship), Santas makes this connection more explicit by suggesting that biophilia is merely one aspect of *philia* that is based on affection and friendship, which might also be reconfigured as a larger context of biotic interconnection between humans and the natural world. Biophilia has also been constructed as an environmental virtue from both individual and collective standpoints (Clowney, 2013). In this way virtue may contribute to human flourishing through care about nature as the primary target, which may also help to unite communities for the purpose of conserving biodiversity. Curtin (2009) and Fennell (2012), for example, have used the biophilia hypothesis in tourism scholarship: Fennell (2012) in the context of animal ethics, while Curtin discussed biophilia according to the restorative effects of wildlife on human wellbeing. Humans experience joy and happiness upon witnessing animals in nature.

CONCLUSION

This chapter provided the foundations for an understanding of the basic functions of the ecosphere and why these functions (and services) are so important to secure if we are to achieve sustainability. Ecosphere was chosen as the title of the chapter over biosphere because the former underscores the importance of both the living and the non-living components of the Earth. A significant amount of space was allocated to biodiversity conservation (referencing SDGs 14 and 15) because for many it is the most pressing environmental issue that we are now confronting – even more serious than climate change. The services that ecosystems provide for us are estimated to be worth at least double the world's total GDP (trillions of dollars), so if human actions (industrial sites, housing, agriculture) continue to destroy natural places we are only hurting ourselves in the end (Carrington, 2018). Habitat loss, invasive alien species, pollution, population increases and overharvesting of resources are all contributing factors to biodiversity loss. The chapter also introduced climate change, water scarcity, energy and waste management – all identified within the matrix of SDGs (SDGs 13, 6, 7 and 12, respectively). Tourism has an impact on all of these, and it will be the task ahead to intersect tourism and non-tourism studies in these areas for the purpose of building a more complete picture of sustainable tourism.

CHAPTER 2 DISCUSSION QUESTIONS

1. Define biodiversity and list the various main contributing factors (HIPPO) to biodiversity loss.

2. What is an ecosystem service, and why are ecosystem services so important to our economy and wellbeing?

3. Compare and contrast personal values with non-use values.

4. What is climate change and how is tourism a contributing factor?

5. What is water security? List examples of places where water security is problematic.

6. Why is waste management a problem in the airline industry?

7. What is biophilia?

End of Chapter Case Study 2.1 Ecological engagement and ecosystem service in a National Urban Park

In a statement by Parks Canada (2018), Rouge National Urban Park 'is home to amazing biodiversity, some of the last remaining working farms in the Greater Toronto Area, Carolinian ecosystems, Toronto's only campground, one of the region's largest marshes, unspoiled beaches, amazing hiking opportunities, and human history dating back over 10,000 years, including some of Canada's oldest known Indigenous sites'.

Livingstone *et al.* (2018) sought to understand how stakeholder valuation of ecosystem services (ESs), and perceptions of threats to their conservation, can improve planning for urban protected areas. Their study objectives were to examine ES valuations by Rouge National Urban Park (NUP) users as well as perceptions of the impact of the invasive vine *Vincetoxicum rossicum*. These authors also sought to determine how those valuations and perceptions are affected by 'ecological engagement' (EE). The authors conducted a social survey of Rouge NUP users and found that valuation of most ESs was significantly greater for EE users. Interestingly, non-EE users tended to give recreation ('cultural' ES) the highest importance value. Conversely, EE users tended to assign pollination ('supporting' ES), the highest importance. Further, we were surprised to find that 15.2% of EE and 38.4% of non-EE users disagreed or were neutral to the notion that *V. rossicum* is negatively impacting the Park's supporting ESs. Similarly, 32% of EE and 54.1% of non-EE users disagreed or were neutral to the notion that *V. rossicum* is negatively impacting the Park's aesthetic ESs. They conclude that examination of EE can reveal differential ES valuations and perceptions of invasion impact. Additionally, they suggest that such examination can inform conservation management plans and public engagement strategies.

Sources: Livingstone *et al.* (2018); Parks Canada (2018).

CHAPTER 3
INTRODUCTION TO THE NOÖSPHERE

LEARNING OUTCOMES

The focus of this chapter is on gaining appreciation of the importance of the UN Sustainable Development Goals (SDGs) that relate more specifically to the noösphere and their potential for informing tourism theory and practice. The chapter will provide the reader with:

1. A short overview of the remaining 12 UN SDGs, with reference to research both inside and outside of tourism to provide context.

2. A review of some major organizations that relate to these SDGs.

3. The final pieces of information that relate to the book's Conceptual Framework (Chapter 1, p. 26).

4. A focus on human nature and how this knowledge is viewed as essential in better understanding tourists on an atomistic scale, as well as the tourism industry, when it comes to sustainable tourism (ST).

INTRODUCTION

In the Introduction to Chapter 2 we discussed the correct terminology for labelling the Earth's living and non-living properties. The term 'ecosphere' was selected over 'biosphere' because the former was deemed more inclusive of all the living and non-living elements comprising the Earth that humans use for their myriad projects and enterprises. A brief discussion is required here to find the correct terminology to describe these various projects and enterprises – to describe the evolution of human ingenuity and innovation that has been so vital in exploiting the ecosphere for our own instrumental ends. For Wyndham (2000), and many other scholars, the proper term is *noösphere* (derived from Greek *νοῦς*, as nous or 'mind', and *σφαῖρα*, as sphaira or 'sphere') (https://en.wikipedia.org/wiki/Noosphere), translated as the sphere of human thought.

The origins of the term noösphere can be traced back to Vernadsky (1926: Chapter 2), who in turn adopted the term from his colleagues E. Le Roy (1928) and P. Teilhard de Chardin (1955/2008). Vernadsky defined the noösphere as a 'new form of biogeochemical energy, which might be called the energy of human culture or cultural biogeochemical energy' (Vernadsky, 2012: 18). As such, the noösphere is about harnessing nature's energy (mastering the forces of nature, including the discovery of fire as the first great leap forward, agriculture, animal husbandry and new technologies) into a new form of cultural energy. For Wyndham (2000: 87), the noösphere is 'produced and maintained by increasing complexity of human interaction in cultural, social, biological and physical environments'.

Teilhard de Chardin (1955/2018) described the noösphere as a form of consciousness evolving as a thin layer of tissue enveloping the Earth. He felt, for example, that love was a promising form of cultural energy that could unit all living things in generating global connectedness. This view parallels how biophilia ties humans to the natural world (Santas, 2014), or how biophilia stands as an environmental virtue enabling humans both to flourish and to unite communities for conservation purposes (Clowney, 2013). There is presently a renewed interest in the concept of the noösphere as a savoury connection to the internet, with the latter viewed as 'an emergent global realm of human thought' (Peters & Reveley, 2015: 4). The lines moving outwards in Figure 1.3 represent this thin layer of tissue enveloping the Earth. The noösphere has relevance to tourism, as tourism is a global human phenomenon that touches the entirety of the planet from the equator to the poles, from urban to wilderness settings, and from aquatic to the highest terrestrial environments.

Like the previous chapter on the ecosphere, the formula used in the present chapter is to introduce the reader to a number of general issues and their impacts that relate to the sphere of human thought. These follow from the UN SDGs and include poverty, hunger, education, gender equity, working conditions, industry, inequalities, community development, consumption and consumerism, peace and justice, and partnerships. The chapter concludes with a discussion of human nature, which helps us to better understand the actions of tourists and the tourism industry from a philosophically embedded perspective.

GENERAL ISSUES

Each of the remaining SDGs not discussed in Chapter 2 is introduced briefly below for the purpose of establishing a benchmark from which to discuss ST in the remaining chapters. The template is not identical for the discussion of each of the SDGs, but will follow a loose outline around major organizations involved in the area of a particular SDG, definitions, theories, indicators or an example of research taking place outside tourism. Much greater emphasis will be placed on tourism in later chapters.

SDG 1: NO POVERTY

Nelson Mandela observed that poverty is not a natural phenomenon, but rather a condition created by human action. Its eradication will also involve human action, not from the perspective of charity but rather from that of justice (UN, 2014). For many, the eradication of poverty is the single greatest challenge facing the global community, the magnitude of which is easy to imagine given that urban areas are expected to receive in excess of 90% of global population growth in the next 40 years (UNDESA/PD, 2012).

Definitions of poverty have long been contested, with disagreement around four different elements (Laderchi *et al.*, 2003). The first is the space in which poverty ought to be defined. This includes the degree of emphasis that should be placed on different spheres, including social, cultural, resource or political concerns. The second is universality of definitions, or how definitions fit within local or regional contexts and if they fit in cross-cultural situations. The third concern is the objective versus subjective measurement of poverty. Objective measures are captured by statistics, whereas subjective measures are based on value judgements, which opens the door to different agendas and interests of different stakeholder groups. The fourth issue is how to differentiate the poor from the non-poor along poverty lines. Challenges include the justification of specific lines, and whether such lines reflect absolute standards of deprivation. What is clear in the literature on poverty, however, is the need to look beyond simply income as the only measure of poverty towards other critical variables including, for example, education, public health, safe drinking water and proper sanitation facilities (Liu *et al.*, 2015). These variables have been discussed by Waglé (2008) under the category of capabilities, i.e. whether individuals have the capacity to access education, health and so on in securing wellbeing. Wagle also mentions the element of social exclusion as an important poverty variable. Even though individuals may not be income poor or capabilities poor, they may still be impoverished if they are excluded from mainstream economic, political, cultural and civic activities – a topic that should be of interest to tourism scholars.

More specific to ST is the notion that sustainable development (SD) may only be truly realized in the absence of poverty (Lui *et al.*, 2015). Achieving SD in developing countries is therefore necessary because evidence suggests that poor people in such regions may not embrace stewardship when the economy does not support their immediate basic needs such as shelter and food. Education about the human and ecological value of SD is critical, because the poor may not embrace SD if it is perceived to threaten their livelihoods (Cobbinah *et al.*, 2015).

Some of the basic facts and figures reported by the UN Environment Programme (UNEP, 2018b), and associated examples of progress (Box 3.1), on poverty are as follows:

- 650 million people still live in extreme poverty;
- about one in five persons in developing regions lives on less than US$1.25 per day, with 80% of these people living in South Asia and sub-Saharan Africa;
- 11% of the world's population live in extreme poverty, down from 28% in 1999;
- every day, an average of 42,000 people had to abandon their homes due to conflict.

Box 3.1 Progress of SDG 1 in 2018

While extreme poverty has eased considerably since 1990, pockets of the worst forms of poverty persist. Ending poverty requires universal social protection systems aimed at safeguarding all individuals throughout the life cycle. It also requires targeted measures to reduce vulnerability to disasters and to address specific underserved geographic areas within each country.

The rate of extreme poverty has fallen rapidly: in 2013 it was a third of the 1990 value. The latest global estimate suggests that 11% of the world population, or 783 million people, lived below the extreme poverty threshold in 2013.

The proportion of the world's workers living with their families on less than $1.90 per person a day declined significantly over the past two decades, falling from 26.9% in 2000 to 9.2% in 2017.

Based on 2016 estimates, only 45% of the world's population were effectively covered by at least one social protection cash benefit.

In 2017, economic losses attributed to disasters were estimated at over $300 billion. This is among the highest losses in recent years, owing to three major hurricanes affecting the USA and several countries across the Caribbean.

Source: UN Report of the Secretary-General (2018a)

The depth of research in the area of poverty is voluminous. Some of this work investigates the extent to which foreign aid reduces poverty. Researchers have found that aid packages that are larger in size do not necessarily translate into greater gains in poverty reduction. This is due to the altruistic tendencies of those who are in charge of handling this aid. Higher levels of embezzlement have been detected as well as an increased propensity to relax efforts in addressing poverty problems because of the size of the aid project (Bourguignon & Platteau, 2017). Bigger aid packages often result in a lower degree of proportionate poverty reduction impact than smaller packages that appear to have a more direct impact because the money is used for its intended purpose.

In tourism, there are significant challenges in eradicating poverty in developing countries, even though there are several mechanisms in place such as pro-poor tourism and Fair Trade tourism that are designed to provide assistance. The former is based on formal justice and fairness and is said to generate net benefits for poor people through strategies such as expanding business and employment opportunities for the poor, capacity building and training, mitigating environmental impacts of tourism on the poor and promoting participation (Roe & Urquhart, 2002). The latter focuses on fair production and trade practices between developing and developed countries, as well as fairness in the distribution of benefits and strong cooperative partnerships (Cleverdon & Kalisch, 2000).

However, as Bauer (2017) observes, economic priorities still take precedence in tourism with the rich having power over the poor. The system is not geared towards enhancing the capacity for communities to secure self-determination and equal partnership, and the equal distribution of benefits. Altruistic forms of tourism like pro-poor tourism and Fair Trade tourism are ineffective because they have been appropriated by the tourism industry in the pursuit of corporate interests (Higgins-Desboilles, 2008).

SDG 2: ZERO HUNGER

Closely related to the concept of poverty is hunger, which is a developing world problem but also problematic in the developed world. The FAO, International Fund for Agricultural Development and World Food Program (2015) report that of the 815 million people (10.7% of the world's population) who are chronically undernourished, 11 million people are undernourished in the developed world – a small percentage of the total. Poverty is a main cause of hunger, but hunger is also a contributing cause of poverty in a cyclical relationship (World Hunger Education Service, 2018).

One of the greatest challenges in the future will be how to feed 9–10 billion people by the year 2050 and, at the same time, reduce the impact that we have on the natural world in areas such as greenhouse gas (GHG) emissions, losses to biodiversity, land use change and the loss of ecosystem services (Smith & Gregory, 2013). Other challenges relate to the alarming increase in the rate of child obesity as a result of poor dietary choices, and how marketing is identified as a contributing factor to this issue (Rayner *et al.*, 2008), and the sheer number of people who are undernourished.

Several gains have been achieved in the eradication of poverty. For example, UNEP (2018c) reports that economic growth and increased agricultural productivity, especially in Central and East Asia and Latin America and the Caribbean, have reduced by a half the number of undernourished people on the planet over the course of the last 20 years. While impressive, such statistics often mask the huge disparities that continue to exist globally. UNEP (2018c) reports that:

- one in nine people in the world today is undernourished;
- a quarter of children suffer from stunted growth; in some developing countries, it is as high as one in three;
- if women farmers had the same access to resources as men, the number of hungry people in the world could be reduced by 150 million;
- Asia is the continent with the hungriest people, two-thirds of the total;
- since the 1990s, some 75% of crop diversity has been lost from farmers' fields.

There are two main factors that are often addressed in defining hunger. First is the actual state of hunger, which is defined as a short-term physical discomfort as a result of chronic food shortage or, in severe cases, a life-threatening lack of food (NRC, 2006). The second refers to the food insecurity aspects of hunger, i.e. the lack of access that people have to food, which is said to be the most important direct cause of hunger (Smith *et al.*, 2006).

In addressing both of these factors, several organizations have established indicators or indices to measure hunger. The World Bank focuses primarily on caloric energy consumption. If an individual consumes fewer calories than the minimum number recommended by the WHO, he or she is classified as undernourished (Smith *et al.*, 2006). The FAO index measures 'hunger as the fraction of the population with per capita energy consumption below standard nutritional requirements' (Smith *et al.*, 2006: S103). The Global Hunger Index is a composite index developed around three indicators – the proportion of a population undernourished by FAO standards; the prevalence of underweight children under five years of age as established by WHO standards; and the under-five mortality rate as set by UNICEF – with all three indicators used to set an average score (Aiga, 2015; IFRPI, 2013).

However, these indices have been criticized because they fail to reference other important factors such as distribution issues, the failure to account for food and health shocks, and unreliable data (Masset, 2011; see also Nah & Chau, 2010). In an effort to build greater capacity in the fight against hunger, Masset argues for: (1) campaigns and information to generate political will (conditional on the institutional environment); (2) political will translates into policies (conditional on the political environment); (3) policies translate into programmes on the ground (conditional on resources); and programmes reduce hunger (conditional on capacity).

Targets for SDG 2 include ending hunger for all people, ending malnutrition, expanding agricultural productivity, maintaining genetic diversity of seeds and plants, increasing investments, correcting trade restrictions and adopting measures to ensure proper functioning of food commodity markets. Target 2.4 reads:

> By 2030, ensure sustainable food production systems and implement resilient agricultural practices that increase productivity and production, that help maintain ecosystems, that strengthen capacity for adaptation to climate change, extreme weather, drought, flooding and other disasters and that progressively improve land and soil quality.

The sole indicator for this target is: 'Proportion of agricultural area under productive and sustainable agriculture' (https://sustainabledevelopment.un.org/sdg2).

While these are vital objectives which all must be addressed, there will need to be a greater focus on the critical habitats and non-human animal dimensions of this goal (this SDG could also have been placed in the previous chapter because of its link to agriculture and land practices). Chapter 7 will provide greater scrutiny of this dilemma, but it is worth noting at this juncture that livestock production in South and Central America to serve the culinary (more like fast-food) interests of North Americans has a huge effect on tropical forests. For each North American hamburger, 'the environmental cost is half a ton of rainforest … Expressed as forest area, the cost is 67 square feet – more than 6.25 square metres of forest' (Newman, 1990: 142, cited in Fox, 2000: 170). Nibert (2002) adds that with a projected global population of 9 billion people by the year 2050 the production of meat for such a population 'would require even greater levels of deforestation, increase desertification ... add to air pollution ... exhaust freshwater supplies, and compound already critical levels of water pollution' (Nibert, 2002: 115).

The complexity of issues tied to the production and consumption of food is only recently coming to the fore in tourism studies (see, for example, Hall *et al.*, 2003; Hjalager & Richards, 2002). It is only lately that the contents of tourism books on food have included ethical implications tied to the use of animals (Kline, 2018). Sustainability in tourism as applied to food must be bound to a broader agenda beyond just pleasure. In fact, a 20-page Google Scholar search of academic sources on 'tourism' and 'hunger' by relevance yielded no papers with both terms *in the title*, apart from two, both of which focused on gastronomy, or satisfying hunger at good restaurants (e.g. Svard, 2013). While several tourism scholars have addressed poverty, hunger is usually addressed in a more cursory manner. It is encouraging that past work has investigated poverty and hunger together as they relate to the Millennium Development Goals. Two examples found in the same journal special edition include Saarinen *et al.* (2011) and Novelli and Hellwig (2011). This latter paper lists a variety of actions that tourism operators have taken in the eradication of extreme poverty and hunger. Some of these include the building of schools, assisting with new business initiatives to create jobs, and the selling of local goods in safari shops, as well as helping local women to improve the quality of their craft and souvenirs.

The market of tourists interested in vegetarianism and veganism is expanding, as is evident in a recent study in Turkey. Dilek and Fennell studied visitors to the first ever Didim VegFest in Turkey over two days in April 2017, which itself is a marker for changes in tourist food consumption, and asked tourists about the main hotel selection factors that were of particular interest to them. Findings point in the direction of eco/animal-friendly establishments where staff and services are oriented towards safeguarding the natural environment. Scholars are also reporting that vegetarianism is not just dietary, but a lifestyle, which is one of the fastest growing trends worldwide (Wu, 2014). Dilek and Fennell (2018) conclude that given this trend, businesses, especially restaurants and hoteliers, need to diversify their services to accommodate these needs so that vegetarians will not be marginalized because of a moral choice that is both personal and global in nature. In other research, Yudina and Fennell (2013) explored the tourism dimensions of eating animals from an ecofeminist perspective, and argued that there really are no boundaries or limits to what animal is consumed, where and why. This is because food is no longer a necessity but rather a psychological pleasure (Finkelstein, 1989), and embodies culinary experiences that access all of the senses (Crouch, 2000), as postmodernism brings with it the desire to eat exotic foods at the cosmopolitan scale. These are 'pluralized and aestheticized experiences that have fostered new patterns of tourism consumption and the development of new individualized identities' (Everett, 2009: 340).

SDG 3: GOOD HEALTH AND WELLBEING

Health and wellbeing is an area that has made significant progress since the dawn of the new millennium. UNEP (2018d) states that between the years of 2000 and 2013 HIV/AIDS infections dropped by 30%, and over 6 million lives were saved from malaria prevention. Since 1990, preventable child deaths globally declined by more than 50%. The UN, through the SDGs, is committed to ending AIDS, tuberculosis and malaria by 2030, with the purpose of instituting universal health coverage and access to medicine sand vaccines for all. Still, there are challenges as outlined in the following facts and figures by UNEP (2018d):

- globally, each year more than 6 million children die before reaching their first birthday;
- children born into poverty are almost twice as likely to die before the age of five as those from wealthier families;
- 16,000 children die each day from preventable diseases.

Several indices of wellbeing are presently in use as reported by Giannetti *et al.* (2015), but they may be criticized for failing to be inclusive of a more robust definition of wellbeing. In an effort to be more comprehensive in conceptualizing wellbeing, Stanojevi and Benčina (2019) collapsed several existing indices into a more robust scale. Their resultant health and education, political system, economy (HEPE) wellbeing index is based on six dimensions of social indicators, including economy, education, entrepreneurship and opportunity, governance, health, and safety and security, with a total of 50 variables. Other studies have emphasized the need for strong government intervention in striving for universal health care, as well as the resources in place to ensure that the health workforce, especially in developing countries, does not experience a widening gap between the supply and demand of workers (Tangcharoensathien *et al.*, 2015).

Health and wellbeing is a major theme in tourism research. General texts are available on health and wellness tourism (Smith & Puczkó, 2013), as well as targeted works on happiness and subjective wellbeing (McCabe & Johnson, 2013), medical tourism (Carrera & Bridges, 2006), quality of life and wellbeing (Uysal *et al.*, 2016), and through a range of practices and settings including spa tourism (Erfurt-Cooper & Cooper, 2009), outdoor education and bush therapy (Pryor *et al.*, 2005) and remote indigenous community contexts (Tedmanson & Guerin, 2011). Uysal *et al.* (2016) argue that long-term sustainability will depend on the ability of tourism to contribute to quality of life in communities. Making connections between sustainability, tourism and wellness would appear to be a fruitful area of research in the future. We make further observations on this connection in Chapter 10.

SDG 4: QUALITY EDUCATION

Like health and wellbeing, above, the UNDP (2018c) reports that considerable improvement has been made in the area of education. The goal of achieving universal education for children is well on its way, with 91% of children in developing countries enrolled in education in 2015. The number of drop-outs has decreased dramatically, while literacy rates and the number of girls in school have risen. The following facts and figures from UNEP (2018e) shed more light on the challenges that remain:

- about half of all out-of-school children of primary school age live in conflict-affected areas;
- 57 million primary-aged children remain out of school, more than half of them in sub-Saharan Africa;
- 103 million youth worldwide lack basic literacy skills, and more than 60% of them are women;
- in developing countries, one in four girls is not in school;
- globally, six out of 10 children and adolescents are not achieving a minimum level of proficiency in reading and maths.

The connection between education and sustainability has long been recognized as essential in moving the SD agenda forward. McKeown (2000), for example, developed the first Education for Sustainable Development (ESD) Toolkit based on the notion that communities and educational systems needed to be intertwined. In an author's note in McKeown *et al.* (2002), McKeown states that she was inspired to create the ESD Toolkit after the UN Commission on Sustainable Development meetings in 1998 demonstrated a high level of frustration with the development of education and sustainability, despite the universal affirmation of the need for this agenda to move forward. ESD is characterized as follows:

> ESD carries with it the inherent idea of implementing programs that are locally relevant and culturally appropriate. All sustainable development programs including ESD must take into consideration the local environmental, economic, and societal conditions. As a result, ESD will take many forms around the world. (McKeown *et al.*, 2002: 13)

The ESD agenda follows four key themes: (1) improve basic education; (2) reorient existing education to address SD; (3) develop public understanding and awareness; and (4) improve training (McKeown *et al.*, 2002). Around the same time as the ESD Toolkit was developed, other initiatives on sustainability and education emerged. The German BLK '21' programme for education and sustainability was implemented in 1999, following very quickly after the principle of sustainability being implemented as a German state objective in 1994 (de Haan, 2006). This programme was built around several key competencies and forms of learning, as outlined by de Haan (2006). These include:

- competencies in foresighted thinking;
- competence in interdisciplinary work;
- the call for interdisciplinary learning;

- competence in cosmopolitan perception, transcultural understanding and cooperation;
- learning participatory skills;
- competence in planning and implementation skills;
- the capacity for empathy, compassion and solidarity;
- competence in self-motivation and in motivating others;
- competence in distanced reflection on individual and cultural models.

There are several key emerging themes driving the education and sustainability platform forward. Some of these include the notion that education needs to move beyond the simple acquisition of facts to a level that increasingly considers evidence in decision making, that is: the evidence that ESD contributes to a higher quality education (Dixit, 2018); inclusive education for the purpose of removing barriers for all children to have optimal experiences at school (Slee, 2018); access to education (Chiarelli-Helminiak & Lewis, 2018); gender equality (Esteves, 2018); and the role that libraries play in developing countries as a conduit to knowledge attainment on sustainability, such as the training of female farmers on sustainability (Jain & Jibril, 2018).

The importance of education is recognized by Adetiba (2017), who calls for a new model in African higher education that needs to do a better job at driving change. Such a model, Adetiba claims, needs to be built around sustainability, and especially the UN SDGs, because education has the ability to provide solutions for poverty, hunger, poor health services, gender inequality, food insecurity and many other challenges identified by the SDGs. Other scholars in the African context are on side with this need. Offorma and Obiefuna (2017) argue that teachers need to be better acquainted with active learning strategy skills in order to better address the quality education goal of the UN.

The education platform in tourism has been advanced on a number of fronts. The BEST EN (Building Excellence in Sustainable Tourism Education Network) was developed in 1999 with the intent of merging tourism practice with curricula and module development around sustainability (Moscardo, 2016). The Tourism Education Futures Initiative embraces sustainability within the stewardship dimension of their values, as we examine in more detail in Chapter 8 (Padurean & Maggi, 2011). Other recent work specifically involving the SDGs and education comes from Dimitrov *et al.* (2018) on the influence of neoliberal values in university settings. These influences focus on training instead of critical thinking because of the business-oriented approach inherent in neoliberalism. Dimitrov *et al.* (2018) observe that such an approach, if it continues, will seriously challenge and undermine the ability to build and sustain the SDGs in education now and in the future.

SDG 5: GENDER EQUALITY AND SDG 10: REDUCED INEQUALITIES

Gender equality is 'achieved when women and men enjoy the same rights and opportunities across all sectors of society, including economic participation and decision making, and when the different behaviours, aspirations and needs of women and men are equally valued and favoured' (Irish Department of Justice and Equality, 2018). Equality is closely related to equity with the latter defined as 'the process of allocating resources, programmes, and decision making fairly to both males and females without any discrimination on the basis of sex ... and addressing any imbalances in the benefits available to males and females' (CAAWS, 2018). While equality is designed to create the same starting line for everybody, equity provides everyone with the full range of opportunities and benefits – the same finish line (CAAWS, 2018).

SDG 5, to achieve gender equality and empower all women and girls, emphasizes the magnitude of discrimination that takes place globally, even though equality should be a human right. It is an accepted fact that if women and girls are empowered there is a multiplier effect that enhances economic growth and development in all regions of the world, which dramatically increases our chances of attaining a sustainable future (UNDP, 2018: gender equality; see also SDG 8). The following facts and figures are reported by UNEP (2018f):

- globally, women earn only 77 cents for every dollar that men earn doing the same work;
- worldwide, almost 750 million women and girls alive today were married before their 18th birthday;
- up to seven in 10 women around the world experience physical and/or sexual violence at some point in their lives;
- less than 20% of the world's landowners are women;
- only 22.8% of all national parliamentarians were women as of June 2016, up from 11.3% in 1995.

UNEP (2018h) facts and figures on reduced inequalities illustrate that:

- the richest 10% earn up to 50% of total global income;
- at current rates of progress, the World Economic Forum states that it will take 217 years to close the gender gap in employment opportunities and pay;
- on average, income inequality increased by 11% in least developed countries (LDCs) between 1990 and 2010;
- in developing countries, rural women are up to three times more likely to die in childbirth than women living in urban centres;
- for every dollar sent home in 2015, migrant workers paid 7.5 cents in fees, more than double the target rate of 3%.

What often constrains women in attaining the same rights and opportunities as men are formal and informal norms based on cultural history. These may include religious practices in Africa that celebrate female submission as the mark of womanhood, or the stigma around abortion (Ibitayo, 2017). In India, 57% of young women between the ages of 15 and 24 are out of school and unemployed, as compared to 15% of their male counterparts (OECD, 2013). A critical lack of education and job training is cited as a principal driver for such high levels of unemployment among girls and women (Mehra & Shebi, 2018).

There are also massive inequalities that exist in the corporate domain, where businesses have traditionally valued and rewarded the efforts of men over women. A potentially fruitful way of addressing gender equality issues in corporate social responsibility (CSR) is to view women through the lens of feminist citizenship theory for a better and more holistic approach to stakeholder relations. Grosser (2009), following the work of Crane *et al.* (2008), argues that the 'stakeholders as citizens' approach sees corporations 'as new sites of citizenship where stakeholders can be viewed as citizens in relation to the firm' (Grosser, 2009: 294). As society changes along neoliberal lines, the role of the private sector is becoming more important in areas such as regulation and privatization. This also means that the private sector should be a key driver for the protection of citizen rights (Crane *et al.*, 2004), which includes education, community development, health and political spheres of life (Grosser, 2009). And while women often play a vital role in community development as a manner by which to exercise citizenship, these roles are often more informal when it comes to political participation – a point noted above with reference to the under-representation of women as parliamentarians.

Tourism research on gender inequality has established a strong presence in the tourism studies literature in a relatively short period of time. For example, Ferguson (2011) argues that after the first decade of the new millennium, little work had been done in the area of tourism and gender. A paper just four years later by Ferguson and Moreno Alarcón (2015: 401) commences with the observation that 'gender inequality in the tourism sector has attracted widespread attention in recent years'. Munar *et al.* (2015) addressed the issue of gender equality more formally by documenting the large gap that exists between men and women in many critical categories of scholarship within the tourism academic realm. For example, only 13% of the members of the International Academy for the Study of Tourism are female.

SDG 8: DECENT WORK AND ECONOMIC GROWTH

UNEP's (2018g) facts and figures on SDG 8 illustrate that although the number of workers living in extreme poverty has declined and the size of the middle class has expanded to 34% of total employment, there are still many more disparities and inequalities.

- 204 million people were unemployed in 2015, 192.7 million in 2017;
- 470 million jobs will be needed to absorb new entrants into the labour market between 2016 and 2030;
- unemployment among youth (aged 15–24) reached 13% in 2014, nearly three times higher than the rate for adults;

- only 29% of the global population has comprehensive social security;
- in 2017 there were around 300 million workers in extreme poverty, living on less than US$1.90 per day;
- gender equality in the labour force could add US$28 trillion to the global economy by 2025.

The International Labour Organization (ILO, 2018) provides a comprehensive description of the importance of decent work for the dignity and wellbeing of people living around the world, and the campaign has been a central objective of the ILO ever since the organization coined the term 'decent work' in 1999 (Anker *et al.*, 2003). For the ILO, decent work is a catch-all phrase that addresses fair income, opportunities for productive work, workplace security, social protection for family units, personal development, social integration, freedom to express concerns, the ability to participate in decision making and equality between men and women.

The ILO is committed to working within the parameters of the UN SDGs through a number of key policy directives. Some of these initiatives include efforts to eliminate child and forced labour, occupational safety and health action, jobs for peace and resilience, respect for fundamental rights at work, youth employment, decent work for sustainable development, ending poverty, the needs of the LDCs, evidence-based research and the development of a catalogue of SDG tools (ILO, n.d.). The International Training Centre (ITC, 2018) of the ILO offers a number of courses that support SDG8, including 'Opportunities for green jobs in the waste sector', 'Project cycle management in fragile settings', 'Leadership for employment promotion in fragile settings' and 'Effective investment facilitation and sustainable development'.

Attempts have been made to measure decent work through the use of statistical indicators in an effort to quantify change that is taking place globally. However, scholars argue that there needs to be a move away from the collection of data on labour using employment and unemployment statistics alone (Anker *et al.*, 2003). More emphasis, they add, is required in areas that relate more to quality-of-life indicators and person-enabling characteristics. Bridging off the ILO's work, Anker *et al.* (2003) suggest the following indicators to conceptualize decent work: (1) opportunities for work; (2) work in conditions of freedom; (3) productive work; (4) equity in work; (5) security at work; and (6) dignity at work. Decent work has also been applied in the context of value chain development for government and private sectors usage. Value chains include all the various steps involved in bringing a product to market – from the farm to the retail counter. Herr *et al.* (2009) observe that decent work issues exist within value chains because when workers are disadvantaged there are problems in production and transportation, within factories or retail, that take away from the competitiveness of a product.

Tourism research appears to be at an incipient stage in the area of decent work. A working paper by Bolwell and Weinz (2008) suggests that a challenge for the ILO will be to promote decent work principles for the reduction of poverty not only within the niche tourism markets but also for mass tourism. This study demonstrates the crossover that will invariably take place between different SDGs – in this case decent work with poverty reduction. De Beer *et al.* (2014) found that very little work has been done on employment and tourism in Africa in general, and the rhetoric behind decent work by government has thus far not translated into action. They found that although large tourism companies provide more stability for guides, it is the smaller firms that offer only temporary positions with minimal, if any, benefits.

SDG 9: INDUSTRY, INNOVATION AND INFRASTRUCTURE

A review of academic work on industry and sustainability shows a broad array of different industries involved, many of which would not have addressed the notion of sustainability a decade ago. Industries that are readily accessed in general literature searches in the area of sustainability include the concrete and cement industry, oil and gas, construction, gold mining, the automotive industry, coffee, steel, water, food and hospitality. These are signposts that industry is embracing sustainability on a number of fronts. Industry advances in the areas of innovation and infrastructure development are of tremendous value as models or benchmarks of how industry should proceed along sustainable lines in the future.

Innovation is defined as 'large-scale transformations in the way societal functions such as transportation, communication, housing, feeding, are fulfilled' (Geels, 2004: 19). Technology plays an important role in any transformation, and so it is, as Geels observes, the combination of society and technology that is critical in how physical artefacts, organizations, natural resources, science and legislation are intertwined to achieve specific desired functionalities.

Many authors believe that because that there is no alternative to sustainability now and in the future, sustainability initiatives ought to be the key drivers of innovation (Nidumolu *et al.*, 2009). Other scholars view sustainability innovations very much like radical innovations (Schaltegger & Wagner, 2011). A radical innovation is defined as 'innovation that is characterized by creating new-to-the-world markets that are disruptive for both customers and manufacturers' (Markides & Geroski, 2005: 17). Uber is an example of a radical innovation in the way that it disrupts the conventional models of transportation. In fact, universal acceptance of the Uber model has been problematic because of how transformative it has been to other related industries like the taxi industry. Other commentators have written on the benefits of innovation in the marketplace and suggest that sustainability innovation is realized by organizations that seek societal benefits as well as a multitude of stakeholder partnerships in creating institutional stability (Schaltegger & Wagner, 2011).

In a provocative commentary on transforming innovation for sustainability, Leach *et al.* (2012) argue that, as we approach critical thresholds in many different social and ecological domains (e.g. food shortages), transformative innovation is required to chart new pathways to sustainability – including the SDGs. Greater recognition is needed of the importance of inclusive grassroots innovation strategies and how these connect with global politics and initiatives. As such, although consistently pushed to the margins, local knowledge, experimentation and technology are required to solve local and global issues. As observed by Leach *et al.* (2012):

> We believe that there is now a new urgency to (re)connect these two strands of sustainable development in order to find ways to navigate a safe operating space for humanity from the bottom up. Global and multi-scale planetary boundaries and SDGs now, more than ever, need to be met by embracing local action in multi-scale approaches. (Leach *et al.*, 2012: 11)

Innovation must also be built into infrastructure. Infrastructure includes facilities and services such as sewers, power plants, tunnels, bridges and roads, which allow communities, enterprises and society in general to function smoothly. Sustainable and resilient infrastructure 'integrates ESG [environmental, social, and governance] aspects into a project's planning, building, and operating phases while ensuring resilience in the face of climate change or shocks such as rapid migration, natural disasters or economic downturns' (Egler & Frazao, 2016: 22). Egler and Frazao (2016) estimate the world's existing infrastructure to be in the order of US$20–50 trillion, with an expected US$93 trillion future demand over the course of the next 15 years. This new demand far exceeds the level of demand that currently exists. This fact makes it essential that new infrastructure should be built with resiliency and sustainability in mind, not only for ecological (reduced consumption of resources and reduced environmental impact) and sociocultural (to advance social inclusiveness) benefits, but also for competitiveness in a changing world (Egler & Frazao, 2016). Given the debt/deficit constraints that governments are under, large institutional investors have come to the fore in efforts to fill these gaps and must incorporate ESG factors into their development plans (Hebb, 2019).

Innovation also requires taking a step back. Nature should be used as a form of soft infrastructure. This means restoration of environments back to original states, which might provide the capacity for withstanding other natural events. This includes the restoration of wetlands along the Gulf Coast in the USA and in New York City (where wetlands used to exist before development) as buffers against hurricanes (Zolli, 2012).

The tourism industry would not operate to capacity in the absence of these critical services. And, in fact, many tourism destinations are challenged because they lack critical infrastructural services such as proper sewage management systems to accommodate the often large influxes of tourists within destinations (see, for example, Brenner & Guillermo Aguilar, 2002, in the context of regional development challenges in Mexico).

SDG 11: SUSTAINABLE CITIES AND COMMUNITIES

The growing complexity of cities and communities poses significant challenges for achieving SD. The oft-quoted adage of 'Grub first, then ethics' (by the German playwright Bertolt Brecht) suggests that, especially in the face of increasing poverty in some urban contexts, priorities surround feeding one's family first over concerns about environmental stewardship. This means that for a large percentage of city dwellers, and also those in smaller communities, sustainability may not even be a goal. The UNDP (2018) emphasizes some of the real challenges in building sustainability into city and community contexts:

- 3.5 billion people, half of the world's population, live in cities; by 2050, the urban population is expected to reach 6.5 billion, two-thirds of all humanity;
- cities occupy just 3% of the Earth's land but account for 60–80% of energy consumption and 75% of carbon emissions;
- currently, 828 million people live in slums, and the number is rising;
- in 1990 there were ten cities with 10 million inhabitants or more; by 2014 the number of 'mega-cities' had reached 28;
- in the coming decades, 95% of urban expansion will take place in the developing world.

The challenges are readily apparent for the prospects of sustainability in large cities based on the UNDP (2018) facts and figures above. Studies routinely cite the massive environmental footprint that these regions have compared to hinterland areas. For example, the footprint of the city of Bath in the UK is some 20 times larger than its corresponding land area (Doughty & Hammond, 2003), leading the authors to criticize the idea of cities ever being sustainable.

Other scholars, however, see urban regions as being advantaged over smaller communities when it comes to sustainability, because of access to monetary resources, existing infrastructure and human resources. Yedla (2015) contends that by embarking on ecological initiatives and corrective, sustainable policies, tremendous gains in efficiency may be realized. Targeted areas include production, landscaping, ecological-friendly cultures, services provision and sanitation systems (Yedla, 2015). The list of activities in Table 3.1, adopted by several communities, both large and small, in the USA, is an example of some of the sustainability initiatives being undertaken (Svara *et al.*, 2013).

Svara *et al.* (2013) conclude by suggesting that sustainability adoption is low and uneven, with many communities not taking advantage of sustainability innovations to the extent that they should. Only one in six governments is judged to be relatively high in its adoption of sustainability activities. Furthermore, sustainability is not associated with a specific or typical sociodemographic category based on race, class or wealth.

There is an abundance of research on community and community development, with an especially strong tie to tourism studies as one of the most studied aspects in tourism research. Where it is not the direct topic of research, community is mentioned indirectly or tangentially in study after study. The connection between community and sustainability is savoury. In fact, Dangi and Jamal (2016) argue that there are two fertile knowledge domains in tourism that have been evolving side by side: sustainable tourism and community-based tourism. Khazaei *et al.* (2015) argue that community development initiatives need to go further in reaching, engaging with and empowering fringe community groups

Table 3.1 Major sustainability activity areas

Major activity areas	*Average % of activities used*
Recycling	33
Water conservation	28
Transportation improvements	22
Energy use in transportation and exterior lighting	22
Social inclusion	21
Reducing building energy use	19
Local production and green purchasing	18
Land conservation and development rights	15
Greenhouse gas reduction and air quality	12
Building and land use regulations	12
Workplace alternatives to reduce commuting	8
Alternative energy generation	7
Overall adoption rating across all activity areas	**18**

Source: Svara *et al.* (2013)

such as first-generation immigrants, who they characterize as a less engaged community sub-group. If sustainability is about the 'now' and the 'future', these groups may be an important resource for understanding the dynamics of change that inevitably take place in communities. Inclusivity, the authors argue, is a virtue that must be emphasized in moving the community agenda forward (cf. Chapter 9).

SDG 12: RESPONSIBLE CONSUMPTION AND PRODUCTION

A significant challenge for attaining SD lies in how we produce and consume products (Box 3.2). UNDP (2018i) states that there is an urgent need to reduce our ecological footprint because there is simply too much waste of valuable resources in our production and consumption practices. While the vast majority of world citizens have too little to meet their basic needs, those that have more than enough, generally speaking, are wasteful in their production and consumption practices. UNEP (2018i) illustrates that:

1. agriculture is the biggest user of water worldwide, with irrigation using about 70% of all freshwater for human use;

2. 1.3 billion tonnes of food is wasted every year, while almost 2 billion people go hungry or are undernourished;
3. only 3% of the world's water is fresh (drinkable), and humans are using it faster than nature can replenish it;
4. the food sector accounts for around 22% of total GHG emissions, largely from the conversion of forests into farmland;
5. globally, 2 billion people are overweight or obese;
6. one-fifth of the world's final energy consumption in 2013 was from renewable sources.

Bullet points 1 and 4 above shed light on the tremendous impact that the agricultural sector has on world resources – a topic that we treat more extensively in Chapter 7. The problem is compounded by the fact that we are in the midst of a livestock revolution where more people are demanding meat products in both the developed and developing worlds. Delgado *et al.* (2001) found that people in the developed world derive 27% of their calories and 56% of their protein from animal products – 11% and 26%, respectively, for people in developing countries. Table 3.2 demonstrates the growth in demand for meat consumption through to 2020.

Box 3.2 SDG 12: Ensure sustainable consumption and production patterns

Sustainable consumption and production is about promoting resource and energy efficiency, sustainable infrastructure, and providing access to basic services, green and decent jobs and a better quality of life for all. Its implementation helps to achieve overall development plans, reduce future economic, environmental and social costs, strengthen economic competitiveness and reduce poverty.

Sustainable consumption and production aims at 'doing more and better with less', increasing net welfare gains from economic activities by reducing resource use, degradation and pollution along the whole life cycle, while increasing quality of life. It involves different stakeholders, including business, consumers, policy makers, researchers, scientists, retailers, media and development cooperation agencies, among others.

It also requires a systemic approach and cooperation among actors operating in the supply chain, from producer to final consumer. It involves engaging consumers through awareness-raising and education on sustainable consumption and lifestyles, providing consumers with adequate information through standards and labels and engaging in sustainable public procurement, among others.

Source: UNEP (2018i)

Tukker *et al.* (2010) echo the sustainability challenges inherent in food production, but add mobility and energy use as targets to foster change in achieving sustainability because they have the greatest environmental impact. As noted in Table 3.2, not all economies are equal in their use of resources. The developed world with 20% of the world's population is responsible for approximately 80% of the life cycle impacts of consumption. Existing infrastructure and habits are said to be the prime drivers preventing change (Tukker *et al.*, 2008).

Businesses often struggle with the intricacies of sustainability at many key stages of a company's activities, such as manufacturing, products/service development and stakeholder cooperation, because these areas are viewed as being too amorphous. An example of a model designed to better integrate these activities in achieving a unified approach to sustainability comes from Jonkute and Staniškis (2016). Their SUstainable and RESponsible COMpany model for SCP (SURECOM) incorporates engineering, management and communication tools, industrial processes, products and services and stakeholder input, and involves consumers as partners in an integrated management system.

Table 3.2 Actual and projected meat consumption by region

Region	*Annual growth of total meat consumption*		*Total meat consumption in million metric tonnes*		
	1982–1994	*1993–2020*	*1983*	*1993*	*2020*
China	8.6	3.0	16	38	85
Other East Asia	5.8	2.4	1	3	8
India	3.6	2.9	3	4	8
Other South Asia	4.8	3.2	1	2	5
Southeast Asia	5.6	3.0	4	7	16
Latin America	3.3	2.3	15	21	39
West Asia/North Africa	2.4	2.8	5	6	15
Sub-Saharan Africa	2.2	3.5	4	5	12
Developing world	5.4	2.8	50	88	188
Developed world	1.0	0.6	88	97	115
World	2.9	1.8	139	184	303

Source: Delgado *et al.* (2001)

Other studies at the atomistic or individual scale establish that women, as compared to men, live more sustainably and leave a smaller global ecological footprint. Johnsson-Latham (2007) attributes this to the fact that, among other aspects, the consumption patterns of men are different from those of women, with men, especially rich men, having greater access to resources and power and an associated increase in the degree of mobility and transport opportunities. Men on average consume more than women in several major categories, including eating out, alcohol and tobacco use, transport and sport. Furthermore, a study on values and sustainability found that individuals who are respectful and compassionate towards others, who see society as an extended family and who wish to alleviate the suffering of others, were found to have higher levels of sustainable consumption than others not possessing these traits (Sharma & Jha, 2017). This has led marketers to produce sustainable variations on existing brands, which are no longer categorized as niche products but rather as mainstream.

Scholars also argue that there needs to a better connection between sustainable consumption and de-growth, two areas that share similar challenges in SD. The concept of strong sustainable consumption focuses on the 'appropriate levels and patterns of consumption, paying attention to the social dimension of wellbeing, and assessing the need for changes based on a risk-averse perspective' (Lorek & Fuchs, 2013: 36). Such an approach gets away from too much of a focus on improving the efficiency of consumption through technological means only, by paying attention to more contemporary issues around sustainability such as distributive justice.

Examples of production and consumption linked to tourism will be numerous throughout the course of this book. Chapters 6 and 7 on the tourism industry are cases in point, where examples of tools and CSR will highlight how tourism products are produced, consumed and monitored in the pursuit of more sustainable ends.

SDG 16: PEACE, JUSTICE AND STRONG INSTITUTIONS

While some countries enjoy peace and prosperity, others are trapped into perpetual sequences of conflict marked by torture, crime, sexual violence and the stripping of rights of many of the most vulnerable populations. UNEP (2018j) illustrates some of the deep-seated problems in the area of peace, justice and strong institutions:

- every minute, nearly 20 people are displaced as a result of conflict or persecution; at the end of 2016, the total number of forcibly displaced persons was 65.6 million;
- 603 million women live in countries where domestic violence is considered not to be a crime;
- there are 10 million stateless people around the world who have been denied a nationality and related rights;
- corruption, bribery, theft and tax evasion cost developing countries US$1.26 trillion per year.

The connection between peace and sustainability is a topic of considerable interest. Cairns (2000), for example, argued at the turn of the century that, while peace heightens the sustainable use of the planet, war impedes it, and also that 'peace among humans is a necessary precursor to sustainability'. This is the first of several principles that Cairns organized in a declaration of world peace and sustainability. These principles are not unlike those that have defined the essence of Green Cross, which is an organization committed to building a global culture of peace and sustainability (Likhotal, 2007).

The challenges inherent in conflict-prone regions are particularly thorny, especially for SD as noted above through the UNDP facts and figures. Even in post-conflict regions, the task of moving in the direction of sustainability is ominous as a result of the lack of resources. Samuels (2005) contends that the link between strong institutions, sustainability and post-conflict regions requires transformative action on three different fronts. First, conflict-prone societies must move from a culture of violence to one that institutes political mechanisms via elite negotiation and social dialogue. Secondly, good governance structures of a democratic nature need to be implemented to prevent future conflict. Thirdly, sustainable institutions, i.e. those of a long-term nature, need to be embedded for long-term success.

Justice is well placed alongside peace and strong institutions in the movement towards a more sustainable tourism industry because of the sheer number of ethical issues that have rights and justice implications tied to them. One of the most comprehensive treatments of justice comes from Rawls (1972), who argued that just societies are marked by: moderate equality, where people, in theory, cannot impose their will on others; moderate egoism, where cooperation and fairness are the rule; and moderate scarcity. Rawls argued that it is this last condition that becomes problematic in the fair treatment of people and groups, where cooperation is easy in a climate of abundance, and rather more challenging if resources are scarce. Rawls also talked about intergenerational justice and the obligations or duty that we have to future citizens of the planet. His 'just savings' principle is to be implemented as a tool for 'constraint on the rate of accumulation … Each age is to do its fair share in achieving the conditions necessary for just institutions and the fair value of society' (Rawls, 1999: 263). Mihali and Fennell (2015) used Rawls' notions of justice to configure a system of trading tourism rights between wealthy and impoverished members of the world community in a system that would be more equitable in consideration of those who have the means to travel internationally and those that do not.

However, even though there ought to be a natural bond between justice and sustainability, at least in theory, there are several practical and theoretical challenges that prevent this bond from solidifying. These challenges lie in a series of vexing issues that are complicated by multiscalar, multigenerational and multidimensional characteristics of sustainability (Klinsky & Golub, 2016). Social contexts are subject to change, negotiations on long-term issues like climate change, for example, are time sensitive because of the nature of political cycles, and environmental justice issues are subject to social, political and economic forces.

Other scholars believe that the discourse on justice and rights ought to be embedded in new paradigms of thought that contest the dominant grip that corporate interests have on the pace and direction of global development (Shiva, 2016). The privatization of public goods, land grabs and collusion between corporations for organizational benefit are just a few practices that emphasize organizational and individual interests over other social and ecological interests. Shiva's (2016) *Earth Democracy* is a movement that emphasizes care and gratitude for the Earth and all of its inhabitants over greed and consumerism, and it is premised on reclaiming the commons along with the rights of nature and of human beings in the pursuit of a just future. This new movement is immersed in what Shiva refers to as multiple 'clashes of civilization' over what Aristotle referred to as *chrematistics*, loosely translated as the gaining of wealth or money making, sometimes through illicit or unfair means, and *oikonomia*, or the art of obtaining sufficient property for living a good life (Dierksmeier & Pirson, 2009).

SDG 17: PARTNERSHIPS FOR THE GOALS

Following on from the focus on justice and strong institutions above, this final SDG places emphasis on the commitment to build effective global partnerships through cooperative engagement. There are several targets in the areas of finance, technology, capacity building, trade, policy, multi-stakeholder partnerships and data monitoring and accountability that are important to address and achieve in this SDG. For example, among other targets, SDG 17 is striving to (UNDP, 2018):

- assist developing countries in attaining long-term debt sustainability through coordinated policies aimed at fostering debt financing, debt relief and debt restructuring, as appropriate, and address the external debt of highly indebted poor countries to reduce debt distress;
- promote the development, transfer, dissemination and diffusion of environmentally sound technologies to developing countries on favourable terms, including on concessional and preferential terms, as mutually agreed;
- enhance international support for implementing effective and targeted capacity building in developing countries to support national plans to implement all the SDGs, including through North–South, South–South and triangular cooperation;
- significantly increase the exports of developing countries, in particular with a view to doubling the LDCs' share of global exports by 2020;
- encourage and promote effective public, public–private and civil society partnerships, building on the experience and resourcing strategies of partnerships;
- by 2020, enhance capacity-building support to developing countries, including for LDCs and small island developing states, to increase significantly the availability of high-quality, timely and reliable data disaggregated by income, gender, age, race, ethnicity, migratory status, disability, geographic location and other characteristics relevant to national contexts.

Governance is an essential element in developing partnership initiatives, and this holds true for partnerships in sustainability. McAllister and Taylor (2015) describe governance for partnerships in sustainability as necessary for cooperation in the design and implementation of sustainability policies. These authors argue that partnerships cannot be expected to radically change governance tactics, but are more helpful in nudging governance towards including a greater diversity of stakeholder interests in policy initiatives. An example that will be examined in greater detail later in this book is how animal interests in tourism are starting to bubble to the surface. The argument is that animals should now have a 'seat' at the table as bona fide stakeholders when it comes to defining the principles and practices of the tourism industry. This is a new and innovative idea that will no doubt challenge the conventional mindset when it comes to the manner in which we practise tourism, especially in how we may or may not be able to step outside the anthropocentric way in which the world works.

HUMAN NATURE

The final feature of the Conceptual Framework (Figure 1.3) to explain before moving forward with ST in the rest of the book is the aspect of human nature as a core component of the noösphere – the sphere of human thought. Human nature is a broad canvas with traditions in evolutionary biology, psychology and neuropsychology, theology, sociology, economics, politics and philosophy, among others, and it involves the basic characteristics that are important in influencing or dictating the natural tendencies of human beings along cognitive, emotional and behavioural lines. Philosophers, for example, have been preoccupied with the importance of reason in living the good life, which has often been based on the pursuit of pleasure over pain. Chinese philosophers, including Hsun Tzu, felt that our human nature was basically evil. And psychologists (Pinker, 2002) and evolutionary biologists (Dawkins, 1989) have examined the altruistic and self-interested tendencies of humans in general and in specific cohorts.

We have elected to focus on a few aspects of human nature that are derived from both natural and social science research, and that are manifest in tourism studies research and closely connected disciplines. A brief foray into biology, particularly theories of inclusive fitness and reciprocal altruism, provides us with ultimate explanations of causation, rather than proximate ones (Crouch, 2013). In the former case, it is the evolutionary forces that dictate why we act the way we do, and in the latter case it is the immediate environmental or physiological/psychological factors that dictate behaviour. This section will also gravitate towards ethics and some philosophical concepts that help us to better understand the choices that we make as tourists and service providers, and then on to values, attitudes and behaviour. A more comprehensive section on pro-environmental behaviours addresses the challenges of translating values and attitudes into action. The section concludes with a focus on the environmental/sustainable citizen.

SELF-INTEREST AND ALTRUISM

An example of proximate causation as it applies to self-interest follows from the work of Newswander *et al.* (2017), who consider the nature of self-interest as a regime value both in American history and in present-day society. They observe that a 'regime' is a 'normative standard that extends beyond one's own place and time' (Newswander *et al.*, 2017: 553). Regimes, they argue, may be centred on power and interest, and may be political, social or religious in their orientation. In considering how self-interest is a regime value in US society in contemporary times, Newswander *et al.* (2017) trace the history of the word 'interest' back to the earliest days of the USA. The authors cite George Washington, who wrote: 'a small knowledge of human nature will convince us, that, with far the greatest part of mankind, interest is the governing principle; and that, almost every man is more or less, under its influence' (Newswander *et al.*, 2017: 553). The reason why the first generation of US citizens were led by interest as a normative moral standard is because they wanted to suppress the basest forms of human passion including greed, avarice, venality and lawlessness, among other forms. As such, the pursuit of self-interest was instituted rather like a constitutional order which would sanction some acceptable normative pursuits, while suppressing the worst of these. These interest-centred principles helped to define the boundaries of self-interest in America as a 'unique set of regime of [sic] values' (Newswander *et al.*, 2017: 553), based on elements such as liberty and the ownership of property. Why Newswander *et al.*'s (2017) work has value is because it recognizes that a transition has taken place away from these fundamental regime values as pillars of an emerging society, towards ones that represent personal preferences over the collective. Self-interest is now reducible to cost–benefit calculations on economic terms, with the elevation of these baser human impulses, noted above, that serve to degrade the chance to reach higher human potential. Individualism has now morphed into a different beast as agents strive to maximize their own personal desires at the expense of a broader public ethos.

Ultimate explanations of self-interest and altruism are centred around the work of two evolutionary biologists, W.D. Hamilton and Robert Trivers. Hamilton (1964) provided the strongest theoretical argument for why we choose to act altruistically towards individuals who are related to us over those who are not. Altruism is defined as behaviour that promotes the fitness (wellbeing) of another organism at a cost to the fitness of the agent (Kitcher, 1993). Giving one's seat to an elderly person on a bus is an example of altruism. There is a cost to the donor of the seat (energy and comfort) and a benefit to the elderly person, who now does not have to endure standing for a prolonged period of time. Hamilton reasoned that we are more prone to kin selection (i.e. helping our own relatives, also referred to as the theory of inclusive fitness) because kin share our genes and natural selection has hard-wired us to ensure that our genes, even if they are housed in a relative, persist in successive generations. So, in the event of two people drowning, one being your sister and the other your sister's friend, and given that you only have enough time and energy to save one person, the theory of inclusive fitness argues that you would save your sister because she shares your genes. 'Blood is thicker than water', as the saying goes.

What this means in tourism is that there is likely to be a much stronger chance that altruism (benefits) will be extended to one's own family over non-kin working in the same organization. Hires and promotions may go to relatives over other, perhaps more deserving, workers who are unrelated to the boss. The technical term for this type of behaviour is nepotism, and it is theoretically substantiated, at least from the biological side of things, by inclusive fitness (see Fennell, 2018a, for a more detailed description of this theory in tourism).

Following on from the work of Hamilton, the biologist Robert Trivers (1971) developed the theoretical premise explaining why non-kin choose to be, or not to be, altruistic towards others. Trivers' theory of reciprocal altruism (RA), or the theory of 'You scratch my back, I'll scratch yours', provides the best explanation as to why this phenomenon takes place. Trivers argued that natural selection has endowed us with:

- *A complex regulating system*: Given the unstable nature of human altruism, natural selection will favour a complex psychological system that regulates not only each individual's own cheating and altruism tendencies, but also the detection of these characteristics in others.
- *Friendship and the emotions of liking and disliking*: The tendency to like others and to act altruistically towards these others will be selected as the immediate emotional reward motivating altruistic behaviour and the formation of partnerships. Selection will also favour liking those who are themselves altruistic.
- *Moralistic aggression*: Injustice, unfairness and a lack of reciprocity motivate human aggression and indignation.
- *Gratitude, sympathy and the cost/benefit ratio of an altruistic act*: Humans are selected to be sensitive to the costs and benefits of altruistic acts. The greater the need state, the greater the tendency for an individual to be altruistic, even to strangers.
- *Guilt and reparative altruism*: Cheaters (defined as those who fail to reciprocate) pay for their transgressions by being cut off from future acts of aid. The cheater should be selected to make up for his transgression and to show convincing evidence that s/he does not plan on cheating in the future. Selection would favour reparative gestures.
- *Subtle cheating: the evolution of mimics*: Once friendship, moralistic aggression, guilt, sympathy and gratitude have evolved to regulate the altruistic system, selection will favour mimicking these traits in influencing others to one's own advantage (hypocrisy and mimicking sympathy and gratitude).
- *Detection of the subtle cheater: trustworthiness, trust and suspicion*: Selection favours the ability to detect and discriminate against subtle cheaters, as well as sham moralistic aggression.

Selection favours distrusting those who perform altruistic acts without the emotional basis of generosity or guilt.

- *Setting up altruistic partnerships*: Since humans respond to acts of altruism with feelings of friendship that lead to reciprocity, one such mechanism might be the performing of altruistic acts towards strangers, or even enemies, in inducing friendship.
- *Multiparty interactions*: Selection favours the formation of norms of reciprocal conduct. This would stem from learning how to deal with cheaters, and developing rules of exchange.
- *Developmental plasticity*: Selection favours the ability for RA to grow, learn and be adaptive under a number of different circumstances.

There is widespread acceptance of the theory of RA, as well as Hamilton's theory of inclusive fitness above. RA explains how cooperation evolved in small, stable and dependent societies: small because very few people lived in these families or clans; stable in the sense that very few left or joined the clan; and dependent in that the individuals of this small community relied on each other to help one another. Trust and cooperation in these small communities were built and solidified over time. But far from being a theory of altruism alone, people are often altruistic to others in order to secure benefits from these same people down the road – a form of cooperation in which in individual helps himself or herself by helping others (Trivers, 1971). As outlined above, cheating (the failure to reciprocate) serves to truncate these friendships, demonstrating how powerful this theory is in explaining how these cooperative relationships are made and lost.

The foregoing discussion shows that altruism has many sides. Jensen and Meckling (1994; see also Jensen, 1994) investigated this fact by arguing that we should be careful when we assume that those who sacrifice their time, resources and energy for others – true altruists – are in fact perfect agents. There is ample evidence pointing to the fact that although people have altruistic intentions and motives, which may translate into altruistic behaviour, they often engage in non-rational, counter-productive behaviour that is harmful to themselves and others (a point made below in the discussion on akrasia). The same holds true in reverse. Self-interest does not always mean that people are not at times altruistic. Jensen and Meckling (1994) make two valid points. The first is that people are, in the end, self-interested, and as such they will encounter conflicts in cases where cooperation is required between parties. This conflict can take place within families, partnerships, organizations and any other social arrangements. The second is that human nature, and therefore human behaviour, is dualistic. Agents are both rational and non-rational in their behaviours and this translates into contradictory actions. But while these contradictions exist, they also note that these imperfections are not deviations, but rather fundamental elements of normality itself. As such, evolution, the authors observe, has endowed humans with a brain that blinds them from perceiving and rectifying actions that can cause emotional pain in others.

Fennell (2006) reasoned that because cooperation and trust are central outcomes of RA and these take time to build and refine, cooperation might be challenged in tourism because of the one-off nature of tourist–service provider interactions. It is sometimes a more profitable strategy to cheat others in increasing one's own fitness (financial wellbeing). Examples of tourists being cheated in far-off destinations, especially in highly time-sensitive situations (i.e. when the tourist is an excursionist or has little time at an attraction or at their accommodation) are numerous because there is no shadow of the future (Axelrod, 1984). This means that agents (in this case tourists and service providers) do not see each other on a regular basis. If they did, this fact alone would be instrumental in lessening the propensity for one to cheat the other because of more frequent interaction.

In other related work, Paraskevaidis and Andriotis (2017) investigated two types of altruistic motivations of members of two voluntary tourism associations in Greece. The first, true altruism, was defined as altruism without the expectation of return rewards later on, while the second, reciprocal altruism, was defined as acting (at a cost to the agent) but with the expectation of return rewards down the road. The authors found that reciprocal altruism, more than true altruism, was the principal impetus for volunteering time and energy to increase the standard of living with the community. More specifically, individual and organizational volunteer efforts for the purpose of improving tourism development in this example of an emerging destination were done under the premise that such work would translate into personal and collective gain. This corroborates Fennell's (2018; see also Coghlan & Fennell, 2009) original contention that there is a healthy dose of self-interest built into volunteer experiences.

ETHICS

The range that RA has as a shaper of human nature is demonstrated in the fact that many scholars believe that it is the foundation of barter, the prisoner's dilemma, the tragedy of the commons, and ethics (Pinker, 2002; Ridley, 2003). While Hamilton's inclusive fitness explains altruism at the family group level, RA explains the importance of cooperation as human social groups extended outwards from family to society. As Ehrlich (2000) observes, ethics evolved in these small, stable dependent communities, as individuals quickly understood the rules of engagement between members, and natural selection rewarded individuals for certain unselfish traits that had benefits to the group (Mayr, 1988). Moral philosophy, Ehrlich (2000) continues, evolved when the taken-for-granted rules of small societies were insufficient to guide behaviour, with the need for more formalized principles and laws. There are tensions that exist within communities and societies because we are all spectators, advisors, judges, critics and instructors as well as agents of morality (Frankena, 1963) in deciding what is right and wrong for ourselves and for the group. Added to this is the fact that we must constantly compete with conflicting demands along the lines of greed, power, sex and so on, as biological and social beings (Fennell, 2018a).

Ethics is defined as the 'rules, standards and principles that dictate right conduct among members of a society or profession. Ethics are based on moral values' (Ray, 2000: 241). These rules, standards and principles that dictate right conduct may be applied in all walks of life. Examples include medicine, business, marketing, engineering and the environment. Business ethics, for example, is concerned with how organizations adhere to the rules, standards and principles that are derived from within the organization, but questions arise regarding the extent to which these rules and standards are reflective of broader societal rules. When businesses march to the beat of their own drum and focus on organizational self-interest at the expense of other human groups or the environment, there are often conflicts.

Environmental ethics has played an important part in identifying how businesses and other human endeavours have placed the interests of people over planet. The environmental ethics field has taken shape over the last 40 years, drawing on the seminal works of Aldo Leopold (1949/1989), Rachel Carson (1962), Lynn White (1967) and Garrett Hardin (1968). The principal thrust of this field is recognizing the entities that occur in nature that are worthy of moral standing, and the associated duties that must follow from this standing (Nash, 1989). If nature has intrinsic value on the basis of ecosystem services, scientific importance and so on, as argued by Rolston (1981), we ought to move away from treating it instrumentally, i.e. as a means solely to our own ends.

Curry (2011) provides a good overview of the spectrum of environmental ethics theories from the perspective of three rather distinctive categories: light green or shallow anthropocentric ethics; mid-green or intermediate ethics; and deep green or deep ecocentric ethics. Light green or shallow anthropocentric ethics includes those theories (this category embraces both weak and strong versions of anthropocentrism) that emphasize the interests of humans over all other entities in nature (Humphrey, 2000). There is indeed reason to protect elements of the natural world, both living and non-living, if these entities contribute to human wellbeing (Biro, 2002).

The mid-green category of environmental ethics places more emphasis on the interests of individual animals. This perspective includes a number of different theoretical perspectives including biocentrism, i.e. the belief that *all life* has inherent value, which is held *equally* in all constituent parts of nature (Taylor, 1986). The other principal group of ethical perspectives in the mid-green category includes those that pertain to animal liberation. Singer's (2009) focus on utilitarianism and Regan's (2004) focus on animal rights are noteworthy. In the former case, we ought to place value on equal consideration of interests, or the notion that we should not place precedence on the marginal interests that people gain in using animals through practices such as experimentation, factory farms, sources of entertainment and commerce in tourism, if these uses cause undue pain and suffering. Human pleasure, therefore, should not outweigh the interests animals have in avoiding suffering. In the latter case, animals are said to have inherent value as subjects-of-a-life because they possess the same sorts of capacities that humans do: sentience, consciousness, intentionality and so on.

The final category, deep green or deep ecocentric ethics, places value not on individual members of the natural world but rather on ecosystems as whole systems. Examples include deep ecology (Naess, 1984), the Gaia hypothesis (Lovelock, 1979), Aldo Leopold's land ethics (Leopold, 1949/1989) and, more contemporarily, ecofeminism and care ethics (Adams, 1993; Held, 2004). Whereas biocentrists place value on all living things, as noted above, ecocentrists place intrinsic value on all entities of the natural world, living and non-living, because these entities contribute to the interconnected relations that are essential in building dynamic natural systems. So, for example, Leopold's (1949/1989) land ethics is based on 'A thing is right when it tends to preserve the integrity, stability, and beauty of the biotic community. It is wrong when it tends otherwise' (Leopold, 1949/1989: 224). It is morally acceptable, therefore, to use elements of the natural world (e.g. hunting and fishing) as long as this use does not disrupt the stability and harmony of the whole system (Fennell, 2013).

At the heart of environmental ethics is the dichotomy between anthropocentric viewpoints on the one hand, and alternative viewpoints on the other. One of the earliest attempts to conceptualize these differences comes from O'Riordan (1981). His views on ecocentrism versus technocentrism were important in illustrating emerging alternatives to the conventional anthropocentric viewpoint. In this case technocentrism, or the feeling that environmental problems can be solved by technology and scientific thinking, is contrasted with ecocentrism (Table 3.3).

In tourism studies, scholars have addressed issues of overuse of the environment by contrasting instrumental versus intrinsic values and use. Holden (2005), for example, argues that tourism practice has not undergone a moral shift in human behaviour to the point where intrinsic value takes precedence over instrumental value. We still use environmental resources as means to our own particular ends in ways that are inherently damaging to ecological systems, as noted in Chapter 2. Holden feels that this moral shift in tourism needs to be driven by the market. And so it follows that the market will need to make considerable shifts in what they purchase and how much they purchase. And further, there are philosophical questions about human nature that pull us in the direction of self-interest instead of regard for other humans and the environment. We now look to philosophy to address these tendencies before moving on to values, attitudes and behaviours.

Akrasia

In the previous section on altruism, Jensen and Meckling (1994) suggested that agents will at times engage in non-rational behaviour that maybe harmful to themselves or others because they are often more self-interested than altruistic in their actions. Fennell (2015) examined this type of thinking in a tourism capacity through the concept of *akrasia*. Akrasia can be traced back to the Greek philosophers Socrates and Aristotle, both of whom felt that people do irrational things in their pursuit of pleasure over pain. Akrasia is defined as 'a deficient

Table 3.3 Ecocentrism versus technocentrism

Ecocentrism		***Technocentrism***	
Modern technology and existing mainstream economic strategies seen as fundamentally flawed		Technical solutions to environmental problems can be found by appropriate application of scientific thinking	
Deep ecology	*Soft reliance soft technology*	*Environmental managers*	*'Cornucopians'*
Intrinsic importance of nature to mankind Ecological laws/bio-rights of species and unique landscapes to remain unmolested	Emphasis on small scale and community, and on participation as a continuing education and political function No work/leisure distinction, only personal and communal development	Economic growth and resource exploitation can continue if economic/tax adjustments made, legal guarantees of minimum environmental standards produced, compensation for those adversely affected exist Project appraisal and public consultation properly structured into policy stream	Faith in science providing the basis of advice pertaining to economic growth, public health and safety Suspicion of attempts to widen participation and lengthy discussion in project appraisal and policy review Belief that all impediments to growth can be overcome with scientific ingenuity

Source: O'Riordan (1981)

capacity to contain or restrain one's desires (Mele, 1994: 424), and where the anticipation of pleasure overwhelms good judgement (Adler, 2002). Judgement is a key feature in understanding akrasia. We are being akratic when we know that option 'A' is the right course of action, but instead choose option 'B' because of a weakness of will. For example, we know that eating a dessert after a big meal will not help us achieve our dietary goals but, even so, we choose to eat a large piece of chocolate cake and ice cream because we have succumbed to a weakness of will. Fennell (2015) used akrasia to explain why sex tourists, who know that sex tourism is wrong, still travel to international destinations to engage in these practices. We may experience akrasia when we travel long distances by airplane because we know such travel has environmental impacts. In response to Fennell, Dann (2015) argued that philosophical deliberations on tourism might surely be enhanced through multi- or interdisciplinarity. Arguments on what is rational and irrational, building off the concept of akrasia, may be aided by the consideration of the work of other scholars in a type of '*addo tertium*, or the addition of a third alternative to a previously considered dichotomy' (Dann, 2015: 263). The take-away from Dann's viewpoint is that there is a deep well of knowledge from which to draw insight into the challenges we face in both tourism studies and tourism practice.

Theorists have also discussed the concept of aesthetic akrasia, which has relevance to certain tourism experiences. Thériault (2017) argues that we are irrational if we fail to react in the most appropriate way over forms of art or music. This entails disliking good art or liking bad art, which in the end shows bad aesthetic judgement (Thériault, 2017). The same premise applies: there is a weakness of will which amounts to an irrational judgement or behaviour. Thériault uses the example of the movie *Borat* within one of several categories of aesthetic akrasia. She refers to this category as the 'elephant in the room', where there is the deliberate expression of reprehensible content for the purpose of gratuitous provocation or to express a progressive vision.

The *summum bonum*

Another example of how the wisdom of ancient Greece has recently entered the tourism studies lexicon is through the concept of the *summum bonum*, loosely translated as the ultimate end or the supreme good. For a vast number of scholars, it is pleasure that we are in pursuit of as the supreme good. How this translates into our everyday lives is subject to the types of experiences that we choose to pursue and the level of happiness or pleasure that is derived from these pursuits. For Epicureans, said to be almost wholly hedonistic in their pursuits, pleasure came in the form of gluttony and sensual desires (Crossett, 1961). Other scholars have written that far from a focus on desire alone, Epicurus and his disciples sought moral pleasures around the 'the art or practice of living fully, ideally, happily' (Motto & Clark, 1968: 39).

The tourist experience has been traditionally conceived as one that is different from everyday life. Cohen (1979), for example, argued that tourists derive meaning from attempts to find their spiritual or cultural 'centre' as a zone of absolute reality through travel if they are unable to do so at home. By contrast, Boorstin (1987) felt that the tourist experience was wrapped up in the pursuit of pseudo-events, or inauthentic contrived events that are superficial in nature, which was different from MacCannell (2013) who argued that tourists are motivated to penetrate into back regions of the destination in the pursuit of authentic experiences. Fennell (2019b), however, argued that what lies even deeper than the pursuit of spiritual centres, authenticity and inauthenticity is the need to derive pleasure as the basic ontological platform from which to explain the nature of the tourist experience. Pleasure is the ultimate end that we seek, and tourism, for many, is arguably the chief mechanism to reach the *summum bonum*. As recognized by Leslie (1994: 30), tourism is 'perhaps more than any other aspect of human activity indicative of the state of a society and the level of consumerism. When one takes the whole package into account it is the ultimate in conspicuous consumption.'

Fennell observed that there are two different domains of tourism that represent two very different routes for achieving pleasure and happiness through tourism. One includes ego-centred forms of travel where there is a focus on satisfying the interests of the individual tourist. Examples include sex tourism, cruise line tourism, medical tourism and business travel. Given that cruise line tourism is a way of life for some tourists by virtue of the sheer number of times they participate in this form of travel, this is evidence of the important role that tourism plays for these people in seeking pleasure. The other domain is based on other-centred forms of tourism, which include pilgrimages, volunteer tourism, visiting friends and relatives and justice/solidarity tourism. In this latter domain, it is the interests of other people or entities like the environment that are the motivating factors behind travel, and it is through these events that tourists derive meaning as pleasure.

Interestingly, the task of deciphering the *summum bonum* has fallen to ethics as the most logical province to unveil its unique intricacies as a feature of human nature (Seth, 1896). Ethics, Seth points out, entices us to ask not what is (this is biology's jurisdiction), but rather what ought to be in judging the value of something or someone. It is to values, attitudes and behaviour that we now turn briefly in an effort to connect sustainability with human nature, which in turn will provide a springboard for analysing tourism and sustainability in a different light.

VALUES AND ATTITUDES

Values and attitudes are important dimensions that provide a window into better understanding our likes and dislikes. People's underlying values refer to general life priorities and guiding principles, or they can be assigned values, which refer to priorities or principles assigned to a valued object (Seymour *et al.*, 2010). By contrast, attitudes refer to one's dispositions towards an object after evaluation (Ives & Kendal, 2013). Values and attitudes may be similar according to: how they influence our cognitions and behaviours; how they may be learned and acquired from the same sources (e.g. parents); how they resist change; and how they play upon one another in a reciprocal manner. Table 3.4 illustrates the dissimilarities between values and attitudes.

To Milgrath (1989: 58), values are 'held strongly and generalize readily to many situations', and social values are those that most people in a given society believe to be important. Values make the most sense, Milgrath adds, when they are arranged in hierarchies. These value hierarchies, which are essential in shaping societies, are assembled through trial and error and over longer periods of time – decades and even centuries. Milgrath also disentangles values from preferences and beliefs. Values are different from preferences, which are held less strongly and are not easily generalizable, while beliefs are similar to values because we start with a cognition and then attach a feeling to it. However, beliefs and values differ because:

Table 3.4 Difference between values and attitudes

Values	*Attitudes*
Values help to guide our behaviour.	Attitudes are the response that is a result of our values.
Values decide what we think as for right, wrong, good or unjust.	Attitudes are our likes and dislikes towards things, people and objects.
Values are more or less permanent in nature.	Attitudes are changeable with favourable experiences.
Values represent a single belief that guides actions and judgement across objects and situations.	Attitudes represent several beliefs focused on a specific object or situation.
Values are derived from social and cultural mores.	Attitudes are personal experiences.

Source: iEduNote (2018)

> We tend to believe things that we value and disbelieve things that we do not value. Yet we value things that we do not believe. I have often felt that it would be wonderful to live in a peaceful world, but I do not believe I will ever do so. Similarly, we believe things that we do not value. For example, I believe there was a Nazi holocaust against the Jews, but I totally reject the way the Nazis treated the Jews. (Milgrath, 1989: 59–60)

Milgrath makes a compelling argument for the types of values, and their hierarchy, for a sustainable society. The most important value is the preservation of a viable ecosystem, while the second priority should go with nurturing the proper functioning of society. (This point on a hierarchy of values and priorities is emphasized in our Conceptual Framework in Chapter 1.)

Scholars have identified a series of core values that individuals and organizations hold to be true in defining the essence of their practices and behaviours, and that help these individuals and groups determine what is right or wrong. Kinnier *et al.* (2000) have compiled a list of universal moral values that have been drawn from both religious and secular sources having relevance to our discussion on ST, and that follow from Milgrath's perspective on values, above. These include:

1. Commitment to something greater than oneself (supreme being):
   to seek truth
   to seek justice
2. Self-respect, but with humility, self-discipline, and acceptance of personal responsibility:
   to respect and care for oneself

to not exalt oneself or overindulge – to show humility and avoid gluttony, greed or other forms of selfishness or self-centeredness

3. Respect and caring for others (i.e. the Golden Rule):
   to recognize the connectedness between all people
   to serve humankind and to be helpful to individuals
   to be caring, respectful, compassionate, tolerant and forgiving of others
   to not hurt others (e.g. murder, abuse, steal from, cheat or lie to others)
4. Caring for other living things and the environment.

Studies on environmental attitudes have explored a number of different dimensions. For example, Kiley *et al.* (2017) conducted a study on the public's (Victoria, Australia) perceptions of attitudes towards five terrestrial ecosystems, and found that not all ecosystems were valued the same. In order of preference, respondents favoured wet tropics followed by dry forest, arid woodland/shrubland, heathland and finally grassland. The study, said to be the first of its kind, was deemed important in generating support for the conservation of less appreciated ecosystems. Furthermore, ecological worldviews are said to vary depending on one's socioeconomic characteristics. Halkos and Matsiori (2015) found in a study of environmental attitudes and values associated with marine biodiversity protection that those with higher environmental attitudes have higher non-use motivations, as well as willingness to pay and ethical motivations for the protection of species. Studies have also found that younger and better educated individuals with a liberal orientation maintain higher environmental attitudes than their lower educated and less liberal orientated counterparts (Dietz *et al.*, 1998).

PRO-ENVIRONMENTAL BEHAVIOURS

A related body of literature that focuses on the connection between the natural world and human agency is pro-environmental behaviours (PEB). In tourism, a good percentage of research on PEB has been done in the area of ecotourism, demonstrating mixed results. In the Galapagos Islands, for example, scholars report that targeted interpretation programmes are successful at inducing ecotourists towards greater philanthropy and the intention to donate to organizations that support Galapagos conservation. Powell and Ham (2008) found that 70% of their Galapagos sample of tourists had intentions to support Galapagos conservation in the future. What should be recognized is that the expense of a Galapagos ecotourism experience, coupled with the charismatic nature of the destination itself, may be helpful in pushing tourists towards this type of commitment.

However, while it is certainly the case that even though ecotourism has as one of its prime mandates the education of tourists, often there is little increase in knowledge from ecotourism participation (Markwell, 1998), which is discouraging if ecotourism guides are to be successful at translating education into ethical environmental behaviour. Even in those studies that demonstrate knowledge increases from the ecotourism experience, environmental attitudes and behaviours often do not increase (Beaumont, 2001; Orams, 1997).

PEB and ecotourism have also been investigated using social psychological measures. Ting and Cheng (2014) argue that intrinsic motivation is a good predictor of PEB because of the responsibility that people feel towards protecting the environment. The authors argue that a positive attitude towards PEB is not in itself sufficient to motivate action, which compelled them to argue that moral behaviour responses are subject to social desirability bias. They also note that even though people may know that their behaviours are wrong, this awareness will not necessarily translate into more positive action – a dilemma that we encountered above in the discussion on akrasia (see also Böhler *et al.*, 2006). A companion study by Ting and Cheng (2017) found that student participation on guided ecotourism trips as part of their studies led to positive effects on PEB. They conclude by suggesting that more countries, and especially developing countries, should build experiential learning into their curricula to motivate young adults to adopt stronger environmentally sensitive attitudes and behaviours. Students who have more environmental knowledge are more likely to adopt PEB (Ajaps & McLellan, 2015).

Several studies, like the one on Galapagos above, emphasize the 'intention' to behave in an pro-environmental manner, as well as the actual behaviour itself, either during an event, immediately after an event or long-term after the event. For example, Ballantyne and Packer (2011) note that even though tourists intend to be sustainable in their actions, only a minority of tourists translate these intentions into behaviour. Their study found that while 33% of tourists intended to conserve the environment right after their trip, only 7% had followed through on these intentions with actual behaviour four months after the trip. The same 'drop-off' effect has been found in several wildlife tourism attractions in Australia (Ballantyne *et al.*, 2011).

Miller *et al.* (2015) investigated sustainable destinations from the perspective of large mass tourism urban destinations. They found that PEB reduced when individuals went on vacations (as opposed to their behaviours at home) in reference to recycling, sustainable energy use and green food consumption. By contrast, transportation became more sustainable as tourists used public transport, cycling and walking which increased in the tourist mode (cf. Chapter 7). Furthermore, certain nationalities have been shown to have greater sensitivities to environmental matters. For example, Hjalager (1999) found that German visitors assign greater importance to environmental issues than other dominant nationalities visiting Denmark. Studies, although of a more general nature, illustrate that PEB in individuals are influenced by relevant others such as family and friends (see Grønhøj & Thøgersen, 2017, who show that young people are primarily motivated to act in PEB ways as a result of family norms), which in turn are influenced by PEB norms at the country level (Culiberg & Elgaaied-Gambier, 2016).

Still other studies have found that PEB may be elicited through other-oriented tendencies (Pfattheicher *et al.*, 2016). While it has been shown that concern for the destruction of elements of nature (e.g. trees) can bolster environmental attitudes and the moral duty to help nature (see Berenguer, 2007), Pfattheicher *et al.* (2016) show that having compassion for the suffering of others is also positively related to pro-environmental tendencies. They demonstrate that by inducing the emotion of compassion, the motivation to protect nature will be heightened, which is an effective approach to building environmental tendencies. Similar work on compassion has been undertaken by Weaver and Jin (2016) but in the context of tourism – compassion defined as 'the feeling that arises in witnessing another's suffering and that motivates a subsequent desire to help' (Goetz *et al.*, 2010: 351). Weaver and Jin argue that aspirational forms of tourism like just tourism, hopeful tourism and enlightened mass tourism may attain sustainable outcomes if they are fortified by compassionate responses from others.

THE ENVIRONMENTAL SUBJECT

In light of the previous section specifically, and the overall focus of this chapter, we conclude by examining research on individuals referred to as 'environmental subjects', a topic that will resurface in Chapter 11.

Based on his work in northern India, Agrawal (2005) documents the evolution of how some rural residents go through a personal transformation to care about the environment – in this case forestry resources that had been subject to over-utilization by a general populace that had little regard for sustainable extraction. Agrawal argues that environmental subjects can emerge under these circumstances because beliefs and thoughts get framed around the experiences that people have with the environment, with unanticipated outcomes which, in retrospect, may be viewed as inappropriate. These interventions provide the agent with incentives to work on their pre-existing beliefs, i.e. to challenge those beliefs and to incorporate new ways of thinking about the world. This transformation often takes place through involvement in regulatory practices. In the case of northern India, Agrawal found that the state's choice to govern at a distance opened the door for the creation of forest councils as a new form of government. These forest councils introduced intimate government (dispersing rules, more diffuse government involvement, attention to environmental practices), which created opportunities for citizens to participate in local government initiatives and to alter their own subjective belief structures. The end result was the creation of new environmental subjects who voluntarily care about the natural world.

A greater affiliation towards nature has also been demonstrated in the tourism industry. Tzschentke *et al.* (2008) used operators affiliated with the Green Tourism Business Scheme of the Scottish Tourism and Environment Forum to understand the factors that led to the adoption of environmental measures by operators of small hospitality firms. The propensity to become environmentally active resulted from what the authors characterize as a value-driven journey, which has been heavily influenced by the development of environmental consciousness. This consciousness is a result of personal, sociocultural and situational factors. For example, respondents stated that parental education was a big factor in shaping their green tendencies, and that the greening of their businesses was a reflection of the prior greening of them as individuals. It would seem that the greening of the individual is especially important as other studies have reported that there are two primary factors that dictate environmental behaviour: a moral disposition and high income (Dolnicar, 2010).

Several studies have documented the steady move away from or alienation from a connection to the natural world (refer back to the discussion in the previous chapter on biophilia). Foremost among these is Pyle (1978, 1993), who wrote of an 'extinction of experience' whereby the direct personal connection with nature is essential in building an emotional tie with nature. There are several reasons why this connection is eroding. Some attribute it to the higher percentage of people now living in urban environments (White *et al.*, 2018; Zhang *et al.*, 2014), technology including the internet, and video games and smartphones which contribute to more sedentary lifestyles (Ballouard *et al.*, 2011), while others credit it to the over-programming of children and the micromanagement that goes along with it (Hofferth, 2009). There are obvious benefits to biodiversity conservation if people connect with nature, but there are also psychological, physiological and societal wellbeing benefits that go along with this connection (Miller, 2005; Pyle, 1993). Studies of children in urban primary schools have shown that exposure to short-term environmental education programmes, such as bird feeding and monitoring projects, lead to enhanced understanding of biodiversity, with the willingness on the part of many children to continue involvement (White *et al.*, 2018).

Wells and Lekies (2006) used Bronfenbrenner's (1995) life course perspective to investigate the likelihood that children exposed to nature in their youth would have a closer tie to the natural world as adults. The life course perspective examines the lives of individuals as sets of interlaced trajectories that, when assembled together, tell the story of a life, i.e. each person has a life path that is comprised of various categories including work, health, recreation, family and so on, that together tell the story of that individual. Wells and Lekies (2006) studied 2004 adults aged 18–90 living in urban areas of the USA and found that childhood participation in 'wild nature' (hiking, camping, fishing, playing in the woods and so on) was positively related to environmental attitudes and behaviours in adult life. Children who engaged in these sorts of activities before the age of 11 were more likely to express pro-environmental attitudes and to engage in PEB.

CONCLUSION

This chapter was organized around two main components. The first included a discussion of the remaining SDGs. These SDGs were grouped together at this juncture because they reflected more of a focus on the noösphere as compared to those selected for Chapter 2. The noösphere was defined as the sphere of human thought or the complexity of human culture that plays such an integral part in harnessing nature's energy. Partnerships, justice, institutions, education, health, community development and industry, and all the other

SDGs discussed here, are shaped by the intricacies and complexities of people and institutions caught up in struggles defined by politics, power, economics and access to resources.

The second main component of the chapter sought to better understand human action through a discussion of human nature. Key perspectives in evolutionary biology, including inclusive fitness and reciprocal altruism, provided deep theoretical explanations as to why we can be altruistic and cooperative on the one hand, and self-interested on the other. Philosophy was also consulted to aid in exploring the workings of human nature. Akrasia, or weakness of will, illustrates why we are often irrational in the choices we make, and the discussion on *summum bonum* demonstrates how tourism may be the supreme good or end that we seek as an important mechanism by which to attain pleasure. The chapter concluded with a discussion of pro-environmental behaviours and the environmental citizen. While some studies report that interventions (tourist guides and interpretation) may lead to changed behaviours down the road, the vast majority of studies illustrate that while there may be the intention to adopt more environmentally sound behaviours from tourism experiences, these often do not emerge.

Furthermore, there appear to be triggers at various stages and through various events that can transform people into environmental citizens. Especially with youth, greater exposure to nature leads to a higher likelihood of embracing environmental attitudes and behaviours. As we proceed even further down the road of a more technologically centred world driven by information and communication technologies that pull us a way from nature, the task of drawing us back into nature, the expression of biophilic tendencies as noted in Chapter 2, seems ominous.

CHAPTER 3 DISCUSSION QUESTIONS

1. What is the noösphere and how does it relate both to tourism and the SDGs?

2. What are some indicators and indices of hunger, and on what basis have they been criticized?

3. What is the Education for Sustainable Development Toolkit, and how could it be applied in a tourism context?

4. What is the International Labour Organization's decent work campaign about, and what are some of its policy directives?

5. Why is innovation so important to the future of sustainable tourism?

6. What is reciprocal altruism, and how does it inform tourism theory and practice?

End of Chapter Case Study 3.1

This case study is an example of the complex set of relationships and controversies that exist over the keeping of wild tigers in a captive environment for tourism purposes. What follows is a transcript of the work of Erik Cohen (2019) on the case of the fall and reincarnation of Thailand's Tiger Temple, which was closed because of animal rights and animal welfare activists, and through the efforts of the Thailand government, principally the Department of National Parks (DNP), but subsequently reopened as a zoo. In the end the DNP and animals liberationists ended up being divided as to how to keep the animals. It is an example of Curry's mid-green level of environmental ethics, where the individual interests of animals are taken into consideration through a range of animal ethics theories, such as welfare and rights.

Thailand as well as several other Asian countries is home to a number of illegal facilities such as zoos, theme parks and farms that hold animals captive. One of the most famous of these establishments is the Thailand Tiger Temple which is located in Wat [Temple] Pa Luangta Bua Yanasampanno, a Buddhist forest monastery, which was established in 1994 in the province of Kanchanaburi in the western part of Thailand. Upon Cohen's first visit to the temple in 2002 there were eight tigers. By 2016 the number of tigers had risen to 147. As Cohen writes, the temple had virtually turned into a theme park where tourists could take selfies with the tigers who were rendered less ferocious because of various measures by the monks to make them docile by administering drugs to the tigers (Cohen, 2013). In the name of Buddhist compassion, the animals were thoroughly abused by the monks for purposes of entertainment and commerce. While tourism numbers continued to increase year after year, as well as the entrance fees, several controversies emerged that were religious, legal and ethical in nature. These include:

- *Religious*: It is considered inappropriate and an infringement of Buddhist law for a Buddhist temple to provide commercialized entertainment to tourists, unrelated to Buddhism. However, while this might be an opinion held by some members of the public, it is remarkable that the Sangha authorities apparently did not call upon the temple to desist from such commercial activities.
- *Legal*: In Thailand wild tigers are considered state property; hence it is illegal for individuals or establishments to keep them without an official license. Keeping the tigers in the temple was considered an infringement of the law, which prompted the authorities to confiscate them and remove them from the temple.
- *Ethical*: The keeping and exposure of captive wild tigers for a tourist public incited animal rights and welfare activists to increasingly attack the temple for mistreatment of the animals and to demand their removal from the temple.

Cohen summarizes the case study in the following way:

> The DNP and the animal rights and welfare activists differed in their views of the status of the animals and the purpose of their removal from the temple. The DNP perceived the tigers as 'property', to be handled by humans as any other object. It approached the retrieval of the tigers primarily as a legal issue, a matter of law-enforcement. Since the tigers were state property, they had to be returned to their owner, the state. But the DNP was not opposed in principle to the use of tigers for tourism, as seen from its readiness to grant a license to the temple's new zoo, while some of its representatives talked about selling or renting the confiscated animals back to the temple. The wellbeing of the tigers was for the DNP of secondary significance. In fact, the welfare of the animals in DNP's facilities worsened rather than improved. The animal rights and welfare supporters differed from the DNP in their approach: they perceived the animals as subjects (rather than mere property) and were primarily concerned with an ethical issue: to stop what they perceived as animal exploitation and abuse in the temple's tiger exhibit, and set the tigers free (although they had no clear conception of the concrete circumstances in which such freedom will be realized). The freedom and welfare of the tigers, rather than their legal standing, was thus their principal concern. So when the Golden Tiger zoo received a license from the DNP, and announced its pending opening, animal activists strongly opposed it. Yet, in sharp contrast, the activists did not show much concern for the conditions under which the tigers were held in the DNP facilities after their 'liberation', and did not protest publicly against their indefinite confinement to small cages in the DNP facilities. This indicates that for the animal activists the saliency of the issue was reduced, once the tigers were moved from the public limelight into a remote place, far from the public eye.

This case study thus draws attention to an often overlooked question in the fight against trafficking in wild animals and their exploitation for tourism: the low concern for the fate of these animals after they have been freed from the clutches of traffickers and tourist entertainment entrepreneurs. As this case study shows, authorities and animal rights and welfare activists tend to focus on cases of animal abuse when they are in the spotlight. But the suffering of 'saved' animals in obscure holding facilities deserves equal attention by both animal rights and welfare activists, as well as animal ethics proponents and tourism researchers.

Source: Cohen (2013, 2019)

CHAPTER 4

SUSTAINABLE TOURISM IN ACTION

LEARNING OUTCOMES

The importance of this chapter lies in the organization of a number of essential approaches that are geared around sustainability concepts and tools, both of which focus around the conceptualization of sustainability and its practice. The chapter is designed to provide the reader with:

1. Issues around the implementation of sustainable tourism.

2. An understanding of what a sustainability concept is, and examples of these.

3. An understanding of what sustainability tools are, and examples of these.

4. How social and ecological systems research has entered the tourism lexicon and what possibilities it has for the implementation of sustainability in tourism.

INTRODUCTION

In Chapter 1 we discussed the history and origins of sustainable development and sustainable tourism, as well as fundamentals and definitions. Here the focus switches to many of the entrenched criticisms around ST before venturing into a discussion of some more tangible concepts and tools of ST. ST is fine in theory, indeed it is a savoury construct, but how this translates into practice is a very different question. Because of these challenges, a range of approaches have been developed to better accomplish implementation ends. We follow the work of Schianetz *et al.* (2007b) who developed a comprehensive overview of the place of concepts and tools in sustainable tourism, and so we organize the second and third sections of this chapter according to their overview (with some inclusions and exclusions of their work). Schianetz *et al.* (2007b: 372) argue that a '*concept* is an idea of how to achieve sustainability'. Examples of concepts include the commons, carrying capacity, ecotourism, the precautionary principle, environmental management systems, codes of ethics, and certification and ecolabelling (combined here because of their close connection). A *tool*, on the other hand, 'is something that typically consists of a systematic step-by-step assessment procedure and/or a computational algorithm that is used to implement a concept' (Schianetz *et al.*, 2007b: 372). Tools that will be discussed in this chapter include sustainability indicators, ecological footprint, life cycle assessment, environmental impact assessment, auditing and reporting. While a range of concepts are said to be important in the implementation of sustainability in tourism destinations (see Lee, 2001), it is the tools that are essential in implementing these concepts, according to Schianetz *et al.* (2007b). A short section is included on social and ecological systems science, as new and innovative ways in which to conceive and implement sustainability in tourism.

CRITICISMS OF SUSTAINABLE TOURISM

In Chapter 1 we discussed that even though sustainability has been an important topic in tourism studies for years, very little has changed in its practice. The issue at hand, according to Moscardo and Murphy (2014), is in the way tourism academics have conceptualized ST. Butler (1991) detected this problem early in the ST discourse by arguing that curbing tourist numbers, changing the tourist type, changing the resource for resistance and educating people involved in the industry stand as our best chances at becoming more sustainable.

Hunter (1995) argued that ST is too tourism centric and parochial and, as such, stakeholders are unable to plan and manage the industry consistent with the tenets of ST (see also Hardy & Beeton, 2001). There are issues of scale at hand, Hunter notes, especially when planning takes into consideration only specific entities like a single resort, when multi-sectoral planning integrated more comprehensively both inside and outside tourism is demanded (see Wall, 1993, in the context of Bali, Indonesia). Hunter advanced two alternative conceptual models. Model 1, total immersion, has ST development nested inside a broader notion of ST. Model 2, partial immersion, has ST development half inside the broader SD paradigm and half outside it. The implications of each are far reaching. In Model 2 the door is left open for various aspects of the tourism system, e.g. transportation, to ignore ST principles, when other industry sectors are adopting these principles.

Others argue that both industry and conservation movements are using sustainability to justify their policies and practices, creating an inherent weakness (McKercher, 1993b). Tourism as a business emphasizes economic development, whereas conservation advocates protection of resources. Can we have it both ways? Conflict over the proper use of resources will continue to be at the heart of disagreements with no end in sight. One merely has to look at just about any tourism impact dilemma to see how sociocultural and ecological factors are compromised in the face of development needs. It is impossible to imagine any type of tourism activity that is developed and operated without reducing the quantity or quality of natural resources in that location (Welford *et al.*, 1999).

Moreover, the term sustainable tourism has been criticized as lacking integrity, not much more than a buzzword or marketing gimmick emphasizing sensibility, sensitivity and sophistication (Butcher, 1997; Wheeller, 1997). Indeed, ST has been criticized as being another empty cliché, with questions surrounding its operationalization, its measurement, and consensus on its definition. Further, some believe that trying to produce definitions of sustainable tourism is dangerous because general definitions can connote the impression of simplicity in what is a complex area, while some descriptions may be irrelevant, misleading and ever-changing (Bramwell *et al.*, 1996).

Butcher (1997: 37) makes the extraordinary statement that 'sustainable tourism is a concept with little to offer the tourist'. He takes this position based on the belief that the new moral imperative in tourism threatens to suppress hedonism as tourists become increasingly regulated in what is acceptable behaviour and what is not acceptable. So, 'Hedonism, once a virtue of tourism, becomes a threat. Caution and wariness are characteristics of the new tourism' (Butcher, 1997: 35). Additionally, Butcher argues that the sustainability rhetoric has become the bane of the lesser developed countries who badly need development in order to overcome debilitating political, social and economic constraints. The choice to preserve cultural and ecological artefacts over the development needs of the impoverished is deemed irresponsible from Butcher's perspective. This perspective has been embraced by other scholars who believe that ST ought to be configured according to poverty alleviation through the pro-poor tourism agenda (increasing access for the poor to tourism benefits), and through ecotourism with its focus on the reduction of environmental impacts (Neto, 2003). This may mean that different interpretations of ST may be appropriate for developed and developing countries. For instance, poor, developing countries may emphasize an economic imperative, whereas other, stronger interpretations of ST are based upon a 'Western environmentalism' (Munt, 1992).

Muller (1994) suggests that the objective of ST is to influence the following factors: economic health, the subjective wellbeing of locals, unspoilt nature and the protection of resources, healthy culture, and the optimum satisfaction of guest requirements. Thus, the desired situation is balanced tourism development in which no one element, whether it be environmental protection, visitor satisfaction or economic health, predominates over the others. Still other authors (Hunter, 1997) argue that the concept of balancing all goals is unrealistic, since such competing aspects are often traded off and priorities emerge which skew the decision making in favour of certain aspects – as noted above by Munt (1992).

Nonetheless, the dominant perception of ST is a destination area tourism/environment system in balance, where none of the above aspects can be allowed to dominate. Although the notion of balance in ST is attractive, questions need to be addressed such as what protecting the resource base really means. As well, some studies interchangeably use the words 'protection', 'conservation' and 'preservation', although each has a very different meaning. Often, no reference is made to particular resources (renewable or non-renewable natural resources), exhibiting a further lack of detail and clarity (Hunter, 1997). Hunter (1997) states that perhaps the most appropriate way to perceive ST is not as a narrowly defined concept reliant on a search for balance, but instead as an overarching, adaptive paradigm within which several different development pathways may be legitimized according to circumstances. Put simply, there may be a need to consider factors such as supply, demand, host community needs and desires, and a consideration of impacts on environmental resources. As well, location-specific factors such as environmental characteristics and existing tourism developments should be considered. Thus, Hunter (1997) surmises that ST research would benefit from a closer inspection of the broader SD literature, which demonstrates a greater flexibility in determining potential development pathways. It is also suggested that ST research could benefit from a more penetrating appreciation of the complexities inherent in human–environment interactions. This would allow for more detailed analyses of the interactions between economic sectors, the level of precaution to be adopted in environmental management (Fennell & Ebert, 2004), potential environmental management techniques, and the extent to which these should be utilized depending on the degree of efficiency sought in the utilization of resources. Table 4.1 provides a breakdown of specific concerns that have been voiced by Moscardo and Murphy (2014), following from Getz (1986) and Moscardo (2011) on the left-hand side of the box, and Liu (2003) on the right-hand side of the box.

Newer approaches to ST are therefore required to overcome these deficiencies in the nature of how tourism is planned, developed and managed from a sustainability perspective. For example, Moscardo and Murphy (2014) develop an alternative approach to tourism sustainability based on quality-of-life factors. These include the complexity of the tourism system at local and global levels, the incorporation of responsible tourism principles, and that tourism should be viewed as one tool among many in striving for sustainability.

IMPLEMENTATION

Although ST has received a great deal of support globally, difficulties remain in the practical implementation of relevant policies applied to both mass and alternative forms of tourism. Hunter and Green (1995) argue that a number of interrelated difficulties exist, including ignorance of tourism's impacts, the diversity of interests and attitudes, and the diversity of the tourism industry itself. It has become evident that many tourists, operators and government authorities still remain ignorant of the potential impacts of tourism on the natural world, even though the long-term viability of the industry depends on the maintenance of high environmental quality. Even if these problems become better appreciated, there still remains the dilemma of convincing those in the industry to embrace ST practices. Clearly, changing static attitudes is not easy, as many developers need to be coerced into implementing ST policies, especially when their profit motive is geared towards ensuring short-term gain rather than long-term conservation (Hunter & Green, 1995).

Table 4.1 Criticisms of sustainable tourism

Moscardo and Murphy (2014)	*Liu (2003)*
1. A narrow focus on specific projects, rather than considerations of tourism as a whole.	**1.** Little effort has been placed into tourism demand at the destination level, with often unfounded assumptions that a sustained level of visitation will occur.
2. Limited attention given to tourism impacts.	**2.** Resource sustainability is often discussed only from the perspective of preservation and conservation, with little recognition of the dynamic and complex nature of resources, which evolve according to the needs of society.
3. A focus on economic factors with occasional limited acknowledgement of environmental issues.	**3.** Little effort has been placed into the emphasis on intergenerational equity according to the fairness of benefits and costs that ought to be distributed across stakeholder groups.
4. A failure to consider how tourism would interact with and affect other activities at a destination.	**4.** More of a focus should be placed on the economic benefits of communities from tourism while maintaining their cultural identities.
5. The naïve adoption of business strategic planning as the dominant framework for tourism planning and, as a consequence.	**5.** Tourism destination and regional decision makers in tourism often seek ways to limit tourism growth through the implementation of carrying capacities, but with little success.
6. The placement of market or tourist needs and expectations as the core drivers of tourism planning, giving destination residents a very limited role, if any.	**6.** Greener or softer forms of tourism like ecotourism, alternative tourism and responsible tourism have been advocated to achieve sustainable outcomes, but with little success.

At the onset of the new millennium, the jury was still out with regard to the success behind turning principles into practice. Welford *et al.* (1999) argue that there is indeed a great deal of rhetoric associated with the concept of ST and rather less guidance on how to operationalize the idea. Swarbrooke (1999) notes that emphasis needs to shift from strategy generation to implementation, because there are many ST strategies in destinations but as yet few examples of successful initiatives. Butler (1998) argues that the overwhelming appeal of ST lies in the generality of the concept while the true costs of the implementation of the concept have never been fully articulated. Butler (1998) explains that where the costs are perceived to be a reduction in development, fewer tourists, less employment and lower income (i.e. mass tourism), then the concept is not supported enthusiastically. In such scenarios, ST is interpreted in terms of economic sustainability, wherein the primary concern is with maintaining the long-term viability of the economy as opposed to sociocultural and ecological viability. Butler (1998) further states that where it has been adopted in the tourism industry, ST has been accepted for three reasons secondary to its rationale: economics, public relations and marketing. For instance, encouraging tourists to conserve power and water can cut utility costs while making guests feel that the accommodation operator is concerned about environmental wellbeing and ST principles, thus garnering good public relations. In the marketing of ST, the tourism industry has achieved success in promoting the concept and its principles perhaps more in line with organizational self-interest than anything else.

The key challenges to implementing SD in tourism, according to Berno and Bricker (2001) can be reduced to three main points. The first includes the nature of the tourism industry and tourism products, which do not conform to classical definitions of 'industry'. The industry (transport, attractions, facilities, food and beverage, accommodation) is amorphous and diffuse and it is difficult to understand and organize all of these on the same plane. The second includes the fragmented fashion in which decision making is made in tourism, which takes place at local, state, national, regional and international levels with little coordination and collaboration. The third includes the conflicting interests that take place among a broad spectrum of different stakeholder groups. These groups include the public sector (governments), the private sector (industry), the not-for-profit sector (e.g. NGOs), tourism associations, Aboriginal communities, the media, host communities (locals) and tourists.

SUSTAINABILITY CONCEPTS

This section introduces the reader to a number of concepts that have been used liberally in the past, not only in tourism and sustainable tourism, but also more generally. These concepts explain how natural resources have been used and misused, and thus the vexing issues of how best to use and manage resources in socially and ecologically responsible ways. A recurring theme among these concepts is the notion that the social and ecological elements of this use can never be separated into discrete entities, i.e. they inform each other. As such, all have important implications in understanding limits of use, and will be useful in describing ST at various junctures throughout the course of the book.

THE COMMONS

Issues over the utilization of common pool resources have received a great deal of attention in the academic literature, especially in economics where the seeds of the concept appear to have first germinated. Huntsman (1944) argued that fisheries depletion boils down to economics, while Gordon (1954) reasoned that common pool resources are difficult to manage because they yield no economic rent – there is no legal entitlement to resources like the marine environment. Commercial fishermen, for example, are free to fish waters wherever and whenever they wish, with the potential – indeed the result in many cases – that fisheries (e.g. cod) have become decimated over the years.

Garrett Hardin's (1968) seminal essay on the tragedy of the commons pushed the discussion even further by turning his attention to human nature. Hardin reasoned that farmers would increase sheep stocks on common lands until such time as the resource could no longer support economic activity. There is thus an element of self-interest at work in this theory which explains how collective demand in the face of diminishing supply leads to the ruin of all. Easter Island is a classic example of this phenomenon where the overutilization of resources led to the ultimate demise of the island's entire population. Jared Diamond (2005) documents the conditions under which collapse has taken place in Easter Island and other similar environments. He lists several contributing factors, including deforestation and habitat loss, soil problems, water management problems, overhunting, overfishing, introduced species, overpopulation and the increased per capita impact on people. Four more recent factors contributing to collapse, according to Diamond, include climate change, the build-up of toxins, energy shortages and overuse of the Earth's photosynthetic capability.

Despite the tremendous gains that Hardin made in advancing common pool resource theory, he has been criticized for failing to appreciate the importance of institutional arrangements in determining the outcome of resource utilization. Theorists such as Ostrom (1986) argued that the tragedy of the commons, or the steady and predictable degradation of natural resources like fisheries, under certain conditions, may be thwarted through the implementation of formal (i.e. policies) and informal (voluntary codes of ethics) rules that dictate and control use. Common pool resources are defined by Ostrom (1990: 30) as any 'natural or man-made resource system that is sufficiently large as to make it costly (but not impossible) to exclude potential beneficiaries from obtaining benefits from its use'. Institutional arrangements are critical factors that determine the success or failure of resource management efforts, and they may be bureaucratic in nature (top down), community-based (bottom up) or a combination of many different management types (Tang, 1991).

While it is beyond the scope of this work to venture into the very deep waters of institutional arrangements, natural resources usage may be classified according to the following categories, each with their own differentiating factors: total open access (e.g. flying or boating), quasi open access (e.g. ocean fisheries), public open access (e.g. forests and lakes), community property (e.g. resources management in developing countries), common property (e.g. resources management in developed countries), and private (Stabler, 1996). Factors include the extent to which access is open to all, whether they are owned, in which domain they are owned (private or public), whether or not property rights are attached, the existence or not of management institutions, the nature of rules governing management and whether they are enforceable, and agency, information and transaction costs, i.e. the cost of making resources exclusive. Tourism impacts often result from the fact that many of the resources the industry depends upon for success are common pool resources such as beaches, mountains and forests, where there is no regulatory agency responsible for the management of these places, and they can be used free of charge (Twining-Ward, 1999). The overuse of these resources for tourism purposes is common, and leads not only to the diminished quality of the resource base, but also to the quality of the experience (Healy, 1992). The problem with common pool resources, as outlined in the definition by Ostrom above, is that there is often no incentive in maintaining or even increasing supply, so 'free riders' are able to reap benefits without putting more back into the maintenance of the resource. Resources, therefore, tend to be overused and under managed (Healy, 1992), which has led tourism theorists to argue that tourism should be viewed as an extractive industry in the same way as mining or fisheries are (Garrod & Fyall, 1998; McKercher, 1993b).

The village of Kiwengwa on the east coast of Zanzibar in Tanzania is an example of a commons-related problem in tourism. Gössling (2001b) writes that the open-water fisheries utilized by the local people of Kiwengwa were viewed as common property. Governance was based on extracting as much of the resource (fish) as local people wanted, as only a few families used the resource. Self-governance was therefore the norm. What changed the property rights regime in Kiwengwa was an increasing local population and rising tourist numbers, in association with the introduction of new technologies such as snorkel mask and spears, which altered traditional use systems; that is, traditional extraction methods for the collection of prawns, gastropods, bivalves and octopuses fell out of use and were replaced by newer approaches, employed by far more users of the resource. As tourism has become a major factor in the modernization of Kiwengwa, rapid changes in the fabric of the relationship between local people and the natural world have also changed and contributed to a cycle of ecosystem degradation. These changes are illustrated in Figure 4.1, based on the following characteristics observed by Gössling (2001b: 449), where tourism has:

1. given rise to individualism and focus on personal economic benefit;
2. encouraged the abandonment of traditional resource-use systems;
3. contributed to turning local natural resources into commodities;
4. spread the idea that resources can be replaced by imports;
5. directly and indirectly imparted a negative effect on the local ecosystems; and
6. turned the village into an emerging centre of resource allocation on an industrial basis

Figure 4.1 shows how capital investment in tourism from external organizations and entities will contribute to increased local incomes while at the same time increasing population growth based on migrants, staffing and tourists. This will in turn lead to the rising demand for natural resources and, later, overexploitation leading to degraded ecosystems. The impact to the community is a weakening of traditional resource use, a loss of indigenous knowledge and a turning to new modes of production that tend to exacerbate the problem, i.e. moving away from a traditional economic structure and entering into the global market economy.

Gössling's (2001b) study shows us that tourism has the potential to transform property rights regimes from primary sector interests to tertiary. This type of evolution is also demonstrated in changes taking place in Goa, India, from Portuguese colonial rule to Indian control in 1961 (Noronha *et al.*, 2002). The new political institution based on democracy and electoral politics, including the small nature of Goa, tourism's demonstration effect and the desire to cross class lines, created pressure to move away from agriculture and into tourism. Noronha *et al.* (2002) argue that the over-specification of rights created a climate

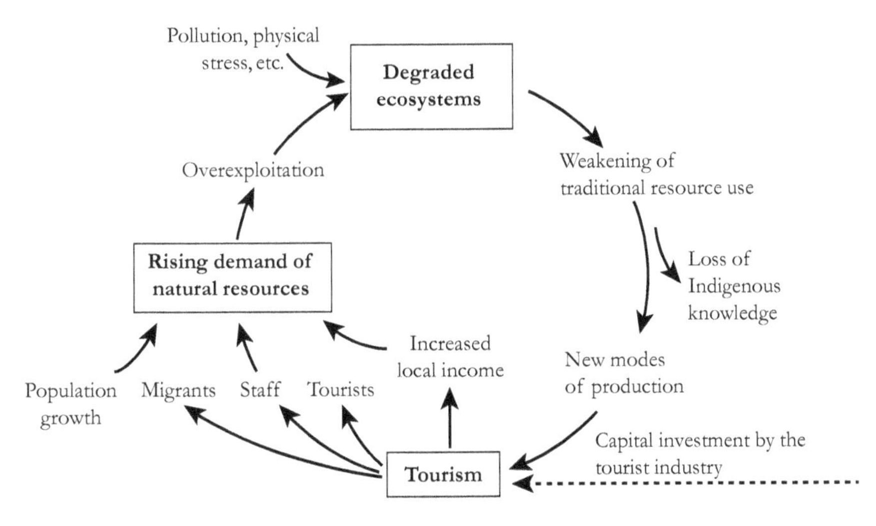

Figure 4.1 The cycle of ecosystem degradation
Source: After Gössling (2001b)

of individual over communitarian values and opened the door to a market culture that had transformative consequences.

CARRYING CAPACITY

The point raised in the introduction about the inability to separate social and ecological factors is not new. It was recognized in the first studies on recreational carrying capacity during the mid-1960s by both Lucas (1964) and Wagar (1964). Wagar wrote that:

> The study ... was initiated with the view that the carrying capacity of recreation lands could be determined primarily in terms of ecology and the deterioration of the areas. However, it soon became obvious that the resource-oriented point of view must be augmented by consideration of human values. (Wagar, 1964: i)

Fundamentally, carrying capacity measures how much tourist use of an area can be accommodated over time before a level of unacceptable damage takes place that reduces the quality of the experience and the site itself. Several types of carrying capacity have been discussed in the tourism literature; the first three are drawn from the Federation of Nature and National Parks of Europe (FNNPE, 1993), while the fourth is taken from the general literature.

1. *Environmental or ecological carrying capacity*: The degree to which an ecosystem, habitat or landscape can accommodate the various impacts of tourism and its associated infrastructure without damage being caused or without losing its 'sense of place';
2. *Cultural and social carrying capacity*: The level beyond which tourism developments and visitor numbers adversely affect local communities and their ways of life;
3. *Psychological (or recreational) carrying capacity*: The level beyond which the essential qualities that people seek in the protected area (such as peace and quiet, few other people, few signs of human developments) would be damaged by tourism developments;
4. *Physical carrying capacity*: Often referred to as the number of people able to fit on a specific human-made structure like a football stadium, airplane, or bus. There is a fixed limit, after which the capacity of the structure has been overshot.

One of the major problems in assessing carrying capacity is determining the specific indicators used to measure limits of acceptable and unacceptable use (McMinn, 1997). That is, how is it possible to ascertain if a proposed development is in fact sustainable and how can it be measured? While some elements of the environment are easy to measure (e.g. volume of tourists that an area can physically accommodate without detriment to the environment), other elements, particularly those associated with the social environment, are extremely difficult to measure. As well, McMinn (1997) points out that any one tourist site is likely to have many different factors, all with their respective carrying capacities and many with extremely subjective criteria. Such factors have different values attached to them depending on the elements as well as the person making the judgement. Thus, the measurement of carrying capacity can be vague when considering the many different factors in the location, the philosophical and ethical issues involved in determining how the physical limits are established, and who is entrusted with the responsibility of making the decisions.

The vast majority of studies on carrying capacity share the concern that even if one can measure the physical, ecological and social/psychological capacity of an area, how should it be put into practice? Each locality is different in terms of geography, ecosystems, social structure and economy, and tourist demand. Thus, it is unlikely that the carrying capacity will be the same in any two places, so its measurement and application must be site specific while at the same time being very difficult to forecast (Swarbrooke, 1999). As a result, when it comes to implementing the system as an inter-organizational instrument, the carrying capacity concept has not been particularly useful. Only for well-defined areas (provincial parks or national parks) has it been possible to develop the tool to such an extent that it can be classified as an effective integrated management system (Hjalager, 1997).

The most widely used application of the carrying capacity concept in tourism studies is the tourist area life cycle model developed by Butler (1980). Butler argued that increases in tourist numbers to a destination over time could lead to eventual decreases if the carrying capacity of the destination is breached. Characteristics of overuse of destinations include lack of investment in facilities that are in need of upgrading (e.g. hotels), changes in the nature of tourists as prices drop, and investment in new properties external to the destination region.

Other scholars have used the basic premise behind stage-based models like Butler's to argue that ecotourism in developing countries can be positive in the latter stages of development. Favourable outcomes and positive attitudes result from increases in quality of life of residents benefiting from tourism throughout the life cycle of tourism development (Hunt & Stronza, 2014).

Carrying capacity has also been widely discussed from the perspective of ST where, according to Saarinen (2006), there are distinct traditions of sustainability in tourism studies based on limits of growth. The resource-based tradition emphasizes the importance of carrying capacity in the formative days of tourism and sustainability, and the idea that specific numbers could be formulated in efforts to avoid negative ecological impacts. The activity-based tradition complicates the resource-based tradition owing to the fact that tourism is a dynamic activity. There are too many intervening variables around different activities in time and space that complicate efforts to control numbers according to ecological and sociocultural factors. The community-based tradition underscores the relationship between activities and the resource base and the various participants involved in the tourism landscape. Tourism as a dynamic force is continually being constructed and reconstructed through various phases of planning, development and management.

THE PRECAUTIONARY PRINCIPLE

An emerging norm that has been embraced in some jurisdictions and sectors more readily than in others is the precautionary principle (PP). Conceived in Germany during the 1970s (Boehmer-Christiansen, 1994), the PP is 'a culturally framed concept that takes its cue from changing social conceptions about the appropriate roles of science, economics, ethics, politics and the law in pro-active environmental protection and management' (O'Riordan & Cameron, 1994: 12). Caution and a conservative approach to how we engage with the natural world are two fundamental concepts needed to guide human use of the natural world (Myers, 1993). According to VanderZwaag (1994: 7), the PP is framed around:

- a willingness to take action (or no action) in advance of formal scientific proof;
- the cost-effectiveness of action, i.e. some consideration of the proportionality of costs;
- providing ecological margins of error;
- the intrinsic value of non-human entities;
- a shift in the onus of proof to those who propose change;
- a concern with future generations; and
- paying for ecological debts through strict/absolute liability regimes.

Although the PP has been adopted at international levels, VanderZwaag (1999) argues that often the intent of these global perspectives is to be general when being more specific in policy and practice is essential. VanderZwaag (1999) uses the example of the 1992 Earth Summit as an example, where Principle 15 states that:

> In order to protect the environment, the precautionary approach shall be widely applied by States according to their capabilities. Where there are threats of serious or irreversible damage, lack of full scientific certainty shall not be used as a reason for postponing cost-effective measures to prevent environmental degradation. (Van Dyke, 1996: 10)

Phrases such as 'according to their capabilities' are not helpful, because of the situational and relative notions of what is acceptable or not. This means that reducing uncertainty (Dovers & Handmer, 1995), risk (Bodansky, 1994), and better understanding the costs and benefits in our use of natural resources (Scott, 2000) are all important. Science nested within pressure from public and private sectors is also essential in the production of data that protect human health and the environment as a first priority (CBD & UNEP, 2001). There are many challenges pertaining to how or if experimental research can be assembled into predictive models showing how the natural world behaves over time (Kaiser, 1997).

The PP has direct relevance to sustainability because of its demand that we think in expanded periods of time, the emphasis on restraint, and the need for the present generation to think and act in ways that conserve resources and maintain health so that these later generations will benefit and thrive (deFur & Kaszuba, 2002). For the tourism industry, this means that operators, organizations, scientists and other stakeholder groups need to exercise precaution in their short- and long-term actions and be subject to external scrutiny in the form of audits or other mechanisms in maintaining appropriate standards of practice (Kirstges, 1995). In order to better plan, develop and manage the tourism industry, Tickner and Raffensberger (1998) offer the following six steps:

1. *Define the general duty to take precautionary action.* This involves the adoption of a corporate or industry-wide duty to take precautionary action in the face of scientific uncertainty where there is a threat to human health or the environment. The concept of human health could be expanded to include an assessment of how tourism developments, for example, have transformative impacts on the ecology and customs of local communities.
2. *Set aggressive goals/vision for achieving sustainability (backcasting).* This step involves the establishment of clear and measurable goals by which to drive innovative best practices within the industry. The establishment of vision statements, which cut across sectors of the tourism industry and provide the mechanism by which to develop effective goals and objectives, should be likened to a road map that provides a way to the future.
3. *Assume responsibility for demonstrating the safety of products and processes.* Tourism industry stakeholders involved in the planning, development and management of the tourism industry must demonstrate the safety of their operations before engaging in such activities. The choice of technology, materials and products is paramount in demonstrating that their activities have a limited impact on people and natural resources.
4. *Create criteria for decision making under uncertainty.* Companies will need to create and adhere to criteria that will guide decision making where harm to people or natural resources is a possibility. Indicators of sustainability and other such tools will need to be employed for the purpose of determining what type of evidence to weigh in assessing impacts and how to do it.
5. *Use tools for implementing precautionary, preventative approaches.* There are numerous tools for carrying out precautionary policies, including: (a) clean production and pollution prevention; (b) corporate product phase-outs; (c) strict standards within the firm; and (d) alternative assessments. The recent focus on environmental management systems will provide direction in this regard.
6. *Use the 'polluter-pays' principle.* This principle, which places responsibility on the shoulders of the offending party, demands that such parties pay the costs of the damage they cause. One mechanism that has garnered a high level of support is assurance bonding. Companies are required to pay a premium before undertaking a project, which is based on the worst potential damage that might occur from the development. If no damage occurs, the bond is returned to the developer.

As tourism continues to increase in magnitude in urban, rural and wilderness areas, and in both terrestrial and aquatic environments, there is a pressing need to incorporate precautionary measures to ensure that the industry is not taking more at the expense of human and ecological health. In referencing Confucius who said that 'the cautious seldom err', it follows that precaution needs to be exercised in tourism even before harm can be demonstrated (Fennell & Ebert, 2004).

REGULATION VERSUS VOLUNTARY INITIATIVES (CODES OF ETHICS)

Regulations are the rules established by the state or other regulatory agencies to prescribe proper conduct by members of the jurisdiction. For example, regulation of the airline industry by government has taken place since the Air Mail Act of 1925 and included, in various forms, the control by government of fares, routes, the entry of airlines into the market, and safety, as well as flight schedules. Deregulation of the US airline industry in 1978 gave more control back to airlines in many of the aforementioned areas except safety (McDermott, 2017).

An early foray into the connection between tourism, regulation and sustainability comes from Forsyth (1995), who examined the business attitudes of the UK's outgoing tourism industry. Forsyth found that several companies felt that, although embracing sustainability would be good for business, and many had actually adopted certain measures of sustainability (e.g. donations to local charities, 'ecotips' and codes of ethics), operators felt that it was host governments that were most responsible for ST. Of a sample of 69 interviewees from business, 63.8% felt that host governments alone were responsible for implementing ST, followed by 30.4% who felt that operators and host governments were responsible, followed by tourism associations (4.4%), and then operators alone (1.4%). These results provide scope into the extent to which operators feel a lack of responsibility in implementing ST. Although operators stand to gain most from tourism in the form of economic benefits, access to foreign lands and utilization of common pool resources, another sector or domain should provide leadership. Organizational and individual self-interest in a culture of competition must surely be the rationale for such behaviour – a topic explored more at length in Chapter 3.

The tourism industry typically chooses self-regulation over governmental regulation because it affords industry stakeholders greater flexibility in service provision (Buckley, 2012). A good example of a group that has embraced self-regulation is the International Association of Antarctica Tour Operators (IAATO). This organization not only self-regulates according to visitor interaction with animals (e.g. how close tourists can get to animals), but also on who and where the 40-odd companies that operate cruises in Antarctica can access landing sites through the course of the tourist season (Marshall, 2016). The work of the IAATO is also a good example of how codes of ethics are implemented in tourism. The IAATO's 'Guidance for Visitors to the Antarctic' outlines general as well as specific categories to maintain the ecological integrity of the continent. The general category has four guidelines. These are:

- Walk slowly, occasionally stopping to give wildlife the time and space they need.
- As a general principle, keep noise to a minimum and avoid approaching birds and animals any closer than 5 metres/15 feet; in some instances even this may be too close, so watch the birds' and animals' behaviour as you approach and stop, or retreat, if they show signs of disturbance.
- Be aware of your location relative to your fellow visitors – making sure you do not surround animals or cut off their route to the sea.
- Heed the advice of your guides; they want you to gain as much as possible from your experience while treading softly on this unique environment. (IAATO, 2019)

Codes of ethics are said to be one of the most important expressions of an organization's philosophy. They induce members of an organization to think about their actions in a philosophical manner and to do their duty as moral individuals in the context of the expressed values of the organization (Fennell & Malloy, 2007). More formally, the code of ethics is said to serve three general aims: (1) to establish the moral values recognized by a company; (2) to communicate the company's expectations to employees; and (3) to demonstrate to employees and the public that the company operates within specific ethical parameters (Montoya & Richard, 1994: 713).

Goodall and Cater (1996) argued that although codes of ethics have a place in tourism, and by their sheer number they have an important place, there is little possibility that they will make tourism sustainable. One of their main reasons to downplay codes in reaching sustainability is because businesses are seeking stronger operational guidance for making better practical changes, and so other measures are needed to push the sustainability agenda. Certification and ecolabels are a further step in this direction.

CERTIFICATION AND ECOLABELS

Certification

Certification schemes have been widely used in tourism and ST because of their emphasis on the reduction of ecological and sociocultural impacts. Such programmes hold the industry accountable for its actions, and for those organizations that choose to adopt certification standards there are benefits in the form of leverage with demand. Tourists who are becoming increasingly savvy in their selection of ethical tourism operators often choose operators who have adopted certification schemes over those that have not. Font *et al.* (2003: 213) define certification of sustainable tourism (and ecotourism) as 'the process of providing documented assurance that a product, service or organization complies with a given standard'. Piper and Yeo (2011: 281–282) developed a more detailed overview of certification and related concepts:

- *Certification*: A procedure by which a third party gives written assurance that a product, process or service is in conformity with certain standards.
- *Certification body*: An organization performing certification, sometimes referred to as the certifier or the certification agency. The certification body may use an existing standard or it may set its own standard, perhaps based on an international and/or normative standard.
- *Certification label*: A label or symbol indicating that compliance with specific standards has been verified. Use of the label is usually controlled by the standard-setting body.
- *Standards*: Documented agreements containing technical specifications or other precise criteria to be used consistently as rules, guidelines or definitions, to ensure that materials, products, processes and services are fit for their purpose. Standards include environmental standards, organic standards, labour standards, social standards and normative standards. Environmental standards are standards for materials, products and production processes to ensure that negative impacts on the environment are minimal or kept within certain limits.

Font *et al.* (2003; see also Font & Sallows, 2002) document the consultation process that led to the development of the Global Sustainable Tourism Council (GSTC), spearheaded by the Rainforest Alliance. The GSTC is a not-for-profit organization that has as its mission to act as the main global force in the certification of various types of tourism operations (Box 4.1). More specifically, the GSTC has established a set of criteria and

Box 4.1 The Global Sustainable Tourism Council

The GSTC is an independent and neutral organization, legally registered in the USA as a 501(c)3 non-profit organization that represents a diverse and global membership, including UN agencies, NGOs, national and provincial governments, leading travel companies, hotels, tour operators, individuals and communities – all striving to achieve best practices in sustainable tourism. It is a virtual organization without a main office, with staff and volunteers working from all six populated continents. Financial support from donations, sponsorship and membership fees allows them to provide services at low cost and to create, revise and make available the GSTC Criteria.

The GSTC establishes and manages global sustainable standards, known as the GSTC Criteria. There are two sets: Destination Criteria for public policy makers and destination managers, and Industry Criteria for hotels and tour operators. These are the guiding principles and minimum requirements that any tourism business or destination should aspire to reach in order to protect and sustain the world's natural and cultural resources, while ensuring that tourism meets its potential as a tool for conservation and poverty alleviation.

The GSTC Criteria form the foundation for the GSTC's role as the global accreditation body for certification programmes that certify hotels/accommodations, tour operators and destinations as having sustainable policies and practices in place. The GSTC does not directly certify any products or services, but it accredits those that do.

Source: Adapted from GSTC (2019a)

performance indicators for key players in the tourism industry, including hotels, tour operators and destinations. The criteria and performance indicators for hotels can be found at GSTC (2016).

Costa Rica's Certification for Sustainable Tourism (CST) run by the Costa Rica Tourist Institute (ICT) is a concept that provides competitive advantage for participating firms while at the same time guaranteeing homogeneity and quality of goods and services. There are four categories of assessment in satisfying the criteria for certification:

1. *Physical-biological parameters*: Evaluating the interaction between the tourism company and the surrounding natural world.
2. *Infrastructure and services*: Evaluating the management policies and operations of the company along with its use of infrastructure.
3. *External clients*: Evaluating the interaction of the company with its clients in terms of how much it allows and invites the client to be an active contributor to the company's policies and sustainability.
4. *Socioeconomic environment*: Evaluating the interaction of the company with local communities and the population in general.

Within each broad sector, companies are measured with a series of questions based on a scale that ranges from 0 to 5, with 5 being outstanding. A hotel at Level 1 of sustainability scores between the 20th and 39th percentile, suggesting that the firm is meeting basic criteria with reference to sustainability. By contrast, a hotel that measures 5 is in the 95%+ percentile and has scored high on all sustainability measures (Anywhere Inc, 2018). Vasconcelos-Vasquez *et al.* (2011) have illustrated the advantages and disadvantages of the CST scheme (see Table 4.2).

Studies on sustainability certification at the producer level (bananas, coffee, fish products, forest products and tourism) indicate that producers feel that there are more economic benefits than environmental benefits of such schemes (Blackman & Rivera, 2011).

Table 4.2 Advantages and disadvantages of the Certification for Sustainable Tourism (CST)

CST advantages	*CST disadvantages*
Can be applied to any kind of firm.	
The cost is accessible for any firm in Central America and for firms all over the world.	Competes with other similar certifications.
It is an external certification, performance is measured.	Ignores some legal elements.
It takes into account the social environment to determine sustainability.	Focuses mainly on the environment.
It is government awarded, which adds value to this certification.	The assessor is not internationally recognized.

Source: Vasconcelos-Vasquez *et al.* (2011)

Using the example of tourism specifically, Rivera (2002) found that certification led to price premiums in hotels in Costa Rica, suggesting positive economic benefits. Rivera and de Leon (2004) found that certification of ski resorts in the USA weighed more heavily in the direction of environmental costs than improved environmental performance. Dziuba (2016) found through the implementation of econometric modelling that hotels complying with certification standards had lower maintenance costs than hotels that did not, representing 19.09% of the average costs incurred by all the hotels included in the study.

Ecolabels

Ecolabels are an important aspect of a company's environmental management strategy in the way in which they reduce the level of disconnect between consumers and producers regarding the environmental qualities of their products (Delmas & Lessem, 2017; Piper & Yeo, 2011). Consumer awareness, therefore, is forcing more sustainable practices and creating a critical cycle between consumers and producers, driven by eco-labelling (Miranda-Ackerman & Azzaro-Pantel, 2017). Ecolabels provide reliable information to consumers about the environmental impact of products or services in the expectation that this information will influence their purchasing decisions.

In some rural tourism regions, guesthouse entrepreneurs are unwilling to adopt ecolabels as an expression of European values. Such is the case in Bucovina, Romania, where sustainable use of energy, water and soil are not present in environmental policies. This led Stanciu *et al.* (2015: 515) to conclude that 'the common tourism policy of the guesthouses in the analysed area is pragmatic, risky, uncompetitive and uncorroborated with the European (EU) vision and requirements'. In other studies, fish farming operators (a supplier to the tourism industry) and other stakeholders in the industry express several different perceptions about the use and viability of ecolabel schemes. For example: optimists argue that there is a large market for ecolabelled fish; improvers argue that more research is needed on ecolabels and fish farming; pragmatists say that there is a market for ecolabelled fish but it is small; and sceptics note that certification does not equal sustainable, not does it address the many ecological concerns generated by the aquaculture industry (Weitzman & Bailey, 2018).

ENVIRONMENTAL MANAGEMENT

An environmental management system (EMS) is 'a collection of internal policies, assessments, plans and implementation actions … affecting the entire organization and its relationships with the natural environment' (Darnell *et al.*, 2008). Although EMSs typically vary across a number of dimensions, all of them are characterized by (Coglianese & Nash, 2001):

- having an environmental policy or plan;
- internal assessments of the organization's environmental impacts;
- having quantifiable goals to reduce impacts;
- programmes and resources to train staff;
- checking implementation of strategies through an audit process to ensure that chosen goals are being achieved;
- correcting problems and deviations from chosen goals through a review process.

Scholars argue that the commitment to an EMS ought to be firmly embedded within the operational procedures of organizations so that protecting the natural environment becomes second nature (Shireman, 2003) and there is improvement in business and environmental practices (Curcovic *et al.*, 2000). The International Organization for Standardization (ISO) has a family of different frameworks, and reports that as of 2018 there were more than 300,000 certifications of ISO 14000 in 171 countries (ISO, 2018). The 14000 family, referred to as 'Environmental Management', is characterized as follows:

> ISO 14001:2015 sets out the criteria for an environmental management system. It maps out a framework that a company or organization can follow to set up an effective environmental management system. It can be used by any organization regardless of its activity or sector.
>
> Using ISO 14001:2015 can provide assurance to company management and employees as well as external stakeholders that environmental impact is being measured and improved. (ISO, 2018)

In tourism it is often the larger tourist operations that have well-developed EMS programmes in attempts to achieve ST objectives (Pigram, 1996). Thus, ST faces the challenge of raising concern for the environment among the numerous smaller establishments and encouraging the spread of environmental best practice at all levels of tourism activity. Convincing these sectors to step towards sustainability may call for a range of incentives, or even sanctions. Improved measures of environmental performance that contribute to better environmental and sociocultural accounting are required, as is the push for accountability from many of the major players in tourism such as the biggest tourism firms, as noted above, as well as governments and international organizations such as the UNWTO.

This discussion indicates that in order to move towards the goal of ST, a greater level of participation is required on both the supply and demand sides. Carey *et al.* (1997) point out that the supply side must take greater responsibility for the planning, organization and implementation of a consistent and sustainable tourism policy which integrates the public and private sectors. Although some organizations seek to achieve the objectives of environmental sustainability by implementing an EMS, they often do not address many of the broader social, cultural and community aspects of sustainability (Welford *et al.*, 1999).

ECOTOURISM

If there is one thing that characterizes tourism, in both theory and practice, it is the preoccupation with nomenclature: we love the development of new terms, presumably because of our penchant for exposing the vulnerable underbelly of older terms and for competitive advantage. A simple overview of the literature will uncover dozens of different tourism types, with many newer forms embracing the alternative side of the ledger. Examples include green tourism, geotourism, environmental tourism, nature-based tourism, adventure tourism, rural tourism, cultural tourism and too many others to name. Here, we briefly focus on just one type of alternative tourism – ecotourism – and illustrate how overlap between ecotourism and other forms of tourism maybe confusing not only to operators but also to tourists.

Ecotourism is defined as 'Travel with a primary interest in the natural history of a destination. It is a form of nature-based tourism that places ... emphasis on learning, sustainability (conservation and local participation/benefits), and ethical planning, development and management (Fennell, 2014: 17). This type of tourism involves direct experience with the natural world, and it is the resource base or natural capital stock (ecosystems) that ecotourists are in pursuit of. As observed by Tyler and Dangerfield (1999), Orca ecotourism in Norway is made possible because of the fjord ecosystem that leads to an abundance of prey species (e.g. salmon) that holds densities of killer whales sufficient in size to support a whale-watching industry. The number of people who seek tourism experiences that link directly with low-impact use of the resource base, environmental conservation and sustainable activity (i.e. ecotourism) has steadily increased since the late 1980s (Fennell, 2015). As a relatively new form of tourism, ecotourism by nature should be sustainable.

The company Adventure Travel, founded in 2001, argues that the ecotourism industry has not gone far enough because the concern is only about nature and not about the people who live in the places Adventure Travel visits and who must benefit from the industry.

A further flaw observed by Adventure Travel is that ecotourism is watered down by greenwashing – ecotourism operators pretending to be something in theory (ethical, nature-based, responsible and so on) that they are not in practice. With the misrepresentation taking place in ecotourism, many organizations are using 'responsible tourism' (RT) as the term of choice (Russell & Wallace, 2004).

These observations, however, do not necessarily hold true, as there are plenty of ecotour operators that do not greenwash and that actively support the interests of local people, especially through efforts to allow local people to benefit in many different ways: control of decision making, economic benefits, ecological benefits, etc. Furthermore, there is considerable overlap between companies who are defined by different labels (ecotourism, green tourism, sustainable tourism, ecotourism and so on) but end up doing many of the same things. This example also shows how tourism operators often search for niche opportunities in order to influence a travel public that may not be educated in the subtleties in or differences between related types of tourism. And so we would be remiss in failing to identify that RT, too, has its critics. While some are staunch advocates of this form of tourism and have made good strides in meshing social and environmental needs (Discover Corps, 2018), others argue that RT cannot be the answer because there are always winners and losers, i.e. the costs and benefits are not distributed evenly (Wheeller, 1991). Thinking otherwise, according to Wheeller (1991: 92), is 'dangerously misleading'. Ecotourism is discussed in greater detail in Chapter 9.

SUSTAINABILITY TOOLS

FROM PRINCIPLES TO INDICATORS

Moving forward from definitions and basic concepts, several groups have developed principles and indicators of ST. In general, principles are 'essential elements of the areas ... which help to elaborate the meaning of objectives', while indicators are 'measurable states which allow the assessment of whether or not associated criteria are being met' (NRI, 2002). Criteria are more robust because they demand that something must be measured, not just stated as aims or attitudes in the case of principles, i.e. there must be some meaningful form of measurement in the operationalization of these principles to make sure practice is following on from theory (Garrod & Fyall, 1998). But as MacGillivray and Zadek (1995) point out, indicators are criticized because they privilege some aspects of sustainability while excluding others because they reflect the interests of their creators. A case in point is the sustainability indicators developed by Roberts and Tribe (2008), which are based around four dimensions: economic, management/institutional, environmental and sociocultural. Roberts and Tribe identify: five issues/themes and 14 performance indicators for the economic dimension; only two issues/themes and five performance indicators for the management/institutional dimension; eight issues/themes and 22 performance indicators for the environmental dimension; and six issues/themes and 13 performance indicators for the sociocultural theme.

An example of principles used in ST is *Beyond the Green Horizon: Principles for Sustainable Tourism* (Eber, 1992), sponsored by Tourism Concern and the World Wildlife Fund. This publication focused on 10 principles for ST. These are: (1) using resources sustainably; (2) reducing over-consumption and waste; (3) maintaining diversity; (4) integrating tourism into planning; (5) supporting local economies; (6) involving local communities; (7) consulting stakeholders and the public; (8) training staff; (9) marketing tourism responsibly; and (10) undertaking research.

The Tourism Industry Association of Canada (1995), together with the National Round Table on the Environment and Economy, developed guidelines for ST practices with an emphasis on multi-level cooperation and environmental protection. This document contains guidelines for tourists, the tourism industry and related associations, tour operators and government factions, among others. While too large to present here in its entirety, categories of guidelines for tour operators include: (1) policy, planning and decision making; (2) the tourism experience; (3) the host community; (4) development; (5) natural, cultural and historic resources; (6) conservation of natural resources; (7) environmental protection; (8) marketing; (9) research and education; (10) public awareness; (11) industry cooperation; and (12) the global village. In this last case, awareness needs to be heightened about the role that tourism ought to play in 'promoting international understanding and cooperation'.

The International Institute for Sustainable Development (IISD, 1992) developed national and local/hotspot indicators for the proper sustainable management of tourism as shown in Table 4.3.

A more comprehensive set of indicators was developed for the World Tourism Organization (see Manning, 1996; UNWTO, 1996), designed to enable the tourism sector to manage environmental and sociocultural costs, but also to better integrate with regional planning initiatives. These indicators are of a number and scale that would be easily adopted (implementable and understandable), and they should be heterogeneous such

Table 4.3 Indicators for the sustainable management of tourism

National indicators	***Local/hot spot indicators***
Area protected	Destination attractiveness index
Endangered spaces	Site stress index
Cultural protection	Consumption
Travel intensity	Ratio of tourists to clients
Use intensity	Development density
Key resource consumption	Percentage of foreign-owned facilities
Ratio of tourists to residents	Environmental quality
Health/social impacts	

Source: IISD (1992)

Table 4.4 Core indicators of sustainable tourism

Indicator	*Specific measures*
Site protection	Category of site protection according to IUCN[a] index
Stress	Tourist numbers visiting site (per annum/peak month)
Use intensity	Intensity of use – peak period (persons/ha)
Social impact	Ratio of tourists to locals (peak period)
Development control	Existence of environmental review procedure or formal controls over development of site and use densities
Waste management	Percentage of sewage from site receiving treatment (additional indicators may include structural limits of other infrastructural capacity on site, e.g. water supply, garbage)
Planning process	Existence of organized regional plan for tourist destination region (including tourism component)
Critical ecosystems	Number of rare/endangered species
Consumer satisfaction	Level of satisfaction by visitors (survey based)
Local satisfaction	Level of satisfaction by locals (survey based)
Composite indices	
Carrying capacity	A. Composite early-warning measure of key factors affecting the ability of the site to support different levels of tourism
Site stress	B. Composite measure of levels of impact on site – its natural and cultural attributes due to tourism and other sector cumulative stresses
Attractiveness	C. Qualitative measure of those site attributes which make it attractive to tourism and which can change over time

Note: [a]IUCN: International Union for the Conservation of Nature.

that they could be applied in different contexts. The two tables are built around two different types of indicators. The first are core indicators (Table 4.4), and the second are ecosystem-specific indicators (Table 4.5), suggesting that ST ought to be situational around the specific demands of different types of natural environments.

Indicators have also been developed for specific ST purposes. Roberts and Tribe (2008), as noted above, developed indicators for small tourism enterprises, while Choi and Turk

Table 4.5 Ecosystem-specific indicators

Ecosystem	*Sample indicators*
Coastal zones	• Degradation (% of beach degraded, eroded) • Use intensity (person per metre of accessible beach) • Shoremarine fauna (number of key species sightings) • Water quality (faecal coliform and heavy metals counts)
Mountain regions	• Erosion (% of surface area eroded) • Biodiversity (key species counts) • Access to key sites (hours' wait)
Managed wildlife parks	• Species health (reproductive success, species diversity) • Use intensity (ratio of visitors to game) • Encroachment (% of park affected by unauthorized activity)
Unique ecological sites	• Ecosystem degradation (number and mix of species) • Ecosystem degradation (% area with change in cover) • Stress on site (number of operators using the site) • Number of tourist sightings of key species (% success)
Urban environments	• Safety (crime numbers) • Waste counts (trash amounts, costs) • Pollution (air pollution counts)
Cultural sites (built heritage)	• Site degradation (restoration/repair costs) • Structure degradation (precipitation acidity, air pollution counts) • Safety (crime levels)
Cultural sites (traditional communities)	• Potential social stress (ratio average income of tourists/locals) • In seasonal sites (% of vendors open year-round) • Antagonism (reported incidents between locals/tourists)
Small islands	• Currency leakages (% loss from total tourism revenues) • Ownership (% foreign/non-local ownership of tourism businesses • Water availability (costs, remaining supply) • Use-intensity measures (at scale of entire island as well as for impacted sites)

Source: Manning (1996)

Table 4.6 Top three sustainability indicators for each community tourism dimension

Dimension	*Sustainability indicator*
Economic	• Availability of local credit to local business • Employment growth in tourism • Percentage of income leakage out of the community
Social	• Resident involvement in tourism industry • Visitor satisfaction/attitude towards tourism development • Litter/pollution
Cultural	• Availability of cultural site maintenance fund and resources • Type and amount of training given to tourism employees (guides) • Types of building material and décor
Ecological	• Air quality index • Amount of on-site erosion • Frequency of environmental accidents related to tourism
Political	• Availability and level of land zoning policy • Availability of air, water pollution, waste management policy • Availability of development control policy
Technological	• Accurate data collection • Use of low-impact technology • Technology for benchmarking

Source: Choi and Sirakaya (2006)

(2006) developed sustainability indicators to measure community tourism development through a Delphi technique involving 38 tourism scholars. After three rounds of discussion, the scholars arrived at a list of 125 indicators categorized as political (32), social (28), ecological (25), economic (24), technological (3) and cultural (13). Although too numerous to mention in their entirety here, the authors identified the top three sustainability indicators for each dimension. These are shown in Table 4.6.

ECOLOGICAL FOOTPRINT

The concept of an ecological footprint (EF) was first introduced by William Rees (1992), and further advanced by Wackernagel and Rees (1996). As described by Wang *et al.* (2017: 2), the 'ecological footprint is an estimate of the area of biotically productive land and water that are appropriated exclusively to produce the natural resources used and assimilate the wastes generated'. In less detail, the EF 'is a quantitative measurement describing the appropriation of natural resources by humans' (Hoekstra, 2008: 10).

Environmental footprints are numerous and include carbon, water, energy, emissions, nitrogen, land, biodiversity, phosphorous, fishing-grounds, waste and human footprints (Čuček *et al.*, 2012). Several applications of these will be examined in coming chapters.

Used in a tourist context, the EF, or rather the 'touristic ecological footprint' (TEF; Hunter, 2002), is defined and operationalized on a number of different levels. These include complete sector (the entire industry), component sectors, products, destination areas, temporal and outlets. The EF for the entire tourism industry, i.e. the sum of all tourism-related activities, has been estimated to be 10% of the planetary EF given that the tourism industry accounts for approximately 10% of world revenue. Such estimates are difficult because of the nature of tourism itself along the lines of diversity and overall consumption patterns. Hunter (2002) argues that it may be more rational to focus on more specific aspects of tourism like transportation. A transit-related TEF, at a destination or within a region, with a compendium of these measures, would be helpful in understanding the magnitude of the sector on ecology. Hunter's third suggestion, product, might focus on, for example, a one- or two-week vacation at a specific resort at a mass tourism destination, while the product TEF could be measured according to the information in Table 4.7.

Hunter queries whether an all-inclusive resort (mass tourism destination that is close by) would be more taxing on the resource base than a week's vacation to a far-off destination that is supposedly alternative in its design (e.g. trekking in the hills of Nepal). There are also questions about the TEF for tourists as compared to local people, from a product standpoint but also from the destination more generally, as well as questions about the

Table 4.7 Contributions of the product TEF associate with a 'typical' foreign holiday package

Tourist zone	*Examples of contribution to total product TEF*
Source area	• Purchases made specifically for the holiday (e.g. clothing, camera) • Travel to the airport
Transit area	• Travel in the air (outbound and return) • Food and beverages consumed during flight (outbound and return)
Destination area	• Travel while on holiday (e.g. organized tour, travel in hired car) • Purchases made (e.g. clothing, gifts) • Food and beverages consumed (e.g. meals at hotel, restaurants) • Water consumed (e.g. drinking, swimming pool) • Waste products (e.g. sewage, food packaging) • Energy requirements (e.g. lighting, air conditioning, heating)

Source: Hunter (2002)

assessment of the TEF for a tourist as compared to his or her EF at home. We have an impact on the planet in our daily lives at the point of origin. How this compares to the TEF when we travel in terms of trade-offs is something that needs further investigation. The TEF might also be assessed from a temporal standpoint according to seasonality, i.e. high-season demands far exceed low-season demands, and certain outputs should be measured for individual businesses, like hotels. Comparing the per capita outputs of these businesses, e.g. room sizes, per capita water usage, number and use of televisions, as Hunter claims, would allow the more environmentally aware tourists to select units that had lower demands than ones with higher demands.

More recent studies on the EF in tourism have evolved into a family of different footprint methods, as noted above. Beyond the traditional TEF, there is the tourism carbon footprint (TCF) as well as the tourism water footprint (TWF), all of which are effective tools for quantitatively assessing the level of impact that tourism has on the ecosystems of specific destinations. As such, EF analysis can estimate regional sustainability by measuring natural resource consumption against the ecological carrying capacity of the destination (Wackernagel *et al.*, 2002).

Wang *et al.* (2017) summarized a number of studies on the TEF and found that tourism transportation energy consumption represented the greatest contribution to the footprint, accounting for 59–97% of it. Examples include: Hawaii, with transportation representing 69% of the TEF, accommodation 25%, recreation 6% (Konan & Chan, 2010); New Zealand, with transportation representing 73% of the TEF, accommodation 17%, recreation 10% (Becken *et al.*, 2003); Switzerland, with transportation representing 87% of the TEF, accommodation (10%), recreation 1% (Perch-Nielsen *et al.*, 2010); and the Seychelles, with transportation representing 97% of natural resources used (Gössling, 2002). Zhang *et al.* (2017) provide an example of the TWF in the Mount Huangshan region of China. They report that the total TWF, which included green (rainwater used in production of goods), blue (surface and ground water consumed) and grey (wastewater from all sources except toilets) water, was approximately 10.19 million m^3/year, or about 3.39 m^3/day or 3387 L/day. Tourist sewage and food production were the two main sources of water consumption.

LIFE CYCLE ASSESSMENT

Kiddee *et al.* (2013) argue that there are several tools that may be implemented in order to track and manage persistent problems such as e-waste. One of the most important of a range of tools is life cycle assessment (LCA), which Kiddee *et al.* define as a tool to design environmentally friendly devices for the purpose of minimizing e-waste, and to identify environmental impacts such as carcinogens, climate change ozone layer effects, ecotoxicity, acidification, eutrophication and land use impacts to improve the environmental performance of products (see also Belboom *et al.*, 2011). The key feature of LCA, therefore, is the tracking of environmental impacts created in the various stages of a product's development and use (its life). This often includes the use of LCA in a cradle-to-grave capacity, where a product is tracked right from the initial stages of resource extraction for the development of the product, through product use, and to its final destination in a landfill site.

A recent example of a timely issue in environmental affairs is the use of onshore and offshore wind turbines and the creation of greenhouse gas emissions (GHGs). Wang *et al.* (2019) report that offshore wind turbines have higher GHG emissions than onshore units because of the floating platform used for the former to fix turbines at sea. There is more energy required in the production of these fixed platforms, which in turn creates a greater energy demand. The analysis also suggests that wind turbine factories and wind turbine farm sites should be as close as possible to one another to minimize increases in energy demand and GHG emissions.

Other tools mentioned by Kiddee *et al.* (2013) that help to understand the impact that our actions have on the natural world include multi-criteria analysis (MCA) and material flow analysis (MFA). MCA is a strategic decision-making tool to solve complex multi-criteria problems that include qualitative and quantitative elements of the problem, used mostly in the area of solid waste (see also Garfi *et al.*, 2009). MFA analyses the route of materials, like e-waste, that flow into recycling sites or waste disposal units, in space and time. It links sources, pathways, and the intermediate and final destinations of materials (see also Brunner & Rechberger, 2004).

AUDITING

Auditing is the on-site verification activity, such as an inspection or examination of a process or quality system, to ensure compliance to requirements. An audit can apply to an entire organization or might be specific to a function, process or production step (ASQ, 2018). Audits can be performed internally or within the organization (first-party), or externally as a second-party audit (on a supplier by or on behalf of a customer), or as a third-party audit from an independent organization where the firm is in pursuit of a certification, recognition or award (ASQ, 2018). An example of a third-party auditing system in the pursuit of certification is the international Organization for Standardization (ISO 9001) programme which has standards for quality management systems regarding products that meet the requirements of customers and regulatory bodies.

Environmental auditing is a process that organizations undertake to determine whether they are in compliance with regulatory requirements and environmental policies and standards (Pigram, 1996). More formally, the environmental audit is:

> A management tool comprising a systematic, documented, periodic and objective evaluation of how well organisations, management and equipment are performing with an aim of helping to safeguard the environment by: (a) facilitating management control of environmental practices; (b) assessing compliance with company policies, which would include meeting regulatory requirements. (UNEP/IEO, 1989: 100)

Beyond measuring environmental impacts, the environmental audit also heightens employee awareness of environmental problems and issues, provides greater understanding of cost-saving opportunities, configures award structures for employee achievements in environmental stewardship, fosters the establishment of environmental training and allows for the efficient use of resources (Diamantis, 1998).

CORPORATE SOCIAL RESPONSIBILITY AND REPORTING

Corporate social responsibility (CSR) has been a topic of research interest since the 1950s. But it was Carroll (1979) and later Carroll (1991) who provided much more scope into the concept from a management perspective. Carroll (1991) argues that companies are socially responsible when they meet four discrete classes or stages of responsibility: economic, legal, ethical, and discretionary or philanthropic. In more detail, these are (Carroll, 1979: 500):

- *Economic*: Before anything else, the business institution is the basic economic unit in our society. As such it has a responsibility to produce goods and services that society wants and to sell them at a profit. All other business roles are predicated on this fundamental assumption.
- *Legal*: Just as society has sanctioned the economic system by permitting business to assume the productive role, as a partial fulfilment of the 'social contract', it has also laid down the ground rules – the laws and regulations – under which business is expected to operate. Society expects business to fulfil its economic mission within the framework of legal requirements.
- *Ethical*: Ethical responsibilities are ill defined and consequently are among the most difficult for business to deal with. In recent years, however, ethical responsibilities have clearly been emphasized – although debate continues as to what is and is not ethical. Suffice it to say that society has expectations of business over and above legal requirements.
- *Discretionary*: Societal expectations do exist for businesses to assume social roles over and above those described thus far. These roles are purely voluntary, and the decision to assume them is guided only by a business's desire to engage in social roles not mandated, not required by law and not even generally expected of businesses in an ethical sense. Examples of voluntary activities might be making philanthropic contributions, conducting in-house programmes for drug abusers, training the hard-core unemployed or providing day-care centres for working mothers.

CSR is defined as

> the responsibility of an organization for the impacts of its decisions and activities on society and the environment, through transparent and ethical behaviour that contributes to sustainable development, including health and welfare of society, takes into account expectations of stakeholders, is in compliance with applicable law and

> consistent with international norms of behaviour and is integrated throughout and practiced in an organization's relationships. (ISO, 2010; see also Steiner, 1972)

The growth of CSR is due to the perceived failure of governmental regulation in the face of privatization and globalization pressures, a shift in values of the Western citizenry (media and civil society) and the revolution in information technology (Hartmann, 2011). The insistence that corporations consider the social and environmental impacts of their practices is evident in the belief on the part of managers in the global and consumer goods sector (as of 2011) that CSR is one of the top priorities (Consumer Goods Forum, 2011). CSR has relevance to our discussion on ST because, as observed by Connell (2000), social responsibility is now recognized as an essential component of ST management. Social responsibility is widely used in the private commercial realm and is expressed more generally as CSR.

Several CSR publications can be found in the tourism literature, with many of these either of a general nature or focusing on specific sectors. Kalisch's (2002) work on behalf of Tourism Concern emphasizes the importance of large tourism firms having to shoulder the responsibility for implementing sustainability measures in their operations. Miller (2001) argued that CSR has been adopted in the UK tourism industry because of lack of control and resources, to gain market advantage and positive public relations, to save costs, and for legal and moral reasons. In reference to this last reason, Miller argues that some large companies are adopting CSR in an effort to move beyond the commercial into the altruistic. In this vein, Hudson and Miller (2005) focused on the responsible marketing practices of the heli-ski operator, Canadian Mountain Holidays, particularly in how this organization seeks to find a balance between communication to clients and responsible action. There is a danger in not walking the talk and the perception on the part of clients that such communication amounts to greenwash. Researchers also view RT not only in the context of competitive advantage – and in this there is an extrinsic motivation for RT – but also in the context of protecting the rights and interests of tourists (Epuran *et al.*, 2017).

Reporting has been in practice for many years and has now become a staple feature of CSR practices. The use of reporting as a feature of CSR appears to be on the rise. For example, the number of CSR mandates, globally, increased from 130 to 250 between 2013 and 2016 (Carrots & Sticks, 2016), and over 90% of the world's largest companies regularly publish sustainability reports (KPMG, 2017). The introduction of EU Directive 2014/95/EU now requires large companies to report on the environmental and social impacts of their operations, which has raised the profile of CSR reporting to a whole new level (Sassen & Isenmann, 2018). Gulenko (2018) defines CSR reporting and mandatory CSR reporting as 'the issuance of a broad report on companies' CSR issues', and 'a binding requirement by a legal or financial institution (e.g. government or stock exchange) that mandates certain companies to publish a CSR report', respectively. The practice of CSR reporting has been dominated by the for-profit sector over the public sector, the latter of which has been characterized as 'patchy' (Ball, 2005), even though the public sector may be able to achieve better results because of its uniqueness (Ball & Bebbington, 2008). Much like auditing, organizations may pursue reporting for the rewards it offers in terms of competitiveness and positive public image (Bebbington *et al.*, 2009).

Three of the largest and most frequently used CSR reporting guidelines include the Sustainability Reporting Standards of the Global Reporting Initiative (GRI), Integrated Reporting of the International Integrated Reporting Council, and the Sustainability Accounting Standards Board. Alonso-Almeida *et al.* (2014) note that the energy sector has been an active user of the GRI in an effort to be seen as more sustainable, given how visible the sector is as a polluter at the international scale.

The GRI developed in 1997 is an independent international organization which has been a pioneer in sustainability reporting, and is the first and most widely used methodology for the reporting of sustainability practices. Taken from their homepage (GRI, 2018), their goal, vision and mission are as follows:

> Goal: businesses and governments worldwide understand and communicate their impact on critical sustainability issues such as climate change, human rights, governance and social wellbeing. This enables real action to create social, environmental and economic benefits for everyone. The GRI Sustainability Reporting Standards are developed with true multi-stakeholder contributions and rooted in the public interest.
>
> Vision: A thriving global community that lifts humanity and enhances the resources on which all life depends.
>
> Mission: To empower decisions that create social, environmental and economic benefits for everyone.

The GRI operational framework includes four different categories: (1) universal standards applicable to every organization (foundation, general disclosures, management approach); (2) economic standards (economic performance, market presence, indirect economic impacts, procurement practices, anti-corruption, anti-competitive behaviour); (3) environmental standards (materials, energy, water, biodiversity, emissions, effluents and waste, environmental compliance, supplier environmental assessment); and (4) social standards (employment, labour/management relations, occupational health and safety, training and education, diversity and equal opportunity, non-discrimination, freedom of association and collective bargaining, child labour, forced or compulsory labour, security practice, rights and Indigenous peoples, human rights assessment, local communities, supplier social assessment, public policy, consumer health and safety, marketing and labelling, customer privacy, and socioeconomic compliance) (GRI, 2018).

Criticisms of CSR reporting are numerous. Examples of issues include intentionally narrow definitions of sustainability used by the corporate sector, failure to fully complete reporting, striving towards eco-efficiency instead of tougher actions, and the failure to report on ecological limits that will affect operations (Milne & Gray, 2007). Problems with mandatory approaches to CSR reporting possibly fall within the realm of intrinsic versus extrinsic motivation. There may be an increase in the number of topics covered in CSR reports, but the quality of reporting has not improved (Gulenko, 2018). Others have found that CSR reporting by businesses maintains a largely instrumental and economic approach to the natural world (Milne *et al.*, 2009). A comprehensive list of challenges that firms have encountered in CSR reporting has been organized into four main categories (Brand *et al.*, 2018). These include: (1) value chain (e.g. how to define and depict the value chain of companies); (2) stakeholder orientation (e.g. how stakeholders can help to determine if companies have an impact concerning a specific topic); (3) materiality (e.g. how to include information about the impact of a company concerning sustainability topics in a materiality assessment); and (4) target group orientation (e.g. how a company can know whether or not a certain target group perceives the reporting as credible).

The use of reporting in tourism research is in its infancy. One of the few studies in this area by de Grosbois and Fennell (2011) investigated 150 of the largest hotel companies in the world according to their carbon footprint. The authors found that only a very small number of the largest hotel companies report their carbon footprints, and in a patchy way. Medrado and Jackson (2016) investigated the reporting practices of US hospitality and tourism firms in the area of CSR. They found that most firms used the GRI as a standard guideline for reporting. They also discovered that different industry sectors reported in different ways. For example, accommodation firms report on CSR/sustainability far more than food and beverage and the cruise line sectors. The information most frequently disclosed was on performance related to water use, energy conservation and waste generation. Less information was disclosed on worker compensation and work–life equilibrium.

SOCIAL-ECOLOGICAL SYSTEMS RESEARCH

Natural resource science took a radical turn before the end of the previous century based on the new science of complex social-ecological systems. This new science is an amalgam of three disparate disciplines: ecology, economics and physics (Schoon & Van der Leeuw, 2015), the latter of which was most essential in laying down the scientific structure for understanding complexity. The field of complex social-ecological systems is an excellent example of how interdisciplinarity can advance science into new and exciting areas of study. While Holling (1973) is often credited with theorizing how ecosystems move between multiple stable states, and indeed he was the essential figure in ecology and a burgeoning social-ecological systems camp, it was the physicist Prigogene and colleagues who set the stage for this radical move (Prigogene *et al.*, 1972). In general, Prigogene *et al.* (1972) discovered that while some systems work towards a state of equilibrium (positive entropy), other systems demonstrate negative entropy, or a state of thermal disequilibrium described as far-from-equilibrated systems. These systems are characterized as evolving entities because they increase in complexity as they move from one equilibrated state to another through a series of internal fluctuations. These internal fluctuations consistently push the system outside its normal range so that 'there is seldom any way of using our knowledge of a system's past behaviour to predict its future sequential unfolding or evolutionary end-point' (Reed & Harvey, 1992: 363; see also Fennell, 2004, for a lengthier discussion of this research).

Following from Prigogine *et al.* (1972), complex socioecological systems are marked by dynamic processes and reciprocal feedback mechanisms, with a substantial exchange of energy and materials across boundaries (Berkes *et al.*, 2001). This type of thinking is now being used in several different disciplines and sectors and will be addressed in examples used throughout the course of the book. There are, generally speaking, three camps that have emerged out of the complex social and ecological systems manner of thinking: technology, resilience and transformation.

TECHNOLOGY, RESILIENCY AND TRANSFORMATION

The first camp focuses on how technology may be used as a vehicle to secure sustainable futures. There is little doubt that technology has contributed significantly to our present unsustainable problems at all scales, but there is the belief that it must also play a vital role in solving these problems. This broad area includes, for example, biofuels, energy technologies, fuel cells, green chemistry, renewable energy technologies, sustainable agriculture, sustainable building and infrastructure development, and sustainable transportation (see Ray & Jain, 2011, for a comprehensive overview of technology utilization for the treatment of safe drinking water). The organization FirstCarbon Solutions (2018) argues that new innovative solutions in technology are essential in achieving sustainability for the future. FirstCarbon lists the following examples of how technology is making a difference:

- *New Delhi's smart city*: The New Delhi Municipal Council (NDMC) announced its plans for installing solar panels on top of public schools and willing households. It is working with the Delhi Electricity Regulatory Commission to come up with billing rates to be used when the transition happens. While plans for the project are not yet finalized, officials are hopeful that they will be able to generate at least 1.7 MW of electricity from schools alone (Indian Express, 2015).
- *The Netherlands' liveable wind turbine*: A Dutch design firm unveiled its proposal for a wind turbine that also functions as an apartment block and hotel. The project, called the Dutch Windwheel, will generate electricity from both wind and water because of its planned location along the Rotterdam waterside. The team in charge of the project is still in talks with the city council and developers, and are eyeing 2020 to begin construction (Frearson, 2015).
- *Perth's wave project*: The Carnegie Perth Wave Energy Project was created in order to fully utilize the powerful waves in Western Australia. The buoy-like structures created

by CETO Technologies will be placed in the water to generate power from the waves while collecting water to be desalinated. The energy and clean water generated from the project will be sold to the Australian Department of Defense for its naval base (Renner, 2015).

- *Johannesburg's kinetic walkway*: Pavegen, a London-based clean tech company, developed flooring that generates energy via people's footsteps (i.e. kinetic energy). The company partnered with Samsung to install a walkway in one of Johannesburg's shopping malls. While it primarily provides energy for Samsung's interactive data screen, it also collects energy to be distributed to underdeveloped communities in South Africa (Singh, 2015).

The second camp, resilience, continues to be an important concept in social and ecological systems thinking, and it has been the topic of recent major publications in tourism (Butler, 2017; Cheer & Lew, 2019; Hall *et al.*, 2017). Resilience is 'the ability of a system to prepare for threats, absorb impacts, recover and adapt following persistent stress or a disruptive event' (Marchese *et al.*, 2018: 1275). Resilience emphasizes the ability of systems to 'bounce back', recover and adapt after disturbances of one magnitude or another. In a detailed overview of the similarities and differences between resilience and sustainability, Marchese *et al.* (2018) note that scholars have taken one of three different pathways in considering both terms: (1) resilience as a component of sustainability, (2) sustainability as a component of resilience, or (3) resilience and sustainability as separate (with greater general support for this latter view) (McPherson, 2014). Sustainability has been conceptualized as prioritizing *outcomes* around environmental and social concerns like food security, justice and equity, i.e. a focus on the future (see Prosperi *et al.*, 2016), whereas resilience has priorities found in *processes* and functionality during and directly after events (see Park *et al.*, 2013). In other words, sustainability is about the processes required to place the world back into balance, while resilience aims at ways in which to manage this imbalanced world (Zolli, 2012).

Pigram (1990) made early reference to resilience in tourism studies by arguing that irreversibility is a function of the spatial and temporal patterns of impacts to a resource, and how resilient the resource is, as well as the scope of managerial response to such impacts. Simmons (2013) has shown that tourism in New Zealand has proved to be extraordinarily resilient over the course of five decades. Since 1959, when the first modern jet-propelled aircraft arrived in New Zealand, the industry has doubled every 10.5 years, reaching 2.6 million international visitors in 2012. Tourism attractions and destinations can be more resilient through technical improvements and innovations, business management decisions and shifts in behaviour (Jopp *et al.*, 2010). In the context of the ski resort municipality of Whistler in British Columbia, Canada, experience has allowed the region to respond to a number of shocks and stressors that continually challenge similar places along many lines. Sheppard and Williams (2017) document how several governance system innovations buffered Whistler against significant stressors like the 2008 financial crisis and economic slowdowns, as well as changes in municipal leadership. Resiliency came in many forms, including changes in community energy use, community and corporate performance indicators and community open houses. In a companion article, Sheppard (2017) writes that the governance and community characteristics of a resilient resort destination (Whistler) are:

1. passion, pride and a can-do attitude;
2. strong and effective governance;
3. shared vision, values and single-mindedness;
4. strong partnerships, collaboration and community involvement;
5. community and corporate memory;
6. taking care of social issues;
7. a well-resourced community; and
8. a sense of place and connectedness.

The third domain in social and ecological systems research is transformation, which is defined as 'The capacity to transform the stability landscape itself in order to become a different kind of system, to create a fundamentally new system when ecological, economic, or social structures make the existing system untenable' (Folke *et al.*, 2010: 3). The transformation domain has used the concept of leverage points to explain why other systems of thought have been relatively unsuccessful at instituting positive change in the area of sustainability. Leverage points are 'places within a complex system (a corporation, an economy, a living body, a city, an ecosystem) where a small shift in one thing can produce big changes in everything' (Meadows, 1999: 1). Meadows contends that while people working within systems are often pushing very hard for progress, they are often pushing in the wrong direction in making things worse. The key, Meadows contends, is finding the correct leverage points (Meadows identifies 12 important places to intervene in systems as leverage points, the last and most important of which is to have the power to transcend paradigms).

Abson *et al.* (2017) adopted Meadows' perspective on leverage points as a tool to better explain sustainability transformation. These authors argue that we continue to follow unsustainable development routes because we have not found the correct leverage points to move us in different directions: science and technology become path dependent, creating lock-ins and lags (Westley *et al.*, 2011) and fail to employ the correct methods in attacking the fundamental causes of unsustainability. As such, there is a penchant towards what the authors view to be 'quick fixes' in addressing sustainability issues rather than the pursuit of more appropriate mechanisms – we choose proximate instead of ultimate drivers of change, a topic of discussion in the previous chapter in the section on human nature. These proximate drivers include a range of what Abson *et al.* (2017) refer to as weak or shallow interventions. Examples cited by the authors include policy changes that focus on financial incentives or setting targets within existing structures such as carbon pricing that fail get to the root cause of problems. More powerful areas for intervention, according to Abson *et al.* (2017), are centred around three leverage domains: (1) reconnecting people to nature; (2) restructuring institutions; and (3) rethinking how knowledge is created and used in the pursuit of sustainability. An example of a deeper intervention in domain (1), above, includes the notion that pro-environmental behaviours alone will not lead to a more sustainable lifestyle. There is also the need to change institutional structures in society that make it easier for people to behave in more sustainable ways. *Intent*, the authors note, needs to be better coupled with *design* or the 'methods and means that help us to get there' (Abson *et al.*, 2017: 35). For Westley *et al.* (2011), transformation will be as a much a social innovation as it is a technical one, and will involve a major shift in institutions and governance in cultural, political and economic spheres.

Linking concepts of institutional failure with leverage points and tourism, Hall (2011b) argues that although ST represents a success given how liberally it is infused in industry, government and academia, it is also a failure because of its inability to pull back from a development model based on growth. The problem is that key stakeholders in tourism are unwilling to recognize policy failure. ST continues to evolve, Hall (2011b) adds, from first-order change that is characterized by the setting of policy instruments and sustainability indicators, to second-order change marked by a more balanced approach to SD. A third or alternative way, termed 'de-growth', 'slow' or 'steady-state' tourism, is emerging based on external pressures from a number of different sectors. This steady-state configuration is one that is based on qualitative expansion rather than pure aggregate quantitative growth, the latter of which currently exists in most scenarios by damaging natural capital (Hall, 2009, 2010b, 2011b; see also Daly, 2008). Not unlike the Jevons paradox or Jevons effect (William Stanley Jevons was an English economist in the 19th century concerned with the increase in coal-burning factories, and the associated increase in demand), where technological progress will not reduce consumption as resource consumption rises due to an increase in demand, policies of economic growth as the main objective will not result in conservation of the natural world (Czech, 2006; Hall, 2010b). As we walk the thin line between technological progress and conservation, the steady-state economy appears to be the only rational approach to satisfying both sides of the equation (Czech, 2006). This preoccupation with growth is itself a leverage point, where there are benefits as well as costs (Forrester, 1971). These costs include environmental impacts, poverty, hunger and a range of other disturbances. Forrester argues that growth is needed in some contexts, but so is slow growth and de-growth in other contexts.

CONCLUSION

Following Hunter (1997), Fennell and Ebert (2004) and several other tourism scholars, we argue that contemporary ST needs to be conceived within a broader SD framework in order to move the agenda forward. This broader agenda will hopefully open up new learning pathways and emphasize existing ones, which in turn should allow scholars and practitioners to gain an appreciation of the magnitude and complexity that sustainability poses in human-ecological systems and interactions.

The formula moving forward in this book is to look more specifically at the many destinations in which tourism takes place (e.g. islands, polar regions, tropical rainforests), as well as the many sectors that comprise the fabric of the tourism industry (e.g. food, accommodation, transportation), and critically analyse these in light of the platform of knowledge presented in these first four chapters. This includes a continued focus on the UN SDGs, along with research on cooperation and self-interest, consumerism, technology, biodiversity, ecological footprints, certification, corporate social responsibility and other tools and concepts.

CHAPTER 4 DISCUSSION QUESTIONS

1. List and discuss some of the main criticisms of sustainable tourism.
2. Discuss some of the challenges to the implementation of sustainable tourism.
3. How are commons issues related to the concept of carrying capacity?
4. What is the difference between regulation and voluntary codes of ethics?
5. How can ecotourism and alternative tourism be used to inform sustainable tourism?
6. What is the ecological footprint and how can it be used in tourism?
7. What is corporate social responsibility? Illustrate how it is linked to reporting.

End of Chapter Case Study 4.1 Cities, tourism and the tragedy of the commons

This case study makes reference to the problems that can emerge when self-interest rears its ugly head in the face of limited or diminishing resources. The tragedy of the commons was discussed earlier in this chapter as a theoretical position that explains why individual interests lead to collective ruin. In this case study by Dans, the topic is Airbnb which, even though utilizing private properties, there is an interpretation of these properties as a common good. There are questions regarding the sustainability of the Airbnb initiative if everybody wishes to rent out their properties for personal gain. The consequences to the accommodation sector would seem to be front and centre in questions concerning what is best for cities in reference to municipal taxes, personal and corporate revenue and city planning. Here is what Dans has to say on the matter.

The impact of an application such as Airbnb on the development and planning of cities, especially those attractive to tourists, have been the object of study and controversy for some time, while anybody who has recently tried to rent a property in such cities.

The application created by Brian Chesky, Joe Gebbia and Nathan Blecharczyk eight years ago has made it extremely easy to rent properties to tourists, and many home owners have been unable to resist the temptation. The attractive areas of cities, since the development of mass tourism, have been under heavy pressure and, in many cases, have turned into theme parks, with business oriented only towards tourists. As more and more homeowners in these areas turn to Airbnb and similar applications to fetch higher profits, less affordable properties remain available to long term residents, driving them out of their neighborhood.

The process of touristification, known for decades in cities such as Venice, is significantly aggravated by making it possible for anybody to let their property out. The initial idea of the founders of Airbnb, two young people who did not have the money to pay the rent on their San Francisco home and so decided to rent one of their rooms to conference delegates, has in many cases been taken over by companies that buy up entire buildings and let out properties through the platform, intermediaries that manage third-party properties, along with an entire industry to service the sector. In San Francisco, Airbnb's offices were occupied by demonstrators in November 2015, prompting City Hall to implement measures such as a register of rental properties, along with restrictions on the number of nights per year and the number of properties an individual can offer through the platform. Since then, the company has seen protests in other cities out of the 65,000 it operates around the world.

This is a textbook case of the tragedy of the commons: several individuals, motivated only by personal interest and acting independently but rationally, end up destroying a limited public resource, even though none of the parties benefit from that destruction.

Even though the main resource in this case is private property, the resource that generates the problems can be interpreted as a common good: the pool of properties that make up the rental offer of a particular city or neighborhood. It is hard to convince the owners of properties in the center of a city that they cannot maximize the profitability of their property by whatever method they deem appropriate within the law; at the same time, if everybody in a given area chooses to rent out their property through Airbnb it's not going to be sustainable, however much economic activity it generates. Do certain tourism models kill neighborhoods? Of course, it can create instability… but this was already happening, in some areas, long before Airbnb. To blame Airbnb alone is not to have looked at the question in detail.

Prior to Airbnb, short-term rentals took place in a legal vacuum, generally at the lower end of the economic scale, and often giving rise to both disappointing experiences and a submerged economy. The platform offers accommodations appealing to a wide range of people – from low-cost to luxury, while the user experience is improved by a peer-rating system, making it easier to control the resulting economic activity.

At the same time, as we have seen, this democratization of short-term rentals for property owners in an attractive area has become a problem.

The usual way to deal with these kinds of situations has been through regulation, and this is happening in some cities. But regulation must meet different criteria: it cannot just respond to the interests of the hotel industry, nor impose so many restrictions that it drives the activity underground. A balance has to be found between understanding the wealth that tourism creates, and the need for tourism to be sustainable, while at the same time taking into account specifics such as seasonality.

The solution is to reevaluate a resource, tourism, which has become the economic motor of many countries and regions, in the context of a technology that cannot be halted.

Source: Dans (2017)

CHAPTER 5
THE SUSTAINABLE DESTINATION

LEARNING OUTCOMES

This chapter focuses on the sustainable destination, outlining the key features of destinations that can promote, or inhibit, sustainability. The chapter is designed to provide you with:

1. An understanding of how we can define destinations.

2. The ability to identify the key components of a sustainable destination.

3. An awareness of the key features of a destination that promote or inhibit sustainability.

4. A disciplined approach to the analysis of destinations as networks and complex adaptive systems.

5. An understanding of how the tourism area life cycle helps us to understand sustainable destinations.

INTRODUCTION TO THE SUSTAINABLE DESTINATION

This chapter focuses on the tourism destination as a fundamental unit of analysis for sustainability, particularly focusing on the UN Sustainable Development Goals (SDGs) 14 and 15 on life below water and life on land. This is because destinations, their images and their digital traces do not only attract tourists, motivate visits and therefore energize the tourism system, but they are also at the sharp end of tourism, at the same time suffering and benefiting from visitation. The richness and variety of destinations are key drivers of the success of tourism and they demonstrate a complex pattern across the world where tourists are attracted to the unique, the exotic and the vulnerable – hence the need for a sustainability approach. And this is needed now more than ever, as the pleasure periphery reaches ever more distant and remote locations, including the Polar regions. In other words, a sustainable approach is required because the destination is where the consequences of tourism occur – whether positive or negative – and it is therefore the focus for the planning and management of tourism, all wrapped into a framework of sustainability. This approach is increasingly based on partnerships, as we will see in this chapter, and aligns with UN SDG 17 on partnerships for the goals.

Clearly the components of the tourist destination can be effective only if careful planning and management deliver a sustainable tourism (ST) product, and in so doing ensure that one or more of the components does not surge ahead of the others. It is clear that the concept of sustainability demands a long-term view of tourism and ensures that consumption of tourism does not exceed the ability of a host destination to provide for future tourists. In other words, it represents a trade-off between present and future needs. In the past, sustainability has been of low priority compared with the short-term drive for profitability and growth, but as pressure has grown for a more responsible tourism industry, it is difficult to see how such short-term views on consumption can continue. Indeed, once the principle of the sustainable destination has been accepted, the benefits are clear (https://www.europarc.org):

- measurable economic, social and environmental benefits from well-managed ST; and
- strengthened relations with local tourism stakeholders and the wider tourism industry.

Destinations are part of complex environmental and social systems where the relationships between the various consequences of tourism are only just beginning to be understood. There is no doubt that with concerns for climate change we will move to quadruple bottom-line sustainability for destinations: environment, the economy and the community, with the fourth element being carbon. Indeed, throughout this book we stress the vital need for tourism to embrace the low-carbon economy and wherever possible contribute to reducing climate change.

DEFINING THE DESTINATION

Despite the fact that the destination can be considered alongside the tourist as the basic unit of analysis in tourism, defining destinations has generated controversy, with some arguing for a geographical definition while others see it as an idea or a concept that cannot be confined within geographical boundaries. Effectively the destination is a focal point for the generation and delivery of tourism products and experiences and the implementation of sustainable management planning and policy and provides the facilities and services to meet the needs of the tourist. Destinations are both tangible as physical spaces, and intangible as they generate images, expectations, digital traces and memories. In an attempt to generate some clarity in the debate, the UNWTO held a Think Tank in 2002 to establish a definition, although this proved more difficult than they expected. The following definition resulted from the meeting:

> A tourism destination is a physical space in which tourists spend at least one overnight. It includes tourism products such as support services and attractions and tourist resources within one day's return travel time. It has physical and administrative boundaries defining its management, images and perceptions defining its market competitiveness. Destinations incorporate various stakeholders often including a host community, and can nest and network to form larger destinations. Destinations can be on any scale, from a whole country, a region, or an island, to a village, town or city, or a self contained centre. (UNWTO Destination Think Tank, December 2002)

This definition is more of an all-encompassing list of destination attributes and already seems rather dated and, from the point of view of sustainability, not very useful. For this chapter it is more helpful to identify what a sustainable destination looks like. Here there are two approaches. First, the European Charter for Sustainable Tourism in Protected Areas (https://www.europarc.org) has outlined the key components of a sustainable destination. These are:

- protection of the natural and cultural heritage;
- participation by all stakeholders through the creation of a sustainable destination forum;
- effective partnership working;
- planning to prepare and implement a ST strategy and action plan, including monitoring and evaluation; and
- to realize the environmental, social and economic benefits of everyone working more sustainably.

A much more detailed framework for the development of a sustainable destination is provided by the Global Sustainable Tourism Council (see Table 5.1).

These two organizations provide a clear-sighted view as to just what makes a destination sustainable and many of these components are discussed elsewhere in this book. The remainder of this chapter analyses the particular features of the destination that can hinder, or promote, a sustainable approach. We believe that there are five such features:

1. the notion of inseparability;
2. the tourism context;
3. the stakeholder approach.
4. the network gaze; and
5. a complex systems approach.

THE NOTION OF INSEPARABILITY

A fundamental issue for sustainable destinations is the fact that tourism is consumed where it is produced, as visitors have to be physically present at a destination to experience tourism. This has led to the term ‘overtourism’ in some destinations where it is felt that there are simply too many tourists for the community and the locale to cope with – examples here include Venice, Barcelona and Dubrovnik. This has resulted because tourism, by its very nature, is attracted to the unique and fragile parts of the world, and so destinations, their environments and communities are vulnerable to tourist pressure and may suffer alteration. This is exacerbated by the fact that visitor pressure is often concentrated seasonally in time and at specific popular locations, creating ‘hot spots’ of tourist pressure. As a result, it is imperative that destinations are planned and managed sustainably to cope with visitation. As we will see later in this book, there are tried and tested techniques to

Table 5.1 Global sustainable tourism criteria for destinations

Section A: Demonstrate sustainable destination management
A1 Sustainable destination strategy
A2 Destination management organization
A3 Monitoring
A4 Tourism seasonality management
A5 Climate change adaptation
A6 Inventory of tourism assets and attractions
A7 Planning regulations
A8 Access for all
A9 Property acquisitions
A10 Visitor satisfaction
A11 Sustainability standards
A12 Safety and security
A13 Crisis and emergency management
A14 Promotion
Section B: Maximize economic benefits to the host community and minimize negative impacts
B1 Economic monitoring
B2 Local career opportunities
B3 Public participation
B4 Local community opinion
B5 Local access
B6 Tourism awareness and education
B7 Preventing exploitation
B8 Support for community
B9 Supporting local entrepreneurs and fair trade

Section C: Maximize benefits to communities, visitors, and culture; minimize negative impacts
C1 Attraction protection
C2 Visitor management
C3 Visitor behaviour
C4 Cultural heritage protection
C5 Site interpretation
C6 Intellectual property
Section D: Maximize benefits to the environment and minimize negative impacts
D1 Environmental risks
D2 Protection of sensitive environments
D3 Wildlife protection
D4 Greenhouse gas emissions
D5 Energy conservation
D6 Water management
D7 Water security
D8 Water quality
D9 Wastewater
D10 Solid waste reduction
D11 Light and noise pollution
D12 Low-impact transportation

Source: GSTC (2019b)

deal with the problem of 'overtourism'; indeed, while the term itself might be new and fashionably controversial, in fact the problem has been with us for many decades and links to the fundamental concept of 'carrying capacity'.

CITIES

We introduced SDG 11 and cities in Chapter 3. Sustainable Development Goal 11 relates to 'sustainable cities', and if the 20th century belonged to nation states, then the 21st century will be dominated by metropolitan areas. The UN HABITAT (2016) *World Cities Report* states that two-thirds of the global population is expected to live in cities by 2030 and produce as much as 80% of the global gross domestic product (GDP). Urbanization provides an opportunity to achieve the SDGs, particularly SDG 11, which is to make cities inclusive, safe, resilient and sustainable. The world's cities occupy just 3% of the Earth's land, but account for 60–80% of energy consumption and 75% of carbon emissions. Rapid urbanization is exerting pressure on freshwater supplies, sewage, the living environment and public health. But the high density of cities can also bring efficiency gains and technological innovation while reducing resource and energy consumption.

Box 5.1 Barcelona and overtourism: Balancing the needs of residents and tourists

Introduction

Barcelona has taken a number of strategic initiatives to balance the needs of both visitors and residents. The brand Barcelona has been linked to outstanding economic and tourism success in the last few decades. The city has become a leading destination for leisure, business and cruise tourism. Its name is included in any international ranking as a cosmopolitan, attractive, innovative metropolis. However, the promotion of Barcelona as a tourist destination 'is not a recent trend, but it has been an essential target for local authorities since the International Exhibition in 1888' (Cócola, 2014: 22).

The beginning of the modern tourism growth of Barcelona dates back to 1992 (Ajuntament de Barcelona, 2010). The city hosted the Olympic Games, which launched its renewed image worldwide and allowed huge city brand exposure. 'The Games were a unique and indispensable marketing instrument in bringing about the Barcelona we now enjoy today' (Duran, 2002: 6). The Turisme de Barcelona Consortium was created for the Games in order to determine the main guidelines for planning tourism growth. Despite the good intentions of the city's planners, the reality was that uncontrolled and mass tourism began to affect residents' lifestyles and their perception of tourism over the next few years.

The new tourism strategy

By 2015, after local elections, a new Municipal Action Plan was developed which included a commitment to drafting a Strategic Tourism Plan for the 2016–2020 period. The plan involved all the stakeholders in a participative process 'to establish a local agreement for the management and promotion of responsible and sustainable tourism' (Ajuntament de Barcelona, 2015: 5). Several commitments were made: signing a responsible tourism city charter; obtaining the biosphere destination certificate; the declaration of the vision for responsible tourism; and joining the World Charter for Sustainable Tourism +20. The first step was to carry out a strategic diagnosis, identifying future challenges and goals to address and action proposals. The goals were as follows (Ajuntament de Barcelona, 2016: 4–5):

- to prepare a roadmap for Barcelona's tourism policies to 2020, based on a participatory diagnosis;
- to generate public debate and shared knowledge on tourism and its effects, through an analysis of the current situation and anticipated future scenarios; and
- to integrate the planning approaches towards tourism in the city.

The measures to implement the plan are subdivided into the following main programmes (Ajuntament de Barcelona, 2016):

1. *Governance*: Aiming to reinforce municipality leadership, ensuring stakeholder participation and efficient coordination between public administrations. This programme also adapts the work of the Barcelona Tourism Consortium to better integrate the tourism marketing with local policy.
2. *Linking shared knowledge* with decision making, deeper strategic understanding and richer public debate about tourism and city.
3. *'Barcelona destination'*: To build an economic, social and environmentally sustainable destination that goes beyond administrative boundaries. The destination must be open and innovative, welcoming its visitors while guaranteeing the residents' quality of life. Tourism assets and products must be adapted to sustainability criteria, empowering local providers and promoting singular cultural values.
4. *Mobility*: Evaluating tourism transport needs to enable a coordinated mobility plan for any kind of user (including tourists, cruise passengers and day trippers), promoting more coordinated, sustainable mobility. Accommodation is also considered a core issue as major hotel growth and new providers require a solid, coherent regulation which avoids and/or reduces undesired gentrification.
5. *Managing urban spaces* under a cross-cutting, integrated vision such that residents' lives will not be dramatically affected by visitors, and the latter will enjoy the true essence of the city. The spaces considered in this plan include both the non-tourism districts as well as crowded places. Finally, this programme also includes a plan for accessible tourism.

6. *Communication and hosting* to transmit the diversity and complexity of the city, in order to showcase new possibilities beyond the crowded, iconic attractions. In addition, the improvement of tourist information services, working closely with private influencers and adapting new technologies to communicate in real time with the visitors.
7. *Taxation*: The use and distribution of existing tourism taxes will be revised.
8. *Financing measures and regulation*: The balance between the costs and benefits of tourism is unclear, and the municipality wants to increase the city's tourism return on investment. The incomes from tourism must be socially redistributed and partly used to reinvest in the destination.

The challenge for the city of Barcelona is twofold: on the one hand, fighting against the loss of good brand reputation and competitiveness, as potential visitors may feel the city is losing its essence; and, on the other hand, managing residents' negative perception towards 'overtourism'. As such the case of Barcelona is a paradigmatic example of the sensible and sustainable balance of city tourism. Barcelona has been regarded as a model of tourism development by all its competitors. However, the steady growth of visitors and positive financial inputs might have been hiding a social conflict that needed to be readdressed.

Sources: Ajuntament de Barcelona (2010, 2015, 2016); Cócola (2014); Duran (2002)

The migration of the world's population to cities has significant implications for tourism demand, supply and governance in the future. This has been heightened by the public opposition to tourism in major European cities such as Barcelona, Amsterdam and Venice. The view is that residents are being squeezed out by spiralling rents, the rise of Airbnb, the influx of cruise visitors and the bad behaviour of tourists who take over the centres of cities in the summer months. Clearly, then, the balance between residents and tourists has shifted in some of these cities. García-Hernández *et al.* (2017) discussed the issue of tourist pressure in European city environments, and specifically the historic centre of Donostia-San Sebastián in Spain. They identified a series of tourism issues faced by residents and these are summarized in Table 5.2.

Box 5.2 Sustainable tourism in smaller towns and rural communities

Of course, tourism does not only occur in large cities. Indeed it could be argued that tourism is more visible in smaller towns and communities and it is here that the impacts of tourism can noticeably affect lifestyles. In these smaller communities, tourists tend to be domestic and visiting friends and relatives. For these communities, the development of ST must recognize that every place is different with its own unique community and atmosphere. ST, when introduced in a phased and gradual manner and with local control, is seen to bring significant benefits as it acts as an agent of change. These benefits include:

- economic diversification;
- empowerment of local communities;
- enhanced quality of life;
- a sense of community solidarity and pride;
- capacity building in terms of destination resilience; and
- environmental conservation.

To ensure the sustainable development of tourism in small communities, particularly those close to heritage sites, UNESCO (n.d.) outlines three principles:

1. Talk and listen to the host community and local businesses.
2. Identify and communicate sustainable local economic opportunities.
3. Empower the host community by telling their story.

THE TOURISM CONTEXT

There are a range of issues related to the features of destinations and how they function which impact on the delivery of a sustainable approach. These include inherent local stakeholder conflict and the entrenched interests of small town politics, where the destination itself is the focus and its relationship with the wider world is missed. More specifically, from the point of view of sustainability, the tourism context at the destination can be problematic for a variety of reasons:

> First, destinations are dominated by small and medium-sized businesses (SMEs). Typically such enterprises are individually or family owned, often lacking management expertise and/or the training needed for sustainable approaches. At the same time they lack the means for investment in sustainable practices and may not fully understand the need for it – unless it is directly relevant to their operation. In addition, the very competitive nature of SMEs leads to a lack of trust and collaboration, both of which are important to the development of a sustainable destination. Finally, in such businesses, the rapid turnover of employees – and indeed the SMEs themselves – works against the continuity of relationships needed for the development and understanding of the sustainable destination. (Weidenfeld *et al.*, 2009)

Table 5.2 Negative impacts of tourism identified by neighbourhood groups

Residence and everyday life	Rising land prices and consequent eviction of local residents; neglect and lack of investment in infrastructure and services for local users; increased police control; growth in the number of traders moving into the neighbourhood to the detriment of its residential function and the resulting alterations to everyday life (for example, opening hours).
Public space	Noise; occupation by open-air terrace bars and tables; billboards and display units; excessive overcrowding in some streets and consequent mobility problems; poor waste management; growing privatization of space.
Culture and identity	The commercialization of culture; banalization (in the form of souvenirs and experiences); loss of local languages on the streets; loss of typical food and drink.
Local economy	Commercial homogenization and disappearance of traditional local traders; colonization by foreign brands; rise in unstable and uncertain employment in the hospitality industry; threat to the neighbourhood's social and economic fabric due to the growth of tourist rental property business.

Source: García-Hernández *et al.* (2017)

Secondly, poor human resources practices in the tourism sector include the employment of 'non-standard' labour such as seasonal and part-time workers, resulting in the high labour turnover referred to above. As a result, tourism labour at the destination tends to be poorly qualified. These processes (1) work against the development of sustainable destination by mitigating against the continuity of knowledge transfer, and (2) work against the development of resilience at the destination, leaving it more vulnerable to change, as we will see below.

Thirdly, delivery of the tourism product at the destination is complex with myriad providers involved, including many different types of organization from the private sector to the public sector and, indeed, the host community itself (see Agarwal, 2012, for examples of fragmentation impacting on strategy implementation). This fragmentation of the tourism product and its delivery across various providers leads to poor coordination for a sustainable approach and its adoption across the sector and underscores the need for a partnerships approach as in UN SDG 17 (EC, 2014).

Finally, much of the planning and management at the destination is undertaken by the public sector, often through a dedicated agency such as a destination management organization (DMO). While some DMOs are excellent, others are dependent on local political agendas for their continuity of funding, contrasting with the need for a long-term approach if sustainability is to be taken seriously. Also, destinations do not exhibit the lines of responsibility found in, say, private sector organizations, which means that DMOs have no authority over destination stakeholders and must operate by persuasion and influence to deliver a sustainable agenda.

THE STAKEHOLDER APPROACH

A useful way of viewing the features of destinations and their impact on the adoption of a sustainability approach is by looking at the fact that tourism destinations comprise a mosaic of different actors that we can term stakeholders. A truly sustainable destination will recognize that it must satisfy all of its stakeholders in the long term. This can be achieved by a strategic planning approach that balances a marketing orientation focused on tourists with a planning orientation focused on the needs of local people. The Case Study on the New Forest in the UK at the end of this chapter exemplifies this approach. We must also remember that the tourist experience is made up of a series of small *encounters* with many stakeholders and that these encounters strongly influence the success or otherwise of the visit. In every destination there are several stakeholders which have a wide range of both compatible and conflicting interests, including sustainability:

- *The host community* is the most important stakeholder as they live and work at the destination and provide local resources to visitors. A sustainable destination will involve the local community in decision taking, and ensure that tourism does not bring unacceptable impacts on the local people and their homes.
- *Tourists* are looking for a satisfying experience, through properly segmented and developed products. Increasingly, they not only seek a high quality of service but also a well-managed and organized sustainable destination – they will shun those destinations that are overtly unsustainable.
- *The tourism industry* is to a large extent responsible for the existing development of tourism and the delivery of the tourism product. The industry can be thought of as polarizing between global and niche players. The global players tend to be multinational and well-resourced with capital, expertise and power and increasingly they show leadership in terms of destination sustainability (see Chapter 6). Niche players are traditionally small, family-based enterprises lacking capital, expertise, qualified human resources and influence at the destination as noted above, and here sustainability is more of a challenge.

- *The public sector* sees tourism as a means to increase incomes, stimulate regional development and generate employment. The public sector is an important stakeholder, often taking a destination leadership or governance role in terms of sustainability and owning key parts of the tourism destination.
- *Other stakeholders* include pressure groups, chambers of commerce and power brokers within the local, regional or national community.

Holden (1999) uses the example of skiing in the Cairngorms of Scotland as an example of the differing priorities of destination stakeholders. The key issue that has given rise to these differing priorities is the push to halt skiing for a period of 20 years to allow the environment to return to a natural state. Here conservationists argue the need for alternative economies around conservation programmes designed to improve the ecology of the region, set against opposition from the Cairngorms Partnership, which is a private sector led organization that operates in the Cairngorms National Park (https://visitcairngorms.com/membership). Holden argues that there is often little will on the part of the ski industry, and those who stand to gain through commercial skiing enterprises, to move towards the conservationist view. This position is said by Holden to be a conservative interpretation of sustainability where instrumental values and economic benefits take precedence over nature's intrinsic value.

THE NETWORK GAZE

We can take the idea of a destination made up of many stakeholders and think of destinations as loosely articulated networks of these organizations, with the organizations acting as nodes, and their relationships as the flows around a network. The nature of tourism demands collaboration and partnerships to deliver a sustainable destination and this happens through the creation of inter-organizational networks. This is stressed under UN SDG 17 and this section of the chapter shows how this can be achieved using a network approach. Viewing destinations as networks allows us to visualize their network structures in terms of the positions of the organizations and the relationships between them and how they can influence sustainability. In turn, these structures can be measured, calibrated, compared and classified. This involves the use of network analysis to understand how destinations function and allows the structure of destination networks – and their potential shortcomings and inefficiencies in creating sustainable destinations – to be identified.

The 'network gaze' facilitates sustainability by showing how knowledge, information and consensus can be generated, trust built and knowledge exchange facilitated at the destination (Scott, 2015). To deliver sustainability across a destination network, the configuration of the network and individual organizations' positions within the network are key. Reagans and McEvily (2003) state that the knowledge transfer necessary for sustainability is facilitated by social cohesion among the network members and is as important as how strong the linkages are between them. Braun (2004) adds to this, arguing that as well as network position, there is a need for the destination members to have trust in, and engagement with, the network. This can be a real inhibitor to sustainability among SMEs, as we saw above.

Effective governance and management of the destination network is another important factor in facilitating sustainable destinations. This is because active participation in formal or informal networks is a common way of acquiring knowledge in tourism, and hence understanding of sustainability principles (Baggio & Cooper, 2010). Here social relationships play a critical role in these 'knowledge networks', requiring participants to emphasize the management of relationships as well as the management of processes or organizations (Beesley, 2005). Governance of destination networks ensures that knowledge is not lost when a member leaves, and it also ensures that all members work towards the same sustainability goals. However, we must remember that, at the end of the day, networks are simply relational structures and do not have a purpose. For this we must turn to the notion of communities of practice.

Here, the idea of communities of practice (COPs) takes the network gaze further by explaining the *purpose* of a destination network – such as the development of sustainability. The concept of COPs recognizes that the process for destination innovation and strategy takes place within communities and not among individuals. A COP can be thought of as a group of individuals who develop a shared way of working together and engaging in common activity to accomplish a common purpose, in this case sustainability (Schianetz *et al.*, 2007a). As such, COPs share a repertoire of history and concepts where trust and collaboration are important dimensions for effectiveness and for delivering strategic advantage for the destination. It is this dimension of trust and collaboration that sees a possible departure from way in which destinations function, as noted above. As this chapter shows, trust, or mistrust, is central to the effective implementation of sustainability in tourism destinations.

COMPLEX SYSTEMS

We introduced the systems approach in Chapter 4. It takes the network gaze one step further by thinking of a tourism destination as a complex system whereby the components are autonomous agents (such as businesses or community groups) who interact at the destination, but who pursue their own objectives – hence the difficulties faced by DMOs in managing and coordinating the destination. The challenges faced for the sustainable destination are therefore twofold:

1. the agents in the system do not act with the overall goals of the system – in this case sustainability – in mind; and
2. we do not know how stable these systems are and, if they are subject to change, how they will respond.

Hall *et al.* (2017) define a system as:

> a group of elements organised such that each element is, in some way, either directly or indirectly interdependent with every other element. (Hall *et al.*, 2017: 16)

We can think of four further characteristics of tourism destination systems:

i. Destinations are open systems as they interact with elements and influences external to the destination, hence creating a further challenge for sustainability.
ii. The destination system is made up of a series of subsystems – for example, the distribution networks of accommodation or the public sector marketing system.
iii. The causes behind some of the destination system linkages are understood but many are not, as for example with the complex relationship between the environment and tourism at the destination.
iv. Destination systems are subject to 'feedback'. Here Hall *et al.* (2017) provide the example of an eco-tourism destination where its very success leads to further tourists visiting and the loss of the original purpose of ecotourism.

An example of destinations as open systems is provided by Trathan *et al.* (2015), who examined external human threats to penguin colonies and what must be done to mitigate these threats. They identified a variety of threats:

- harvesting adult penguins for oil, skin and feathers and as bait for crab and rock lobster fisheries;
- harvesting of penguin eggs;
- terrestrial habitat degradation; marine pollution;
- fisheries by-catch and resource competition;
- environmental variability and climate change; and
- toxic algal poisoning and disease.

Using the example of the Galapagos Islands, Trathan *et al.* (2015) argued that although tourists themselves do not cause significant damage to penguin populations, the infrastructure built to develop penguin tourism may have impacts. They argue that the impacts on breeding penguin colonies is unknown because of the wide variety of species studies, the locations of penguins and the levels and types of human activity to which penguins are exposed.

The nature of destinations as complex systems effectively means that they have to be conceptualized and understood as a whole; it is impossible to break them down into constituent parts without losing the integrity of the system. This works well for sustainable destinations, where a 'whole of destination' approach is more effective than a piecemeal approach to management.

A key question for sustainability is how the destination system responds to environmental uncertainty, change and surprise – in other words, how the destination will adapt in response to change. Hall *et al.* (2017) define adaptation as the ability of an organization to survive the conditions of its changing environment. This leads to the idea of complex adaptive systems, which can be thought of as a dynamic network of many agents acting in parallel, constantly acting and reacting to what the other agents are doing (Waldrop, 1992: 144). Here, complex adaptive systems operate to general laws which allow for the improved operation of the overall system, and the analogy with the transition of a system towards sustainability is clear.

For complex adaptive systems, the idea of resilience is important as it recognizes the consequences of a system reacting to change, or disturbances from external influences, such as climate change. If the resilience built into the system is poor, then it will result in a move away from sustainable practices and management to deliver 'unwanted and undesirable economic, environmental and social consequences' (Hall *et al.*, 2017: 25). To avoid this it is important that the destination can exchange knowledge around its network and develop a continuous and collective learning that acknowledges uncertainties and allows for timely adjustment of planning and management strategies.

Here, Karatzoglou and Spilanis' (2010) work on adaptive resource management and destination environmental scorecards is indicative of the progression of tourism and sustainability research on island environments. These sorts of approaches are of immediate importance because sustainability is challenged by operators who are in denial of their environmental impacts, are ignorant of environmental legislation, are inefficient at self-regulation and management and lack access to important resources that would help them mitigate their impacts (Hillary, 2000). These authors argue that the adaptive resource management paradigm is valuable as an overarching conceptual framework because: (1) it empowers individuals and communities to make local decisions; (2) people are held accountable for their use of resources; (3) it encourages participation in policy development and networking among the users of the commons; (4) reduction of uncertainty may be achieved through protection from free-riders; and (5) it encourages the development of governance systems for the commons as a whole to establish rules and norms of collective behaviour in protecting the rights and interests of users and non-users (see Briassoulis, 2002).

Box 5.3 Destination adaptation to climate change to 2030

Introduction

UN SDG 13 focuses on climate action – the focus of this case study. The case takes a real, and current, example of an island in the Adriatic – the Croatian island of Mali-Lošinj – and examines the realities of adapting the complex system of tourism on the island to climate change. The process involves a blend of climate science and tourism marketing. By combining these approaches one can see which tourism activities will be supported by the climate in the future – and therefore which tourism products should be developed and which should be phased out as the strategy is developed. This is a scientific and accurate approach to deliver a sustainable approach to adaptation.

Strategies for climate change adaptation

Adaptation to climate change is a process by which strategies aim to moderate, cope with and take advantage of the consequences of climate events. Over and above their traditional functions of destination marketing and product development, destination management organizations are increasingly expected to take on a central role in destination-level climate adaptation and mitigation strategies. This is the case for the Croatian island of Mali Lošinj, where the tourist board has taken the lead in working with professional climatologists to predict future climate scenarios and to estimate what products can be supported in the future.

The island of Mali Lošinj

Mali Lošinj is a Croatian island in the northern Adriatic which has been awarded a number of sustainability prizes (https://www.visitlosinj.hr; VisitLošinj, 2019). It has a Mediterranean-style climate and is famed for its wellness products based on the local flora and atmosphere. Tourism on the island is well established and dates back to the late 19th century when it was a winter resort, renowned for the quality of its air for visitors with respiratory problems. Access to the island is by road and ferry, by private boat or yacht, or by private plane to the small airport (a new one is planned). Most visitors arrive by road. A major issue for tourism on the island is seasonality, with the season beginning in April/May and finishing in September/October. This is partly climatic but also strongly influenced by cultural factors, such as school holidays, custom and fashion.

The island's main products are:

- beach tourism;
- hiking, climbing and walking;
- flora and fauna;
- marine tourism – dolphin watching, diving, sailing, water sports, sport fishing and angling;
- hunting;
- a significant archaeological underwater park;
- health, wellness and aromatherapy based on bio diversity and aerosols.

Many of these products are promoted on, and depend on, the island's favourable climate – sunshine, air quality and temperature and light winds. The island also promotes its land and marine biodiversity supporting many of the tourism products. Again this is dependent on the climate.

Tourism climate indices

The relationship between weather and tourism activities is highly dependent on the kind of activity that is assessed, with beach tourism, hiking, marine tourism or wellness tourism each requiring different weather conditions to allow for comfortable activity. Tourism climate indices calculate the various elements of the weather – such as precipitation, wind, temperature, humidity – and combine them to provide a composite index for, say, the ideal conditions for beach tourism. Tourism climate indices use three distinct aspects of climate that are relevant to tourism:

1. The *thermal component* is primarily physiological in nature and determines the 'comfort' of tourists.
2. The *physical component* refers to climate features, such as rain and wind, which may cause 'physical annoyance'.
3. The *aesthetic component* represents climate features (e.g. sunshine) that may influence the 'tourist's appreciation of a view or landscape'.

For a proper assessment of the suitability of climate and weather conditions for tourism purposes, the use of composite measures, which combine the three elements above, is most efficient. Effectively, then, a tourism climate index determines the most 'ideal' weather conditions for each type of tourism activity. However, surprisingly, tourism climate indices have remained in the realm of climatologists and have not been used to develop future policy, planning or market strategies for tourism. And yet we can forecast future climate conditions reasonably accurately and these can be run through the tourism climate index model to show how the destination can support, or not support, particular tourism activities and products in the future. In other words, tourism climate indices should provide the basic data for any climate change adaptation exercise.

The approach to adapting the tourism system of Mali Lošinj to future climate conditions took the climate for 2011 and then forecast the climate to 2030. The two climates were then run through the climate indices to see what products would be supported by the weather conditions in each of the two periods.

Table 5.3 summarizes the difference in the number of optimal, acceptable and unacceptable days for a range of tourism activities on Mali Lošinj between 2011 and 2030. It is clear from the table that the tourism system of Mali Lošinj must adapt to climate change by responding to the shifting seasonal pattern of products that will be supported by the weather conditions in 2030, i.e. the spring season delivers the most optimal days for a range of activities, shifting the peak season away from high summer.

Table 5.3 Change in the number of ideal and acceptable days by activity in the period 2011–2030 at 06 UTC and 12 UTC

Activity	*Year*		*Winter*		*Spring*		*Summer*		*Autumn*	
Beach	8.4	8.0	0.0	0.2	1.8	3.2	5.4	4.2	1.4	0.4
Cycling	−1.6	15.0	0.0	2.8	3.6	8.4	0.0	1.2	−5.2	−2.0
Sightseeing	13.0	12.6	0.8	4.2	10.2	7.2	1.4	−1.2	0.8	0.0
Golf	−1.8	9.4	0.0	2.2	3.2	6.2	−1.6	2.2	−3.6	−1.0
Football	10.2	10.2	1.2	5.6	9.0	4.0	−0.2	0.8	0.4	−0.6
Motor boating	10.2	7.8	0.0	0.6	3.8	9.2	5.4	2.0	1.0	−4.2
Sailing	5.8	6.4	0.0	−0.4	1.8	9.6	3.6	−0.8	0.6	−2.0

Source: Cavlek *et al.* (2019)

THE EVOLVING DESTINATION

A sustainable approach to destinations involves taking a long-term perspective as we saw in the Mali Lošinj case study. As markets develop and change, destinations have had to respond in terms of their tourist facilities and services, for example, through adaptation to climate change. A more formalized representation of this idea is expressed by Butler's (1980) tourist area life cycle (TALC) (see Figure 5.1). This states that destinations go through a cycle of evolution similar to the life cycle of a product (where sales grow as the product evolves). Simply put, numbers of visitors replace sales of a product. Obviously, the shape of the TALC curve will vary, but for each destination it will be dependent on factors such as:

- the rate of tourism development;
- access;

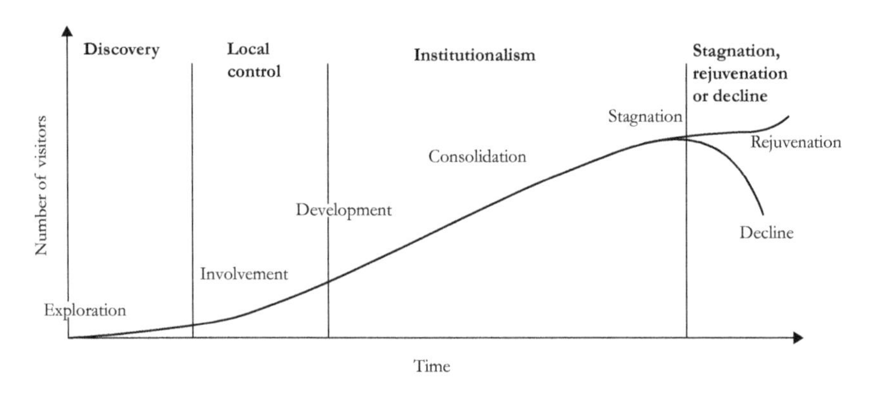

Figure 5.1 The tourist area life cycle
Sources: After Boniface *et al.* (2012); Butler (1980)

- government policy;
- market trends; and
- competing destinations.

Each of these factors can delay or accelerate progress through the various stages. Indeed, development can be arrested at any stage in the cycle, and only tourist developments promising considerable financial returns will mature to experience all stages of the cycle. In turn, the length of each stage, and of the cycle itself, is variable. At one extreme, instant destinations such as Cancun (Mexico) (https://www.visitmexico.com/en/main-destinations/quintana-roo/cancun) move almost immediately to growth; at the other extreme, well-established destinations such as Scarborough, England (www.scarborough.co.uk/) have taken three centuries to move from exploration to rejuvenation. There are clear lessons for sustainability here as, with good planning and management, destinations can influence their trajectory both up, and down, the life cycle or even arrest their development at a particular stage. Effectively, there are different strategies that can be adopted at each stage of the life cycle to achieve this. Cooper's (1995) work on the strategic planning of the offshore islands of the UK clearly demonstrates the strategic priorities at each stage (see Table 5.4).

Here, Weaver's (2018) work on the Southeast Asian island of Timor-Leste adopts the life cycle approach. Timor-Leste is an early-stage destination which might act as a model of sustainable development 'from scratch', taking advantage of its peripherality in building innovation, distinctiveness, autonomy, amenability and other characteristics. Weaver (2018) states that there are four key elements necessary to enable Timor-Leste to realize success as a sustainable destination:

1. marketing that draws on social media accounts of visitors and that builds positive awareness of the potential for peak experiences in the destination;
2. beach-centred resorts that feature best practices;

Table 5.4 Destination strategy by life cycle stage

Competitive position	*Stages of industry maturity*			
	Embryonic	***Growth***	***Mature***	***Ageing***
Dominant	Fast grow Start-up	Fast grow Attain cost leadership Renew Defend position	Defend position Attain cost leadership Renew Fast grow	Defend position Focus Renew Grow with industry
Strong	Start-up Differentiate Focus Fast grow	Fast grow Catch-up Attain cost leadership Differentiate	Attain cost leadership Renew, focus Differentiate Grow with industry	Find niche Hold niche Energize Grow with industry Harvest
Favourable	Start-up Differentiate Focus Fast grow	Differentiate, focus Catch-up Grow with industry	Harvest hang-in Find niche, hold niche Renew, turnaround Differentiate, focus Grow with industry	Retrench Turnaround
Tenable	Start-up Grow with industry Focus	Harvest, catch-up Hold niche, hang-in Find niche Turnaround Focus Grow with industry	Harvest Turnaround Find niche Retrench	Divest Retrench
Weak	Find niche Catch-up Grow with industry	Turnaround Retrench	Withdraw Divest	Withdraw

Source: Cooper (1995)

3. a community-based tourism pole that emphasizes visitation in the agricultural hinterland; and
4. a protected areas pole which draws on the principles of the ecotourism concept proposed by Fennell and Weaver (2005), where visitors act as biodiversity stewards of these natural areas.

We can see that one particular benefit of the TALC is as a framework for understanding how destinations and their markets evolve and the sustainability of the destinations at each stage (see Butler, 2015). Indeed, it could be argued that an understanding of the cycle aids the development of community-based and sustainable tourism strategies at the early stages of the cycle. To implement such approaches in later stages may be inappropriate and is certainly more difficult. In other words, tourist destinations are dynamic, with changing provision of facilities and access matched by an evolving market in both quantitative and qualitative terms, as successive waves of different numbers and types of tourists with distinctive preferences, motivations and desires populate the resort at each stage of the life cycle.

The TALC has many critics, in part drawn by its very simplicity and apparent deterministic approach. Some argue that, far from being an independent guide for decisions, the TALC is determined by the strategic decisions of management and is heavily dependent on external influences. However, as a framework within which to view the development of destinations, albeit with hindsight, and as a way of thinking about the interrelationship of destination and market evolution and hence sustainability, it provides many useful insights.

These insights have been utilized by Lim and Cooper (2009) by linking the life cycle approach to the complex systems approach. They have developed a management model to deliver sustainability in island tourism destinations by developing a set of sustainable indicators that allow for the identification of the life cycle stages. Each set of indicators can then be optimized and monitored for sustainability at each life cycle stage.

CONCLUSION

This chapter has identified the key components of the sustainable destination and analysed how the very nature and make up of destinations can promote or inhibit that goal of sustainability. The key features of a sustainable destination relate to protection of the natural and cultural heritage, participation by all stakeholders through partnerships, and the development and actioning of a sustainability plan. The chapter has then analysed the inherent nature of destinations and how this can support sustainability. These components include, first, the concept of inseparability, whereby tourism is produced where it is consumed and has given rise to the contemporary, and controversial, phenomenon of 'overtourism'. We must also recognize that destinations comprise myriad stakeholders all with their own objectives, which often conflict – a true sustainable destination has mechanisms in place to resolve these conflicts. In terms of analytical approaches, both network analysis and systems thinking help us to understand how destinations function and how they can be configured to support sustainability. Finally, the chapter has reviewed the evolution of destinations through the tourism area life cycle concept and explained how this helps us to understand sustainability at each stage of the cycle.

END OF CHAPTER DISCUSSION QUESTIONS

1. Identify the key components needed to assemble the definition of a sustainable destination.

2. What do you see as the main management approaches to solving the problem of overtourism?

3. How does a network approach help us to understand the functioning of sustainable destinations?

4. Taking a destination that you know well, map the key stakeholder groups and identify their objectives. How can differences be reconciled?

5. How can resilience be built into a destination system?

End of Chapter Case Study 5.1 Managing stakeholders in forest destinations

The VICE model of destination management: The New Forest National Park, UK

This case study outlines an innovative approach to managing the varying and conflicting objectives of stakeholders at a destination, with the overall objective of delivering sustainability. This technique is known as the VICE approach to destination management and was developed in the UK's New Forest – ironically, despite the name, it is one of the most ancient forest landscapes in the country. It is a good example of the partnership approach that underpins UN SDG 17.

The New Forest

The New Forest is an environmentally sensitive area with a unique landscape in the south of England. The whole area is under severe pressure, not only from tourism and recreation but also from other developments such as housing and transport, and as a result it was designated as a National Park in 2005. The New Forest landscape comprises areas of open heathland, interspersed with woodland. Its status as Crown land has protected it from development over the centuries. The New Forest is a significant natural resource that faces many competing demands. Recreation and tourism create major impacts on both the resource and the local community, although the economy does benefit. This has been complicated by National Park designation. The National Park covers an area of 56,651 ha, designed to conserve the New Forest landscape, flora and fauna and to promote its enjoyment by visitors.

Tourism and transport in the New Forest

The New Forest is a very popular destination for both day and overnight visitors. Visitor pressure in the Forest arises from the fact that it is easily accessible through the national motorway network and that over 15 million people live within a 90-minute drive. Tourism in the New Forest is estimated to:

- support almost 2500 jobs in the area;
- contribute £72 million annually to the local economy; and
- attract 13.5 million day visitors and almost 1 million overnight visitors.

A key issue is the management of traffic. Most visitors arrive by car, and a comprehensive traffic management plan includes a 40 mph (65 km/h) speed limit on unfenced roads, and the use of landscaped verges, ditches and ramparts to prevent off-road parking. Visitors are directed instead to over 150 designated parking zones. Other forms of transport include horse riding, cycle hire (although mountain bikes have caused damage in certain areas), horse-drawn wagon rides and regular bus and coach services. These alternatives to the car are coordinated in a series of networks in an attempt to reduce the number of car-borne visitors to the Forest.

Destination management

Managing the New Forest as a sustainable destination is critical given the combination of an environmentally sensitive area and the large number of visitors. The New Forest is fortunate in having developed a visionary approach to destination management. Destination management in the New Forest is based on the principle of partnership, hence the title of their management strategy – *Our Future Together* (NFDC, 2003). This followed an earlier strategy, *Making New Friends*

(NFDC, 1996), and a consultation document, controversially entitled *Living with the Enemy*, which mapped out the challenges for the New Forest tourism industry (NFDC, 1994). A further complication in the management of the New Forest is the plethora of agencies, committees and other bodies involved in the management of the Forest under its National Park status, underscoring the need for a partnership approach. The lead tourism agency is the New Forest Tourism Association (NFTA), formed in 1989 to promote the New Forest as a quality year-round holiday and business destination.

The New Forest destination management approach has become known as the VICE model (see Figure 5.2), based on:

1. visitors;
2. industry;
3. community; and
4. environment.

The model is 'built into' the destination rather than 'bolted on', and stresses the interdependence between the four elements. It has turned a problematic set of

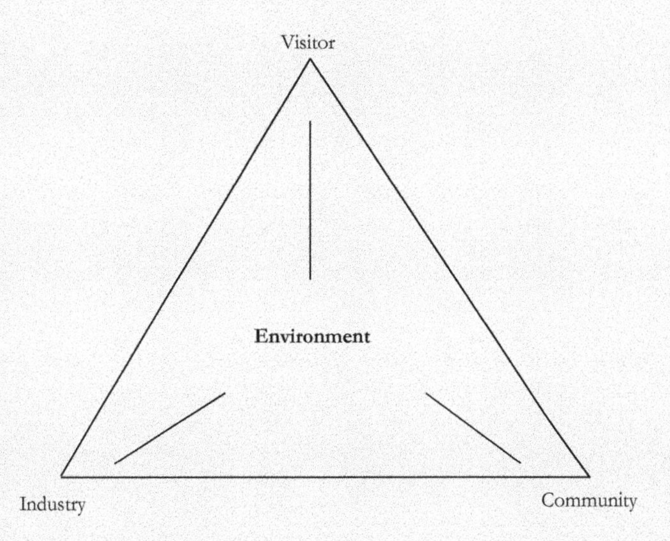

Figure 5.2 The VICE model
Source: After New Forest Tourism

relationships into mutually beneficial partnerships by smart communication and by partnership building with stakeholders. The strengths of the approach are:

- developing a clearly identifiable destination and brand;
- strong leadership from the NFTA;
- ensuring the continuity of stakeholders;
- a long-term commitment to the strategy;
- a positive local profile for tourism;
- a strategy and administrative structure that work together;
- securing trust – changing the culture; and
- good regional and sub-regional relationships.

Beyond the New Forest

The VICE model has proved so effective that other UK regions have adopted its guiding principles for destination management. Cornwall has embedded the VICE model into its local sustainability plans while the Cotswolds region is using the VICE model to evaluate sustainability principles across the region (see Cornwall Destination Management Organisation, 2007 and Cotswolds Tourism Partnership, 2009).

Discussion questions

1. Draft a communication plan to help the New Forest Tourism officer convince local residents that tourism is of economic benefit to the Forest.

2. Other destinations are now adopting the VICE model. How transferable is this approach to other destinations?

3. A key issue for the Forest is the dominance of day visitors and the relatively small number of overnight visitors (around 3 million nights annually). As a result the tourist authorities in the New Forest are concerned that much of the economic benefit from tourism is lost. Draw up a strategy to increase the number of overnight visitors in the Forest.

Sources: Cornwall Destination Management Organisation (2007); Cotswolds Tourism Partnership (2009); Crouch and Ritchie (2012); NFDC (1994, 1996, 2003, 2019); Forestry England (2019); New Forest National Park (2019); NFTA (2019); thenewforest.co.uk (2019); Visit England (2019)

CHAPTER 6
THE TOURISM INDUSTRY (1)

LEARNING OUTCOMES

This chapter focuses upon the sustainability initiatives and issues of three sectors of the tourism industry – hospitality, transportation and attractions. Each of these sectors has specific issues – for example, the air transport industry is an emitter of major greenhouse gases (GHGs), while the hospitality sector is comprised mainly of small and medium-sized enterprises (SMEs) which brings its own challenges in terms of implementing sustainability. Nonetheless, there are common themes that can be detected across the three sectors, including the imperative to engage with local communities, energy, water and waste reduction, and the need to address climate change. The chapter is wide ranging and therefore covers a range of the UN's Sustainable Development Goals (SDGs), specifically SDGs 3, 5, 6, 7, 8, 12 and 13. The chapter is designed to provide you with:

1. An awareness of the sustainability issues facing the tourism industry.

2. An analysis of initiatives being taken to address sustainability across the tourism industry.

3. An understanding of the role of government, the private sector and other stakeholders in addressing sustainability in the tourism industry.

4. An awareness of the challenges of implementing sustainability initiatives in the tourism industry.

5. An awareness of the many examples of good practice across the hospitality, transportation and attraction industries.

HOSPITALITY

INTRODUCTION

The provision of overnight accommodation, food and beverages for tourists partly defines the tourism industry and delivers the most significant economic benefits and spend to the destination when compared to, say, a day visitor. Hospitality comes in many different forms, ranging from condominiums through resorts and conference centres to guesthouses, homestay, visiting friends and relatives, and restaurants and bars. Environmental issues for the hospitality industry have come to the fore since the 1990s and are of increasing concern to consumers (see Hawkins & Bohdanowicz, 2011). For example, surprisingly, the accommodation industry is a significant source of carbon emissions, as well as being perceived as a heavy user of energy and water. Major accommodation developments have to undergo an environmental impact assessment, while many properties adopt environmental auditing of their operations. Of course there is controversy over how the industry has addressed these issues, with some using 'green' initiatives as a major marketing point of differentiation. Accor, for example, have state-of-the-art environmental management systems and publish their green credentials (Accor, 2018).

The magnitude of the accommodation sector in tourism has been addressed through the sustainable tourism (ST) discourse. A key question is whether specialist alternative forms of accommodation are indeed more ecologically benign than traditional forms. Moscardo *et al.* (1996) tackled this question and found that there are many social, cultural, economic and ecological factors that need to be weighed in deciding if one form is ecologically 'better' than the other. In many cases, specialist sustainable forms of accommodation are not in fact holistic in terms of the broad range of elements that would make them truly sustainable. Furthermore, traditional forms of accommodation may be more ecologically benign, especially in urban settings, because of their energy, waste and infrastructure design and practices. The authors also broached the question as to whether specialist forms of accommodation should replace traditional mass or conventional forms. The practicalities of this would be daunting, so the alternative is to get these large-scale enterprises to move towards less destructive practices (de Kadt, 1992). One particular example of a hotel that has embraced these principles is Jakes Hotel in Jamaica, examined in the mini case study in Box 6.1 and an excellent example of the implementation of SDG 12 on responsible production and consumption.

Box 6.1 Jakes Hotel, Jamaica

Introduction

Jakes Hotel in Jamaica (jakeshotel.com) is one of the world's leading hotel advocates of sustainability, not simply through recycling and encouraging guests not to use all their towels, but rather by being fully involved and immersed in its local community and environment. Set on Treasure Beach on Jamaica's south coast, the hotel began as a restaurant in 1991 and now comprises a collection of 30 rooms, cottages and villas, each individually designed. The hotel offers nearby access to golf, fishing, caves and waterfalls, adventure tours, ecotourism and dolphins. The guests are fully involved in the hotel's sustainability mission with a $1 per night levy and the opportunity to visit the various projects.

A sustainable company

As a company, Jakes Hotel has a distinctive view of sustainability. According to the website, 'sustainability is not just about eco-friendly practices, it is as much more about cultural preservation and maintaining what is unique about our community … we believe sustainability is an interactive system between our community and the environment, where each element is cared for and nurtured, so that we can continue to occupy this special place in the world with only positive impact' (Jakes, 2019).

The company aims to be a model for future sustainability initiatives in the Caribbean. Its foundation – BREDS Treasure Beach Foundation – works with the local community, supporting education, sports, cultural heritage and emergency healthcare. The Foundation is led by volunteers and has completed community projects including repairing the roof and building classrooms at local schools and building houses for the disadvantaged.

One of the Foundation's major projects is the development of the BREDS Treasure Beach Sports Park. The principle behind the Park is that sport can bind a community together. The Park was inaugurated in 2010 and when completed it will be a 15-acre park with cricket pavilion, regulation-size soccer pitch, children's playground, and a site for weddings and other functions such as retreats and workshops.

As well as working with the local community, the hotel also is concerned for the local environment, funding projects to protect the local vegetation and ecology of Treasure Beach, rounding off a set of sustainable initiatives that embed the hotel within the destination.

Source: Jakes (2019)

There are a number of initiatives that have been established to promote sound environmental behaviour by accommodation companies, reflecting SDG 12 on responsible production and consumption and SDG 7 on affordable and clean energy:

- The International Tourism Partnership (www.tourismpartnership.org) provides advice to accommodation companies on responsible business including energy consumption and greening the supply chain.

- Hotel Energy Solutions (HES, 2019) is an energy management toolkit available for accommodation properties, supported by international agencies including the IHRA and the UNWTO. The initiative deals with energy strategies including saving energy costs and advice on installing renewable energy systems, claiming savings of 20% on energy costs with simple operational changes. It is the subject of a case study in Chapter 10.
- Considerate Group hoteliers (2018) is an association of independent hoteliers that encourage and support good socially responsible environmental policies and practices in their properties. Initiatives include publications on 'green sourcing' and saving energy costs.
- We introduced the Global Sustainable Tourism Council (GSTC) in Chapter 4. This is an independent and not-for-profit organization representing a diverse and global membership including hotels. The goal of the GSTC is to achieve best practices in ST through the design of a set of criteria for sustainable practices in tourism. For hotels, the criteria cover four major headings:
  A: Demonstrate effective sustainable management.
  B: Maximize social and economic benefits to the local community and minimize negative impacts.
  C: Maximize benefits to cultural heritage and minimize negative impacts.
  D: Maximize benefits to the environment and minimize negative impacts:
    D1 Conserving resources.
    D2 Reducing pollution.
    D3 Conserving biodiversity, ecosystems and landscapes.

The GSTC criteria provide a useful framework for the discussion of sustainability initiatives and approaches in the hospitality sector.

EFFECTIVE SUSTAINABLE HOSPITALITY MANAGEMENT

In this section we consider three major approaches to 'whole of system' sustainable hospitality management:

1. corporate social responsibility;
2. sustainable hospitality management systems;
3. sustainable supply chain management.

Corporate social responsibility

We introduced the concept of corporate social responsibility (CSR) in Chapter 4. Hughes and Scheyvens (2016: 469) are clear that many large hospitality businesses have CSR initiatives that advance environmental, economic and social sustainability. For these businesses CSR is often linked to cost savings and the reputation of the business. They go on to say that CSR initiatives 'typically include donations to schools and hospitals, water and energy saving initiatives, environmental protection, sourcing of local produce and equal opportunity recruitment and training opportunities' (Hughes & Scheyvens, 2016: 471). The environmental and social consequences of CSR are now monitored and reported through annual reports, on company websites and through social media (see, for example, Accor, Marriott and Hilton). It is clear that CSR initiatives are growing although in the early years of the 21st century hospitality was slow to embrace the concept. For example, in a 2007 study of the CSR practices of the top ten hotel companies worldwide by Holcomb *et al.* (2007), their general conclusion was that only 40% of hotels provided mention of social responsibility in their vision and mission statements, whereas 60% reported having a diversity statement. CSR now takes place in both large and small hotel properties and groups, but it is the largest companies that have provided the greatest degree of leadership in this area as part of their business strategies (Abram & Jarząbek, 2016).

In order to counter the charge of 'greenwashing', the measurement and reporting of CSR initiatives is central to a sustainable approach to hospitality. For example, Medrado and Jackson (2016) investigated the reporting practices of US hospitality and tourism firms in the area of CSR. They found that such reporting is in its infancy, and that most firms used the Global Reporting Initiative (GRI, 2018) as a standard guideline for reporting. They also discovered that different industry sectors reported in different ways. For example, accommodation firms were stronger on CSR/sustainability than the food and beverage and cruise line sectors. The information most frequently disclosed was on performance related to water use, energy conservation and waste generation. Less information was disclosed on worker compensation and work–life equilibrium.

Researchers have also investigated CSR from the demand side. Kucukusta *et al.* (2013) queried 150 guests of four- and five-star hotels in Hong Kong in order to identify the CSR factors that were most important for adopting these factors. They found that respondents identified five main factors: community, policy, mission and vision, workforce, and environment. Environment and mission and vision held the greatest predictive power in explaining the preferences around stay, willingness to pay, service quality and brand image. The authors argue that environment and mission and vision ought to be the most important areas communicated to potential guests.

Sustainable hospitality management systems

CSR has prompted a growing awareness of the importance of sustainability for hospitality operations. As a result, a number of over-arching sustainability management systems have been developed to assist the sector. In this section we consider four key initiatives:

1. the IHRA 'Evolution' system;
2. Considerate Hoteliers' Con-Serv system;
3. IHG's Green Engage system; and
4. Accor's PLANET 21 approach.

International Hotel and Restaurant Association

The international industry association – the International Hotel and Restaurant Association (IHRA; www.ih-ra.org) – has developed the 'Evolution' initiative in response to demand from its members for a comprehensive sustainability management approach (IHRA, 2014). The approach:

- measures resource use;
- manages and reduces energy costs;
- provides reporting and disclosure to clients; and
- reduces waste and increases operating profit.

'Evolution' can be managed from a desktop and does not require technical expertise for its use. It delivers a robust audit trail for energy, water, waste, chemicals, refrigerants and GHGs and places the individual property in control of its own data.

Considerate Hoteliers

The private sector company Considerate Hoteliers has developed a similar system for use by accommodation units (consideratgroup.com). Their system, Con-Serv, is ISO 50001 certified as an energy management system. The system is enabled by a dynamic technology platform which has been tailored to capture data relevant to the hospitality industry, including electricity, heat, water, waste and outsourced laundry. It allows properties to identify inefficiencies and improve performance. It provides comprehensive, automated reports that work for individual units or multisite operations. The reporting system compiles cost, consumption and carbon emissions data and benchmark performance.

InterContinental Hotels Group

The Green Engage system developed by the InterContinental Hotels Group (IHG; www.ihg.com) is an online environmental sustainability system which measures the impact that a hotel has on the environment, with over 200 green solutions to help hotels reduce energy, water and waste (Băltescu, 2017). The system has four levels as outlined in Table 6.1.

More specifically, the system:

- Sets and tracks property-specific reduction goals for carbon, energy, water and waste.
- Uses actual data to provide customized environmental performance benchmarking, taking into account hotel location, brand and outfitting.
- Recommends over 200 Green Solutions and provides case studies and implementation plans to reduce the impact of IHG hotels on the environment. This aligns to the UN SDG 12 on responsible consumption and production. The tool also demonstrates the cost savings that can be achieved by hotels when they implement the solutions.

Table 6.1 IHG's Green Engage system

Level	
1	Hotels have completed 10 best practice solutions to set them up for success and support them through activities that provide immediate energy and costs savings. These include actions such as tracking consumption data, setting up a property green team and installing energy-efficient lighting in guest rooms.
2	Hotels have really begun to see the benefits of sustainability on the property, and have taken steps to go above and beyond the basics and implement solutions such as sustainable purchasing and ingraining sustainability into the hotel operations.
3	Hotels have mastered the foundations of sustainability, and are embarking on large projects such as installing energy-efficient appliances and sustainable site management.
4	Hotels are leading hotels in the environmental sustainability area. They demonstrate leading and innovative approaches to being sustainable.

Source: IHG (2018b)

- Supports hotels to create environmental action plans and targets.
- Can achieve energy savings of up to 25% on average for those hotels achieving Level 3 certification, making the IHG hotels more cost-effective to operate and ultimately allowing them to improve the value of service offered to guests.
- Automatically feeds sustainability information about the IHG hotel to clients. In 2016, 54% of IHG business accounts asked for information such as carbon footprint and waste.
- Allows guests to make better informed purchasing decisions. By linking to the hotel booking site, guests can see what level of certification in the IHG Green Engage system the hotel has achieved (ltescu, 2017: 86–87).

Accor PLANET 21

Yilmaz and Yilmaz (2016) use the example of Accor Group and its CSR programme entitled 'PLANET 21', which covers 21 CSR initiatives from seven different domains. These initiatives are illustrated in Table 6.2. One of the key features of Accor's approach is the detailed and comprehensive research that has been done to calibrate the social and environmental footprint of their business operations (Accor, 2016a, 2016b). The research allows Accor to identify the percentage of impacts that are derived from their main activities (Table 6.3).

Table 6.2 Seven pillars and associated goals of PLANET 21

Pillar	*Goal*
Health	Ensure healthy interiors
	Promote responsible eating
	Prevent disease
Nature	Reduce water use
	Expand waste recycling
	Protect biodiversity
Carbon	Reduce energy use
	Reduce CO_2 emissions
	Increase the use of renewable energy
Innovation	Encourage eco-design
	Promote sustainable building
	Introduce sustainable offers and technologies
Local	Protect children from abuse
	Support responsible purchasing practices
	Protect ecosystems
Employment	Support employee growth and skills
	Make diversity an asset
	Improve quality of work life
Dialogue	Conduct business openly and transparently
	Engaged franchised and managed hotels
	Share commitment with suppliers

Source: Accor (2013a)

Sustainable supply chain management

Many of the major hospitality companies have embraced sustainable supply chain management as part of their 'whole of system' approach to managing and certifying

Table 6.3 Impacts of the main activities of hotels

Impact	*1st Contributor*	*2nd Contributor*	*3rd Contributor*	*4th Contributor*
Energy consumption	Onsite energy use (75%)	Air conditioning (12%)	Laundry (7%)	Food & beverage (6%)
Water consumption	Food & beverage (86%)	Onsite water use (11%)	Air conditioning (2%)	Onsite energy use (1%)
Waste production	Construction & renovation (68%)	Onsite energy use (26%)	Operating wastes (5%)	Air conditioning (1.1%)
Greenhouse emissions	Onsite energy use (66%)	Food & beverage (14%)	Air conditioning (12%)	Employee travel (8%)
Water Pollution	Food & beverage (94%)	Onsite water use (4%)	Laundry (2%)	Air conditioning (0%)

Source: Accor (2016a, 2016b)

sustainability. Font *et al.* (2008: 260) observe that sustainable supply chain management, which they describe as 'the trend to use purchasing policies and practices to facilitate sustainable development at the tourist destination' is especially important for tourism operators who act as intermediaries between suppliers and tourism demand. Communication is a key feature, the authors contend, between sustainability and quality in increasing industry and market awareness for marketing and training, within all sectors of the industry.

Xu and Gursoy (2015) undertook a comprehensive review of the literature on hospitality supply chain management towards a more sustainable approach. Their review is divided into three main categories, which reflect the three main pillars of sustainability: the environmental dimension of sustainable supply chain management; the social dimension; and the economic dimension.

Singh Verma (2014) writes that green supply chain management has received growing attention around the world as consumers increasingly ask questions about the environmental impacts of the products they purchase. Typically these questions relate to manufacturing processes, supply chains, carbon footprints and recycling programmes. Following Hervani *et al.* (2005), Singh Verma (2014) developed a conceptual framework for analysing green supply chain management practices in India's hospitality industry, based on green procurement, green design, green manufacturing, green operations and reverse logistics, and waste management. Table 6.4 uses the example of green procurement, which is defined as environmental purchasing consisting of involvement in activities that include the reduction, reuse and recycling of materials.

Table 6.4 Green procurement practices in the Indian hospitality industry

The Taj Residency
• All the hotel's serviettes, tissues, toilet tissue and paper towels are made from 100% recycled paper.
Hotel Bangladore
• All spa beauty products are natural and preservative free.
• The hotel uses eco-labelled weedicide and fungal, rodent and insect killers.
ITC chain of hotels
• Many properties operate on wind energy – the largest self-owned wind farms by any hotel chain in the world.
• LED lights are used in the guest rooms and public areas for efficiency.
• 60% of room stationery and consumables used are either sourced locally, certified or with recycled content.
• Low-volatile organic compound paints and certified wood are used in guest rooms and public areas.
Lemon Tree chain of hotels
• LED lighting in public areas.
• Cen Tra Vac chillers in air conditioning systems that reduce the energy consumption to 0.75 KW/ton.
Oberoi Udaivil
• The resort uses a traditional and natural lime wash for all interior and Udaipur exterior walls.
• Bottled drinking water is made domestically at a centralized facility by Oberoi hotels which reduces greenhouse gas emissions from imports.
• Solar panels on rooftops are used to heat the resort's pools.
• Waste heat is recaptured from the chillers and used to heat hot water.
Our Native Village Hotel Bangladore
• All soaps and shampoos in the rooms are 100% natural, chemical free and handmade exclusively for the hotel.

Source: Singh Verma (2014)

MAXIMIZE SOCIAL AND ECONOMIC BENEFITS TO THE LOCAL COMMUNITY AND MINIMIZE NEGATIVE IMPACTS

Whitbread, the UK pub and restaurant chain, has created 'Force for Good – a New Direction' to guide values and to provide a clear code of conduct and robust ethical principles to underpin the company's operation (Whitbread, 2018). Part of the initiative focuses on strengthening the communities where they operate and reflects SDG 8 on decent work.

In addition, the company is committed to the wellbeing of its employees and the creation of 'decent work', including upskilling, recruiting and empowering employees and providing opportunities for people with no employment, education or training. They are encouraged to volunteer in their local communities and fundraise for charities (including Great Ormond Street Hospital Children's Charity in London, and Costa's dedicated charity, the Costa Foundation). In 2012, Whitbread created WISE (Whitbread Investing in Skills and Employment), a recruitment, training and educational scheme, to speed up, scale up and join up Whitbread's engagement with the education system. Of course, this investment makes business sense as it ensures a pipeline of talent to enable planned growth. Whitbread also ensures that these principles operate throughout the value chain by supporting suppliers to invest in developing the skills of their own workforces. In terms of workforce, most major hospitality companies have now initiated diversity and inclusion in their operations and the supply chain, and are adopting human rights principles in their HR operations. This includes being aware of the dangers of modern slavery and human trafficking. Marriott International (2017a), for example, has developed three key goals in this area:

1. By 2025, 100% of associates will have completed human rights training, including on human trafficking awareness, responsible sourcing and recruitment policies and practices.
2. By 2025, Marriott will enhance or embed human rights criteria in recruitment and sourcing policies and work with the sector to address human rights risks in the construction phase.
3. By 2025, to promote a peaceful world through travel by investing at least $500,000 in partnerships that drive, evaluate and elevate travel and tourism's role in cultural understanding.

Another hospitality company, the Starbucks Corporation (www.starbucks.com) has developed an open-source approach to supplies, involving over 2 million farmers, investing in their communities and purchasing to increase the prosperity and resilience of the communities and sharing farming knowledge and expertise. Since 2005, the Starbucks Foundation has awarded origin grants to support farming families in coffee- and tea-growing communities (Starbucks Corporation, 2018).

MAXIMIZE BENEFITS TO THE ENVIRONMENT AND MINIMIZE NEGATIVE IMPACTS

1. Conserving resources

The hospitality sector is a voracious consumer of resources, particularly of energy and water.

The importance of the environment is also recognized by hotel executives in the USA, who stated that the highest performing CSR initiatives were in energy, waste and water management. The greatest benefits of CSR implementation to these executives included cost savings and branding-related (e.g. reputation and image) outcomes (Levy & Park, 2011).

There are a number of initiatives to reduce consumption, including measures to tax the user. Here, the well-cited example of the problems of implementing measures to control over-exploitation of resources is the Balearic accommodation eco-tax law, which 'is a tourist, extra-fiscal, finalist, earmarked and direct tax' (Ariño, 2002: 172). Every tourist client visiting an accommodation unit in the islands pays tax directly.

Attempts to reduce consumption of resources are subject to performance monitoring and certification. For example, Nelson (2010) investigated the accommodation sector in Australia that has been eco-certified by Ecotourism Australia. She found that although energy is a key issue contained within eco-certification guidelines in Australia, less than half of the 50 accommodation units studied provided any information on energy on their websites. Those that did discussed issues around reducing energy use, improving energy efficiency, using alternative forms of energy and offsetting emissions. Nelson concludes by suggesting that more emphasis is being placed on facilitating nature-based experiences in these enterprises than on minimizing environmental impacts (see also Lai & Shafter, 2005; Warnken *et al.*, 2004).

Nonetheless, there are many examples of organizations that are attempting to reduce consumption. It has been identified that food production has a large environmental impact, considering the energy, water and habitat loss involved. The hospitality sector wastes large amounts of food and in 2017 the World Wildlife Fund (WWF) partnered with the American Hotel & Lodging Association in a project to reduce food waste in hotels. The project initiated a number of pilot programmes along the food waste supply chain, including measuring food waste outputs, improving employee training programmes, creating menus designed to limit food waste and raising awareness with customers (www.worldwildlife.org/initiatives/foodwaste).

In terms of water usage, IHG has established a water stewardship scheme and between 2013 and 2017 achieved a 5% reduction in water use per occupied room in water-stressed destinations (www.cr.hilton/environment). Hilton has launched water commitments applying a value chain approach across hotel operations, supply chains and communities. The water commitments include:

- context-based water pilot programmes in collaboration with WWF to promote stewardship in high water risk areas in the USA, South Africa and China;

- use of low-temperature laundry technologies that can deliver 40% water reduction and 50–75% energy savings with every wash;
- a global water risk assessment for all hotels using WWF's Water Risk Filter;
- launching a water-awareness training video for employees; and
- reducing water consumption by 20.0% per square foot since 2008.

In terms of reducing energy use, Hilton has developed a number of initiatives including:

- a high level of documented energy reduction (22.7%) across US hotels;
- a comprehensive lighting retrofit programme, reducing power by as much as 92% in many areas;
- installation of LED bulbs;
- upgrading building automation systems to improve system programming to control energy usage;
- replacing conventional light switches with switches that have built-in occupancy sensors and a built-in LED nightlight in guest bathrooms; and
- replacing thermostats in rooms with ones with a built-in occupancy sensor.

2. Reducing pollution

Through its sustainability initiatives, the hospitality sector is working to reduce pollution from its operations. This includes waste material as well as GHG emissions (see SDG 13 on combating climate change). For example, Marriott (2017b) has set a goal to reduce GHG emissions per square metre by 30% from 2016 to 2025. Marriott is looking to increase its commitment to renewable energy, including on-site generation. The company is installing low-carbon or renewable energy systems, such as geothermal, wind and solar. Starbucks is also looking to invest in renewable energy with direct investments in new geographically relevant renewable energy projects including both solar and wind farms.

The issue of plastic pollution is one where take-away hospitality operations have come under considerable criticism for the single-use packaging they use, and also the use of plastic drinking straws. Starbucks, for example, owns and operates over 28,000 stores worldwide and has developed the company's Greener Stores Framework, with six key standards including recycling for coffee cups, use of strawless lids and the consequent infrastructure needed, discounts to customers who bring their own cups, and waste management tactics.

3. Conserving biodiversity, ecosystems and landscapes

We introduced the concept of biodiversity in Chapter 2. For accommodation, the IUCN (2008) has identified the various stages of an accommodation unit's life cycle and outlined the implications for biodiversity at each stage, reflecting SDG 15, life on land. At the planning stage it is important to take care in the siting and design of the unit to avoid 'biodiversity-sensitive areas'. Similarly, local design and care with the use of building materials will determine the 'physical footprint' of the accommodation unit. During construction, land clearance can be a significant issue, as well as the materials used and the amount of water and energy taken from the local environment. The housing of workers and disposal of waste can also have major impacts upon biodiversity. Once the accommodation is operational, the level of energy, water, food and other resources are key issues. In addition, the disposal of waste, the management of the estate and purchasing choices all have to be managed with conservation of biodiversity in mind. Here care has to be taken to source sustainable food supplies, manage the estate planting and build a strong relationship with the local community.

The IUCN (2008) has identified a number of business benefits for accommodation units to conserve biodiversity by implementing good environmental practices in their operations.

- playing to environmentally concerned consumers;
- reducing costs;
- improving the quality of the destination;
- improving employee productivity and sense of responsibility to the environment;
- securing a hotel's 'licence to operate' by demonstrating a sense of responsibility; and
- attracting investment from socially responsible investors.

TRANSPORTATION

Transport has significant environmental implications – particularly in terms of carbon emissions – and it is important that tourists understand the consequences of their transport choices as part of SDG 13 on combating climate change. We deal further with transportation and carbon emissions in Chapter 7. Boniface *et al.* (2012) state that each transport mode has different operational characteristics, based on the different ways in which technology is applied to the four elements of a transport system. Technology determines the appropriateness of the mode for a particular type of journey and each mode has different environmental impacts. Governments and international agencies are pressing to make transport more sustainable. It is now recognized that tourism transport accounts for around 5% of the world's carbon emissions. However, it is estimated that around 80% of these emissions are caused by about 20% of all trips, and that these are mostly long-haul trips (Peeters *et al.*, 2007). Governmental responses are through legislation to encourage cleaner technologies, particularly hydrogen-based transport, in the face of dwindling oil reserves. Passengers using the bullet train in Japan, for example, receive a note of their carbon emissions comparing it to less environmentally friendly forms of transport such as flying.

AIR TRAVEL

Air travel has significant negative environmental impacts, both internationally through its contribution to climate change and locally through impacts on noise and air quality. The true cost of air travel for the airlines and their passengers has been masked by the fact that the airlines, unlike other business enterprises, have been exempt from certain taxes, although this situation is changing with the onset of passenger and environmental taxes. The continued growth of air traffic, particularly long-haul, may not be sustainable given aircraft emissions. For example, it has been estimated that a passenger on a single trans-Atlantic flight contributes as much to global pollution as a motorist driving 16,000 km in a year. The International Air Transport Association (IATA) estimates that air transport accounts for 2% of global manmade CO_2 emissions. Each new generation of aircraft is on average 20% more fuel efficient than the model it replaces and airlines have continued to improve their fuel efficiency performance between 2009 and 2016 (by 10.2% over the period).

However, some argue that this rate of improvement and the initiatives taken by the airlines have not kept up with the growth of demand for air transport. Indeed, the authorities have taken a light touch in the regulation of issues such as emissions, relying instead on the industry to develop more efficient aircraft. While this is happening to a certain extent, the fact that there is no tax on kerosene means that airlines are not pressured to use 'sustainable aviation fuel' which is almost twice the price of kerosene.

Aviation and emissions

Gössling (2002) provides a detailed analysis of the energy costs associated with long-haul flights from developed to lesser developed countries. Focusing on aviation, Gössling illustrates that for a typical two-week vacation in a developing country about 76% of energy demands are used for air travel. However, given that aviation emissions released into the troposphere and lower stratosphere have a larger impact on climate change, because of longer residence times, low temperatures, low background concentrations and greater radiative sensitivity (Schumann, 1994, cited in Gössling, 1999a), they have a greater impact on climate change. Based on these greater atmospheric impacts, Gössling speculates that air travel represents about 90% of the holiday's general contribution to global warming. Complicating these figures is the statistic that only around 5% of humanity participates in air travel (Schallaböck & Köhn, 1997).

The international aviation industry has prioritized the reduction of GHG emissions through the targets set by IATA:

- an average improvement in fuel efficiency of 1.5% per year from 2009 to 2020;
- a cap on net aviation CO_2 emissions from 2020 (carbon-neutral growth); and
- a reduction in net aviation CO_2 emissions of 50% by 2050, relative to 2005 levels.

In 2016 the International Civil Aviation Organization (ICAO) and its Member States developed a global market-based measure framework to guide the industry in meeting the targets set by IATA. In order to achieve these emission targets, IATA has devised a four-fold approach:

- improved technology, including the deployment of sustainable low-carbon fuels;
- more efficient aircraft operations;
- infrastructure improvements, including modernized air traffic management systems; and
- a single global market-based measure, to fill the remaining emissions gap.

Within this framework we can identify a number of specific and concrete actions that are being taken by those involved in air transport (including the passengers). These are:

- carbon offsetting;
- limiting engine emissions;
- airline cabin waste; and
- noise.

Carbon offsetting

Carbon offsetting is an action designed to compensate for emissions by financing a reduction in emissions elsewhere. Offsetting and carbon markets are a fundamental component of global, regional and national emissions reduction policies. In 2016, ICAO adopted a global offsetting scheme to address CO_2 emissions from international aviation – the Carbon Offsetting and Reduction Scheme for International Aviation) (CORSIA; IATA, 2018). Offsets come from different activities including wind energy, clean cook stoves, methane capture and other emissions-reducing or avoidance projects.

However there are a number of issues associated with offsetting. For example, ICAO can only legislate for international air travel and has no authority over national policies. Country-specific actions fall under the scope of the UN Framework Convention on Climate Change and the Paris Agreement. This means that there is a danger of a patchwork of uncoordinated regional and domestic policies for international aviation. Indeed, the success of initiatives such as CORSIA may be jeopardized by countries applying carbon pricing instruments or ticket taxes to address emissions.

Gössling (1999b) discusses the value of afforestation programmes that are designed to offset the loading of carbon in the atmosphere. He contends that these sequestration programmes should be viewed as important only in as far as how they allow for the buying of time for other mitigating strategies to take effect in curbing the effects of global tourism. The reason he is not more supportive of these programmes is because they cannot compensate for atmospheric changes in air chemistry and gas composition, they cannot address the problem of the acceleration in the use of fossil fuels, and they cannot meet the scale of fossil fuel loading in the atmosphere.

Limiting engine emissions

IATA states that the emissions from aircraft engines that affect air quality are nitrogen oxides (NO_x), carbon monoxide (CO), sulphur oxides, unburned hydrocarbons (HC), smoke and particulate matter. Improved engine designs have gradually reduced the emissions of NO_x and CO and have almost completely eliminated emissions of unburned HC and smoke. Aircraft engines have to meet mandatory certification requirements established by ICAO. In addition there are other local emissions from airport activities and the road traffic generated.

Airline cabin waste

We introduced the importance of waste management in Chapter 2. Airlines are adopting schemes to reduce, reuse and recycle cabin waste including single-use plastics and food waste. This is because the airline industry has been cited as problematic when it comes to the handling of waste. Arp *et al.* (2018) investigated the waste management practices of airlines, and found that some airlines transport over 50 million passengers per year, and most of their inflight meals and beverages are served in or wrapped with plastic, such as cups and snacks, and also include other items like plastic-wrapped blankets and headphones. Arp *et al.* (2018) point out that disposal takes place at departure airports, and those regions with poor waste management facilities become overburdened or waste simply ends up in the ocean. Other scholars have studied the local citizen perceptions of poor waste management practices. For those places with high levels of uncollected waste, there is an overwhelming sense that the presence of this waste will be off-putting to tourism and a significant threat to public health (Little, 2017).

Noise

Governments are concerned to minimize aircraft noise. Actions to limit noise include monitoring, limiting flying times and airspace, and legislation in terms of acceptable noise limits. The 2014 EU Environmental Noise Directive requires noise action plans to be drawn up by Member States addressing the main sources of noise, including aviation, with the aim of reducing the impact of noise upon affected populations (EEA, 2016). Progress has been made in reducing the noise made by jet aircraft over the years with an average reduction of 4 decibels per decade. However, this rate is slowing and, bizarrely, some of the new generation of aircraft engines may actually increase noise, specifically the counter-rotating open rotor that is due to enter service around 2030.

Airports

When we think of sustainability, airports are often overshadowed by aircraft operations. Yet airports have a significant environmental footprint and impact on their local communities. In the USA, the Federal Aviation Authority (FAA) has established programmes of noise control, emissions control and master planning at US airports. Master plans fully integrate sustainability into airport planning. These plans are facilitated with FAA grants and include initiatives for reducing environmental impacts, achieving economic benefits and increasing integration with local communities. They use baseline assessments of environmental resources and community outreach to identify sustainability objectives. Similar airport management approaches exist across Europe, augmented by certified environmental or quality management systems.

Airline sustainability in action

Airlines are now adopting many of the initiatives outlined in the above sections, and in the Norwegian Airlines Case Study in Chapter 1. In the section below we outline the environmental and community action of two airlines, one in Australia and one in Europe.

Virgin Australia

Virgin Australia is a budget airline that is taking innovative approaches to environmental policy (www.virginaustralia.com). The company estimates that emissions associated with its jet fuel account for 98% of their total emissions footprint. This is being tackled in three ways:

- trialling sustainable aviation fuel;
- introducing fuel efficiency programmes, monitoring and reducing fuel use by using single-engine taxiing; and
- cooperation with key partners such as weather forecasters to enable fuel savings.

Other sustainability initiatives by Virgin Australia include:

- *Preserving Tasmanian forest*: 30,000 tonnes of CO_2 emissions are offset annually, helping to preserve more than 28,000 ha of forest. In addition, passengers can offset their emissions.
- *Upcycling crew uniforms*: Crew uniforms are donated through the LOOP programme where they are remade into pillows, blankets and teddy bears for charities and shelters.
- *Reducing food waste*: Virgin Australia donates over 8 tonnes of food to OzHarvest each month, providing over 16,000 meals to people in need.

Ryanair

Ryanair is a major low-cost airline carrying over 130 million passengers in Europe (www.ryanair.com). The sheer scale of the airline's operation means that it has a high carbon footprint. To combat this it has developed a comprehensive environmental policy encompassing governance and reporting (Ryanair, 2018). Low-cost carriers such as Ryanair have a business model that facilitates environmental sustainability as they operate point-to-point services, thus reducing noise and emissions, and have to operate in the most fuel-efficient way possible. Ryanair, for example, is ranked as the most efficient world airline in terms of emissions per passenger mile. Ryanair is committed to reducing its global impacts (including emissions) and local impacts of noise and air quality. The airline's chief operating officer has direct responsibility for environmental risks and impacts, reporting directly to the Board. The Environmental Policy has five key pillars:

1. Addressing GHG emissions: Flying point-to point with high load factors using fuel-efficient aircraft. The policy includes single engine taxiing, use of solar power, LED lighting, and use of ground power units instead of the aircraft's power on the ground.
2. Moving towards the use of alternative fuels.
3. Working with designers to deliver a step change in fuel-efficient, next-generation aircraft design including drag and noise reduction.
4. Carbon offsetting as a voluntary scheme for passengers.
5. Elimination of non-recyclable plastic in operations and in the supply chain.

SURFACE TRAVEL

Surface travel is by far the most common form of tourism transport for domestic travel and for short international trips. Of course, the most sustainable forms of land transport are walking, cycling and horse riding and for niche products such as ecotourism, these forms of transport are increasingly popular with dedicated suppliers (Dolnicar *et al.*, 2010).

Road transport

Traditionally, road transport has been a significant polluter, particularly through old diesel engines. Over half of the world's air pollution comes from road transport. However, the 21st century has seen a move away from mainstream road transport companies towards sustainable road transport, particularly in the form of electric or hydrogen-powered cars, with estimates that electric cars will have largely replaced petrol and diesel by 2030 (Prideaux, 2018). For example, in 2018 the city of Madrid introduced a far-reaching new scheme to ban polluting vehicles from the city centre, called 'Madrid Central' (http://urbanaccessregulations.eu/countries-mainmenu-147/spain/madrid-access-restriction). Other measures to reduce the use of cars include:

- park and ride;
- improving access to bus, tram, light rail and train services;
- road pricing or congestion charges;
- taxing cars according to their carbon emissions;

- development of 'zero' emission and hybrid cars;
- development of autonomous electric vehicles to facilitate urban planning and reduce pollution; and
- development of disruptive innovations such as car ownership sharing.

More broad ranging sustainability initiatives are being developed by the technology taxi disrupter Uber (https://www.uber.com/in/en/). Uber is using its data harvested from trips taken to facilitate the transformation of cities into safer, more efficient and more beautiful places by helping urban planners design cities of the future (Uber, 2019). This involves mitigating traffic congestion and improving infrastructure, making driving safer and preventing accidents.

The Greyhound Bus Company estimates that bus travel has the lowest carbon footprint of any motorized mode of transport, with an 85% reduction over the same journey by car for one person (www.greyhound.com). Similarly, bus travel reduces emissions by 55–77% over the same journey by air. All new Greyhound buses are built with the latest emissions-reducing parts and refurbished buses have upgrades to increase engine efficiency and fuel economy. Like many other bus and coach companies, Greyhound are reconfiguring their buses to reduce pollution and emissions. For example their buses use:

- low-sulphur fuel;
- diesel particulate filters; and
- idle management systems.

At their bus stations and offices Greyhound are saving energy, cutting out single-use plastic, implementing environmentally friendly landscaping, recycling cooking oil, recycling electronics, reducing water waste and installing motion-activated lights and motion-activated paper towel dispensers.

Rail transport

Environmental pressures combined with technological innovations such as high-speed trains allow rail transport to compete with aircraft on short- to medium-haul routes. Eurostar, for example, is marketed as the 'green' alternative to flying (www.eurostar.com), while in France, China and Japan there has been considerable government investment in applying new technology to the development of high-speed trains and upgrading trunk lines between major cities. It is estimated that trains are four times more fuel efficient than trucks, reduce highway congestion, lower GHG emissions and reduce air pollution (Canadian National, 2016). New technologies are also reinforcing the green credentials of rail travel, including electric-powered high-speed trains and Maglev technology.

The UK has developed a set of sustainable development principles for rail travel which recognize the need for an efficient, integrated rail system, but which also incorporate new knowledge and thinking about rail travel (RSSB, 2016):

- *Customer-driven*: Embed a culture where dialogue with customers puts them at the very heart of the railway, and where they are able to make optimal travel and logistics choices.
- *Putting rail in reach of people*: Position rail as an inclusive, affordable and accessible transport system through the provision of information and accessible facilities.
- *Providing an end-to-end journey*: Work together with all transport modes to provide an integrated, accessible transport system.
- *Being an employer of choice*: Respect, encourage and develop a diverse workforce, support its wellbeing and actively consider and address the challenges of the future labour market.
- *Reducing environmental impacts*: Operate and improve the business in a way that minimizes the negative impacts and maximizes the benefits of the railway to the environment.
- *Carbon smart*: Achieve long-term reductions in carbon emissions through improved energy efficiency, new power sources and modal shift.
- *Having a positive social impact*: Focus on local impacts and communities through better understanding and engagement.
- *Supporting the economy*: Boost the productivity and competitiveness of the economy through efficient services and by facilitating agglomeration and catalysing economic regeneration.
- *Optimizing the railway*: Maximize rail's capability, build on its strengths and improve efficiency to deliver a transport system that is resilient and offers good value for money.
- *Being transparent*: Promote a culture of open and accountable decision making and measure, monitor and report publicly on progress towards sustainability.

A good example of a railway company that is delivering on sustainability is the Canadian National Railway Company (CN; www.cn.ca). The company has adopted global best practice in reporting and verification of data and aligned its five sustainability initiatives with the UN SDGs. The company bases its sustainability work on innovation through technology and its people (including communities) (CN, 2016, 2018):

- Transitioning to a low-carbon world:
  - decoupling growth from carbon emissions;
  - using renewable fuels;
  - using green energy.
- Innovation for sustainable future – environment:
  - protecting biodiversity and managing land;
  - community eco-grants;
  - conserve resources.

- Innovation for sustainable future – people:
  - become a top employer;
  - employee engagement and innovation;
  - transform teaching and learning;
  - recruit top talent.
- Communities (CN, 2018):
  - prioritize safety;
  - Aboriginal outreach;
  - policing;
  - community development grants.

Innovation and contemporary thinking is also evident in CN's commitment to the circular economy by piloting projects to extend the lifespan of their products through recycling. For example, steel rail tracks and locomotives are reused from the main lines to secondary lines and then at CN's yards, and finally sold to be recycled into new steel products; wooden rail ties are sent for co-generation of electricity.

Box 6.2 Public transport and sustainable tourism

Tourism and leisure travel tend not to be heavy users of public transport, partly due to the fact that routes and schedules are often inconvenient for destinations and do not provide the door-to-door convenience of, say, car travel (Gronau & Kagermeier, 2007). This lack of convenience is particularly acute for rural and peripheral coastal destinations (Le Klahn & Hall, 2015). However, it is important to encourage tourists to be more frequent users of public transport as we transition towards a low-carbon economy. For tourists, public transport is shared with other users such as commuters and it has a role in reducing the use of fossil fuels, encouraging modal switching from the car and reducing congestion, particularly in major tourism cities such as London or Hong Kong (Public Transport & Tourism in Ireland ITIC Dublin).

The key mechanisms for improving public transport to encourage tourism use include:

- integration of transport networks to provide interconnections across terminals and modes of transport (e.g. coordinated bus, rail and ferry links in Sydney, Scotland and British Columbia);
- specific ticketing options for tourists including multi-modal ticketing;
- marketing and promotion, particularly of scenic and themed routes;
- deeper understanding of the barriers to tourists using public transport, including lack of information, too many transfers, complex tourism schedules, comfort, cleanliness and inflexible schedules/routes;

- use of technology and online purchasing and information systems;
- integration of transport routes and schedules with tourism facilities and attractions; and
- development of train connections to support regional tourism development as we see in both China and Spain (Le Klahn & Hall, 2015).

WATER-BORNE TRANSPORT

It is generally accepted that, while shipping is relatively safe and clean compared to other transport modes, the industry does have a significant impact on the environment (WWF, 2012). Shipping is subject to less stringent environmental demands than land-based transport. The international agency with oversight of shipping is the Marine Environmental Protection Committee of the International Maritime Organization (www.imo.org), using legislative instruments, codes and guidance. The IMO is the UN's agency responsible for the safety and security of shipping and the prevention of marine pollution by ships (WWF, 2012).

As well as statutory agencies with oversight of marine sustainability, there are also voluntary initiatives that attempt to go beyond legal compliance with environmental regulation to work towards SDG 14, life below water. For example:

Green Star

Green Star is a scheme from the Royal Institution of Naval Architects (RINA; www.rina.org). The scheme has both a clean sea and a clean air element and has been particularly popular with cruise companies. The scheme helps companies to incorporate low-emission gas turbines, advanced waste management systems, protected fuel tanks and the use of non-toxic anti-fouling hull coatings.

Green Marine

Green Marine (www.green-marine.org) is an environmental certification programme for the North American marine industry. It is a voluntary, transparent and inclusive initiative that addresses key environmental issues through 12 performance indicators. Certification is based on benchmarking annual environmental performance through the programme's self-evaluation guides, external verification and publication of results.

The Sustainable Shipping Initiative

The Sustainable Shipping Initiative (SSI) is a multi-stakeholder initiative that brings together like-minded and leading organizations with shared goals and equal determination to improve the sustainability of the shipping industry in terms of social, environmental and economic impacts (https://www.ssi2040.org). The SSI roadmap provides a clear overview of the macro-environment that a sustainable shipping industry will require by 2040. The six core areas of the roadmap are:

1. proactively contributing to the responsible governance of the oceans;
2. earning the reputation of being a trusted and responsible partner in the communities where shipping interacts;
3. providing healthy, safe and secure work environments, so that people want to work in shipping, where they can enjoy rewarding careers and achieve their full potential;
4. instilling real transparency and accountability within the industry to drive performance improvements and enable better, sustainable decision making;
5. developing financial solutions that reward sustainable performance and enable large-scale uptake of innovation, technology, design and operational efficiencies;
6. changing to a diverse range of energy sources, using resources more efficiently and responsibly, and dramatically reducing GHGs.

Cruising

Cruising has a poor environmental reputation with numerous issues. This is partly because cruising involves large concentrations of tourists at destinations for a small period of time with consequent impacts on the economics, environment and communities of those destinations, particularly in terms of fragile destinations such as the Galapagos or Antarctica and historic cities such as Venice and Dubrovnik (Ponton & Asero, 2018). The ships themselves also create many negative environmental and social issues. These include smokestack emissions, disposal of waste and waste water, sewerage, air pollution, anchor damage on the sea bed and coral reefs, and poor human resources practices (sexual harassment, poor wages and conditions, and discrimination leading to some being termed 'sweatships'). In addition, cruise companies are only now catching up with good CSR and reporting practice. For example, de Grosbois (2016) found that the industry demonstrated limited use of international reporting guidelines and an almost universal absence of third-party assurance of reported information. More specifically, websites were unclear in their presentation of CSR information with little reference to time frames and the scope and sources of information. And while information on environmental and community issues were addressed, industry economic benefits, employment equity, and diversity and accessibility were not at all well represented on the websites.

Nonetheless, as Klein (2011) observes, the cruise line industry can be more sustainable through a focus on the responsible tourism agenda (see also Lück *et al.*, 2010). For the industry, it means building in greater transparency over environmental practices, distributing and sharing benefits more evenly with other shore excursion providers, and better assessing the carrying capacity of ports so that cruise passengers do not overwhelm these areas. In reference to carrying capacity, itineraries need to be fashioned that stay within the limits of the human and ecological conditions that exist at ports.

There are examples of companies that are taking environmental responsibility more seriously. P&O, for example, launched 'Fathom', where passengers worked alongside NGOs and locals on environmental and social projects while the Holland America Line has been a pioneer of sustainable cruising, as we see in the case study in Box 6.3.

Box 6.3 Environmental and socially sustainable cruising – the Holland America Line

Introduction

Cruise tourism has shown remarkable growth since the 1990s. While Johnson (2002) clearly states that the health of the world's oceans is critical to the future of the world, we are seeing an increase in forms of tourism that utilize the oceans as a resource. Here, cruising has become the target of environmentalists, although in the 21st century there have been substantial improvements in their environmental practices. Johnson outlines the environmental consequences of cruising as:

- the coastal infrastructure needs of cruising;
- cruise operations in terms of waste disposal, energy and water consumption and carbon emissions;
- impacts of transferring tourists to the cruise liners, often by air; and
- social, cultural and economic consequences of cruise visitors at the destination.

Johnson (2002) concludes that while cruising allows large numbers of visitors to enjoy the ocean as a resource, if cruising is to be sustainable it needs to:

- implement long-term integrated planning at the international level;
- invest in and promote environmental good practice;
- ensure that destinations are protected from 'mobile' cruise tourism;
- consider profit sharing between cruise line shareholders and host destinations; and
- raise cruise tourists' environmental awareness.

Holland America Line

The Holland America Line is a cruise company at the forefront of implementing environmentally and socially sustainable cruise tourism, conscious of the need for stewardship of the oceans. The Holland America Line (2019) is committed to:

1. safeguarding the health of guests, crew, contractors and shore-side employees;
2. protecting the environment, and using resources efficiently and sustainably;

3. operating all ships and land-based assets safely to prevent damage, injury or loss of life; and
4. ensuring the security of all.

The company is continuing to reduce its environmental footprint through a number of innovative practices including:

- Implementing an environmental management and reporting system to integrate sustainability into all aspects of the company's operations. Here the goals are to reduce fuel consumption, water use and refrigerant releases and implement more recycling. The management system ensures that the company adheres to international regulations for the environment.
- Including environmental duties into on-board employees' work plans.
- Energy and emissions management to ensure the ships are as efficient as possible, including efforts to reduce fuel consumption by optimizing speed and sharing best practice.
- Almost three-quarters of on-board water used is produced on board from seawater and condensation.
- Solid waste is managed carefully and the majority is not hazardous. It is recycled or disposed of on shore, incinerated on board or discharged to sea. Recyclable materials are separated and collected.
- The company is conscious of the need to safeguard biodiversity and so manages discharges from the ships. Other approaches include:
  – ensuring that ballast water, which could introduce invasive species into the ocean, is disposed of onshore;
  – only using sustainable seafood on board;
  – monitoring the paths of marine mammals to minimize striking them; and
  – limiting the time in Antarctica for each voyage.

Sources: Dowling (2006); Holland America Line (n.d.); Johnson (2002); Klein (2011); Lamers *et al.* (2015); Peisley (2006)

One cruise sector that has grown in popularity in the 21st century is river cruising. By 2016 it was estimated that 12 cruise lines were operating 185 river cruise boats in Europe (Prideaux, 2018). Here, the Travel Foundation (2013) has developed a guide to best practice for environmentally sustainable river cruising based on ten principles:

1. *Improve monitoring*: Establish easy-to-follow monitoring systems and continuously provide feedback on performance, in order to improve environmental sustainability.

2. *Reduce energy use* and generate energy from non-renewable fuel.
3. *Save water*: Reduce water consumption by using the best technology available and creating awareness among both guests and staff about water saving practices.
4. *Limit waste production*: Solid waste is one of the largest environmental impacts a river cruise produces.
5. *Foster a sustainability culture*: Sustainability efforts cannot be effective in the long term if they are not genuinely rooted in the company's philosophy.
6. *Define operating procedures*: Environmental considerations and criteria should be systematically integrated into standard operating procedures and clear responsibilities assigned to staff.
7. *Increase cooperation*: Cooperation with other river cruise companies, both formal and informal, can lead to an exchange of experiences, lobbying for better port facilities, negotiating with common suppliers and making progress in other areas of shared interest.
8. *Bring the supply chain in line*: Work with the entire supply chain to minimize impact on the environment.
9. *Incorporate innovation*: Incorporate state-of-the-art technology in fleets.
10. *Minimize absolute impact*: Focus on efficiency when considering sustainability indicators, for example, in terms of 'waste produced per guest night' or 'fuel spent per voyage'.

Finally it is important to recognize the significance of tourism-generated ferry traffic in terms of impact on the environment. Here there are two stand-out examples:

1. The Canadian company, British Columbia Ferries, is committed to preserving and protecting the environment (www.bcferries.com). The company recognizes the touristic value of the coast and has implemented sustainable operations to reduce their environmental footprint. This is done under the umbrella 'SeaForward' programme (BC Ferries, 2019) which is a programme that brings together BC Ferries' existing environmental activities, conservation projects, community investments and new sustainability endeavours. In terms of communities, BC Ferries operates a community investment programme and fundraising efforts designed to support programmes, services and organizations that improve the health and wellbeing of coastal residents.
2. In Scotland, Caledonian MacBrayne (CalMac) is a ferry company that not only acts as a significant form of tourism transport throughout the highlands and islands, but is also a community lifeline (www.calmac.co.uk). Rather like British Columbia, Scotland's environment is a significant tourist resource and this is recognized by the company. CalMac is committed to minimizing the impact on the marine and terrestrial environment and enhancing biodiversity and the quality of Scotland's natural environment (CalMac, 2018). They operate under the requirements of ISO 14001:2015 Environmental Management System, which identifies and mitigates the environmental risk of their operations. They are also in the process of introducing hybrid powered ferries.

ATTRACTIONS AND EVENTS

Visitor attractions and events are the *raison d'être* for tourism: they generate the visit, give rise to excursion circuits and create an industry of their own. As such they are the main motivator for travel, energizing the tourism system and providing tourists with the reason to visit a destination. However, attractions and events also generate consequences for the environment, society and economy, particularly in terms of water and energy, waste and relations with neighbouring communities (see Andersson & Lundberg, 2013). They therefore have to embrace sustainability in all its forms, from the type of transport used to reach them, through the management of local community relationships, to the notion of ethical trading in shops and restaurants. ST demands not only that the visitor receives a satisfying and high-quality experience, but also that the destination is sustainable. Innovative application of visitor management ensures that the increasingly experienced and discerning new tourist does indeed receive a high-quality experience at the attraction or event.

In terms of working towards sustainability, attractions and events are no different from other tourism sectors. Larger organizations are leading the way in sustainable operations by adopting the principles of the four 'Rs' – reduce, re-use, recycle and then replace, focusing in particular on GHG reduction, energy, water, waste, environmental conservation and rescue, community engagement, responsible sourcing and procurement. This reflects SDG 12 on responsible consumption and production, SDGs 6 and 7 on water and energy and SDG 13 on climate action. It involves putting in place management systems with objectives and targets, monitoring and complying with legislation.

STANDARDS

In terms of standards, the event sector has led the way with the adoption of ISO 20121 in 2012 which 'provides the framework for identifying the potentially negative social, economic and environmental impacts of events by removing or reducing them, and capitalizing on more positive impacts through improved planning and processes' (Getz & Page, 2016; ISO, 2012). The standard is premised on the sustainable development principles of inclusivity, integrity, stewardship, transparency, labour standards, human rights and legacy (Raj & Musgrave, 2009).

GREENHOUSE GAS EMISSIONS

All attractions and events use fuels and energy for their operation (see, for example, Mair & Jago, 2010). There are a number of examples where attractions are working towards zero emissions. For example, in North America, the Aquarium of the Pacific in Long Beach, CA was one of the first aquariums to construct a carbon-neutral building. Other examples include Taronga Zoo, Sydney, Australia and the large attractions corporations such as Disney:

1. Taronga Zoo (https://taronga.org.au) is committed to pursuing a responsible model of environmentally sustainable management. Taronga Zoo's sustainability strategy has a primary goal to reduce its carbon footprint by more than 10% by 2020 and to achieve carbon neutrality by 2025. This goal is underpinned by objectives to integrate sustainability into Taronga Zoo's business areas. Taronga has developed a robust strategy to become carbon neutral by 2019. Originally this goal was set to be achieved by 2025; however, Taronga has now recommitted itself to being NCOS certified by 2019. This will be achieved through a combination of:

- reducing energy consumption;
- developing onsite renewable energy;
- embracing ecological sustainable design;
- using carbon offset credits;
- adopting waste management;
- sustainable sourcing and procurement; and
- care with animal nutrition.

2. Disney has focused on calculating its carbon footprint and is committed to reducing GHG emissions. They have set clear targets to reduce emissions and will achieve these by a series of emissions-reduction projects around the company. These include projects at the parks and on cruise ships, including energy efficiency measures, waste heat recovery, lighting upgrades, fuel cells, a geothermal well, employee education, feasibility studies for renewables and investment in forest carbon projects.

ENERGY

We introduced the importance of energy management in Chapter 2. Sustainable attraction operation is seeing many organizations moving towards on-site renewables and developing strategies to cut their energy consumption. For example, in both their Florida and California operations, Universal Studios has been reducing energy consumption through a number of approaches. These include expanding LED lighting fixtures, incorporating energy saving in designing their new rides, in Florida saving energy by using a system that provides cooling to all the buildings on the theme parks, as well as using solar power to pre heat water. This allows the chilled water plants to be operated at their most efficient levels with real-time automatic adjustments to the system based on building loads. The programme is averaging 30% energy savings for chilled water production at the resort. Universal Orlando has also implemented an experimental system that uses thermal solar to preheat water (www.universalorlando.com). An example of the use of renewable energy at attractions is found at SeaWorld (https://seaworld.com), including a new solar array.

WATER

We introduced the importance of conserving water in Chapter 2. Attractions are also reducing water use. Traditionally tourism is a significant consumer of water – Disney, for example, uses 8 billion gallons of water annually. Approaches include:

- recycling of water for irrigation control by monitoring weather conditions;
- waterless urinals;
- using recycled water in chilled water system cooling towers;
- capturing urban runoff and filtering water for reuse;
- water conservation plans and best practices for water management; and
- use of filtering systems to allow grey water to be used.

WASTE

Attractions are attempting to reduce waste and recycle or upcycle existing materials, with some attractions having a recycling facility that separates out plastic, glass and metals. Approaches include the collection of food waste, cardboard and recyclables like metal, glass, plastic and paper, pallet reuse and recycling, textiles collected for waste to energy, and material reuse programmes, as well as bins that encourage customers to pre-sort waste. Universal Orlando Resort, for example, collects food waste from more than 30 restaurants daily, which is placed in an onsite compactor and then sent to an anaerobic digester to generate energy. Disney uses thermal waste-to-energy facilities to manage otherwise unrecoverable waste, and in future aims to divert waste from both landfills and thermal waste-to-energy to achieve a long-term goal of zero waste.

Of course, a contemporary controversy has highlighted the consequences of the growing threat of plastics to our oceans and wildlife. Here, SeaWorld (2019) in the USA is reducing single-use plastics, and in some parks single-use plastic straws and plastic shopping bags have been eliminated.

ENVIRONMENTAL CONSERVATION AND RESCUE

A number of zoos and aquaria work closely with charities, universities, government agencies, conservationists and stranding networks to further conservation and rescue threatened or damaged species. For example, SeaWorld has a 24/7/365 team of animal care, veterinary and animal rescue experts. Their legacy of animal rescue goes back to the 1980s. The company has incorporated education into the visitor experience and uses this to inspire their customers to act in this sphere. Thayer (1990) has observed that aesthetic experiences must be supported by an educational component. What this means is that in understanding an attraction like a landscape, we would have a greater appreciation of that landscape if the experience were supported by a learning or educational message.

Additionally, the SeaWorld & Busch Gardens Conservation Fund has donated more than $16.5 million, supporting over 1200 animal conservation projects on all seven continents, including coral reef restoration, preventing coastal erosion, habitat protection and simply keeping our oceans clean.

Taronga Zoo is also committed to conservation and rescue, and 'protecting and conserving areas of historic or Indigenous significance' is enshrined in their strategy (see Chapter 10 for a section on animal ethics). The Zoo prioritizes ecological sustainability and is committed to reducing the impact of their operation through a process of continual improvement and integration of sustainability objectives. This is achieved through a 360° approach to conservation to inspire and educate visitors and the community. The Zoo organizes environmental days, litter clean-ups, the Taronga Mini Green Grants and environmental workshops in the local area through volunteer teams to care for the environment and engage the wider Taronga community to take action, participate in activities and become positive agents for change.

COMMUNITY ENGAGEMENT

Attractions and events have an impact on their neighbourhood communities and their employees. Community engagement can take a variety of forms and is commonly classified under CSR initiatives. Walt Disney (n.d.), for example, engages with issues as diverse as charitable giving and philanthropy through to human trafficking, forced labour and modern slavery. They are concerned with contemporary debates such as the ILO's 'decent work' agenda, child nutrition and healthy eating in their parks and encourage workplace practices such as volunteering. Taronga Zoo also focuses on its employees, educating them on behaviours and practices. The Taronga Green initiative is responsible for supporting employees in sustainable work practices and promoting sustainability in the workplace.

SOURCING AND PROCUREMENT

Larger organizations are ensuring the sourcing and procurement of sustainable products. This includes life cycle assessment of products such as food supplies, understanding supply chains and managing to reduce environmental and social risk associated with procurement. Many attractions now publicize organic food, Fair Trade suppliers, sustainably sourced food, including cage-free eggs, pork and chicken, locally sourced produce and sustainable seafood.

CONCLUSION

This chapter has outlined the sustainability initiatives and challenges facing the hospitality, transport and attractions industries. The chapter showed how the hospitality industry has been subject to a number of international initiatives to green the industry. It then went on to outline the major pillars of sustainability for hospitality in terms of CSR, 'whole of hospitality' initiatives, sustainable supply chains, community considerations and finally environmental issues. Environmental concerns are a real issue for the transport industry, particularly the aviation sector which is subject to considerable scrutiny for its GHG emissions and noise impacts. The chapter outlined a number of ways in which these are being tackled and detailed the plans of two airlines – Virgin Australia and Ryanair. Finally, the chapter examined the sustainability issues of attractions, including events. Here the main dimensions are GHG emissions, energy, water and waste, In addition, the chapter considered the conservation and rescue work of zoos and aquaria and finished by looking at community engagement and sustainable sourcing and procurement.

END OF CHAPTER DISCUSSION QUESTIONS

1. In class, discuss whether persuading passengers to reduce their annual air miles is a realistic option.
2. What communication messages would you use to persuade a small hospitality business to adopt sustainability initiatives?
3. Examining cruise company websites, to what extent can their environmental statements be accused of 'greenwashing'?
4. What are the main elements of 'green' events?
5. How effective do you feel Disney's sustainability initiatives are?

End of Chapter Case Study 6.1 The Burren and Cliffs of Moher Geopark, Ireland

Introduction

Geoparks are a network of parks with significant environmental value. They are managed by developing strong and sustainable relationships with all the stakeholders involved. The key for this type of attraction is therefore to ensure that the visitor experience is enhanced while also conserving the environmental values of the particular location.

A recent addition to the UNESCO list of Geoparks is The Burren and Cliffs of Moher Geopark in the west of Ireland. The Burren is a complex area, with sensitive issues embracing landscape, history, geology and archaeology. Because of this complexity, the project demands a multi-organization and partnership approach and therefore the management of stakeholders is an essential part of running the Geopark. This approach will ensure 'a cared-for landscape, a better understood heritage, more sustainable tourism, a vibrant community and strengthened livelihoods' (http://www.burrengeopark.ie/).

The Geopark is managed by Clare County Council and supported by a range of national bodies including the Failte Ireland (Ireland's national tourist board), geological agencies, nature conservation and planning agencies and universities. Core funding for the Geopark comes from Clare County Council, the Geological Survey of Ireland and Failte Ireland.

As a tourist attraction, the purpose of the park is:

> To spearhead sustainable tourism that develops and promotes the area as a truly special encounter-rich destination, strengthens the local economy and improves the visitor experience. (Geopark, 2019)

The aim of the Geopark is therefore to become a sustainable, vibrant and world-class attraction. According to the website, the key features of the Geopark are to:

- foster collaboration between all stakeholders to collectively develop and promote the Geopark as a ST destination;
- participate in conserving the natural and cultural heritage in accordance with the international standards;
- ensure high standards of communication and understanding of the unique character of the locale and its stories, emphasizing the particular attributes and strengths of the Geopark;
- build capacity in destination management and stewardship, focusing on enhancing the quality and standards of visitor experiences and tourism products and services;
- optimize tourism's potential as both an economic and a social development tool which benefits hosts as well as visitors;
- create strong economic benefits through product development, marketing and promotion, cost and energy savings, local sourcing and the creation of local employment.

The attraction

The Geopark combines two very different visitor and landscape experiences. The Cliffs of Moher is a spectacular coastal landscape, while the Burren is a larger landscape of limestone terrain. The Geopark encourages active visitation with a learning and experience element. The Geopark's management provides a range of educational and interpretive materials, trails which include food and cycling trails and local guides. The aim is for an immersive experience in the area's natural, cultural and local resources including surfing, kayaking, caving and experiencing local food and local music.

The Geopark is structured around nine Geosites, each with its own management plan and interpretive approach. Low-impact tourism is promoted through Leave No Trace Ireland – a not-for-profit company (http://www.leavenotraceireland.org/).

Integrating tourism into a sensitive landscape

The Geopark is actively working with local tourism enterprises through workshops, training courses and seminars. The training is designed for the nature of tourism on the Burren and specifically aims to increase professional expertise in two specific areas:

1. Reducing environmental impacts

While the aim of low-impact tourism is a laudable one, achieving this in practice is more difficult. The key to achieving the Geopark's objectives is to carefully integrate tourism activity with the natural environment, geology and archaeological features. This is done by working with tourism enterprises to reduce their potential impact and strengthen their capability to use natural resources and become resource and energy efficient by using renewable energy, waste reduction and reducing their carbon footprint. Tourism enterprises will be asked to think about their impacts on the environment and the economy and to align their products with conservation. This approach is very much along the lines of 'ecotourism' and the Geopark is hoping to 'mainstream' elements of ecotourism into more general tourism enterprises.

2. Enhancing economic impacts

The Geopark is also keen to enhance the positive economic impact of tourism enterprises for the benefit of the area. Enterprises are encouraged to cooperate and work with local suppliers and partners, so spreading economic benefit out to the community. Local sourcing can be used as a marketing advantage and builds on the immersive nature of the tourism experience in the Geopark. Here, initiatives such as the establishment of farmers' markets or working with local artists and craftspeople all helps to boost the spend by tourists. The area itself is isolated and most tourism enterprises are small, but also the share of imported materials is low so most of the expenditure stays in the area.

The Burren and Cliffs of Moher Geopark is an excellent case study of how to manage multiple stakeholders to successfully operate a complex tourist attraction in a sustainable way. It also demonstrates clearly the value of working with local tourism enterprises to integrate tourism activity into a sensitive environmental region.

Discussion questions

1. The Geopark is an area of outstanding environmental significance. Does tourism have a role in this type of region or should it be 'de-marketed'?
2. How can the local businesses boost the economy of this remote rural region?
3. What might be the challenges of operating an attraction like the Geopark with so many management interests and agencies involved?

Sources: Geopark (2019); Ireland (2019)

CHAPTER 7

THE TOURISM INDUSTRY (2)

LEARNING OUTCOMES

The objective of this chapter is to introduce the reader to a number of broad issues tied to food, waste management, energy and water in the tourism industry. Examples are drawn from inside and outside tourism for the purpose of demonstrating the direction in which tourism ought to go in the future regarding sustainability. The chapter is designed to allow the reader to:

1. Understand the environmental issues that take place from food production.

2. Know what a food system is.

3. Understand water rights as they relate to the fair and equitable use of water.

4. Have an awareness of the challenges that confront restaurants and hotels with the handling of waste.

5. Understand the concepts and tools that are presently being used to mitigate problems in the area of food, waste, energy and water.

INTRODUCTION

There is a paradigm change well underway that focuses on minimizing the ecological impacts that tourism has on sensitive environments through large-scale projects like hotels, infrastructure and transportation. A case in point is the recent development of Kenya's first 100% solar hotel. St. Ange (2018) writes that the Serena Hotels group is attempting to align all their programmes with the UN Sustainable Development Goals (SDGs), with the Kilaguni Serena Safari Lodge, in association with Mettle Solar OFGEN, as perhaps the best example of how new technologies are transforming the accommodation landscape. The hotel in Tsavo National Park has its own solar power plant for all of its operations that produces 307 kWp along with 670 kWh battery storage as the demand from eco-travellers continues to grow. The benefits of these green strategies for hotels and other tourism establishments include competitive advantage, better financial performance and better environmental performance (Singjai *et al.*, 2017).

The Kilaguni Lodge is just one example of the sorts of initiatives taking place in tourism, which are critical to pushing the boundaries of sustainability in new and innovative ways in an industry that should be a leader in green planning and development. This is reflected in the amount of research taking place on food, energy, waste management and food – the focus of this chapter – in tourism and hospitality, which has experienced strong growth over the last decade. Much of this work focuses on specific aspects in isolation, such as energy or water consumption, while other studies combine these aspects in generating more of a composite measure of resource use (water, energy and waste) in sectors like food and beverage and accommodation (Trung & Kumar, 2005). This chapter discusses these four primary tourism industry components and the impacts that are taking place in these sectors, as well as the concepts and tools that have been developed to mitigate these effects. While food, waste, energy and water were touched on briefly in the previous chapter in the context of hospitality, transportation, and attractions and events, this chapter may be viewed as a continuation of these issues but with a focus squarely on these key sustainability components.

FOOD

There are several key aims of SDG 2 on zero hunger that are at the forefront of changes needed globally in both developing and developed world contexts. Some of these include:

- ending hunger, and ensuring access for all people to safe, nutritious food;
- ending all forms of malnutrition;
- doubling the agricultural productivity and incomes of small-scale food producers;
- ensuring sustainable food production systems;
- increasing investment in agriculture;
- correcting and preventing trade restrictions and distortions in world agricultural markets;
- adopting measures to ensure the proper functioning of food commodity markets.

These measures, the UN argues, will necessitate a 'profound change of the global food and agriculture system' (UN, 2018b). To this end, Smith and Gregory (2013) suggest that the status quo of the current production system is not an option if we are to address the food and ecosystem needs of the future, with the suggestion that radical change is required in both production and consumption in the near future.

FOOD PRODUCTION AND THE ENVIRONMENT

Animal husbandry practices up until World War II allowed animals to live their lives in close proximity to the type of environmental conditions they would have enjoyed prior to domestication. Benton and Redfearn (1996) suggest that there was a skillset, base of knowledge and close affective bond between worker and animal that disappeared by the mid-20th century. They attribute this change to the internationalization of the agribusiness sector, which replaced responsibility and an ethic of care with artificial environments that place a premium on space and high yield, contributing to stress, pathological behaviour and disease – conditions that had to be managed through antibiotics and genetic manipulation.

The environmental impacts of these massive changes on the livestock sector are by all accounts huge. Fox (2000) argues that to keep up the supply with the demand for meat products, the resource base has been seriously compromised. Citing Robbins' (1987) *Diet for a New America*, Fox explains that meat industry operations contribute to large-scale:

> toxic chemical residues in the food chain, pharmaceutical additives in animal feeds, polluting chemical and animal wastes from feedlot runoff in waterways and underground aquifers, loss of topsoil caused by patterns of relentless grazing, domestic and foreign deforestation and desertification that result from the clearing of land for grazing and for cultivating feed, threatened habitats of wild species of plants and animals, intensive exploitation of water and energy supplies, and ozone depletion owing to the extensive use of fossil fuels and to significant production of methane gas by cattle. (Fox, 2000: 165)

In a report by the US Food and Agriculture Organization (FAO, 2006), livestock is responsible for 18% of greenhouse gas (GHG) emissions – a bigger share than transport. The sector is responsible for 9% of anthropogenic carbon dioxide (CO_2), 37% of methane and 65% of nitrous oxide from manure. Livestock production uses 8% of the world's water supply for irrigation, and is the largest source of water pollution mainly from animal wastes, antibiotics, hormones, chemicals, fertilizers and pesticides. Grazing of livestock occupies 26% of the Earth's surface, and the expansion of lands dedicated to grazing is a principal factor in the deforestation of much of the rainforest environments in regions around the equator (FAO, 2006). The production of meat uses water resources that far exceed the demand required for food grown from the soil. One pound of wheat requires only 60 gallons of water (Fox, 2000), while the production of the same unit of beef takes 1799 gallons of water (Foodtank, 2019). The message is that by eating further down the food chain we place far less stress on the ecosphere.

The type of production also has different levels of impact on the resource base. Extensive systems of beef production are based on grazing and forage from grasslands. Intensive systems are different because they house cattle in buildings and use concentrated feeds and home-grown forage. Ogino *et al.* (2016) investigated beef production systems in Thailand, and found through life cycle assessment (animal management through biological activities of cattle, grassland management, purchase of feed and waste treatment) that intensive systems had higher impacts on energy consumption but lower GHG emissions (10.6 kg CO_2) than the extensive system (14.0 kg CO_2). Rivera Huerta *et al.* (2016) also found that intensive systems of beef production (in Mexico) have poorer environmental performance than extensive systems. In particular, these impacts were most marked in reference to climate change, terrestrial acidification and freshwater and marine eutrophication.

The environmental impacts inherent in the production of meat point to certain trade-offs with biodiversity conservation and ecosystem service provision (Holt *et al.*, 2016). Using the example of pesticide use, they note that current changes in EU legislation and risk assessment may have important implications for food production. For example, in the case of the use of herbicides and winter wheat, the trade-off in the protection of ecosystem services will be the loss of yields. Aquatic and terrestrial buffers will reduce crop areas, and alternative herbicides are not effective against some invasive weeds.

The magnitude of the livestock industry has created other problems. The feeding of live animals including bits of discarded dead animals in the form of slop or swill can cause diseases like foot-and-mouth disease (FMD), a viral zoonosis with little effect on human health but with huge implications for the economy. In 2001 the outbreak of FMD in the UK resulted in the death of over 10 million cows, pigs and sheep, most of which were not infected but were unfortunate enough to live in affected areas; hardest hit was Cumbria (Blake *et al.*, 2001). The impact on the economy of the UK from lost tourism revenue alone was huge. Blake *et al.* (2001) employed a micro-regional simulation model to predict the total revenue lost as a result of the outbreak. They estimated at the time that revenue would drop by £7.5 billion, with 21% of this from falls in domestic overnight tourism, 49% from day visits and 31% from reductions in international visits. The authors projected a residual effect over time, with a reduction in tourism expenditures of £4.9 billion in 2002, £1.1 billion in 2003 and £0.5 billion in 2004.

SUSTAINABLE FOOD SYSTEMS

Not unlike other academic fields and disciplines, theorists incorporated the idea of sustainability in food systems research at an early date. Although it is beyond the scope of this book to trace the history of the connection between sustainability and food, work by Dahlberg (1993) on the three 'Es' of ecology, ethics and equity is illustrative of this linkage:

> (1) Sustainability as long-term food sufficiency, i.e. food systems that are more *ecologically based* and that do not destroy their natural resource base. (2) Sustainability as stewardship, i.e. food systems that are based on a conscious *ethic* regarding humankind's relationship to other species and to future generations. (3) Sustainability as community, i.e. food systems that are *equitable* or socially just. (Dahlberg, 1993: 81, cited in Sumner, 2011)

Bridging this set of ideas by Dahlberg, theorists have also made valuable connections between the social economy and local, sustainable food systems. The social economy has been examined in detail by McMurtry (2010), who describes it as economic activity that is not controlled by the state or the market, but rather has its rootedness in the social wellbeing of communities and marginalized individuals. Credit unions and farm cooperatives are examples of these types of arrangements.

Sumner (2012) writes that while there are a number of reasons to develop local sustainable food systems, the primary objective of these systems is to ensure that all citizens of the planet have access to nourishing food that does not, in its production, processing, distribution, consumption and disposal, exceed the ecological limits of the planet (Sumner, 2011). Access to food is critical in any configuration of a sustainable food system, as is food sovereignty (human rights including the right of communities and individuals to choose their own forms of agriculture), and the support for local farmers, workers, local communities and economies, and conservation of the local environment. Sumner (2012) adds that a successful sustainable food system would integrate the local with the global, the latter of which includes fairly traded imports that are non-competing with the local realm in the creation of cooperative networks in a set of 'globally local' arrangements. Yet the challenge of a sustainable food system at local and global levels is a lofty one because, as noted by Wright (2004), there are two key intervening variables that have proven to be insurmountable: population grows to exceed the supply of local resources including food; and the hierarchical nature of all civilizations suggests that wealth becomes concentrated in the hands of a few, ensuring that there is not enough food to go round. Other challenges include food price volatility, shortages of basic commodities, an increase in global rates of obesity along with food-related diseases, as well as land grabbing (Blay-Palmer *et al.*, 2016). These authors argue that the growth of sustainable food systems might better take place through the development of networks of knowledge. Sharing good practices that have grown organically in local communities at a broader scale will bolster cooperative action against the foregoing global pressures.

Pointing out that there is in fact more than enough food to feed the world, Millstone and Lang (2003) place the blame for such disparities and discrepancies squarely in the lap of globalization and neoliberalism:

> Food is no longer viewed first and foremost as a sustainer of life. Rather, to those who seek to command our food supply, it has become instead a major source of corporate cash flow, economic leverage, a form of currency, a tool of international politics, and instrument of power – a weapon! (Krebs, cited in Millstone & Lang, 2003: 11)

This model of accumulation, Sumner (2011: 63) adds, enriches a few while at the same time creating externalities of a social, economic and ecological nature that are steadily 'destroying local economies, devastating individuals, families and communities and degrading the planet'.

FOOD SYSTEM MODELS

The basic model of a food system includes natural resources and societal demands as inputs, and waste emissions and food as outputs. The paradox for Sabaté *et al.* (2016) is that even though food systems are reliant on natural resources, they are also significant contributors to natural resource degradation. These authors propose the implementation of environmental nutrition policies and procedures as a new frontier for public health and as a manner by which to address the need for more sustainable food systems (Table 7.1). Sabaté *et al.* (2016) argue that the environmental impacts of a food system have a direct influence on nutritional outcomes. They use the example of GHGs as an output of the food system that contributes to climate change, which in turn affects the production of food through the reduction of yields and the nutritional content of food, which in turn increases both malnutrition and food insecurity.

Suggestions for change in the connection between agriculture and ecology include the implementation of agroecology, or the application of ecological principles to agriculture in the development of sustainable food systems. This approach includes aspects of transdisciplinarity, productive practices and social movements that move food systems towards ecological sustainability, social equity and greater resiliency (Méndez *et al.*, 2015; Miles *et al.*, 2017). Gliessman (2016) has developed a framework of five different stages required in moving from a conventional food production system to an agroecology one, as follows

- improving system efficiency to reduce the use of conventional agro-chemical inputs and their ecological and social risks (Level 1);
  - substituting more sustainable inputs and practices into farming systems (e.g. many practices included in certified organic agriculture; Level 2: Input substitution);

Table 7.1 Focus of nutrition dimensions

Nutrition dimensions			*Scope of dimensions*
Scientific disciplines covered by each dimension	Biological sciences • Biochemical • Physiological • Medical Social and multidisciplinary sciences • Epidemiology • Anthropology • Political science • Economics • Sociology Environmental sciences • Physical • Atmospheric • Ecology • Geography	Issues addressed	Individuals • Growth • Adequate diets • Nutrition requirements • Disease management Communities • Epidemics of chronic disease • Nutrition deficiencies • Disease prevention • Food security Biosphere • Agricultural practices • Sustainable food systems • Societal

Source: Sabaté *et al.* (2016)

- redesigning farming systems based on ecological knowledge to maximize ecosystem services (Level 3: Farm-scale agroecology);
- re-establishing connections between producers and consumers to support a socioecological transformation of the food system (Level 4: 'Transformative' agroecology); and
- supporting a fundamental shift in global society where ethics, knowledge, culture and economy are rethought and directed towards ecological restoration, social justice and equity in the food system and within all forms of human activity (Level 5: Global transformation to a sustainable society).

AGRITOURISM

The connection between agriculture and tourism is well documented in the literature. This includes research on conceptualizing agritourism as well as an investigation into the social and economic costs and benefits of the industry. Phillip *et al.* (2010) developed a typology for defining agritourism based on five key characteristics derived from whether or not the farm is a working farm, the nature of tourist contact with the enterprise, and the authenticity of agricultural activities. The agritourism typology is as follows:

1. Non-working farm agritourism, e.g. accommodation in an ex-farmhouse property.
2. Working farm, passive contact agritourism, e.g. accommodation in a farmhouse.
3. Working farm, indirect contact agritourism, e.g. farm produce served in tourist meals.
4. Working farm, direct contact, staged agritourism, e.g. farming demonstrations.
5. Working farm, direct contact, authentic agritourism, e.g. participation in farm tasks.

The tourism industry has been effective in the economic restructuring of farms in rural communities. Since the 1970s, farmers in the USA have needed to employ diversification strategies based on rather unconventional means to bolster farm incomes and sustain enterprises. Tourism has become a popular strategy by virtue of its ability to generate jobs and bring in money, with the added benefit of developing coordination and cooperation between farmers, government and local businesses (Wilson *et al.*, 2001). In places like Sicily, for example, rural tourism has become a vital element to the rural community through investment and the creation of jobs, ensuring that agriculture continues to play an important role in the fabric of rural communities (Sgroi *et al.*, 2014), even though it is treated as a secondary economic activity (Weaver & Fennell, 1997). In a study of farms in New Jersey, USA, small farm enterprises were found to realize higher profit impacts through the incorporation of tourism than larger farms (those earning $250,000 or more in annual sales) (Schilling *et al.*, 2014).

Tourism can contribute to local economies in lesser developed countries if backwards linkages are made with different sectors (Telfer & Sharpley, 2008). The agricultural sector is key in this configuration simply because of the diversity of food products that the tourism industry could access in promoting pro-poor tourism (Rogerson, 2006; Torres & Momsen, 2004) and the reduction of the ecological footprint in the production of food.

In a study of the linkages between African safari lodges and agriculture in South Africa, Rogerson (2012) found that there were many factors that prevented a strong relationship. Foremost was the lack of uniform communication between both groups as well as what was characterized as a deep mistrust between local producers and food supply decision makers. Other factors included the inability of producers to supply the quality, consistency and volume required by the lodges, as well as the challenges of long-distance sourcing of products. Rogerson (2012) writes that local agricultural producers should mobilize themselves into groups for the purpose of overcoming these challenges.

Agricultural enterprises that have a tourism component practise sustainability to a much higher degree than non-tourism agricultural units, with a host of sociocultural, economic and ecological benefits (Barbieri, 2013). Regional policy developers would be wise to recognize these benefits in planning for the successful agritourism–sustainability nexus in the future (Broccardo *et al.*, 2017). The direct, hands-on, authentic experience that tourists get from an agricultural lifestyle would help to realize sustainability through the protection of local heritage and tradition, as well as the support of farmers as stewards of the land (Valdivia & Barbieri, 2014).

Other studies have found that fostering sustainable development in the small-scale agricultural enterprises in Tanzania is difficult because of the perception on the part of the hoteliers and restaurateurs that local products are inferior and unreliable (Sanches-Pereira *et al.*, 2017). These tourist establishments prefer imported goods because of convenience and the history of existing relationships. As such, there are few direct supply channels from agricultural producers to the tourism industry and no operating markets. The authors argue that there are four different strategies that can be used to build a more sustainable relationship between industry and farmers. These include:

1. awareness and capacity building to support and implement linkages between tourism and agriculture;
2. start-up drivers or regions that can serve as multipliers on the basis of those suppliers that have been successful;
3. public–private partnerships and destination-level cooperation and action for the implementation of pro-poor tourism initiatives, with emphasis on institutional arrangements and funding to catalyse development;
4. effective promotion of pro-poor tourism and branding.

CONCEPTS AND TOOLS

Following from the template developed in Chapter 4 we identify a number of concepts and tools that are presently being used in sustainability research and that have application to tourism studies. In this section of the chapter, ecotourism, ecolabels, indicators, life cycle assessment (LCA) and corporate social responsibility (CSR) are emphasized.

Advantages can be realized through the production of new products for consumers who are demanding changes in the type and quality of food. In Korea, organic farms are evolving into ecotourism destinations as local farmers seek alternative methods to generate income. Choo and Jamal (2009) analysed 38 websites of organic farmers and found a tight fit between an organic lifestyle and ecotourism through increased consumption of local products, local economic diversification, equitable changes in local lifestyle, attention to cultural heritage, conservation of natural resources, educational programmes for locals and tourists and community participation. This resonates with work by Fennell (2012) on the use of animals in ecotourism operations, and Fennell and Markwell (2015) on Australian ecotourism operators and their choices around organic food supply. Fennell argued that if ecotourism stands as one of the most ecological forms of tourism, this should be reflected in the types of food offered to ecotourists along a continuum: vegan ecotourism, vegetarian ecotourism, domesticated meat (a contentious issue if animals are mass produced) and finally eating bush meat. Fennell and Markwell (2015) found that some of the most highly certified ecotourism operators were not the ones providing meals that were more ethical or sustainable in reference to food-service business practices. It was food and wine from ecolodges and wildlife tourism operators who had higher credentials when it came to ethical food sourcing and food sustainability. Efficiencies in marketing these greener products provide competitive advantage. For example, in the highly competitive wine industry in Italy, Fiore *et al.* (2017) illustrate that marketing green-oriented innovations such as the use of organics, agricultural waste recovery and efficient water use all have proven benefits with a clientele that is now more than ever expecting these sorts of changes in production.

Delmas and Lessem (2017) write that the goal of ecolabels is to reduce the amount of information asymmetry that exists between consumers and producers about the environmental characteristics or attributes of their goods or services. This goal is often not met because of unclear or irrelevant information that ends up confusing the consumer. Ecolabels, therefore, need to communicate clearly their particular attributes to the target audience. Studies indicate that consumers have positive attitudes towards ecolabelled food products; what is less well known, however, is whether this positive attitude translates into behaviour (Isaac *et al.*, 2017). In this study, the authors compared ecolabelled seafood premiums in the market with consumers' willingness to pay. They found that consumers were in fact willing to pay premiums for ecolabelled food in an effort to reward producers who adopted environmental practices (see also Chen *et al.*, 2016). For example, premiums ranged from 24% to 38% for organic aquaculture ecolabelled products compared to 10–13% for ecolabelled Marine Stewardship Council fish, demonstrating the power of organic products. Furthermore, sustainable practices like line-caught fishing generated premiums from between 10% and 25%, and other more unsustainable fishing practices were discounted because of the perception that the quality of the fish was diminished. Even though sectors of the food industry like the seafood industry have developed their own ecolabels and certification systems as a way to market the environmental performance of their products, these are generating confusion in the industry in the USA and internationally (Czarnezki, 2014). Third-party certification bodies are said to be superior because they not only provide verification of this quality but also independence, and therefore credibility, which avoids greenwashing from the first-party approach.

The foregoing suggests that consumers play a vital role in making food more sustainable (Grunert, 2011). What challenges this role is the manner in which sustainability is communicated to consumers. And even if consumers are motivated to support sustainability in food establishments and foods are ecolabelled, barriers exist preventing consumers from using the available information to make choices that are sustainable. The following framework illustrates where these barriers exist within a consumer behaviour perspective (Grunert, 2011: 209):

1. *Exposure does not lead to perception*: Consumers simply do not notice the label, because they are time pressured when shopping and most purchases are made habitually.
2. *Perception leads only to peripheral processing*: Consumers see the label, but do not care enough to make an effort to understand what it means. It still may affect their choices, though.
3. *Consumers make 'wrong' inferences*: Consumers do see the label, make an effort to understand what it means, but draw the wrong inferences. They may end up buying the product, but for the 'wrong' reasons.
4. *Eco-information is traded off against other criteria*: The price may be higher, the taste is not good, and the family prefers something else.
5. *Lack of awareness and/or credibility*: Consumers who want to make sustainable choices may find it hard to carry them out in practice.
6. *Lack of motivation at time of choice*: While consumers have a positive attitude towards sustainability, this attitude is not so strong that it affects behaviours in all situations where sustainability may be a criterion. We can say that consumers 'forget' about their positive attitude to sustainability when making food choices. Such 'dormant' attitudes are a major factor in explaining discrepancies between attitude and behaviour.

A newer incarnation in ecolabel programmes is carbon emissions labelling. Miranda-Ackerman and Azzaro-Pantel (2017) argue that there are critical disconnects between the agricultural stage of food production and other aspects of the food supply chain. Several environmental impact improvements have been made with the former, but not with the latter where there are several inefficiencies. The authors argue that carbon emissions labelling should be integrated along the food supply chain in the development of a green supply chain network design with CO_2 emissions measured along key stages of the product. This includes measuring carbon from the perspective of suppliers, processing, bottling and distribution.

Several indicators have been developed in food production research that should have added value in contextualizing sustainable tourism (ST). Ryan *et al.* (2016) developed generic farm-level sustainability indicators for use in Ireland. These include economic measures (productivity of labour, productivity of land, profitability, market orientation and farm viability), environmental measures (GHG emissions per farm, GHG emissions per kg of output, nitrogen balance and emissions from fuel and electricity) and social indicators (household vulnerability, education level, isolation risk, demographic viability and work–life balance). Researchers have also developed sustainability indicator systems specifically for rural settings for managers and policy makers. A good example of these systems comes from the work of Blancas *et al.* (2011), who have developed an elaborate list of social, economic and environmental indicators that may be collapsed into sustainability indexes to reduce measurement subjectivity.

Goggins and Rau (2016) developed the FOODSCALE assessment tool for measuring a number of key indicators that determine whether food provision is sustainable or not. The scale is based on a points system that ranges from 0 to 100. A maximum score of 100 is possible, which is distributed across 11 different main categories (36 total indicators), which are weighted as 5, 10, 15 or 20, based on an extensive review of the literature on the impact that these categories have on sustainability. This includes commitment to change and how healthy and sustainable the food is for consumers. The assessment tool covers all three pillars of sustainability (economic, environmental and social) as well as the entire food system itself, including production, distribution, procurement, consumption and waste. The 11 indicators and their weighted point scores are as follows.

1. Organic (10 points)
2. Seasonality (5 points)
3. Fairly traded produce (5 points)
4. Meat (15 points)
5. Sustainably sourced seafood (5 points)
6. Eggs (5 points)
7. Water (5 points)
8. Food waste (10 points)
9. Origin of food (20 points)
10. Consumer engagement (10 points)
11. Engaging with small producers and local communities (10 points)

We include a complete overview of the category with the highest weighted number of points (Origin of food) as an example of how the system is organized:

Origin of food (20 points)

Social impacts

- Sourcing food locally/in the region increases food security and resilience to external shocks in the food system
- Links producers and consumers
- Facilitates education
- Protects local food cultures

Environmental impacts

- Reduces long-distance food transportation impacts
- Reduction in energy used for storage
- Protection of biodiversity
- Reduced risk of contamination and disease

Economic impacts

- Contributes to local and rural economy
- Can improve efficiency of delivery systems
- Generates employment in rural areas
- Can reduce procurement costs

FOODSCALE indicators

- Provenance of five key foods to local, regional, national or international origin
- Number of intermediaries between producer and consumer.

Life cycle assessment is being used to measure the environmental impacts of food production. A case in point is the application of LCA to investigate the environmental impact of heavy pig production in Northern Italy from cradle to farm gate, using a functional unit of 1 kg live weight (Bava *et al.*, 2017). The authors found that environmental impacts per kg live weight was a key measurement statistic: pigs slaughtered at a lower weight had lower environmental impacts, as measured through global warming, eutrophication, acidification, the use of non-renewable energy, land occupation, abiotic resource depletion, terrestrial ecotoxicology and ozone layer depletion (the global warming potential being on average 4.25+/− 1.03 kg CO_2 eq./kg live weight). The authors add that environmental efficiencies may be realized through enlargement of the average farm size and improvements in reproductive performance.

LCA was also used to compare six different dietary patterns (omnivorous, vegetarian and vegan) by the method of production (conventional versus organic farming) in determining the environmental impact of each according to carcinogens, respiratory organics, respiratory inorganics, climate change, ozone layer, ecotoxicity, acidification, eutrophication, land use, minerals and fossil fuels (Baroni *et al.*, 2006). A seventh dietary pattern was added, referencing a normal weekly Italian diet (the study took place in Italy) with food from conventional farming methods. The authors found that the normal weekly diet had by far the greatest average environmental impact, followed by omnivorous conventional, omnivorous organic, vegetarian conventional, vegetarian organic, vegan conventional and finally vegan organic. There are several key observations drawn by the authors:

1. A greater consumption of animal products produces greater environmental impacts.
2. Chemical-conventional production methods have a greater environmental impact than organic methods.

3. Beef is the single food with the greatest environmental impact.
4. The production of cheese, fish and milk also has high environmental impact values.

Chapter 4 also discussed the important role that CSR is playing in ST management as a proactive means by which to make an active commitment to society (Lombardi *et al.*, 2015). CSR has an important role to play in the agricultural sector because there is widening gap between the public's perception of farming and food production and the realities of such, which continue to threaten the reputation and legitimacy of these enterprises (Luhmann & Theuvsen, 2017). Following from the research findings and industry practice on the criticisms of CSR in the supply chain of the food industry, Maloni and Brown (2006) developed a comprehensive framework of supply chain CSR in the agricultural industry. A summary of the categories, subcategories and specific elements is included in the following table (Table 7.2).

A thread that continues to weave its way through this book is the important role that large corporations play in sustainable development. An example of an international fast food chain that has had success in the realm of CSR is McDonald's Corporation – despite the magnitude of impact that this company is having on the natural world as a result of the demand for meat products, as noted above. McDonald's is attempting to enhance its reputation through several channels. These include animal welfare, corporate giving, education scholarships, good employment practices, good environmental practices and choosing suppliers that maintain similar social and environmental values (Gheribi, 2017). To this end, Hartmann (2011) argues that there is a need for SMEs to follow the lead of larger corporations in the area of CSR, as well as a need for changes to take place throughout the entire food supply chain. While SMEs, individually, have little power, it is in the aggregate magnitude of these businesses where their social and environmental consequences can be substantial.

Yet the ability and willingness to utilize CSR is a function of various constraints from the regional perspective. In countries such as China, the concepts of CSR and food risk management are not well understood (Zhang *et al.*, 2014). Drivers of change in this area, according to the authors, would include attitude to CSR and risk, management support, training, budget for CSR and FRM, and the implementation of an international standard. In Slovakia, a survey of SME food companies demonstrated that only 42% of respondents were familiar with the term CSR, leading the authors to conclude that the overall situation in Slovakia regarding CSR is not at all optimistic (Ubrežiová & Moravčiková, 2017).

Chapter 4 also introduced the concept of reporting in corporate affairs, which is a practice that continues to gain greater acceptance especially in the commercial realm. Two examples of sustainability reporting schemes in the food and agricultural sectors (FAS) include the G4 Guidelines of the Global Reporting Initiative (GRI) and the Sustainability Assessment of Food and Agriculture (SAFA) of the UN FAO. SAFA is said to be the most important state-of-the-art mechanism in the FAS sector because it utilizes the ISO 14040 family of standards for LCA, the International Social and Environmental Accreditation and Labelling

Table 7.2 Summary of food industry supply chain CSR issues

Category	*Subcategory*	*Elements*
Animal welfare	Humane treatment	Cruelty, handling, housing, slaughter, transport
Biotechnology	Animals, plants	Antibiotics, growth hormones, tissue cultures, genetic testing, recombinant DNA, cloning
Community	Support	Economic development, philanthropy, arts, educational support, job training, volunteering, literacy, health care, childcare, housing
Environment	Conservation	Damage compensation, energy, food miles, forests, farming methods, packaging, resources, species, water, soil
	Pollution and waste disposal	Emissions, waste, manure, water, hazardous materials, organic, herbicides, pesticides, rodenticides, recycling, global warming
Fair trade	Fairness	Fair trade, profit sharing
Health and safety	Safety	Food safety, security, traceability, transportation, disclosure
	Health	Healthy lifestyles, local food sources
Labour and human rights	Compensation	Compensation
	Illegal labour	Captive/forced/bonded labour, child labour, status verification
	Opportunity	Training, education, advancement, regular employment
	Treatment	Accommodations for disabled, discipline/abuse, discrimination, respect
	Worker rights	Legal rights, civil rights, diversity, privacy, collective bargaining, grievances, rights disclosure
	Working conditions	Hygiene, sanitation, healthy, quality, safety, transportation safety, housing safety, training/disclosure, hours

Procurement	Behaviour	Conduct, professional competence
	Purchasing process	Confidentiality/proprietary information, conflict of interest, deception, impropriety, influence, reciprocity, responsibility to employer, power abuse, special treatment
	Legal	Applicable law
	Supplier diversity	Disadvantaged suppliers, minority suppliers, supplier's minority labour/programmes

Source: Maloni and Brown (2006)

alliance (ISEAL) principles, Reference Tools of the Global Social Compliance Programme, and the GRI guidelines, noted above (Hřebíček *et al.*, 2015). For companies/farmers and governments/investors/policy makers, respectively, the SAFA systems allow for:

> Self-assessment for evaluating sustainability at the supply chain level; identifying hot-spots for performance improvement; performing gaps analysis with existing sustainability initiatives for thematic comparison; managing or benchmarking suppliers to improve sustainable procurement.
>
> Informing sustainable development strategies and goals; determining policy, investment and procurement priorities; providing a global guidance on sustainability requisites for the regulation of global supply chains. (SAFA Guidelines, 2013)

The SAFA sustainability reporting polygon includes 21 different themes within four broad areas. These include: governance (corporate ethics, accountability, participation, rule of law, holistic management); environment (atmosphere, water, land, materials and energy, biodiversity, animal welfare); economy (investment, vulnerability, product quality and information, local economy); and social (decent livelihood, fair trade practices, labour rights, equity, human health and safety, cultural diversity). There are 58 sub-themes and over 100 indicators that need to be considered in this reporting scheme (SAFA, 2013).

WASTE MANAGEMENT

As already stated in this book, waste management is an essential element of the environmental management strategies required in the tourism and hospitality sector, with a massive change in these strategies having occurred over the last two decades (Pirani & Arafat, 2014). While waste management is not a stand-alone SDG, it is referenced heavily in SDG 12 – ensure sustainable consumption and production patterns. Key targets in this SDG include: the efficient use of natural resources; halving per capita global food waste; managing chemicals and all other waste through their life cycles; and reducing waste generation through prevention, reduction, recycling and reuse.

The EU (EC, 2010) has developed the mindset that waste should not be viewed as an unwanted burden but rather as a valued resource. The EU's five-step **waste hierarchy** emphasizes the importance of moving up the chain from disposal to recovery, to recycling, to re-use, and finally to prevention, where prevention is elimination of waste before it is created (Omidiani & HashemiHazaveh, 2016). Targets for EU Member States are based on recycling 50% of municipal waste and 70% of construction waste by 2020. Figure 7.1 provides a more comprehensive overview of the waste hierarchy as developed by Radwan *et al.* (2012) and adapted by Pirani and Arafat (2014). It shows that there are several broad-based factors that come into play in how hotel managers should handle their waste. These include economic, legislative, social, marketing, awareness and education, the availability of recycling and composting in the region, the level of support offered by government and the cooperation of manufacturers. These factors play upon the selection of a waste carrier, audits, the type of solid waste management programme implemented, staff training and how best to involve guests in the solid waste management programme.

Waste is a problem in tourism because it is often seen as the most visible impact of a range of different environmental issues. Peak tourism seasons may exacerbate the waste management situation. In the Himalayas, for example, waste is often deposited on the street, in vacant areas and in streams, complicating its proper handling (Wani *et al.*, 2018). Apart from the aesthetic problems that go along with this are the health-related issues to humans from stray dogs, flies and other pests that may cause concern for both tourists and local people. These aesthetic and health issues may ultimately have the effect of curtailing the growth of tourism (Jennifer *et al.*, 2018). Furthermore, many regions do not have the institutional and legal frameworks in place to handle changes in the amount and flow of waste as a result of rapid development and associated increases in tourism.

The lack of institutional structures suggests that even in developing countries there are significant differences in how waste is handled. Dodds and Walsh (2019) studied 56 festivals across Canada and found clear divisions around certain themes. First, festivals in Alberta and British Columbia diverted more waste to disposal sites because of more advanced municipal waste management policies and procedures. They also found that music festivals handled waste more efficiently because of responsible leadership. Thirdly they found that smaller festivals generate more waste per attendee because the smaller venues are constrained by a lack of human and financial resources to handle waste, which in turn demotivates organizers to adopt sustainable practices.

Other studies linking tourism and waste show the importance of using local materials and resources in the creation of several different types of products for tourism purposes, and the limited amount of waste that results from these practices. Vitasurya and Pudianti (2016) found that small-scale traditional craft industries in developing countries could be more sustainable through the minimization of wastes in the production process. The authors found that: (1) the use of local materials for ingredients in crafts reduced and prevented waste; (2) solid waste from animal feed from the tempeh industry (a traditional soy

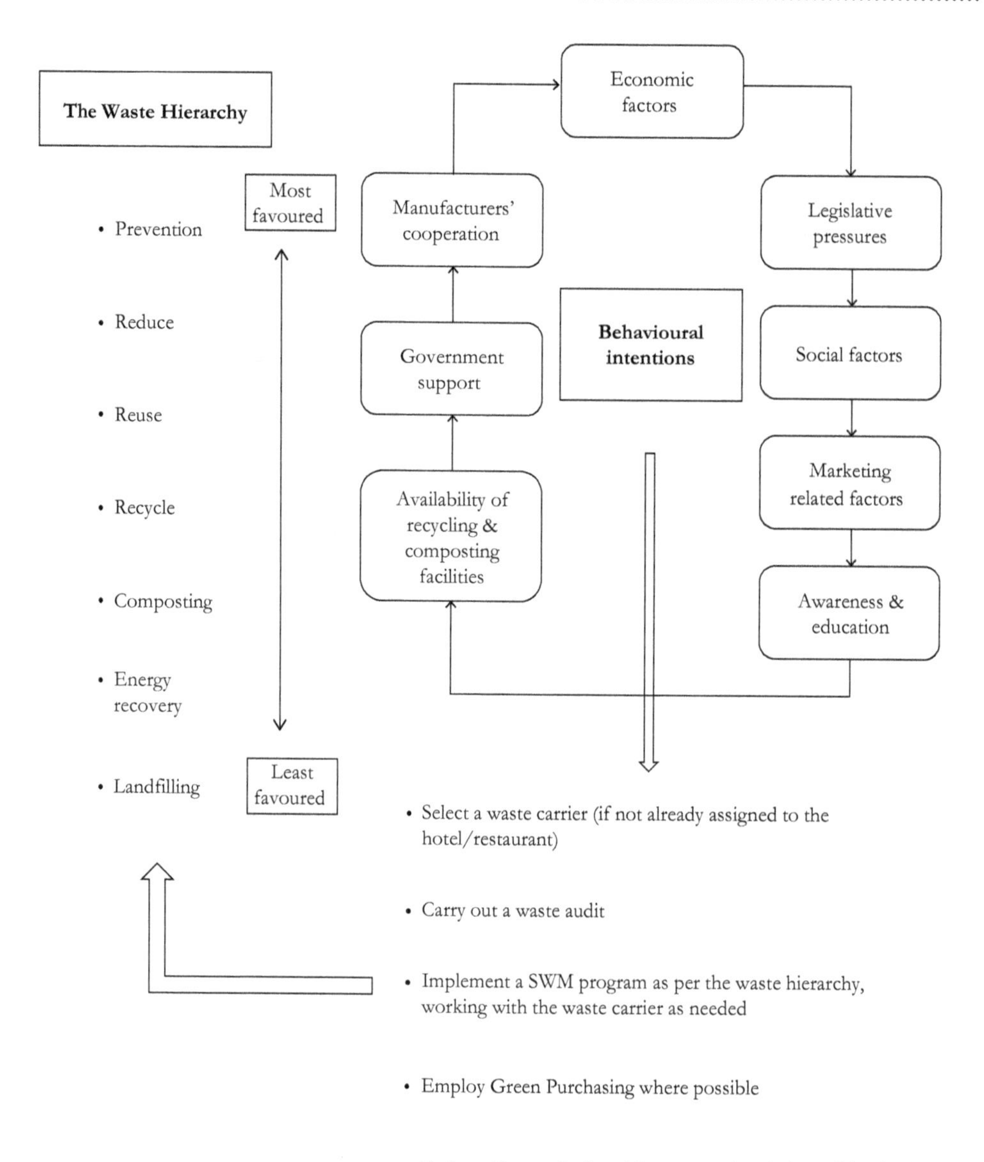

Figure 7.1 Solid waste management practices model for the hospitality sector
Source: After Prani and Arafat (2014), based on Radwan *et al.* (2012)

product) and fertilizer from the bakpia industry (bean pastry) was able to be reused; (3) solid waste from bamboo furniture could be turned into batik bamboo handicrafts; (4) solid waste from coconut shells became charcoal for energy; and (5) communal landfill sites were easily able to handle the minimal amount of waste generated from these activities.

Not unlike other sustainability elements of the tourism industry, it is the accommodation and food and beverage sectors that appear to have the most scholarly work conducted in the area of waste. Tables 7.3 and 7.4 identify the main types of waste generated in the hotel industry. Table 7.3 identifies the non-hazardous forms of waste and Table 7.4 shows the hazardous wastes generated.

Table 7.3 Types of non-hazardous waste in the hotel industry

Non-hazardous waste type	*Components*	*Source*
Household	Food/kitchen waste, used or dirty paper and wrapping, plastic wrapping or bags, composite wrappers	Hotel's different departments
Cardboard	Packaging	Hotel's purchasing and other departments
Paper	Printed documents, brochures, menus, maps, magazines, newspapers	Administration, reception, guest rooms, restaurants
Plastic	Bags, bottles (that did not contain hazardous material), household goods individual portion wrappers for various products	Kitchen, restaurants, bars, guest rooms, administration
Metal	Tin cans, jar lids, soda cans, food containers, mayonnaise, mustard and tomato puree, tubes, aluminium packaging	Kitchen, restaurants, bars, guest rooms
Glass	Bottles, jars, flasks	Kitchen, restaurants, bars, guest rooms
Cloth	Tablecloths, bed-linen, napkins, clothes rags	Kitchen, restaurants, bars, bathrooms, guest rooms
Wood	Wooden packaging, pallets	Purchasing department
Organic waste	Fruit and vegetable peelings, flowers and plants, branches, leaves, grass	Kitchen, restaurants, bars, guest rooms, gardens

Source: Zein *et al.* (2008)

Table 7.4 Types of hazardous waste in the hotel industry

Hazardous waste type	*Source*
Frying oil	Kitchen, restaurants
Mineral oil	Maintenance service
Paint and solvent residues	Maintenance service
Flammable material (gas, petrol, etc.)	Kitchen, garden
Fertilizers and chemicals (insecticides, herbicides, etc.)	Garden
Cleaning chemicals	Maintenance service
Ink cartridges	Administration
IT disks and CDs	Administration, guest rooms
Batteries	Maintenance service, administration, guest rooms
Cleaning chemicals and solvents in dry cleaning	Laundry room
Fluorescent lights, neon tubes and long-life bulbs	Maintenance service

Source: Zein *et al.* (2008)

These two tables combined show the large number of products that contribute to the waste burden in the hotel sector. And waste *is* created. Waste audits indicate that guests, on average, generate about 1 kg of waste per night in the hotel sector, generating about 66 tonnes per hotel per year in the UK (WRAP, 2011). Studies vary considerably on the amount of waste generated in hotels and this is due to a number of different factors relating to the hotel itself as well as the activities and characteristics of guests and employees.

Studies have also documented the variability in amount of waste generated in the restaurant industry. Pirani and Arafat (2014) write that food waste accounts for about 56% of garbage from restaurants and 28% of garbage from hotels. The problem with food waste in landfill sites, they argue, is that it decomposes to create methane, a GHG, and it is financially costly. Curtis and Slocum (2016) illustrate that food represents 40% of all the solid waste generated by resorts. Principato *et al.* (2018) found in a study of 127 restaurants in the Lazio and Tuscany regions of Italy that about 12.9% of food in the preparation stage was wasted because of spoilage or incorrect preparation, while the average amount of waste per customer, estimated by asking waiting staff, was 15.8%. The authors also found that more waste occurred in meat restaurants over fusion and ethnic restaurants. The reason for this is because the former typically gives larger portion sizes that are beyond the needs of the consumer. Additionally, the authors found that there was more food waste in upscale restaurants (€60 or more per plate), attributed to the belief that wealthy clients often order more than they can eat.

Because of the persistent waste problem in restaurants, some establishments are taking an active role in measuring the level of food waste. Sakaguchi *et al.* (2018) found that 65% of restaurants were measuring food waste and 84% are composting inedible food. The most frequently used method to eliminate food waste was to give it to employees. Most restaurateurs were fearful of giving food leftovers as donations because of legal liability.

In the previous discussion on waste hierarchy, the final element of Figure 7.1 illustrated that guests should be allowed to play an active role in the solid waste management programmes of restaurants. This component of the model, getting patrons onside with such programmes, has been investigated in recent studies. Stöckli *et al.* (2018) found that certain types of prompts were helpful in getting restaurant patrons to reduce food waste in restaurants. Two main types of prompts were used (as well as a control group), referred to as information-alone and informational and normative prompts. These were worded as follows:

> *Information-alone*: Food waste happens in the restaurant too. A third of all foods are thrown away. 45% of waste occurs in households and restaurants. Please ask us to box your leftover pizza slices for takeaway to avert food waste.
>
> *Informational and normative prompt*: Our guests expect a reduction of food waste. A third of all foods are thrown away. 45% of the waste occurs in households and restaurants. The majority of our guests expect that the wasting of food is reduced. Therefore, many people ask us to wrap their pizza leftovers. Please ask us to box your leftover pizza slices for takeaway to avert food waste.

The authors found that 25% of the control group patrons asked to have their meal packaged for takeaway, 55% of the patrons subjected to the informational prompt asked to have their meal packaged, and 64% of the group given an informational and normative prompt asked for packaging to take the rest of their meal away. While patrons may ask for a doggy bag to take their food home, it is also suggested that restaurants offer to package a patron's food in efforts to reduce food waste in the restaurant. Other formal mechanisms to reduce food waste in restaurants include (Principato *et al.*, 2018):

- careful ordering and menu planning;
- avoiding spoilage waste by monitoring use-by dates and storage conditions;
- re-using edible food items for making other recipes;
- offering different portion sizes;
- composting food waste.

TOOLS

Two sustainability tools that have significance to waste management and sustainability include waste mapping and materials flow analysis (MFA). Both tools attempt to track, through either graphic or procedural means, the ebb and flow of waste in operations.

Waste mapping

Waste mapping allows hospitality and tourism to 'understand where and how waste occurs, and how much it is really costing them' (WRAP UK, 2013a, cited in Pirani & Arafat, 2014: 326). Waste mapping allows the organization to monitor waste at a property, specifically the type of waste generated, in what amounts and at what particular locations. A waste map is created from these data for each room in the establishment for the purpose of creating an operations strategy that is more efficient (Pirani & Arafat, 2014).

Materials flow analysis

MFA is a tool that allows researchers and managers to better understand solid waste streams in a system. Meylan *et al.* (2018) used this tool to help improve the solid waste management landscape in the Seychelles, a region that is confronted with increasingly higher numbers of tourists. MFA was deemed essential at this stage of tourism development because current systems were unable to accommodate the solid waste management pressure created by this growth. Meylan *et al.* (2018) configured MFA on Mahé according to the following research questions:

- What types of products (classified by material) enter the Seychelles and in what quantities?
- How do materials in each stream exit the Mahé system and in what quantities?
- How are the materials used in terms of their flows between stakeholders?

Figure 7.2 shows that the vast majority of paper and cardboard waste goes to the landfill site with relatively small amounts going to the recycling company. The authors argue that because STAR (the waste removal company) gets paid per ton of waste at the landfill by the government, there is no incentive to divert waste to other facilities for recycling. The MFA graphically demonstrates the direction in which these waste streams are moving and the impacts that are accruing as a result of policy and practice.

ENERGY

In Chapter 2, both non-renewable and renewable energy sources were discussed and contrasted according to energy consumption by type. Non-renewable energy in the form of petroleum and natural gas are the dominant sources followed by coal and nuclear electric energy. Growth in renewable forms like hydroelectric, wind, biomass and geothermal energy is taking place, and their social acceptability is on the rise. Chapter 2 also included

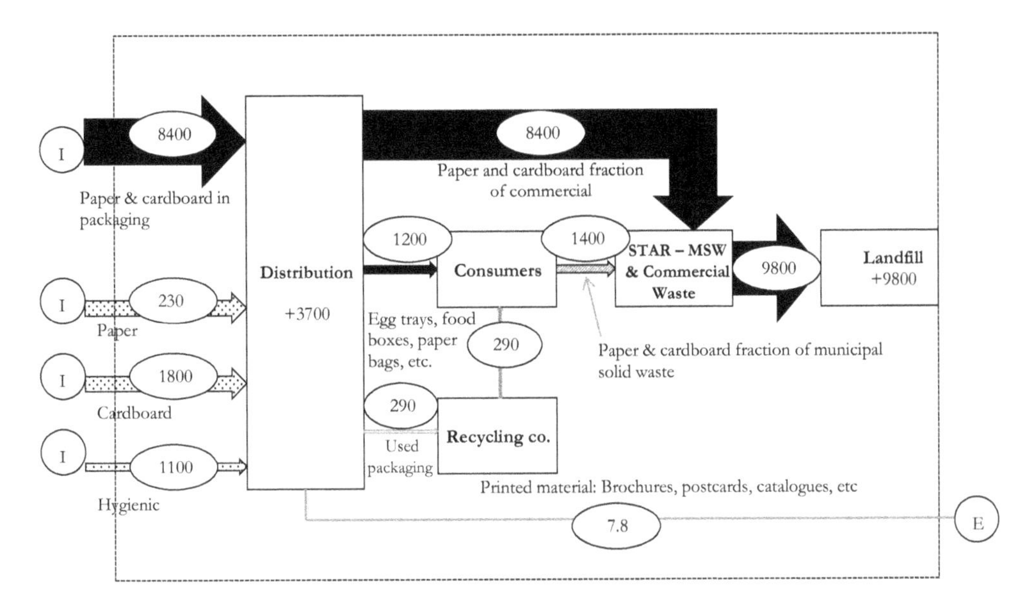

Figure 7.2 System diagram of 2014 material flows of paper and cardboard on Mahé

Notes: Black arrows indicate high data uncertainty; striped average uncertainty; dotted low uncertainty (numbers are in metric tons). I: import flows, E: export flows; STAR: a waste-handling company contracted by the government.

an overview of the importance of access to affordable, reliable, sustainable and modern energy for all (SDG 7). Priorities for the SDG include increasing the share of renewable energies by 2030, universal access to reliable energy, increasing energy efficiency, increasing global cooperation for clean energy, and expanding and upgrading the infrastructure that supplies sustainable forms of energy in developing and developed countries.

The collection of data on the use of new and existing energy sources cannot be overestimated in considering the pros and cons of these sources – hidden as they may be. For example, proponents of ethanol (in the USA ethanol as a fuel is derived mostly from corn) argue that it is a clean-burning fuel source, which has a lower environmental impact than fossil fuels. However, Pimentel (2003) found that ethanol is *not* in fact more environmentally sound because it produces a 29% negative energy balance; that is, 29% more energy in the form of high-grade fossil fuel is required to produce one gallon of ethanol than the energy in a gallon of ethanol. Pimentel lists the impacts from the production of ethanol from corn as taking more feed away from livestock production, causing more soil erosion than any other crop, and requiring the use of more insecticides, herbicides and fertilizers than other crops, all of which contribute to more water and air pollution. The ethical priority for corn, Pimentel argues, should be for food and feed.

Studies on energy use and efficiency in tourism are now appearing more regularly in the literature with a focus on accommodation, transportation and renewable forms of energy. An example of this latter category is wind energy. Some studies on the impacts of wind energy have established that the social impacts can be much greater than the environmental impacts. Saidur *et al.* (2011) found that wind energy is clean, environmentally friendly and ultimately cheaper than other sources of renewable energy. Water consumption, for example, is vastly decreased in comparison to petroleum-based power plants. Its social impacts, however, include visual impacts as well as noise pollution (mechanical noise from the generator and gear box, and aerodynamic noise, which is the noise of air passing over the blades of the turbine), leading to a reduction in property values. Tourists, however, are not reporting social impacts. Carr-Harris and Lang (2018) found that tourists viewed the USA's first offshore wind farm at Block Island, RI not as a deterrent but rather as an attractant. The wind farm has contributed to an 8% increase in occupancy rates and a US$1756 increase in revenue from June through September for accommodation units on the island. Furthermore, other positive impacts include the development of new tours around the wind farm, and the underwater infrastructure that supports the turbines has created better fishing opportunities and, as a result, more business.

Petrevska *et al.* (2018) provide examples of some of the energy eco-efficiency practices of leading hotel properties in the world. Hotel Bordessano in Yountville, CA was LEED Platinum certified in 2015 because, in part, of its rooftop solar array which supplies 50% of the total amount of electricity used by the hotel, with an estimated 6–8 year period to recover the investment. Novotel Jaragua in Sao Paolo, Brazil has an extensive vegetable garden which is organic, pesticide free and solar powered. Supply chain savings are realized as the garden produces food for the restaurant that is of a higher quality and standard than food that would need to be transported to the hotel externally. Finally, the Fairmont Resort Blue Mountains in Leura, Australia saves about 885 kWh/week in the reduction of energy from a 100 kW PV solar system. The system also saves about 155 kg/year of CO_2 for the entire resort. Other studies point out that very simple energy technologies can be used to great effect. For example, hoteliers who installed solar control film on windows in hotel rooms in subtropical regions, reducing the necessity to use air conditioning, saved 155 kWh per room annually. This reduced SO_2 by 920 g and CO_2 by 131 kg, with savings of between RMB 879 and 984 over a period of five years (Chan *et al.*, 2008).

Differences in the standards of hotels in relation to sustainability, as noted above, are a common theme in the literature. Díaz Pérez *et al.* (2019) found in an analysis of five-, four- and three-star hotels in the Canary Islands of Spain that the higher star hotels have energy and water consumption (seawater reverse osmosis systems in place) levels that are triple those of lower quality units, and five times the level of the lowest quality hotels. Furthermore, the emission coefficients were found to be three times higher than coefficients on the mainland of Spain because of the external energy dependence and the need to use water desalination processes, which require a great deal of energy. And while some higher standard hotel properties have eco-certification, there is still a great deal of ground to make up. This is the case in Macedonia, where five-star properties are investing significantly more in energy efficiency programmes than lower standard properties (Petrevska *et al.*, 2018). About half of the five-star hotels have an ecolabel and 60% said they were eco-certified. This contrasts with three- and four-star properties that had much less representation in these areas: 80% of three-star hotels and 52% of four-star hotels do not have eco-certification. A more general conclusion by the authors is that there is 'extremely limited use of alternative sources of energy and new innovative approaches to saving energy consumption' (Petrevska *et al.*, 2018: 403).

Furthermore, the water–energy nexus is an important consideration in understanding the degree of pressure that tourism places on local environments. Yoon *et al.* (2018) found in the case of Benidorm, Spain that six times the amount of energy is required in the use of desalinated water during dry years. As such, it is not water scarcity that is the problem – indeed the sea is abundant with water – but rather the amount of energy required to make water potable. Yoon *et al.* (2018) argue that there are now shifting scarcities taking place in such regions from water scarcity to energy scarcity given the demands placed on the electrical grid. And as this water needs to be purchased, the cost of water is now beyond the reach of some stakeholder groups like farmers for their total needs.

CONCEPTS AND TOOLS

Several tools are now being employed to mitigate the impacts of energy use in the tourism industry. Work is being done on indicators of renewable energy technologies (RET) and carbon reduction technologies, as well as a variety of studies using LCA that investigate the carbon intensities of travel to, from and within destinations. Tourists also have the opportunity to consult carbon calculators to gain an appreciation of the amount of carbon that results from their travel behaviour. Juvan and Dolnicar (2014) discovered that even though there are a variety of calculators, most were found to have limited functionality – they were difficult to use and were perceived to have poor credibility. Only four of 73 calculators reviewed were suitable in figuring the carbon footprint of a vacation.

Michalena *et al.* (2009) have studied the implementation of RET in islands of the Mediterranean in determining whether there is a correlation between renewable energy and ST development. The authors used an assessment framework as part of their methodology based on five main drivers leading to RET implementation and several indicators derived from various sources such as the literature. The drivers and examples of indicators for each driver are shown in Table 7.5.

Teng *et al.* (2012) developed energy conservation and carbon reduction (ECCR) indicators for the purpose of mitigating climate change emissions that stem from the activities of hotels in Taiwan. On the basis of interviews with hotel managers, environmental specialists and government officials, seven categories defined the boundaries of the ECCR framework. These include communication and participation, top management commitment, energy, water, waste, building and purchasing, along with 32 indicators that were weighted according to their value in the proposed framework. The connection between

Table 7.5 Renewable energy technology indicators

Category (driver)	*Indicator*
Processes	• Active participation of local communities • European and international collaboration and networking on renewable energy mechanisms, policies and technologies
Institution and policy	• Strong, constant and effective political support for the RET implementation process • Adequate funding for undertaking actions regarding research on RET
Competence and capability	• Positive impact of RET implementation in the market • Compatibility with other technologies (fossil fuels, etc.)
Technological and research statutes	• Obtaining targets regarding RET implementation at regional scale • Renewable energy production at a low cost
Environment	• Sustainability according to greenhouse pollutant emissions • Sustainability according to other pollutant emissions

Source: Adapted from Michalena *et al.* (2009)

categories, subcategories and indicators is shown in Table 7.6. After a process of prioritizing the indicators it was found that communication and participation was weighted highest by study participants followed by top management commitment. This suggests that implementation is foremost a function of management support and staff engagement, according to the authors.

Other Taiwanese studies on energy demonstrate the degree to which research is interfacing with practice in this island region. Like the study above on Benidorm, scholars are measuring the scope of the environmental impacts of tourism as well as the specific impacts that each tourist is having. A study on the carbon impact of international tourists to Taiwan found, by concentrating on the measurement of ground transportation (without focusing on how tourists got to Taiwan), that accommodation had the highest carbon (CO_2) impact (Tsai *et al.*, 2018). Measures to reduce this impact include the promotion of smaller units like bed-and-breakfast establishments which have lower CO_2 emission coefficients. The authors argue that the development of more convenient public transportation would induce tourists to rely less on cars in efforts to lower the carbon footprint.

Kuo and Chen (2009) used LCA to document tourism impacts on Penghu Island in Taiwan from the perspective of transportation, accommodation and recreational activities. This research included a system boundary built around: type of transportation from origin to Penghu Island (e.g. aeroplane); type of transportation within Penghu Island

Table 7.6 ECCR framework for the hotel industry

Categories	*Subcategories*	*Criteria/indicators*
1. Top management commitment	• Corporate ECCR management policies • ECCR audit measures	1.1 Develop ECCR management plans and policies 1.2 Give designated commissions or department heads responsibility for managing and supervising hotel ECCR performance 1.3 Periodically collect and monitor data for water, energy and waste consumption for ECCR auditing and management
2. Water	• Water conservation equipment and measures • Waste water management and reuse	2.1 Adopt water-saving devices or measures to reduce water consumption 2.2 Comply with government waste water regulations or principles 2.3 Recycle and reuse hotel waste water
3. Energy	• Electricity • Heat • Energy conservation measures	3.1 Adjust the air-conditioning according to the temperature in different climates and during different seasons 3.2 Install linking or auto-sensing systems for power supply 3.3 Use energy-saving lighting devices and avoid using incandescent and halogen light bulbs 3.4 Use energy-efficient refrigeration facilities and measures 3.5 Use heat pumps or heat recovery systems 3.6 Optimize measures to reduce energy consumption in response to weather change and peak usage 3.7 Install energy-efficient monitoring systems

4. Waste	• Waste management • Waste sorting, recycling and reusing	4.1 Reduce waste (e.g. avoid providing disposable amenities and tableware) 4.2 Comply with waste storage principles 4.3 Implement waste sorting (including kitchen waste) 4.4 Implement waste recycling and reuse
5. Building	• Green building materials • Building greening design and measures	5.1 Use environmentally friendly building materials 5.2 Grow plants on the rooftop and at base of building 5.3 Adopt building thermal insulation measures. 5.4 Adopt natural and sufficient lighting in the building 5.5 Ensure that the building has good insulation
6. Communication and participation	• Staff • Tourists/guests • Community/ government	6.1 Inform and educate staff about hotel ECCR policies and practices 6.2 Communicate with guests and request their cooperation in hotel ECCR options 6.3 Promote products and services that support ECCR policies 6.4 Provide guests with information about public transportation or bicycle rent services 6.5 Support ECCR activities hosted by the government and the community

7. Purchasing	• Purchasing policies • Green purchasing practices	7.1 Purchase Energy Star or high-efficiency products and facilities 7.2 Preferentially purchase local or seasonal goods and materials 7.3 Give preference to suppliers who adopt ECCR practices 7.4 Adopt renewable/recycled products 7.5 Purchase environmentally friendly cars for use as hotel vehicles

(e.g. rental car, tour bus); energy intensity and CO_2 emission of different types of accommodation (e.g. hotel, bed-and-breakfast); resource demand and pollution produced within accommodation (e.g. water demand, electricity used, solid waste discharge, wastewater discharge, BOD discharge); and energy intensity and CO_2 emission of different recreational activities (e.g. sightseeing, fishing, rafting). BOD is biological (or biochemical) oxygen demand, which refers to the amount of dissolved oxygen required by organisms to break down organic material based on temperature and time. The environmental loads per tourist on Penghu Island are reported in Table 7.7.

Table 7.7 focuses on two main units of analysis: environmental loads of tourists for an entire trip and environmental loads per day. The results indicate that individual tourists over the course of their trip used 1606 MJ of energy, 34.1 MJ of electricity and just over 600 L of water. Their contribution to air pollution was just over 109,000 g of CO_2, 2660 g of CO, just under 600 g of hydrocarbons, 69.5 g of nitrogen oxides, as well as significant amounts of wastewater discharge contributing to water pollution. The amount of solid waste discharged at 1.95 per tourists per trip appears to be less than in other reported studies in this book.

Filimonau *et al.* (2011) used life cycle energy assessment (LCEA) to measure the carbon footprint of two hotels in Poole, Dorset in the UK. LCEA is a derivative of LCA with a focus on the energy and carbon emissions as the basic units of analysis in the measurement of environmental impacts. This includes the use of a life cycle inventory where energy flows are pinpointed and subsequently quantified with data configured according to GHG emissions standards. Filimonau *et al.* (2011) note that LCEA is not a replacement for LCA, but rather has value as a particular application to energy consumption, especially in gauging the carbon efficiency of buildings. Two broad types of energy had to be identified and measured. The first is operational energy, which includes the ongoing activities of the units that use energy. These include air conditioning, heating, ventilation, elevators, appliances, cooking, catering, refrigeration, water heating, water supply, laundry, wastewater treatment and solid waste generation. The second type, embodied energy, includes the energy needed to construct the building, maintenance of the building and refurbishments.

Table 7.7 Environmental loads per tourist on Penghu Island

Basis	*Indicator*	*Environmental loads per tourist per trip*	*Environmental loads per tourist per day*
Resource demanded	Energy used (MJ)	1606	501.9
	Electricity used (MJ)	34.1	10.7
	Water demand (L)	606.7	189.6
Air pollution produced	CO_2 emission (g)	109,034.0	34,073.1
	CO emission (g)	2659.9	831.2
	HC emission (g)	596.8	186.5
	NO_X emission (g)	69.5	21.7
Water pollution produced	Wastewater discharge (L)	415.6	129.9
	BOD discharge (g)	83.1	26.0
Solid waste produced	Solid waste discharge (kg)	1.95	0.6

Source: Adapted from Kuo and Chen (2009)

Filimonau *et al.* (2011) found that the larger of the two hotels investigated had a much larger carbon footprint from calculations on a 'per one guest night' perspective. The researchers recommend the design of hotel spaces that are smaller and more efficient. For example, this larger hotel had a massive lobby and connected bar as compared to the smaller hotel, and most of this area was empty of people each time the authors made observations. A reduction of the gross floor area, therefore, would decrease the amount of energy needed to heat or cool these spaces. Another recommendation was to switch from electricity to gas for heating, ventilation and air conditioning because the latter would cut GHG emissions by 45%. In contrast, because the smaller hotel undertook regular refurbishment (about every two years), there were several ongoing indirect costs that contributed to higher energy and carbon impacts.

Scholars have also measured the carbon impact of short-haul tourism, suggesting that these trips are better for the environment. Filimonau *et al.* (2014) argue that carbon footprint assessment frameworks are relatively immature because they fail to capture both direct (operational) and indirect (embodied) life cycle elements inherent in the overall carbon footprint, with the hidden or indirect GHG emissions significant in overall accounting (see also Filimonau *et al.*, 2014), as noted above.

In this study, the authors followed the lead of the UK Department of Environment, Food and Rural Affairs (Defra) which has developed a more robust system of accounting for the GHG through conversion factors to estimate the carbon intensity of different products and services (Defra, 2010). Despite the positive changes, Filimonau *et al.* (2014) argue that this system is unable to capture the GHG emissions embodied in non-fuel chain related capital goods and infrastructure. They use the example of coach travel and the industrial processes that must take place in resource extraction, transport, refining these materials, the manufacturing process of the coach, delivery to the final destination and later disposal of the unit. The indirect carbon footprint, they argue, also relates to the energy required in renovations to the coach, refurbishment and maintenance, all of which magnify the carbon requirement for this one vehicle (see also Frischknecht *et al.*, 2007).

The indirect GHG emissions noted above may be accounted for in the LCA tool that is noted here and in other chapters. Although this tool is comprehensive and used in many industry sectors to account for the cradle-to-grave carbon impacts from the development, use and disposal of products, it still has some functional and methodological limitations as described by Filimonau *et al.* (2014), which hinder the precision of its estimates. LCA relies on life cycle inventories such as Ecoinvent (ecological inventories), to include the GHG emissions impacts from a number of products and services, but Ecoinvent suffers from irregular updates. Furthermore, background information on how appraisals are made is often missing, compromising the independent review of carbon calculations which are required in LCA estimates. Finally, Filimonau *et al.* argue that there are difficulties with the definition of what constitutes a short-haul trip, which obviously complicates GHG calculations on the basis of distance. In developing a more holistic approach to documenting the direct and indirect GHG emissions from tourism, a hybrid approach is adopted which combines Defra with LCA (Ecoinvent) 'in order to achieve the highest accuracy of estimates and to demonstrate the share of the "indirect" carbon footprint in the total GHG emissions from different holiday travel scenarios to Southern France' (Filimonau *et al.*, 2014: 633). The system boundary for carbon assessment based on transport from the UK to Southern France, and the associated activities that take place before the trip, during the trip and after the trip are located in Figure 7.3.

Based on the calculations of this model in light of different travel scenarios (car, air travel, train, coach, air + air), Filimonau *et al.* (2014) illustrate that train and coach are the most carbon-efficient modes of travel, producing less than half of the GHG emissions that air travel and car produce. Air travel produces the greatest carbon impacts of all scenarios tested. Airline travel represented 77% of the carbon footprint, followed by car at 71%, coach 43% and rail 40%. Similar findings were identified at the turn of the century by Becken (2001), who identified scheduled coach, motorcycle and shuttle bus as low-energy mode forms of transportation over those that have higher demands like airplane, camper van, ferry and private car.

Furthermore, Filimonau *et al.* (2014) found that the impacts of transportation on the overall carbon footprint fell the longer that tourists stayed at the destination, as tourists involved themselves in other activities and more stays at accommodation units which all accounted for greater uses of carbon. A 14-night stay in Southern France changes the

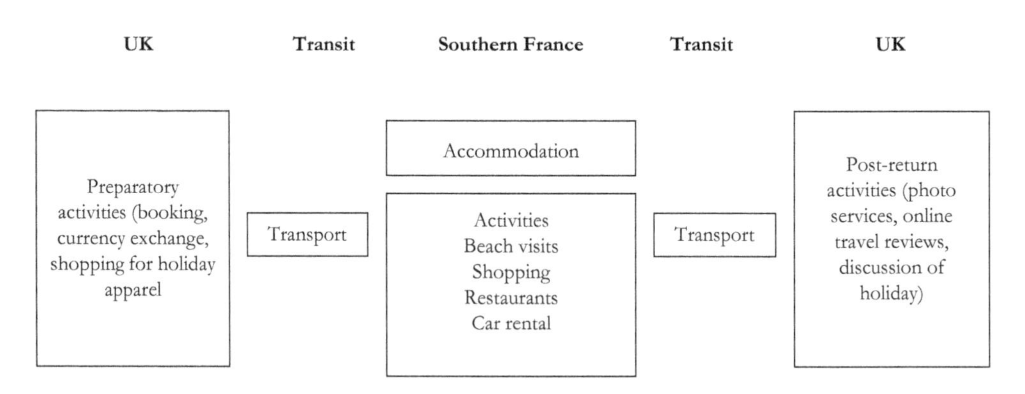

Figure 7.3 System boundary for carbon footprint assessment of holidaying in southern France

Source: After Filimonau *et al.* (2014)

relative share of total GHG emissions for transport accordingly: 65% air, 59% car, 31% coach and 27% rail. The daily carbon intensity loads for different transport scenarios is demonstrated over seven-night and 14-night durations in Table 7.8.

The daily carbon intensity of air and car transport for the longer stay scenario was 70–80% less than the shorter stay, while the changes are not as marked for train and coach which are changed in the order of 40–45%. Longer stays at the destination lead to more eco-efficiency when the metrics are assessed on a daily basis and are not cumulative. The energy bill of international aviation is large, to say the least. Becken (2002) found based on visitor arrival data generated by Statistics New Zealand that aviation is responsible for 27.8 PJ of total energy, which would increase New Zealand national energy use by 6% if such was included in national energy inventories. This use of energy was said by Becken to add 1.9 million tonnes of CO_2 emissions to the atmosphere.

The final piece of Filimonau *et al.*'s (2014) work is the assessment of the indirect GHG emissions based on a more comprehensive overview of resource extraction and manufacturing, as noted above. It is of particular interest to note that train becomes a less eco-efficient mode of travel when these other metrics are configured into the entire analysis, as we also note in Chapter 6. Filimonau *et al.* (2014) observe that:

> While rail is commonly considered as the most carbon-efficient transportation option, the holistic analysis of its carbon impacts shows that its advantage is reduced marginally when the 'indirect' carbon requirements are taken into account. All this emphasizes the necessity to include the estimates of the 'indirect' GHG emissions into carbon footprint assessments of tourism in general and holiday packages in particular. (Filimonau *et al.*, 2014: 635)

Sun (2014) followed the lead of Becken and Patterson (2006) who argued that two approaches could be used in accounting for the CO_2 emissions from tourism. The bottom-up

Table 7.8 Estimates of the 'daily' carbon intensity for different scenarios of holidaying in southern France (kg CO_2-Eq. per tourist)

Scenario	*Duration of stay at the destination*	
	7 nights	*14 nights*
Car	50.9	30.3
Air	52.8	31.3
Air + air	77.1	43.4
Train	20.1	14.9
Coach	25.9	17.8

Source: Filimonau *et al.* (2014)

approach quantifies tourism end-use behaviours and types of energy use and GHG emissions and is best suited to smaller regions where data collection on emissions is manageable. The top-down approach is based on environmental accounting procedures, which allows for the assessment of tourism emissions from a sectoral standpoint within the broader economy. Sun (2014) employed bottom-up and top-down approaches to more consistently measure the carbon footprint of the tourism industry based on a system boundary that included domestic tourism expenditure, inbound tourism expenditure and the spending of locals as outbound tourists, in a framework that measures domestic total carbon effect as well as foreign-sourced effects. The top-down approach was configured around the Tourism Satellite Accounts (TSA) as well as an environmentally extended input–output model (EEIO) to compare supply and demand data on production and consumption. Sun contends that TSA data provides data on tourism expenditures, while the EEIO model quantifies direct, indirect and induced effects of GHG emissions based on tourism consumption (see also Dwyer *et al.*, 2010) in comparing tourism with other sectors of the economy with regard to eco-efficiency. Results indicate that tourism consumption contributed 2.16% to Taiwan's GDP, and represented 3.08% of national GHG emissions. Within tourism, the sectors that produced the greatest amount of CO_2 were aviation (38%), followed by land transportation (21%), shopping (17%) and food (13%). Sun argues that island environments that rely on imported materials and aviation lead to larger emissions outside the country's territory in the development of tourism.

The immediacy of the need for energy efficiency in tourism is so important that if short-term improvements are not made then governments will be forced to impose stronger regulations, carbon or energy taxes, pulling back on tourism investment or controlling the flow of tourists within regions (Wang & Wang, 2018). It is also the role of policy makers, Wang and Wang observe, to stimulate reform in the tourism industry through the development of newer and greener technologies. A well-cited example of the problems of implementing measures to control over-exploitation of resources is the Balearic accommodation eco-tax law, which 'is a tourist, extra-fiscal, finalist, earmarked and direct tax' (Ariño, 2002: 172). Every tourist client visiting an accommodation unit in the islands must pay the tax directly. Although there is a history of protection of natural resources in the Balearics, the implementation of such a tax was met with contempt from those working in the tourism industry, especially the accommodation sector.

WATER

This book has already noted how tourism products are strongly connected to water resource attractions and destinations (Narayana, 2013). Lakes, rivers, seas and oceans are all great attractors to destinations around the world, and properties hold greater value if they are on or adjacent to bodies of water. The cruise line sector has capitalized on the value that water (seas, oceans, lakes and rivers) holds for tourists, but it operates at a cost environmentally (garbage, dredging), socially (some benefit while the vast majority of other community members do not) and economically (tourists eat and sleep on boats, taking money away from the accommodation and food and beverage sectors of the destination). As noted by Klein (2011), cruising can be more sustainable through spreading benefits to local people, greater ecological accountability, and respecting the carrying capacity of ports by limiting the number of passengers disembarking at these places.

Water as an attraction or facilitator, however, is different from water for personal direct consumption, and indirectly as a resource for golf courses or ski hills, and the list goes on. Tourism facilities often overuse water systems to a point where other stakeholder groups are placed at a disadvantage because of the lack of water. The UN SDG 6 on clean water and sanitation stresses the need for improved water quality, water-use efficiency, integrated water resource management, the protection of water-related ecosystems, better sanitation, and capacity-building for better cooperation and support of these initiatives.

Many of these water-related issues have been examined through the discourse on water rights and justice. Risse (2014) suggests that we ought to have a global water compact with an accompanying monitoring body making sure that every human being has access to this vital resource as entitled co-owners. By this he means that humanity collectively owns the Earth (anthropocentric as this sounds or is intended) in a form of common ownership. If individuals were unable to satisfy basic needs because of an absence of access to clean, potable water, a just global society would establish the means to overcome these disparities or barriers. For Arden (2016) the human right to water is surely a 'paradigm universal right', i.e. the right to water must be the same for everyone. And for Hayward (2016) a global right of water ought to be subjected to normative standards that are institutionalized because of the challenges of our time: high ecological demands and sociocultural inequality.

The Universal Declaration of Human Rights and other conventions did not mention the human right to water. This changed in 2002 (contained in Document E/C.12/2002/11) with the Committee on Economic, Social and Cultural Rights which, through its General Comment 15, asserts that, as a legal basis:

> The human right to water entitles everyone to sufficient, safe, acceptable, physically accessible and affordable water for personal and domestic uses. An adequate amount of safe water is necessary to prevent death from dehydration, to reduce the risk of water-related disease and to provide for consumption, cooking, personal and domestic hygienic requirements. (UN Economic and Social Council, 2003: 1)

Risse (2014) lists other international bodies recognizing the importance of water as a right, including the UN General Assembly in 2010 and the Human Rights Council in the same year. Some regions, such as Central Asia, have established intergovernmental bodies to deal with water distribution. The Interstate Commission for Water Coordination is a parity collective body which is jointly managed by member states for water equity, equality and consensus (Javorić Barić, 2016). Some countries have the right to water mentioned specifically in the Constitution (South Africa), while others do not, but maintain these rights through vehicles such as social security rights (Belinskij & Kotzé, 2016).

Gawel and Bretschneider (2017) argue that the right to water is often limited because of a series of hurdles. These include: spatial hurdles, or the distance an individual has to travel to the supply point; temporal, the time that an individual must wait to receive or use water; qualitative, if water quality requires some active intervention to make it usable (e.g. boiling); and pecuniary, which makes reference to the price or storage facilities to use and keep water.

In Bali, Indonesia, Cole (2014) reports that even though the region is blessed with abundant rain, mismanagement of the resource has led to water scarcity (for local people, not tourists) and water pollution. Cole argues that it is not only government that has a legal duty to supply water to all stakeholders in Bali, but business too. The argument is that businesses ought to conduct human rights impact assessments, which would lead to competitive advantage and greater reputational strength. The UN (2011) publication, *Guiding Principles on Business and Human Rights*, outlines principles for general areas of human rights: the state duty to protect human rights, the corporate responsibility to protect human rights, and access to remedy. Businesses can have a direct and significant impact on the right to water through their policies and practices. These are typically derived in three different ways (Gaughran, 2012: 53):

- Where business is involved in the provision of water services.
- Where business is a user of water – and particularly where water is a limited resource and business is competing with other users.
- Where business activities that are unrelated to water per se have an impact on water sources (for example, where industry causes pollution of water systems).

Other studies call for a transformation in the power relationships that exist for the provision of water, from one centred on the technocracy, businesses and policymakers, to one where these groups are informed by civil society in the creation of a new reality where there is access to water for all (Loftus, 2015). Civil society, therefore, is a key driver of change where the most vulnerable segments of global society have voices through these drivers.

TOURISM AND WATER

Water as a focus of research in tourism has followed several courses. Studies have identified the amount of water that tourists use on a per day basis, as well as different forms of tourism like mass tourism, as well as issues of rights and justice around the fair and equitable distribution of water resources for tourists and local people.

Gössling (2001a) investigated the water utilization rates of the tourism industry on the east coast of Zanzibar, Tanzania. He found that the yearly increases in tourism numbers, which are expected to continue to rise in the future, mean that the associated demand for water has shown signs of considerable overuse. Specific impacts include the lowering of the groundwater table, problems for land subsistence, deterioration in the quality of groundwater and the intrusion of saltwater. Recommendations include lowering the average daily consumption of water to about 200 L per day per tourist. This will be difficult, Gössling adds, given that some hotels have per capita demand of around 6000 L per day. Suggestions to reach a more sustainable usage of water include planting more drought-resistant plants as irrigation is a major factor in water usage, importing water via pipeline, and desalination, with the latter two options deemed very expensive. It would seem more appropriate to fix the local water shortage problem as opposed to simply finding more water and supporting unsustainable practices. This case study does indicate that sustainability is situational. Blanket solutions covering numerous resource issues will be difficult to manage. Tourism should thus be developed according to the resource conditions of the destination.

In a study of the per guest night water intensity usage by Becken (2014), it was found that tourists visiting developing countries used much more water than local people; in some cases tourists used more water by a factor of eight times (Fiji and Sri Lanka). Becken also found there was great variability in water use both within and between countries, suggesting that water management initiatives, if implemented, could easily reduce water usage. Furthermore, water-rich countries such as New Zealand were not water-intense users, with relatively high efficiencies that helped to conserve this resource. In these countries there was not such a significant difference between tourist and local community users of water. Yang *et al.* (2011) found that average tourist consumption of water in the Liming Valley of Northwest China is 5.2 m^3 per day, with food production and waste (bath/shower, toilet, wash basin, washing machines, kitchen functions) being the top two uses. The authors found that tourist use of water was greater than usage by local people.

The issue of a more balanced approach to water distribution between locals and tourists has been addressed by Morinville and Rodina (2013) with regard to displaced Bushmen of the Central Kalahari who have been taken from their ancestral lands and denied a proper supply of water. While tourist and mining enterprises were allowed to use water resources in the Central Kalahari Game Reserve, San and Balgalagadi people were denied access and relocated to 'service centres', justified on the basis of the fact that these centres had more amenities and infrastructure. Research by LaVanchy (2017) is another good example of the problems that exist over water sustainability in developing countries where tourism has been given priority. LaVanchy found that in Gigante, Nicaragua, all users are reliant on groundwater, but water removal has been exceeding the capacity of the aquifer to recharge itself. The problem is said to be the tourism industry because international tourism arrivals use water at a much greater rate than the local populations, as noted above. Water 'grabbing' is now a practice of those in power, forcing the marginalized local population into a position of lack of control and access. Many local people must walk great distances to access water that once was much more accessible, which LaVanchy characterized as a 'class of losers' in the political ecology of the demands and normative expectations of global tourism.

The problem of tourism as a heavy water resource user is aggravated by bigger and higher standard enterprises. Hof and Schmitt (2011) observe that the critical water supply situation on Mallorca is being worsened by higher end 'quality' forms of tourism as well as by second homes. In this latter capacity, wealthy urban dwellers have purchased second homes in the mass tourism hinterlands with an accompanying rise in water consumption for outdoor uses, including gardens and swimming pools. Others, including Gabarda-Mallorquí *et al.* (2017), argue that because of their compact footprint and vertical urban configuration, mass tourism establishments can have higher water efficiency characteristics than low-density tourist establishments. In a study of hotels in Lloret de Mar on the Costa Brava in Spain, the authors provide quantitative evidence that larger hotels with greater guest capacity and larger swimming pools use less water per guest night than hotels with fewer beds and smaller swimming pools. This is explained by economies of scale, where water consumption decreases with an increase in guest numbers due to efficiencies in the hotel's water system. As such, higher occupancy rates, longer opening periods, greater technological innovations, maximization of the influence of water-saving devices, efficient management of on-site kitchens and environmental management programmes are all factors that lead to this finding.

Other studies have pointed to the resilience measures employed by the tourism industry to withstand environmental pressures. Dinarès and Saurí (2015) discovered that environmental events like droughts have induced Barcelona hoteliers to adopt technological and behavioural change in terms of water conservation measures (cost-cutting behaviour), with higher star hotels being most active in this regard because they use more water than lower quality establishments. Droughts can have crippling effects on tourism destinations. Such is the case in Benidorm, Spain, which suffered through the drought of 1987, causing significant economic losses with associated damage to the image of Benidorm as a holiday destination (Martinez-Ibarra, 2015). Precipitation in the form of intense rainfall episodes is important for the water supply of Benidorm, and the city is now much more resilient because of the active search for newer surface and groundwater resources.

Sustainability is also about returning to more traditional means by which to use and handle resources. Enriquez *et al.* (2017) investigated how to return to the use of sophisticated pre-modern water cisterns on the island of Santorini in Greece. More recently, inhabitants of the island have been forced to modernize their access to water through the use of desalinization, ship deliveries and well withdrawals because of excessive water consumption from a vastly different population profile in the form of more locals and tourists. The authors conclude by suggesting that the most likely solution is to use the cisterns as decentralized storage reservoirs as part of the broader centralized water infrastructure, and also to use the cisterns as educational devices on the importance of sustainable water harvesting on the island.

CONCEPTS AND TOOLS

This section started off by suggesting that sources of water like seas, lakes and oceans have a great appeal for tourists. Beaches hold considerable appeal by virtue of the sheer number of tourists who use this resource for sea, sun and sand. Beaches are able to accommodate a certain level of demand over time – the carrying capacity – without significant degrees of degradation, but often beaches succumb to too much use because of garbage, gas and oil from motorboats, food, the attraction of problem animals, human waste and improper sewage systems.

The Blue Flag Program (BFP) is renowned as the world's first international ecolabel scheme. Launched in France in 1987, it has as of June 2018 certified 4554 beaches, marinas and ecotourism boats in 45 different countries (Blue Flag, 2019). Beaches are evaluated on the basis of five parameters, including beachfront seawater quality, beach quality (solid waste and wastewater management), drinking water quality, environmental education efforts, and beach safety and administration. Each parameter has a maximum score attached to it, and beaches must obtain a score of 90 out of a possible 100 points (Blackman *et al.*, 2012). Researchers have found that the Blue Flag scheme has an economically significant effect on the development of new hotels, especially new luxury hotels – which in Costa Rica has translated into the construction of 12–19 more hotels per year (Blackman *et al.*, 2012). Research has also confirmed the programme as a positive agent for the improvement of coastal zone management, and the promotion of environmental education among local stakeholders (Creo & Fraboni, 2011).

Styles *et al.* (2015) have developed an innovative model to calculate water and energy savings that may be achieved through the implementation of best practice and performance indicators, for a 100-room hotel and 60-pitch campsite, using data for fitting flow rates, irrigation, swimming pools, kitchens, laundry generation, laundry efficiency, dishwashing and flush volume. The key performance indicators and benchmarks for kitchens and swimming pools are found in Table 7.9.

Table 7.9 Key performance indicators and benchmarks for kitchen and swimming pool area best practice

Best practice	*KPIs*	*Benchmarks*
Optimized dishwashing, cleaning and food preparation	L/cover Management plan L/rack (dishwashers) L/min (pre-rinse spray valve)	• Implementation of a kitchen water management plan that includes monitoring and reporting of total kitchen water consumption normalized per dining guest, and the identification of priority measures to reduce water consumption Installation of efficient equipment and implementation of relevant efficient practices
Optimized pool area management	Ecolabel chemicals Natural pool L/m^2 year L/g$n. kg chemicals/m^2 year kg chemicals/g$n. kWh/m^2 year kWh/g$n. Benchmarking	• At least 70% of the purchase volume of chemical cleaning products (excluding oven cleaners) for dishwashing and cleaning are ecolabelled[a] • The on-site swimming pool(s) incorporate(s) natural plant-based filtration systems to achieve water purification to the required hygiene standard Implementation of an efficiency plan for swimming pool and spa areas that includes: (i) benchmarking specific water, energy and chemical consumption in swimming pool and spa areas, expressed per m^2 pool surface area and per guest-night; (ii) minimization of chlorine consumption through optimized dosing and use of supplementary disinfection methods such as ozonation and UV treatment[a] Optimization of backwash control based on pressure-drop data, use of a pool cover overnight to reduce evaporation and installation of low-flow timer-controlled showers

Source: Adapted from Styles *et al.* (2015)

Notes: [a]Chemical consumption/g.n. benchmark (housekeeping section) also applies.

From a synthesis of all the data, the following results were generated (Styles *et al.*, 2015: 187):

- Benchmarks range from ≤58 L/guest-night on campsites to ≤140 L/guest-night in hotels.
- Achievable water savings amount to 227 and 128 L/guest-night for hotels and campsites, respectively.
- Annual savings for a 100-room hotel amount to 16,573 m^3 water, 209,541 kWh energy and €58,463.
- Best practice across European hotels and campsites could save 422 million m^3 water per year.

In other research, Becken and McLennan (2017) established benchmarks for water, depending on accommodation type (vacation, business, villas), of between 358 and 1338 L/guest-night based on a review of 876 environmentally certified hotels globally. Gössling (2015) proposes the use of new indicators of water consumption in the tourism industry, and a move away from conventional direct use (on-site) water consumption benchmarks. Using the example of the accommodation industry in Rhodes, Greece, Gössling's new performance indicators include both local and global (goods produced elsewhere including food and fuel) water usages for sustainable water management. These indicators include:

Area situation

Indicator 1: Renewable water resources per guest night in peak season.

Planning accommodation

Indicator 2: Area of irrigated land per bed;

Indicator 3: Area of pool per bed;

Indicator 4: Area of solar thermal and PV installed per bed.

Operating accommodation

Indicator 5: Amount of meat and dairy products per guest night;

Indicator 6: Energy use per guest night;

Indicator 7: Share of rooms fitted with low-flow options;

Indicator 8: Kg of laundry used per guest night.

Beyond benchmarks and indicators, the water footprint (WF) has been a frequently used measure to determine the environmental impacts of too much use. Mi *et al.* (2015) studied tourism water usage at the agricultural heritage site of Hani Rice Terraces in Yuanyang County, China, and found that the sewage WF (38.33% of all water usage) and the diet (food) WF (36.15%) were the highest contributors to the overall tourism WF. Transportation usage followed at 21.47% and finally water usage by accommodation (5.05%). Their study indicates that it is the indirect usage of water that is most responsible for enlarging the overall WF. This latter finding has been observed by Gössling *et al.* (2012), who argue that indirect water requirements such as food production, building materials and energy requirements are likely more substantial than direct uses of water in tourism. These authors also conclude that the direct use of water for tourism is much less than 1% of global water consumption, and this figure will not become significant even with tourism's forecasted growth patterns of approximately 4% per year. Notwithstanding, the authors point to the need for continued vigilance around water energy and water management, with a particular focus on policy, management, research and development, as well as education and behaviour change.

Two studies in Spain – a popular destination for collecting environmental data on tourism – show how the WF has been useful in tracking water consumption. Cazcarro *et al.* (2014) report that the total WF of foreign tourism in Spain is 3.74 km^3/year and for national tourism 3.25 km^3/year. The three largest sectors of water use in Spain included: food (1.41 and 1.16 km^3/year for foreign tourism and national tourism, respectively); followed by restaurants, coffee shops and bars with 1.04 and 1.05 km^3/year for foreign tourism and national tourism, respectively; and hotels and bed-and-breakfasts at 0.42 km^3/year and 0.27 km^3/year. The authors conclude by suggesting that water use measurement and governance is especially important in arid and semi-arid countries. In the second study, Cazcarro *et al.* (2016) used input–output analysis to track WFs at both micro (municipality) and meso (regional) levels, allowing the researchers to track water consumption through the entire supply chain. They found that WFs of domestic tourism were highly dispersed geographically, while foreign tourism, being more highly concentrated in time and space, had higher WFs per capita, i.e. higher water intensities per Euro spent by tourists, in the order of 0.1 cubic metres/Euro.

Hadjikakou *et al.* (2013) also emphasize the importance of using both local and global pressure on water resources through their study of five destinations in the eastern Mediterranean (they developed and analysed the WFs of different hypothetical holiday packages). The most important factor in minimizing the WF was the choice of a destination closer to home in association with a largely vegetarian diet. The second most important factor was the choice of a destination, with budget options having a much smaller WF than luxury accommodation options. Given the magnitude of growth in the tourism sector, and the stress already being placed on water resources, the authors argue that water system sustainability will be an important issue in the immediate future.

New approaches to WF analysis are proving essential in examining both touristic and agricultural consumption patterns. Kourgialas *et al.* (2018) developed a new tool for the assessment of groundwater footprint (GF) analysis on the island of Crete, given the importance of agriculture and tourism to this destination. The GF is defined as 'the water budget between inflows and outflows in an aquifer system and is used as an index of the effect of groundwater use in natural resources and environmental flows' (Kourgialas *et al.*, 2018: 381). What is novel about their methodology is that it is able to assess both the quantity and quality of groundwater use. They found that only three of the 11 groundwater systems in Crete showed values that indicate moderate or significant aquifer stress.

Turning our attention to more of a focus on social research, Borden *et al.* (2017) investigated social marketing initiatives promoting water-efficient actions on the part of small/medium-sized tourism enterprises (SMTEs) (interviews with 16 managers), and a subsequent survey on how these initiatives would impact on the behaviour of potential guests ($n = 408$). 'Social marketing is the design, implementation, and control of programmes calculated to influence the acceptability of social ideas and involving considerations of product planning, pricing, communication, distribution, and marketing research' (Kotler & Zaltman, 1971: 5). The authors found that guests were in favour of the types of actions that these SMTEs were unable to provide, such as money-off vouchers and donations to charity, because of their small size (e.g. financial limitations). These enterprises, the authors note, are at a competitive disadvantage to larger firms, which have the resources to promote water efficiency measures, some of which (e.g. Accor) have already implemented these types of programmes. Social marketing campaigns are reported to have made important contributions to ST through behaviour change about environmentally appropriate behaviour. However, such change can only take place in parallel with long-term commitments with private sector corporations (Truong & Hall, 2017).

Earlier in the book it was illustrated that reporting by firms holds tremendous potential for CSR and better environmental management in tourism, but the practice is still in its infancy with significant variability in how firms report and the extent to which they report. Two studies are worthy of consideration in the link between tourism, reporting and water. As noted in Chapter 4, the GRI is being used widely for disclosing information on the CSR of tourism and hospitality firms. Kleinman *et al.* (2017: 343) investigated the water disclosure actions of seven food and beverage corporations using GRI guidelines, and isolated four water indicators: (1) EN8 Total water withdrawal by source; (2) EN9 Water sources significantly affected by withdrawal of water; (3) EN10 Percentage and total volume of water recycled and reused; and (4) EN22 Total water discharge by quality and destination. They found that water consumption and water withdrawal were cited most often in the reports of firms, and that water disclosures vary considerably in content. This variability was the source of some concern for the authors, who argue for a more consistent set of procedures around reporting.

Finally, de Grosbois (2016) found in an investigation of the reporting and commitments of specific CSR goals of the cruise tourism industry that much more work on the part of this sector is needed. The industry demonstrated limited use of international reporting guidelines and an almost universal absence of third-party assurance of reported information. More specifically, company websites were unclear in their presentation of CSR information with little reference to time frames and the scope and sources of information. And while information on environmental and community issues were addressed, industry economic benefits, employment equity, and diversity and accessibility were not at all well represented on the websites.

CONCLUSION

We expect to see a great deal more growth in the paradigm change that is taking place with tourism and sustainability, especially as it applies to the four sectors included in this chapter: food, waste management, energy and water. Tourism scholars should continue to investigate these four important areas, and it would be prudent to forge ahead on both interdisciplinary and transdisciplinary fronts. There is tremendous scope for the intersection of various stakeholders that have vested interests in the social, economic and environmental dimensions of tourism in a variety of different contexts, as we have discussed in other sections of the book. This chapter has also emphasized a number of sustainability concepts and tools, such as ecolabelling, certification, life cycle assessment, waste mapping, materials flow analysis and corporate social responsibility, among others. These instruments must continue to play an important role in finding solutions to some deep-seated environmental issues in these sectors. Leadership must also come form the UN SDGs, which provide an excellent template for moving the sustainability agenda forward in tourism and in general.

END OF CHAPTER DISCUSSION QUESTIONS

1. What are some of the environmental impacts that are derived from the livestock industry?
2. Are consumers willing to pay more for ecolabelled products? If so, why?
3. List some of the hazardous and non-hazardous waste products in the hotel industry.
4. How have wind turbines been used to bolster the tourism industry in Rhode Island, USA?
5. Why is life cycle assessment such a popular sustainability tool in measuring the carbon footprint? Identify some countries where this tool has been used.
6. Identify some of the rights and justice issues associated with to the use of water.

End of Chapter Case Study 7.1 Sustainable tourism: A case study of Nainital, Utterakhand, India

In this case study by Puskar (2011), the tourism industry in the Utterakhand region of India is investigated for the purpose of reconciling increased visitation to this beautiful area against the waste that is created in the wake of such visitation. What is interesting about this case study is that the author is not pointing the finger squarely and entirely at tourists and the tourism industry, but at the actions of local people first and foremost. Proper waste management is primarily a concern for communities to organize and manage, and tourists cannot be blamed if local people are not able to act as good role models.

In Puskar's words …

The tourist is as good as a local person when he visits a place. If a local flouts the rule of the law, the immediate reaction of the tourist is to follow him because he thinks that if the local can do it, so can I. In other words, the actions of the local people are imitated by tourists. If a place is such that nobody follows traffic rules, will a tourist follow them? He realizes that the rules are not stringent in that particular place and starts flouting the rules himself.

Take the case of Nainital, a popular tourist destination of Uttarakhand that was being affected by pollution as a result of the negative impacts of the tourism industry. Nainital is entirely dependent on the tourism industry with little production happening in the town. It is the main livelihood of the people. The people of Nainital realized the ill effects of pollution which would one day threaten the very existence of the town itself. A petition was filed. Following are the excerpts of the Court order:

> 'Nainital, a beautiful butterfly, is said to be turning into an ugly caterpillar. … The growing traffic, with the growth of the town and big turnout of tourists, has contributed much to the environmental pollution. The increased traffic has in its wake brought noise pollution.'

The Court set up an enquiry committee, whose findings were as given below:

- Heavy vehicles like buses on the Mall Road and the bridle paths. They also enter Malli Tal and Talli Tal Bazars. I myself have seen it happening in Nainital, when I used to board a bus for Delhi. It used to be invariably from the Mall Road although the road was not meant to face such heavy traffic as it was already crowded by the tourists taking a stroll in the evening.
- The lake water was found full of human waste and horse dung and other wastes, as already noted. The horse stand having been allowed to be erected near the lake and trotting around the lake being permissible, the report states that horse dung in abundance enters and reaches the lake. The tourists who enjoy boating in the lake throw leftover edibles and polythene bags in the lake.
- Hill cutting and destruction of forests was confirmed. Construction of buildings is going on unauthorized and in a big way. The Commissioner has mentioned about illegal construction of office even by Kuman Mandal Vikas Nigam of the State Government and Lake Development Authority, which constructed several triple-storeyed flats which have been declared as dangerous.

Following the petition, an order was passed by the Bench stating that heavy vehicular movement be banned on the Mall Road, care to be taken that horse dung should not reach the lake, the Horse Stand was subsequently shifted far away from the main town to its present location at Land's End. I was a student in Nainital when this shifting took place. Horse riding was very popular among the tourists in Nainital and much hue and cry was raised; the horse owners pelted stones at the police.

It is sad that the horse owners lost their income but sometimes decisions need to be taken keeping in mind the larger interests of public. I myself have witnessed less dirtiness on the roads; earlier the roads were covered with horse dung which not only spreads pollution but is a primary source of disease too.

A movement called the 'Mission Butterfly' was launched calling for action from the local community of Nainital. It is an integrated solid waste management programme (ISWM) promoting the whole-of-life (cradle to grave) management of solid material wastes with strategies for recycling and minimization and is owned and managed by the community notionally divided into clusters (Swacchtha Samiti) of about 1000 persons (say 250 families).

The key features of the scheme are the well-connected Women's Health Workers (ASHAs) who already operate in the community; the use of dedicated Jumbo Bins (green for bio-degradable, i.e. composting and blue coloured for recycling and disposal); garbage collectors; the provision of services for support and training; and various systems for waste management.

It requires the collection, segregation and management of waste and the use of composting and recycling facilities including privately maintained dry-storage facilities. Income is generated from the collection fees, fees for consultants and service providers, the sale of resultant compost and the recyclables such as plastic, metals and glass.

To encourage wider participation and the evolution of community spirit, membership signs (Green Home logo) are provided to signal household and institutional support. The Mission Butterfly integrated solid waste management (ISWM) programme is managed by the Lok Chetna Manch as the executing agency for a two-year transition period before assumption by the Nainital Nagar Palika Parishad (Municipal Board).

Income generation occurs from fees collected from households, hotels and restaurants, institutions and other waste generators; sales of compost, shredded paper and recyclable waste (e.g. plastics, glass and metals).

The tourists were checked to see if they were carrying any plastic or polythene with them and requested to part with it before entering the town. It is hoped that slowly they would become responsible tourists and while visiting the town again they would not bring any plastic with them. Outside vehicles were refused entry and a new taxi stand was constructed outside the town to reduce the traffic pressure inside the town.

A tourist is as good as the town that he goes to. If you give him clean facilities, a clean town, he will keep it clean. A dirty town polluted by the inhabitants themselves will only invite more pollution from the visiting tourists. Responsible tourism did start in Nainital but the first step was taken by the society of Nainital. We have examples such as the Delhi Metro which is kept spick and span in spite of being used by thousands of locals and tourists. They never throw any litter and follow queues and these are the same tourists who throw waste in other areas of Delhi. Tourism is a wonderful way of earning income and for a town like Nainital which is dependent on the tourism industry, it is pertinent to make the tourist a model tourist so that he also enjoys visiting the town and the town also derives benefits from his visit.

I want to quote the court's order again here:

> *'We part with the hope that the butterfly would regain its beauty and would attract tourists not only in praesenti but in future as well, which would happen if the beauty would remain unsoiled. Given the will, it is not a difficult task to be achieved; the way would lay itself out. Let all concerned try and try hard. Today is the time to act; tomorrow may be late.'*

Source: Puskar (2011)

CHAPTER 8

KEY AGENCIES AND INFLUENCERS OF SUSTAINABLE TOURISM

LEARNING OUTCOMES

This chapter focuses on the key agencies and influencers of sustainable tourism, including the private sector, sustainable citizens and the role of education. The chapter is designed to provide you with:

1. An awareness of the roles and programmes of the key agencies involved in sustainable tourism including those in the public and private sector and pressure groups.
2. An understanding of the changing nature and role of policy for sustainable tourism and the debate surrounding its effectiveness.
3. An analysis of the key private sector organizational arrangements that facilitate sustainability, including 'public–private partnerships' and 'benefit corporations'.
4. An understanding of the influence of the sustainable citizen and their actions.
5. An awareness of the influential role of education for sustainability and the key organizations that serve education for sustainable tourism.

INTRODUCTION

This chapter introduces you to the key agencies and influencers of sustainable tourism (ST). The chapter outlines the key public and private sector agencies and organizations that have been influential in promoting and implementing ST. What is clear is that the private sector now plays a greater and more influential role than in the past and that here ST is a source of innovation. The chapter then goes on to consider public sector policy for tourism and shows how, despite the move to integrated 'whole of destination' policy and governance, some commentators still doubt the effectiveness of ST policies given the continued negative impact of tourism. The influential role of the private sector and in particular organizational arrangements that facilitate sustainability are then considered before moving on to the influential role of the sustainable citizen. The chapter closes with an examination of the key role of education for sustainability and the key organizations involved. The content in this chapter focuses on a number of the UN's Sustainable

Development Goals (SDGs), but in particular SDG 16 which is about building accountable and inclusive institutions and SDG 12 which is about responsible consumption and production patterns.

KEY AGENCIES

INTRODUCTION

There are many organizations involved in both promoting and implementing ST and its principles, all striving towards SDG 16 to build accountable and inclusive organizations. These range from intergovernmental bodies such as the UN World Tourism Organization, through major trade bodies including the World Travel and Tourism Council, to pressure groups and charities such as Greenpeace and the World Wildlife Fund. Each of these organizations has a particular niche in the ST system and attempts to influence the sustainability agenda in different ways. In this section we detail the role and activities of the key agencies.

INTERGOVERNMENTAL BODIES

The World Tourism Organization

The World Tourism Organization (UNWTO, 2019a) is the UN agency responsible for the promotion of responsible, sustainable and universally accessible tourism (see http://www2.unwto.org/). The UNWTO's membership includes 158 countries, six Associate Members and over 500 Affiliate Members representing the private sector, educational institutions, tourism associations and local tourism authorities. The UNWTO is the leading international governmental organization in the field of tourism and as such balances its call for tourism as a driver of economic growth with the need for inclusive development and environmental sustainability. The UNWTO is particularly effective as an influencer in developing a robust research approach to sustainable development.

The Committee on Tourism and Sustainability directs the UNWTO's work on ST. The committee's main role is to monitor the implementation of the programme of work under 'Sustainable Development of Tourism'. The programme of work focuses around four main areas.

Firstly, the UNWTO has developed the Global Code of Ethics for Tourism which is designed to maximize tourism's socioeconomic contribution while minimizing its possible negative impacts (see http://ethics.unwto.org/content/global-code-ethics-tourism).

Secondly, the organization is committed to promoting tourism as an instrument for achieving the UN's SDGs, particularly SDG 12, to ensure responsible consumption and production patterns.

Thirdly, the UNWTO is mainstreaming sustainable consumption and production in tourism through the 'One Planet Sustainable Tourism Programme'. This has the overall objective of enhancing the sustainable development impacts of the tourism sector by 2030, by developing, promoting and scaling up sustainable consumption and production practices. This is recognized as an implementation mechanism for SDG 12.

Fourthly, the UNWTO's Sustainable Tourism Programme (STP) promotes and enables transformation of the tourism sector for enhanced sustainability, through evidence-based decision making, efficiency, innovation, collaboration among stakeholders, monitoring and the adoption of a life cycle approach for continuous improvement. The programme has four key work areas:

1. *Policy*: This work area aims to strengthen ST policy making by the promotion of the integration of ST principles into tourism policies and the legal framework, fostering the implementation of these policies and monitoring progress.
2. *Evidence*: This work area involves the promotion of data sharing and the exchange of information, fostering joint action and tourism stakeholder collaboration at all levels, strengthening technical competences in tourism (future) stakeholders and establishing monitoring frameworks and systems to measure the sector's progress towards sustainability. Here the UNWTO has led the field in terms of observatories (see box below) and sustainability indicators at destination level, and is now working on integrating indicators and monitoring with the Tourism Satellite Account (TSA) system.
3. *Practice*: This work area has the objective of researching and exchanging best practices and experiences to identify existing and effective sustainability tools. It involves the use of integrated tools for promoting ST in both destination and tourism enterprises, promoting research and action on priority issues for sustainability within the tourism value chain and influencing consumers towards sustainable buying decisions and travel behaviour.
4. *Finance*: This work area has the objective of raising awareness of political actors, private investors, donors, developers and operators about the need to establish sustainable financing schemes and an investment-friendly macro-economic policy framework.

Box 8.1 The UNWTO Network of Sustainable Tourism Observatories

Introduction

We introduced the debate surrounding sustainability indicators in Chapter 4. The UN Environment Agency has been working on the use of sustainability indicators since the 1990s (see UNEP, 2017). The overall objective of sustainable indicators is to guide and measure the impact of the move towards sustainable consumption and production patterns worldwide. The approach is informed by the SDGs, particularly SDG 12. In tourism, work on indicators is supported by a variety of initiatives including:

- the European Tourism Indicators System (ETIS);
- the Global Sustainable Tourism Council (GSTC);
- the International Network on Regional Economics Mobility and Tourism (INRouTe); and
- the Measuring Sustainable Tourism (MST) initiative.

The International Network of Sustainable Tourism Observatories

The principle behind indicators is simple. By developing credible indicators of sustainability and monitoring them on a regular basis, destinations receive early warning of potential problems and can take remedial action. The UNWTO (2004: 5) defines sustainability indicators as:

> Information sets which are formally selected for a regular use to measure changes in key assets and issues of tourism destinations and sites.

The tourism work on indicators has been led by the UNWTO who see them as essential instruments for policy making, planning and management processes at destinations. As part of the development and collection of indicators, the UNWTO have established the International Network of Sustainable Tourism Observatories (INSTO; http://insto.unwto.org/). This aims to support ST at the destination level through the systematic, timely and regular monitoring of tourism performance and impact.

The mission of INSTO is to support and connect destinations that are committed to the regular monitoring of the economic, environmental and social impacts of tourism, to unlock the power of evidence-based decision making at the destination level, fostering ST practices locally and globally. The observatories play a strong local role in creating resilient destinations and as such are tasked with engaging all relevant stakeholders through a participatory approach. For example, workshops are organized to design monitoring processes and select the indicators to be collected.

Development of indicators

Each observatory has to monitor a standard set of nine core indicator areas:

1. local satisfaction with tourism;
2. destination economic benefits;

3. employment;
4. tourism seasonality;
5. energy management;
6. water management;
7. wastewater (sewage) management;
8. solid waste management; and
9. governance.

Observatories are also able to monitor additional indicators deemed significant at their destination. All data collected is analysed and reported regularly. Collection of the data allows for the establishment of baselines and targets for each indicator.

The Waikato Observatory

The observatory in New Zealand at Waikato is a good example of the local focus of the network (http://insto.unwto.org/observatories/waikato-region-new-zealand/). This focus is important in New Zealand because policy making is accountable to local residents. Local input into the planning process is expected, a role that is fulfilled by this observatory. The observatory seeks to monitor the activities of tourists, the nature of those activities and the consequences for communities and the environment in addition to the local tourism industry.

Waikato is the first tourism observatory in New Zealand, with a focus on two major tourist destinations in the region – Raglan and Waitomo. The monitoring area covers 25,000 km^2, attracting over half a million tourists and representing almost 10% of the local economy. Fulfilling its local mandate, the observatory aims to become a key research resource for all stakeholders in the local tourism sector. It aims to continuously generate research findings essential for planning and managing the destination and to share good practices with other stakeholders in New Zealand and around the INSTO network. The key monitoring areas are:

- host community wellbeing;
- visitor satisfaction;
- destination economic benefits;
- tourism seasonality;
- development control;
- waste, water and sewage management; and
- housing issues.

Indicator issues

Of course, there are some issues associated with the development of indicators of sustainability (see Miller & Twinning Ward, 2005). These include: lack of consistency in measurement between different destinations and poor coordination across different agencies; the fact that there may be gaps in the indicators and key elements of sustainability are missed; many destinations do not have the qualified staff – or the budget – to monitor indicators; and finally, indicators are often not collected nor communicated to encourage effective decision making. However, the development of the network of observatories should help address these issues.

Sources: Miller and Twinning Ward (2005); UNWTO (2004, 2017, 2019c)

The Organisation for Economic Co-operation and Development (OECD) Tourism Committee

Based in Paris, the OECD represents the governments of the world's leading economies. The OECD has an active tourism committee, formed shortly after World War II to advise countries on how tourism could be used to rebuild shattered economies (www.oecd.org). The committee is in a unique position to serve as an international forum for coordinating tourism policies and actions and it acts as a global forum for discussions of tourism policies. In 2017, the OECD issued a policy statement on ST, calling on governments to recognize that ST requires a 'whole of government approach and a long term planning horizon' (OECD, 2017).

The International Union for Conservation of Nature

The International Union for Conservation of Nature (IUCN) is a membership union formed in 1948 and composed of both government and civil society organizations (see IUCN.org). The IUCN's vision is of 'a just world that values and conserves nature'. Their mission is to:

> influence, encourage and assist societies throughout the world to conserve nature and to ensure that any use of natural resources is equitable and ecologically sustainable.

The IUCN is seen as one of the most influential and effective conservation organizations in the world and as such it provides public, private and non-governmental (NGO) organizations with the knowledge and tools that encourage sustainable development. It has 1300 member organizations and draws on 13,000 experts. The IUCN is the global authority on the status of the natural world and the measures needed to safeguard it. It is organized into six commissions dedicated to species survival, environmental law, protected areas, social and economic policy, ecosystem management and education and communication. The IUCN offers a neutral forum to provide governments and institutions at all levels with the impetus to achieve universal goals, including on biodiversity, climate change and sustainable development. In tourism the IUCN's role is particularly notable in terms of best practices, conservation tools and international guidelines and standards.

The IUCN Programme of work 2017–2020 provides the framework for planning, implementing, monitoring and evaluating the conservation work undertaken with and on behalf of its members. To do this the organization is involved in data gathering and analysis, research, field projects, advocacy and education. The programme has three priority areas:

1. valuing and conserving nature focusing on biodiversity conservation, emphasising both tangible and intangible values of nature;
2. promoting and supporting effective and equitable governance of natural resources; and
3. deploying nature-based solutions to societal challenges, particularly in climate change, food security and social and economic development.

Since its formation, the IUCN has, first, broadened its work beyond conservation ecology to embrace the more general principle of sustainable development and, secondly, focused on influencing the actions of governments, business and other stakeholders by providing information and advice and through building partnerships. For example, the IUCN has a growing programme of partnerships with the corporate sector to promote sustainable use of natural resources. This includes work with Marriott International in Thailand. Nonetheless, despite the need to involve the private sector in its work, this has led to criticism of the organization. In addition, the IUCN's stance on removing Indigenous people from national parks has also drawn controversy.

NON-GOVERNMENTAL ORGANIZATIONS: PRESSURE GROUPS

Greenpeace

Greenpeace is a non-governmental environmental organization created in 1971 by two environmental activists (see https://www.greenpeace.org.uk/). The goal of Greenpeace is to:

> ensure the ability of the Earth to nurture life in all its diversity.

The approach of Greenpeace is unusual as it works through direct action, lobbying and research and, as a result, is regularly featured in the media and comes under criticism, particularly when its methods are illegal or endanger life. Yet this approach is effective and Greenpeace is probably the most visible campaigner for sustainability on the global stage, bringing environmental issues into the public domain. Greenpeace also works in more conventional ways, having consultative status with the UN Economic and Social Council.

The work of Greenpeace focuses around campaigning for a greener and more peaceful world. There are three specific programme areas:

1. climate change;
2. defending the oceans; and
3. protecting forests.

The distinctive feature of Greenpeace is that it is funded by its members and does not accept public sector or commercial funding. It therefore has a passionate recruitment campaign for volunteers and funders, including those who are not afraid to take positive or direct action.

World Wildlife Fund

The World Wildlife Fund (https://www.wwf.org.uk/) is the world's leading independent conservation organization. Their mission is:

> To create a world where people and wildlife can thrive together.

To achieve this mission, WWF is: transforming the future for the world's wildlife, rivers, forests and seas; pushing for a reduction in carbon emissions that will avoid climate change; and pressing for measures to help people live sustainably, within the means of the planet. The WWF's programme of work is focused on six objectives:

1. restoring wildlife;
2. sustaining forests and oceans;
3. keeping rivers flowing;
4. trading sustainable timber and seafood;
5. reducing carbon emissions; and
6. living sustainably.

Friends of the Earth International

Friends of the Earth International (FoEI) was founded in 1969 and is an international network of environmental organizations covering 74 countries (see https://foe.org/). Its main parent body is Friends of the Earth (UK) which is primarily an advocacy group focusing on environmental issues (see https://friendsoftheearth.uk/). Friends of the Earth is a campaigning organization with work focused around forests and biodiversity, food sovereignty, climate justice and energy, economic justice and resisting neoliberalism. In addition, campaigns are planned on issues such as desertification, Antarctica, maritime, mining and extractive industries and nuclear power. However, rather like Greenpeace, the campaigning style of operation has attracted criticism, with some suggesting that the organization is more interested in publicity than working with locals to achieve change.

BUSINESS AND MEMBERSHIP ORGANIZATIONS

World Travel and Tourism Council

The World Travel and Tourism Council (WTTC; https://www.wttc.org/) represents the travel and tourism private sector globally. Its members include over 170 leading travel and tourism companies. While the WTTC is an overtly advocacy-based organization and its focus is on economic growth, one of its priority work areas is sustainable growth. Here, WTTC states that growth in tourism should be sustainable, contributing positively to the communities and ecosystems on which it depends.

The WTTC has identified five key areas for the sustainable growth of tourism:

1. *Climate change*: Raising awareness of the impacts of climate change on the sector as well as minimizing the sector's contribution to it;
2. *Destination stewardship*: Promoting best practice in planning and management, with a focus on public, private and community partnership, to ensure that growth benefits all and to tackle problems associated with the overcrowding of destinations;
3. *The future of work*: Ensuring that the sector is prepared for changing structures of employment and technological developments and that the workforce of the future is fully aware of and skilled to take up the opportunities offered by travel and tourism;
4. *Illegal trade in wildlife*: Promoting industry action to support global efforts to tackle the illegal trade in wildlife; tourism has a unique role to play by providing economic opportunities for communities and an economic rationale for the protection of endangered species; and
5. *Sustainability reporting*: Encouraging and supporting travel and tourism companies to measure, monitor and report their environmental, social and governance activities.

The Pacific Asia Travel Association

The Pacific Asia Travel Association (PATA; https://www.pata.org/) is a not-for-profit association founded in 1951. PATA has a long pedigree of campaigning for the responsible development of tourism in the Asia Pacific region. The Association has a membership comprised of 95 government, state and city tourism bodies, 25 international airlines and airports, 108 hospitality organizations, 72 educational institutions, and hundreds of travel industry companies.

PATA's mission is to:

> enhance the sustainable growth, value and quality of travel and tourism to, from and within the region.

PATA's work on sustainability began in 1974 with the concept of the PATA Development Authority as a leadership vehicle in environmental matters. This became the PATA's current Sustainability and Social Responsibility Committee. PATA's work is organized around three pillars: (i) insightful research, (ii) aligned advocacy and (iii) innovative events. PATA's sustainability strategy has six objectives:

1. generating tools that will allow the sharing of knowledge on good sustainability practices;
2. facilitating knowledge sharing and generating awareness on sustainability issues;
3. advocating for policies related to sustainable development of the visitor economy;
4. advocating for responsible organizational practices;
5. utilizing PATA's network to disseminate key related messages and disseminating awareness and knowledge on sustainability and social responsibility to develop a sustainable and responsible visitor economy; and
6. incorporating sustainability and social responsibility into all facets of PATA's strategy and activities.

In addition, PATA has partnered with the Asia Pacific Economic Cooperation (APEC) to develop a Code for Sustainable Tourism (http://sustain.pata.org/about/apec-pata-code/). The code urges PATA members and APEC Member Economies to:

- conserve the natural environment, ecosystems and biodiversity;
- respect and support local traditions, cultures and communities;
- maintain environmental management systems;
- conserve energy and reduce waste and pollutants;
- encourage a tourism commitment to environments and cultures;
- educate and inform others about local environments and cultures; and
- cooperate with others to sustain environments and cultures.

The Global Sustainable Tourism Council

The Global Sustainable Tourism Council (GSTC) has already featured in this book in Chapters 4 and 5. The GSTC is an independent and not-for-profit organization representing a diverse and global membership which includes UN agencies, NGOs, national and provincial governments, travel companies, hotels, tour operators, individuals and communities (https://www.gstcouncil.org/). The goal of the GSTC is to achieve best practices in ST.

The GSTC establishes and manages global ST standards – the GSTC Criteria. The criteria act as the global baseline standards for sustainability in travel and tourism for both industry and destinations. The criteria are used for education and awareness raising, policy making for businesses and government agencies and other organization types, measurement and evaluation, and as a basis for certification. The criteria are arranged in four pillars:

- sustainable management;
- socioeconomic impacts;

- cultural impacts; and
- environmental impacts.

They come as two specific types of criteria:

1. destination criteria for public policymakers and destination managers; and
2. industry criteria for hotels and tour operators.

By administering the criteria, the GSTC has assumed the role of a global accreditation body for certifying industry organizations and destinations.

The World Business Council for Sustainable Development

The World Business Council for Sustainable Development (WBCSD) had its origins in 1992 at the Rio de Janeiro Earth Summit (https://www.wbcsd.org/), and has a membership of over 200 international company CEOs. The WBCSD works to achieve the UN SDGs through a set of six programmes:

1. circular economy;
2. cities and mobility;
3. climate and energy;
4. food, land and water;
5. people; and
6. redefining value.

Members commit to making their knowledge, experience and people available to achieve the goals of the WBCSD and to transparently report on their environmental performance. In return, members benefit from the latest knowledge on sustainability, sustainability tools and participation in policy development.

Deutsche Gesellschaft für Internationale Zusammenarbeit GmbH

As a federal enterprise, the Deutsche Gesellschaft für Internationale Zusammenarbeit GmbH (GIZ) supports the German government in achieving its objectives in the field of international cooperation for sustainable development, including over 50 tourism projects (https://www.giz.de/en/html/index.html). These projects are implemented on behalf of the Federal Ministry for Economic Cooperation and Development (BMZ). Many of the tourism projects have an element of environmental protection and resource management, with the aim of tapping into new resources of income for people living on the periphery of nature reserves.

The International Tourism Partnership

We introduced the International Tourism Partnership (ITP) in Chapter 6. It promotes responsible business within the hotel industry by engaging with the world's leading hotel companies (https://www.tourismpartnership.org/). The Partnership is a non-competitive platform that shares best practice, offering practical products and programmes and facilitating collaboration in the hotel sector. It offers hoteliers access to sustainability information and resources at no cost and includes the online magazine www.greenhotelier.org.

Key issues that the partnership is working on include human trafficking, sustainable supply chains, carbon emissions (including the Hotel Carbon Measurement Initiative, now used by over 21,000 hotel properties globally), water conservation, fair labour standards and youth unemployment (including the Youth Career Initiative, an innovative hotel-based solution to youth unemployment).

PUBLIC SECTOR INFLUENCERS: POLICY FOR SUSTAINABLE TOURISM

INTRODUCTION

Traditionally, government has taken a leading role in implementing ST development, although of course it is the responsibility of all stakeholders to ensure sustainability. The public sector, particularly at the local level, is critical for the coordination, regulation and facilitation of ST. For ST, the role of government is important because:

- the sector is fragmented and a coordination role is needed for ST;
- the sector lacks leadership and this is a role that government can play to influence sustainability;
- there is often a need for some public funding for sustainability initiatives, for example in terms of overseeing eco-labelling as with the European Commission's eco-labelling scheme (https://ec.europa.eu/environment/ecolabel/);
- many vulnerable elements of the destination – coasts, wetlands, small islands or the built heritage – are in public ownership; and
- government has the mandate to regulate, plan and legislate.

Government was slow to act in addressing sustainable development in tourism until the 1992 UN Earth Summit in Rio. A key outcome from Rio was Agenda 21, signed by over 180 governments (www.un.org/esa/sustdev/agenda21.htm). Agenda 21 was a commitment on the part of governments to address the issue of development and the environment across a range of activities. For tourism, an important feature of Agenda 21 is the focus on implementation at the local destination level. The principles of Agenda 21 were reaffirmed at Rio+20 and reflect contemporary thinking on sustainable development in two ways:

1. While tourism is the focus of many Agenda 21 plans in, for example, resorts, small islands and heritage towns, tourism also forms an integral part of other Agenda 21 initiatives, stressing the need for 'whole of destination' management.
2. Secondly, projects involve not only government but also NGOs (such as the WWF) and the private sector, stressing the need for all stakeholders to be involved.

TOURISM POLICY FOR SUSTAINABLE TOURISM

Despite increasing private sector involvement in sustainability, the market alone cannot facilitate sustainable development and government interventions are needed. A key role of government is the determination of tourism policy. Tourism policy is a macro-level instrument that looks to the long term and, by showing the intention of government, it provides a clear sense of direction for the tourism sector. We can define define tourism policy as a all the actions carried out under the coordination of public administrations with the objectives of achieving previously defined aims in the processes of analysis, attraction, reception and evaluation of the impacts of tourism flows in a tourism system or destination.

The key aims of tourism policy are to create competitive destinations, to ensure that the tourism sector functions efficiently, and to deliver benefits, including income and employment, to the government's stakeholders (Scott, 2015). Tourism policy therefore sets the priorities and administrative framework for ST and allows for coordination with other policy areas such as, say, transport or the environment; indeed, Pforr (2004) states that the sustainability paradigm is a key driver of tourism policy. However, theorists have pointed out that the lack of a political will to plan for tourism over the long term will in the end create significant problems (see Page & Thorn, 2002). Hall (2011b) supports this view, stating that policy for ST presents a paradox. It can be said to have been successful as it has been increasingly adopted across all tourism sectors, yet it also represents a policy failure as negative environmental, social and cultural impacts occur at the hand of tourism. The UNWTO's (2018a) review of the integration of sustainability into tourism policies is also cautious, suggesting that the policies themselves are often vague on sustainability, falling back on voluntary initiatives and lacking data. Governments therefore need to take a leadership position in planning for sustainability so that decisions can be made in a more systematic manner (Page & Thorn, 2002), effectively championing a rebalancing from a dominance of economic thinking towards environmental and social considerations (Butler, 2018).

However, it is clear that in order to provide an effective policy framework for ST, a new approach is needed (see OECD, 2017). This approach involves a shift in thinking in terms of the role of policy, the role of government and the involvement of other stakeholders. This new way of thinking includes:

- *a strategic approach to policy*, encouraging a long-term view and a framework to guide the actions of government and other stakeholders;
- *a whole of government approach* to ensure the fragmentation of the tourism sector is fully represented across all departments and that there is coordination both horizontally and vertically in terms of policy decisions;
- *coordination* across relevant stakeholders as a first step towards a more collaborative and inclusive approach to policy; this includes engagement with the private sector;

- *collaboration and partnerships* across all relevant policy stakeholders to allow for dialogue and information networks to agree a binding approach (Pforr, 2004);
- *decentralization and community involvement*, essential for a truly sustainable approach whereby policy decisions are taken closer to the end user, often utilising an approach known as networked governance to enable greater participation of all stakeholders in decision making; and
- *engagement with the knowledge economy* to ensure that information flows efficiently within the policy community and that knowledge based innovation is facilitated.

Box 8.2 Demarketing

The challenges of balancing use and conservation in heavily used but sensitive spaces such as national parks has prompted some public management agencies to consider social marketing approaches. Social marketing is an approach used to develop activities aimed at changing or maintaining people's behaviour for the benefit of individuals and society as a whole. For national parks, behaviours can be influenced in a number of ways including:

- limiting group size or having visit quotas as happens in the Galapagos and some heritage sites in Egypt;
- capping visitor numbers – or indeed banning visitors altogether for a period (see, for example, the six-month closure of Boracay beach in the Philippines for repair and restoration);
- introducing entrance fees; and
- using media such as education or interpretation to influence behaviour.

Another approach for these special places is demarketing. For marketers, demarketing may seem counter-intuitive as it is premised on the idea of discouraging customers, particularly by targeting certain undesirable groups such as the mass tourism segment in Bermuda. Demarketing has been used in Australia where the nature-based tourism industry is so popular that it clearly represents one-third of all travel and tourism to Australia (Buckley & Sommer, 2000). National parks play an important role as the supply to meet this level of nature-based demand, as the most important tourist destinations in the country (Worboys & De Lacy, 2003).

The Three Sisters rock formation of Blue Mountains National Park near Sydney, Australia receives over 2.8 million visitors per year, with the suggestion that specific attractions, and the park itself, would have less pressure if certain demarketing measures were adopted. Armstrong and Kern (2011) categorize these measures as:

- product (e.g. limiting activities either seasonally or entirely; limiting activities by restricting the areas in which they can be conducted);
- distribution (e.g. limiting the overall capacity of camping and accommodation facilities; developing a 'park full' strategy to encourage other destinations);
- price (e.g. introducing or increasing prices/user fees; discouraging/stopping price discounting practices); and
- promotion (e.g. ceasing/decreasing promotion of a product; highlighting the environmental degradation that will occur if too many people frequent the area).

INTEGRATED TOURISM GOVERNANCE

As this book has stressed, successful ST takes a 'whole of destination' approach, and policies (i) draw on a wide base of participation, including the local community and destination stakeholders; and (ii) are coordinated with other economic sectors. This is reflected in contemporary trends in tourism policy and governance, particularly the concept of integrated tourism governance. We deal with governance for protected areas in Chapter 9. Integrated tourism governance involves forging partnerships between the public sector and other destination stakeholders to create a more holistic approach to governing the destination. As such it works towards SDG 16 which is about building partnerships for sustainable development. This is supportive of sustainability because decisions are being taken closer to the industry, other public sector organizations and the host community – an approach that integrates public sector services closer to the point of delivery. The integrated approach to tourism governance is a trend that can be seen across the world, from Canada to Scotland and New Zealand. It is based on the concept of destinations being made up of a relatively stable set of public and private sector actors who are linked to each other and defined by their relevant administrative boundaries. The linkages allow communication for information, communication, expertise, trust and other policy resources, all influential in promoting ST. Of course, in the past, stakeholders were in touch with each other, but the difference here is the involvement of the public sector who pump resources into the network and make it a more formal arrangement. Above all, though, the involvement of the government makes the networks more democratic and all stakeholders are given the opportunity to engage and participate in decisions.

Box 8.3 The policy and planning challenges of sustainable tourism in Bulgaria

Introduction

Bulgaria faces real challenges in terms of diversifying away from mass tourism towards more sustainable forms of tourism. This case study outlines the policy challenges for the country as it seeks to move away from mass tourism and to transform its tourism sector.

Tourism in Bulgaria

Bulgaria's tourism is dominated by major mass tourism developments along the Black Sea coast and in the mountains. These resorts are characterized by high-volume/low-value tourism. Yet the country has great untapped tourism potential in its unspoiled natural and cultural heritage which includes over 600 spas and mineral springs, nine UNESCO World Heritage Sites and great scenic landscapes and national parks – 5% of the territory is protected.

Diversification

The priority for Bulgarian tourism is to diversify away from mass tourism at the coast and in the mountains. Mass tourism brings a range of problems including dependence on a few markets, geographical concentration, high seasonality, low occupancy, acute price competition and a dependence on overseas tour operators. By diversifying its products to include ST sectors such heritage and cultural tourism, nature-based tourism and food and wine tourism, Bulgaria would be able to solve many of these problems. It would also encourage the development of high-quality products, disperse the benefits of tourism to rural and remote areas and attract high-yield visitors.

The tourism policy challenges

Diversification towards ST must be part of an overall policy and strategies directing the future of Bulgarian tourism – indeed, the country's alternative varieties of tourism are a perfect complement for mass tourism. According to the OECD (2007), policy and planning initiatives will be required to achieve diversification.

Policy recommendations

The OECD (2007) has made a number of recommendations to facilitate diversification towards ST:

- Due to the nature of the Bulgarian political system, tourism power is concentrated at the local level with the regions being particularly poorly funded and ill-equipped to coordinate tourism. Strengthening of regional powers is therefore needed, reflecting current thinking on tourism policy.
- There is also a need for government agencies to coordinate their activities as has not happened to date. In other words, structural changes in the Bulgarian political system are needed to create a transparent and inclusive policy formation process.
- There is a need to develop the technical capacity of public sector personnel in tourism.
- SMEs will be the main medium for diversification. Policy should therefore be designed to support SMEs to innovate and encourage product development in such a way that tourism spend and employment can be captured locally. Policy support for SMEs must therefore be broadly based and multi-layered to include financial incentives, entrepreneurship development and training of the workforce.

This case shows how the public sector can intervene strategically to reposition the tourism sector of a country towards sustainable principles. The challenge for Bulgaria will be to implement these ideas using contemporary approaches of integrated governance and policy development.

Sources: Bulgarian State Tourism Agency (2006); Ministry of Tourism of the Republic of Bulgaria (2016); OECD (2007)

THE INFLUENCE OF THE PRIVATE SECTOR ON SUSTAINABLE TOURISM

Traditionally, it has been the public sector that has led on thinking and innovation for ST, and indeed sustainability more generally. However, in the second decade of the 21st century sustainability has become a driver of innovation in the private sector, including tourism companies. Companies have begun to recognize emerging challenges such as climate change, human trafficking, loss of biodiversity and food security and they are exploring how these issues can be managed. Involvement of tourism companies in sustainability is based on three principles:

1. *Business-driven sustainability*: An increasing view that sustainability makes business sense, with evidence that simply engaging in significant corporate social responsibility (CSR) projects can provide an uplift in performance of more than 10%. Accor's report on CSR states clearly that 'sustainability is not just about doing good: rather it also generates genuine business value' (Accor, 2015: 24).
2. *Stakeholder involvement*: Recognising that it is important to involve all those in the supply chain as well as the consumer of tourism themselves.
3. *Data-driven decision making*: A noticeable trend has been the development of toolkits to help companies measure their carbon emissions and other metrics. Accor is a leader in this respect with extensive research on the impact of its operations.

A number of tourism companies have now embedded sustainability firmly into their operations, with sustainability key performance indicators (KPIs), monitoring and reporting built into their annual performance review. Tourism companies that have taken this approach include:

- TUI
- British Airways
- Virgin Atlantic
- P&O
- Carnival Cruise Lines
- Whitbread

In addition to individual companies becoming involved in ST, there are two key business arrangements that are increasingly being used for ST. These are:

- Benefit corporations; and
- Public–private partnerships.

BENEFIT CORPORATIONS

Benefit corporations represent a significant shift in corporate thinking towards the environment and indeed inclusivity and sustainability more generally. In the USA, a benefit corporation is a type of for-profit corporate entity that includes a positive impact on society, workers, the community and the environment – in addition to profit – as its legally defined goals. Benefit corporations are proud of their values and use them in their marketing. Directors of a benefit corporation have to consider the impact of the actions of the corporation on employees, customers, the community and the local and global environment, as well as shareholders. These impacts are reported in an annual benefit report using a respected third-party standard. The ambitions of the benefit corporation movement are clear from the B Corp (2018) Declaration of Interdependence (https://bcorporation.net/):

> We envision a global economy that uses business as a force for good. This economy is comprised of a new type of corporation – the B Corporation – which is purpose-driven and creates benefit for all stakeholders, not just shareholders. As B Corporations and leaders of this emerging economy, we believe:
>
> - That we must be the change we seek in the world.
> - That all business ought to be conducted as if people and place mattered.
> - That, through their products, practices, and profits, businesses should aspire to do no harm and benefit all.
> - To do so requires that we act with the understanding that we are dependent on another and thus responsible for each other and future generations.

One example of a tourism benefit corporation is the Australian-based adventure tourism Intrepid Group (intrepidtravel.com). As one of the first carbon-neutral tour operators, Intrepid has decided to move beyond the usual sustainability tourism schemes and certification to become a fully fledged benefit corporation. The marketing benefits are clear as taking on benefit corporation status communicates a company that benefits destination communities and environments and represents a shift away from short-term profit. It also puts the company into a sector leadership position, setting a benchmark for the aspirations of other tourism companies.

PUBLIC–PRIVATE PARTNERSHIPS

As this book has shown, ST development projects involve multiple goals and many stakeholders, from the tourism industry to government and the local community. In order to encompass these multiple interests, 'public–private partnerships' (PPPs) are increasingly being used strategically as a tool for ST product development and enhancement, as well as for research, marketing and promotion. They have their roots in the fact that ST demands changes in product development and marketing on behalf of both government and the private sector. As a result, governments are increasingly turning to the private sector as partners to develop and implement ST.

Aims of PPPs

Each PPP is different, but we can identify a number of aims that are common across all PPPs in ST (APEC, 2002):

- putting in place a regulatory regime;
- agreeing sustainability indicators;

- agreeing certification criteria;
- agreeing public funding principles;
- research;
- environmental education and training programmes; and
- fair and non-discriminatory taxes with revenue allocated to sustainability.

Types and benefits

There are a variety of models of PPPs from straightforward social collaborations designed to improve the tourism experience to major infrastructure and project development projects. The size of the project will determine the type of agreement, with larger projects requiring a more formal contractual arrangement. Partnerships between the local community and the tourism sector are particularly significant as they provide opportunities for community involvement and participation in tourism (Tonge *et al.*, 1995). PPPs provide a range of benefits for the development of ST (APEC, 2002):

- public–private tourism partnerships facilitate sharing knowledge, expertise, capital and other resources (Bramwell & Lane, 2008);
- involvement by all relevant stakeholders;
- decision-making power is passed to all stakeholders;
- involving all stakeholders increases acceptance of policies so that implementation is easier;
- more constructive and less adversarial attitudes result from the partnership; and
- improved coordination of policies and actions.

The role of the private sector

In the past, industry has been reluctant to participate in public sector activities, fearing that they may be committed to financing the project. However, there is now a realization that for the private sector there are a number of key advantages in being part of a PPP. Of course, the potential gains of the partnership must match the investment involved as businesses assess the potential for increased tourism revenue through provision of a more attractive, secure or accessible destination. Private sector involvement in ST can include (Smith, 2017):

- marketing and promotion;
- product development;
- infrastructure development/renewal;
- attraction development/renewal/diversification;
- enhanced productivity and service;
- community development/renewal through community engagement;
- cultural and heritage protection;
- environmental protection/enhancement;

- working with small and micro enterprises to build management skills, market development and technology transfer; and
- working with governments to establish an enabling framework for the achievement of sustainable development.

The role of government

Of course, governments also play a key role in these partnerships as certain activities can only be provided by the public sector. These include (Smith, 2017):

- developing tourism strategies and plans to support sustainability;
- contracting major infrastructure projects;
- granting project approvals, permits and licences;
- evaluating contracted services;
- approving payment for contracted services;
- acting as a regulator;
- coordinating between government departments; and
- Facilitating knowledge exchange between companies to help industry develop more sustainable operations and products.

According to Smith (2017), the public sector can deploy a number of available tools to help create a viable revenue stream for private investors in a major project. These include:

- the ability to collect user fees (e.g. for a cruise ship port);
- free (or reduced cost) use of government land for a specified period;
- government-furnished utilities;
- off-site infrastructure or access improvements;
- tax incentives;
- zoning exemption or relaxation;
- exclusivity of or restrictions on approval of competing facilities within a specified distance and/or time; and
- assistance with workforce training and development.

Issues with PPPs

Despite the popularity of the PPP arrangement to support ST, there are cases where the arrangement is not appropriate, fails or outlives its usefulness. According to APEC (2002), this is often due to problems with the collaboration between partners who are either unwilling to cooperate in the project fully, or who dominate the process. Similarly, if only a small number of stakeholders agree to participate, this creates an issue in terms of inclusivity. Sometimes, too, the partners only have a token involvement and avoid tackling real problems. Another issue is that involving a large number of partners can be both expensive and time consuming and may lead to fragmentation of decision making.

THE SUSTAINABLE CITIZEN

The final section of this chapter focuses on the sustainable citizen – a topic that we introduced in Chapter 3, which is linked to SDG 12 on sustainable consumption. Here the key issues are understanding the nature of the sustainable citizen, their willingness to pay for sustainable practices and the role of consumer pressure to encourage innovations in sustainable practices. It is also important to understand how tourism as an activity differs from other forms of consumption. Ottman (1993) defines green consumerism as:

> individuals looking to protect themselves and their world through the power of their purchasing decisions. (Ottman, 1993: 3)

However, when tourism products are purchased, the consequences of this decision are different because tourism is 'inseparable'; in other words, it is produced and consumed at the destination, so ignoring reports on the impacts of tourism at a specific destination means that the tourist visiting such a place must be faced with this dilemma first hand. Consumers of other products, Miller (2003) claims, can switch off concern because these products are produced physically in other places, so the consumer will not see the environmental and social costs associated with this production.

Sustainable citizens actively seek out ST products and engage in environmentally friendly tourism behaviour. They are willing to buy products that have a reduced impact on the environment compared to their conventional alternatives. Sustainable citizens have been one of the main forces behind the development of ST, placing pressure on the industry to change its practices. They have now become a particular market segment as operators and destinations try to understand them and develop appropriate products.

ANTECEDENTS

A number of studies have shown that early exposure to nature in childhood is likely to foster a positive attitude to the environment. In Chapter 3 we highlighted work by Wells and Lekies (2006) who used Bronfenbrenner's (1995) life course perspective to investigate if children exposed to nature in their youth would have a closer tie to the natural world as adults. Wells and Lekies (2006) found that when children participated in what was referred to as 'wild nature' (e.g. fishing, playing in the woods), there was a positive correlation with stronger environmental attitudes and behaviours in adult life. The message in this research is clear. By exposing children at an early age to nature activities and experiences, there is a higher chance that they may embrace pro-environmental attitudes and practise pro-environmental behaviours.

ATTITUDES VERSUS BEHAVIOUR

The contrast between a consumer's favourable attitude towards environmentally responsible behaviour and their actual buying behaviour is referred as the 'attitude–behaviour gap'. Here, Budeanu's (2007) work on sustainable tourist behaviour is important because it places into context many of the challenges and opportunities for the development of a more sustainable tourism industry. Budeanu found that even though tourists expressed positive attitudes towards sustainability in tourism, very few matched their attitudes and intentions with actual behaviour – purchasing responsible tourism products, choosing transportation options that are more environmentally friendly and adopting responsible travel behaviours at the destination. Budeanu argues that there are several tools that should be implemented in moving towards more sustainable travel behaviour. These tools include increasing the costs of environmentally destructive behaviour (and one would expect socially and culturally destructive behaviour), decreasing the costs of choosing good environmentally friendly options, increasing education to tourists, educating tourists about environmentally destructive behaviours, and using resources in a more efficient manner. Miller (2003) supports Budeanu's view by suggesting that although tourists make decisions based on social, cultural and environmental considerations in the purchase of tourism products, there is an element of selfishness built into this altruism.

EDUCATION AND COMMUNICATION INFLUENCERS

We introduced the importance of education for sustainability in Chapter 3 under SDG 4. There is no doubt that education and communication are key to changing the attitudes of the industry and its consumers, and thus behaviour towards ST. Indeed, SDG 4 focuses on education. The issue is complex, however, as sustainability is an abstract concept and requires a means of 'making it real'. Giddens (2002) stated the issue clearly in the context of global warming:

> The dangers faced by global warming are not tangible, immediate or visible in day to life. Therefore many sit on their hands and do nothing. (Giddens, 2002: 151)

There are three possible approaches to communicate ST:

1. public education;
2. communicating with the tourism industry; and
3. education for ST.

In terms of public education, a new wave of public awareness programmes is needed that make ST a reality in people's lives. The question is: How can we involve individuals so that they feel more responsible for learning and acting in a sustainable way? Here, it is important to develop deep engagement and understand the audience and persuade them that ST is about 'them'. For example, Loyau and Schmeller (2017) found that education of the public, including tourists, is essential before the implementation of biodiversity conservation measures. Here the US approach to climate change is informative as exemplified by Abbasi's (2006) report on Americans and climate change. The report came up with a series of recommendations on how education can change attitudes towards climate change, including:

- educate the gatekeepers – editors, the media and journalists;
- design a new 'vision for energy';
- create a well-funded public education programme; and
- improve understanding of climate change in schools.

A similar approach is required in order to engage with the tourism sector and communicate the concept of sustainability. How do we reconcile an industry that focuses on growth with a sustainable stance and practice? Common approaches here are to work through industry associations to reach SMEs, while for larger corporations the channel for communication is through human resources training and awareness programmes. Tzschentke *et al.* (2008) used operators affiliated with the Green Tourism Business Scheme of the Scottish Tourism and Environment Forum to understand the factors that led to the adoption of environmental measures by operators of small hospitality firms. The propensity to become environmentally active or involved resulted from what the authors characterize as a value-driven journey, which has been heavily influenced by the development of environmental consciousness. This consciousness is a result of personal, sociocultural and situational factors. For example, respondents stated that parental education was a major factor in shaping their green tendencies. The greening of their businesses was a reflection of the prior greening of the individual.

EDUCATION FOR SUSTAINABLE TOURISM

Education for ST focuses on educating and inspiring attitudes and behaviours towards ST and equipping future generations with the knowledge and skills to put it into practice. Development of a ST curriculum is a starting point here but it also begs a number of questions. What should be the curriculum content? Is there a core of knowledge about ST that every curriculum should include? What should be the balance between general concepts of sustainability and the tourism focus? And how is success measured in ST education?

Henry and Jackson (1996) provided an early overview of what a sustainable curriculum ought to look like in the development of a better educated tourism workforce, based on three levels. These include:

1. *Concepts/philosophies*:
    - political values/ideologies
    - eco-philosophy and environmentalism
    - planning theory

- social theory
- cultural theory
- economic theory
- management theory

2. *Policy orientations*:
   - ecological policy
   - cultural development
   - economic development
   - managerial policy
3. *Skills/competencies*:
   - environmental interpretation
   - cultural interpretation
   - sustainable management and planning practices

UNESCO recognizes that education for sustainability is a critical enabler for sustainable development (https://en.unesco.org/gap), supported by the SDGs. In particular, Target 4.7 of SDG 4 on education specifically addresses education for sustainability and related approaches. UNESCO's Global Action Programme (GAP) focuses on education for sustainable development and seeks to generate and scale up education to accelerate progress towards sustainable development. The GAP aims to contribute substantially to the 2030 agenda, through two objectives:

1. reorienting education and learning so that everyone has the opportunity to acquire the knowledge, skills, values and attitudes that empower them to contribute to a sustainable future; and
2. strengthening education and learning in all agendas, programmes and activities that promote sustainable development.

As part of its education for sustainable development, in 2010 UNESCO developed a series of online multimedia modules for schools based on sustainable development, one of which was on ST. The module covered the characteristics and objectives of ST through a series of case studies. The objectives of the module were:

- to appreciate the benefits and problems arising from various forms of tourism, especially in terms of social equity and the environment;
- to develop a critical awareness of the ways in which tourism can enhance the welfare of people and protect our natural and cultural heritage;
- to promote a personal commitment to forms of tourism that maximize rather than detract from sustainable human development and environmental quality; and
- to plan ways of teaching about ST.

There are two significant education networks that promote ST.

1. The Tourism Education Futures Initiative

The Tourism Education Futures Initiative (TEFI; http://tourismeducationfutures.org/) is a forward-looking network designed to inspire, inform and support tourism researchers, educators and the tourism sector to 'passionately and courageously transform the world for the better'. TEFI is concerned with ST but not exclusively so. TEFI takes a progressive and distinctive approach tourism education, 'committed to a type, scale and form of tourism that is both sustainable and just, and that values people and planet alongside economic outcomes'. To achieve this, TEFI seeks to facilitate knowledge co-creation between the tourism sector and academics.

2. Building Excellence for Sustainable Tourism – an Education Network

Building Excellence for Sustainable Tourism – an Education Network (BEST EN) is a longstanding collaborative network of tourism academics and practitioners (https://www.besteducationnetwork.org/). BEST EN provides a platform for information, exchange of ideas, knowledge creation and building links and creating partnerships between academics as well as within the tourism industry focused on ST. BEST EN operates through a number of activities, including a website for information exchange (BEST EN, 2019), teaching materials and case studies and think tanks – which are themed on particular issues.

CONCLUSION

The content in this chapter focuses on a number of the UN SDGs, but in particular SDG 16 which is about building accountable and inclusive institutions, SDG 12 which is about responsible consumption and production patterns, SDG 17 on partnerships for sustainable development and SDG 4 on education.

This chapter has introduced the key agencies and influencers of ST. The chapter opened with a review of the key public and private sector agencies and organizations that have been influential in promoting and implementing ST. Here some of the global intergovernmental agencies such as the UNWTO have been influential in promoting the sustainability agenda. What is clear also is that the private sector now plays a greater and more influential role than in the past and that here, ST is a source of innovation. There are also influential pressure groups and lobby organizations in the system such as Greenpeace. The chapter then considered public sector policy for tourism and showed how, despite the move to integrated 'whole of destination' policy and governance, some commentators still doubt the effectiveness of ST policies given the continued negative impact of tourism. The influential role of the private sector and in particular organizational arrangements that facilitate sustainability were then considered, including the contemporary 'benefit corporation' initiative as well as public–private partnerships. The chapter then moved on to consider the influential role of the sustainable citizen, before closing with an examination of the role of education for sustainability and the key organizations involved.

END OF CHAPTER DISCUSSION QUESTIONS

1. Using internet sources, choose a public–private partnership or benefit corporation and show how its organizational arrangement facilitates ST.
2. Taking one organization from each of the public, private sector and pressure groups, evaluate their effectiveness in promoting sustainable tourism.
3. What should be the actions of a true 'sustainable citizen' when travelling?
4. Examining the educational course that you are currently pursuing, evaluate its sustainable tourism content.
5. Why do you think destinations find it difficult to implement the European Tourism Indicators Scheme (ETIS) (http://ec.europa.eu/growth/sectors/tourism/offer/sustainable/indicators_en)?

End of Chapter Case Study 8.1 The Accor Group: Influential leadership in sustainability

Introduction

Accor is one of the world's major international hotel chains. The company is influential in leading the sector in terms of its thinking and operations in sustainability. At Accor, sustainability is wired into KPIs, reporting, and every aspect of the company's operation. It is not treated as an add-on or as 'greenwashing'. Accor's commitment to sustainability dates back to the mid-1990s but it is the more recent initiatives and research that mark the company out as an influential leader in the field. These include a set of reports, toolkits and research which we summarize in this case study.

Accor aims to 'create a virtuous circle that benefits its "ecosystem" – comprising employees, customers, partners, local communities, in which hotels are implanted … creating value for its operations' (Accor, 2011: 2). In other words, all stakeholders are involved, including hotel guests and suppliers, which is essential if Accor is to reduce its environmental impact. But also and perhaps more importantly, Accor is honest in its statement that it is involved in this field to boost its 'competitiveness'.

Accor established an Environment Department back in 1994. This is now the 'Sustainable Development Department' which:

- initiates sustainability projects for the Accor Group; and
- assists in the implementation of sustainable development for both accommodation operations and support services (such as HR, purchasing or marketing) through communication of good practice and development of projects.

Accor has developed a coherent suite of tools for sustainability, each underpinned by research:

Charter 21 and PLANET 21 Management Tool and Reporting System

Charter 21 was introduced in 2005 when the Accor Hotels Environment Charter recommended 65 actions that Accor properties could utilize to reduce their environmental footprint. These actions included recycling glass, recovering rainwater and using eco-labelled products. The reporting side of Charter 21 was achieved by hotel managers providing an annual report of their Charter 21 actions.

As Accor's thinking developed, Charter 21 was updated in 2011 and became part of the new PLANET 21 strategy, reflecting the UN's 'Agenda 21' initiatives. This involved updating indicators for the 65 actions and the addition of items relating to social responsibility (for example, the use of Fair Trade products and the organization of staff training on health and wellbeing). PLANET 21 is made up of seven pillars, with 21 commitments implemented by the company across 92 countries:

1. *Health*
   a. ensure healthy interiors
   b. promote responsible eating
   c. prevent diseases
2. *Nature*
   a. reduce water use
   b. expand waste recycling
   c. protect biodiversity
3. *Carbon*
   a. reduce energy use
   b. reduce Co2 emissions
   c. increase use of renewable energy
4. *Innovation*
   a. encourage eco-design
   b. promote sustainable building
   c. introduce sustainable offers and technologies
5. *Local*
   a. protect children form abuse
   b. support responsible purchasing practices
   c. protect ecosystems

6. *Employment*
   a. support employee growth and skills
   b. make diversity an asset
   c. improve quality of work life
7. *Dialogue*
   a. conduct business openly and transparently
   b. engage franchised and managed hotels
   c. share commitment with suppliers

Earth Guest Research

A key contribution of Accor to sustainability research in the hotel sector is their ground-breaking research reports into the environmental and socio-economic footprint of its guests and the company – 'Earth Guest Research' – available for all to share. The research is the basis for communication to the rest of the industry and for training and e-learning modules, clearly demonstrating Accor's leadership in this field. Three key reports have been released:

1. Sustainable Hospitality is a study of guest attitudes to sustainability (Accor, 2011). The key idea is to understand the impact of its guests from booking to checking out and beyond. There are four key research findings:
   a. all hotel guests feel concerned about sustainable development;
   b. high expectations concerning concrete actions in four areas – energy, water, waste and child protection;
   c. guests consider themselves an essential link in the chain of sustainable development;
   d. hotel guests declare that they are ready to act and change their behaviour.
2. The Accor Group's Environmental Footprint is a multi-criteria life cycle analysis (Accor, 2016a). The study set out to assess the impact of the Accor group on the environment and, by focusing on the total life cycle, it looks back to the impact of suppliers – such as dairy farms – on the environment. There are three key findings:
   a. carbon and energy are the first pointers for progress for the group;
   b. food purchases account for most of the water consumed and contaminated;
   c. building sites are a critical link in the waste production chain.

3. AccorHotels' socioeconomic footprint (Accor, 2016b) is a detailed audit of the social and economic impact of Accor's operation. The two key findings are:

a. 4.1 jobs are supported in the rest of the world economy, per AccorHotels employee, of which on average 70% are in countries where the Group operates;

b. the total business generated directly, indirectly and secondarily by AccorHotels in the world economy contributes €22.4bn, of which 83% on average stays in the country in question.

In order to achieve and manage their environmental mission, Accor have developed a suite of tools, reporting, audits and initiatives. These include:

- *The OPEN tool*
  The OPEN tool was released in 2005. It acts as an internal management tool for the Accor Group's hotel managers, allowing them steer the implementation of PLANET 21, track changes in energy and water use and manage waste production. For the Sustainable Development Department the OPEN tool allows them to have a global picture of sustainability metrics across the group.
- *Environmental data external audit*
  In 2007 Accor initiated an external audit of their environmental data and also its level of reporting to see if improvements could be made. The audit examined monitoring of Charter 21 activities and led to the integration of a sustainable development component in Accor's quality audits. From then on, the audits were expanded to cover water and energy consumption data, GHG emissions and social information.
- *ACT-HIV initiative*
  Also in 2007, Accor launched an international programme to fight HIV/AIDS, for use by hotel managers.
- *The Environmental Sustainable Development Reporting Protocol*
  Accor developed a Sustainable Development Reporting Protocol in 2008 to allow them to 'clarify and improve' the organization of environmental and social information.

Innovation

The sustainability initiatives and research outlined above have allowed Accor to be an influential innovator in sustainability in many ways:

- establishing hotel refurbishment and construction standards;
- renewable energy and solar panel initiatives;

- installation of low-energy light bulbs;
- tree planting schemes;
- water flow regulators;
- fair trade practices; and
- organic waste composters.

Discussion questions

1. How would you communicate Accor's leadership in sustainability to a hotel guest in the property?
2. Why do you think Accor has invested so much in research and development for sustainability?
3. How would you communicate the environmental imperative to a small family-run guest house?

Sources: Accor (2011, 2013b, 2016a, 2016b, 2018)

CHAPTER 9
PROTECTED AREAS, ECOTOURISM AND SUSTAINABILITY

LEARNING OUTCOMES

This chapter fits within the structure of the book by providing a link to two very important aspects of sustainability in tourism: parks and protected areas as popular places for tourism (supply), and ecotourism and related forms of tourism (demand), where visitors and tourism businesses implement aspects of sustainability. The chapter is designed to provide the reader with:

1. An understanding of the extent of the terrestrial and marine protected areas systems around the planet, as well as some challenges to the management of these areas.

2. Key linkages between protected areas and the UN Sustainable Development Goals (SDGs).

3. The IUCN's categories of protected areas, with World Heritage Sites as an example of a category in this system.

4. The defining characteristics of ecotourism, including a nature-based connection, sustainable development (communities and conservation), education and ethics.

5. How ecotourism relates to other forms of nature-based tourism.

INTRODUCTION

> The more people that visit the forest, the more revenue can be earned for local communities and forest management, ... We are firmly of the opinion that conservation success is driven by economics and direct benefits to local people.

This remark was made in reference to the struggles that conservationists experienced in protecting the rare and highly endangered red colobus monkey in Tanzania. After 40 years of research and deliberation, the Mogombera Nature Reserve, 10 km^2 in size, is now a reality despite several years of land ownership issues, threats of deforestation, and ownership of the land by a sugar company (Dasgupta, 2019). This little case study represents the important connection between protected areas, tourism, community development, non-governmental organization (NGO) involvement, government and local involvement and interests in efforts to conserve precious natural resources. The establishment of protected areas is not a uniform process. And it is essential that local people are able to realize the benefits from such areas in the competition for scarce resources. Ecotourism is seen as the principal solution in this case for the sustainable management of the region, which must balance social and cultural needs with preservation.

This chapter explores many of the challenges that exist with the sustainable management of protected areas. There are human and natural resources issues and challenges at stake, as noted above, that are intensified in view of the massive amounts of habitat that are lost each and every year and a global population base that continues to escalate. Protected areas remain one of the most important weapons that we have in our arsenal to preserve as much biodiversity as we can; thus increasing the number of protected area globally and improving on how these areas are managed is a top priority.

PARKS AND PROTECTED AREAS

A protected area as defined by the International Union for the Conservation of Nature is a 'clearly defined geographical space, recognized, dedicated and managed, through legal or other effective means, to achieve the long term conservation of nature with associated ecosystem services and cultural values' (IUCN, 2019a).

The World Database on Protected Areas provides an updated data set of protected areas located around the world (Protected Planet, 2019b). This publication illustrates that the number of protected areas globally continues to increase, with approximately 15% of the total land area of the world under protection. At the end of 2018 there were 15,345 marine protected areas, covering 7.44% of the ocean and representing an area of 26,937,555 km^2 (Protected Planet, 2019a). Over 8 billion people visit the terrestrial parks of the world yearly, with more than 80% of these visits taking place in North American and European protected areas, generating over US$600 billion/year in direct in-country expenditure (Balmford *et al.*, 2015). Some key findings of the Protected Planet Report (UNEP-WCMC, IUCN and NGS, 2018), include:

- There is insufficient protection of areas of importance for biodiversity and ecosystem services through systems of protected and conserved areas; however, significant progress has been made in the protection of Key Biodiversity Areas in coastal areas.
- Systems of protected areas are now covering a wider range of ecosystems, with particular improvements in marine areas; however, the protection of offshore oceans and freshwater ecoregions is lagging behind.
- Equitable governance and management of protected areas is a key aspect of Aichi Target 11 (see Appendix 9.1). Although there are several methodologies and a framework for understanding equity in protected areas, assessments have been scarcely implemented.

- Connectivity between protected areas is key to maintaining the viability of populations and ecosystems. Metrics to measure connectivity at the global level have been developed and reveal that about half of the global protected area network is connected.
- Looking forward, governments and other stakeholders will shortly review options for a post-2020 global biodiversity framework. Spatial conservation efforts are critical to the conservation of biodiversity and sustainable development.

PROTECTED AREA CATEGORIES

Fundamental to the establishment and management of protected areas is what has been referred to as the dual mandate. On one hand, parks and protected areas should actively preserve important natural features (recognizing that preservation is saving *from* use, while conservation is saving *for* use), and on the other hand they should make allowances for human use. There is thus a continuum between use and preservation through a variety of different types of protected areas that accommodate a range of different human priorities.

The IUCN (2019b) has developed a recognized global standard of protected area types, which provides a unified approach to the creation of these protected areas and the types of planning, management and development that ought to take place in these areas. The categories are organized according to their level of protection or preservation, with strict nature reserves and wilderness areas as the first categories, and with protected areas with sustainable use of natural resources as the last category, allowing for a greater degree of use than other categories:

Ia. *Strict nature reserve*: Strictly protected areas set aside to protect biodiversity and also possibly geological/geomorphic features, where human visitation, use and impacts are strictly controlled and limited to ensure protection of the conservation values. Such protected areas can serve as indispensable reference areas for scientific research and monitoring.
Ib. *Wilderness area*: Usually large unmodified or slightly modified areas, retaining their natural character and influence without permanent or significant human habitation, which are protected and managed so as to preserve their natural condition.
II. *National park*: Large natural or near-natural areas set aside to protect large-scale ecological processes along with the complement of species and ecosystems characteristic of the area, which also provide a foundation for environmentally and culturally compatible, spiritual, scientific, educational, recreational and visitor opportunities.
III. *Natural monument or feature*: Set aside to protect a specific natural monument, which can be a landform, sea mount, submarine cavern, geological feature such as a cave, or even

a living feature such as an ancient grove. They are generally quite small protected areas and often have high visitor value.

IV. *Habitat/species management area*: Aim to protect particular species or habitats and management reflects this priority. Many Category IV protected areas will need regular, active interventions to address the requirements of particular species or to maintain habitats, but this is not a requirement of the category.

V. *Protected landscape/seascape*: A protected area where the interaction of people and nature over time has produced an area of distinct character with significant, ecological, biological, cultural and scenic value, and where safeguarding the integrity of this interaction is vital to protecting and sustaining the area and its associated nature conservation and other values.

VI. *Protected area with sustainable use of natural resources*: Conserving ecosystems and habitats together with associated cultural values and traditional natural resource management systems. They are generally large, with most of the area in a natural condition, where a proportion is under sustainable natural resource management and where low-level non-industrial use of natural resources compatible with nature conservation is seen as one of the main aims of the area

These categories are represented in a variety of different protected area types and sub-types. For example, the province of Ontario in Canada has a classification system that includes six different park types. These include recreation, cultural heritage, natural environment, nature reserve, waterway and wilderness class parks, with emphasis placed on different levels of human use and preservation among all six classes. Nature reserve and wilderness class parks emphasize higher levels of protection, as noted above in the IUCN categories, while recreation class parks emphasize a greater degree of human use and are often close to large centres of population to facilitate outdoor recreation. Apart from the several types of state and provincial parks, national parks, wildlife preserves and so on, there are several examples of well-recognized international protected areas that have global significance. These include World Heritage Sites, Biosphere Reserves, Global Geoparks and Ramsar sites, with World Heritage Sites and Global Geoparks explained in more detail below.

WORLD HERITAGE SITES

The World Heritage Convention (UNESCO) is one of the most recognized programmes in existence for the protection of cultural and natural assets around the world. At the end of 2018 there were 1092 properties recognized as World Heritage Sites (WHSs), with 845 of these as cultural sites and 209 as natural sites: 54 sites were in danger of being delisted; two were delisted; 38 were mixed (cultural and natural); 37 of these sites were transboundary; and 167 state parties are signatories of the World Heritage Convention (UNESCO,

2019a). Sites that are included on the list 'must be of outstanding universal value' and satisfy at least one of the following ten selection criteria (UNESCO, 2019b):

1. to represent a masterpiece of human creative genius;
2. to exhibit an important interchange of human values, over a span of time or within a cultural area of the world, on developments in architecture or technology, monumental arts, town-planning or landscape design;
3. to bear a unique or at least exceptional testimony to a cultural tradition or to a civilization which is living or which has disappeared;
4. to be an outstanding example of a type of building, architectural or technological ensemble or landscape which illustrates (a) significant stage(s) in human history;
5. to be an outstanding example of a traditional human settlement, land-use or sea-use which is representative of a culture (or cultures), or human interaction with the environment, especially when it has become vulnerable under the impact of irreversible change;
6. to be directly or tangibly associated with events or living traditions, with ideas or with beliefs, with artistic and literary works of outstanding universal significance (the Committee considers that this criterion should preferably be used in conjunction with other criteria);
7. to contain superlative natural phenomena or areas of exceptional natural beauty and aesthetic importance;
8. to be outstanding examples representing major stages of earth's history, including the record of life, significant ongoing geological processes in the development of landforms, or significant geomorphic or physiographic features;
9. to be outstanding examples representing significant ongoing ecological and biological processes in the evolution and development of terrestrial, fresh water, coastal and marine ecosystems and communities of plants and animals;
10. to contain the most important and significant natural habitats for *in situ* conservation of biological diversity, including those containing threatened species of outstanding universal value from the point of view of science or conservation.

In reference to the link between WHSs and the tourism industry, Landorf (2009) writes that even though World Heritage status is not interrelated with tourism growth, WHSs must implement management plans to moderate or diminish the impacts of tourism. In planning and managing for sustainable heritage tourism at these sites, Landorf identifies two key themes that must be implemented in sustaining site significance: a long-term and holistic planning process, and multiple stakeholder participation in planning (see also Gullino *et al.*, 2015). These are carried forward in the comprehensive WHS management plan evaluation instrument developed by Landorf, as shown in Table 9.1.

In a review of six WHSs in the UK, testing the WHS site management evaluation instrument, Landorf found that these sites lacked mechanisms to integrated more holistic

Table 9.1 World Heritage Site management plan evaluation instrument

Evaluation dimensions	*Assessment items*
Situation analysis	1. Tangible heritage characteristics are described 2. Intangible heritage characteristics are described 3. Land use and ownership patterns are identified 4. Demographic characteristics are identified 5. Economic characteristics are identified 6. Economic benefits of heritage are identified 7. Heritage tourism activities are identified 8. Capacity of tourism infrastructure is identified 9. Visitor numbers, length of stay and value are identified 10. Integration with other planning processes is identified
Strategic orientation	11. The time dimension reflects a long-term orientation 12. Broad-based economic goals are identified 13. Broad-based environmental goals are identified 14. Broad-based social/community goals are identified 15. Broad-based heritage development goals are identified 16. A range of strategic alternatives are identified and evaluated 17. Specific objectives are developed that support goals 18. Specific objectives are based on supply capability 19. Specific objectives target equitable economic distribution 20. Specific objectives are quantifiable and measurable
Community values and attitudes	21. Local community values and attitudes are identified 22. Critical issues for residents are identified 23. Community attitudes to heritage are assessed 24. The quality of life in the local community is assessed 25. The vision aligns with local community values and attitudes 26. The relationship between stakeholders is detailed 27. Relevant government agencies participated in the process 28. Government agencies influenced strategic directions 29. Relevant non-government agencies participated in the process 30. Non-government agencies influenced strategic directions 31. Local businesses and residents participated in the process 32. Local businesses and residents influenced strategic directions 33. Relevant visitor groups participated in the process 34. Relevant visitor groups influenced strategic directions

Source: Landorf (2009: 63)

planning approaches to sustainable development including tools to better engage with local community stakeholders. More success was achieved in the area of long-term, goal-oriented planning within these sites.

In related work, Badia and Donato (2013) investigated the how Italian WHSs (Italy has the most sites at 47 as of 2012) were being measured through performance indicators in reference to any gaps that exist between theory and practice. WHS management plans must have six key elements: (a) a thorough shared understanding of the property by all stakeholders; (b) a cycle of planning, implementation, monitoring, evaluation and feed-back; (c) the involvement of partners and stakeholders; (d) the allocation of necessary resources; (e) capacity building; and (f) an accountable, transparent description of how the management system functions (Badia & Donato, 2013: 25). Badia and Donato found that only 25 of the 47 WHSs had a management plan, and only 10 of the 25 had measureable indicators. Only one of the 25 had a monitoring system that assessed the results achieved by the site/organization. It was concluded that there is an absence of a managerial culture in WHSs as well as a lack of accountability.

The End of Chapter Case Study on Hadrian's Wall identifies some of the challenges identified by Landorf (2009) and Badia and Donato (2013) above. Cooperative arrange-ments must be formally established in working towards a shared goal in conservation. Other studies corroborate this finding by suggesting that it is important to clarify both the capacity and the mission of organizations that would wish to partner with protected areas organizations like the National Park Service (NPS). Some potential partners (non-profit or for-profit enterprises) may have either a strong capacity to fit within the goals of the NPS or a weak capacity, and this is matched up with an understanding of a potential part-ner's mission which may be weakly or strongly aligned with the NPS (McPadden & Margerum, 2014). It is potentially a waste of resources (time, money and human resources) to partner with organizations that neither have the mission nor the capacity that align with the primary agency's goals.

TANGIBLE AND INTANGIBLE HERITAGE

In the formative days of heritage protection it was the physical or concrete structures of human processes (along with natural phenomena) that were recognized as holding herit-age value. This tangible heritage includes 'a physical dimension that requires specific tech-nologies of preservation or conservation and raises questions about integrity and authenticity of the site, often from a historical perspective' (Conway, 2014: 145–146). However, there is also recognition of the value of intangible heritage as being imperative in representing the past, present and future of human societies. This latter form of herit-age is 'much more fluid and raises even more questions about the value of conserving a particular form of the heritage over time when the cultural context in which it is produced has changed' (Conway, 2014: 146). It is the 'the practices, representations, expressions, as well as the knowledge and skills, which communities, groups and [...] individuals recognize as part of their cultural heritage' (UNESCO, 2003, Article 2.2).

Conway (2014) reports on the tangible and intangible heritage found in the rock art and ranching communities of the Sierra de San Francisco site in Mexico's Baja California Peninsula. The rock art murals is an example of WHS status, consisting of 320 sites found in seven caves of the region, dated to as early as 2800 BCE. The intangible heritage is represented in the ranching culture of the region, as remembered past experiences, past practices, technology and the oral history of the region. In helping to sustain this heritage – as there are concerns that these cultural traditions are disappearing – tourism is being intensified to expand the role of more peripheral cohorts within the broader community. Ideas include botany tours led by women, animal herding, crocheting, cheese production, craft production, and food preparation for the trip to see the murals.

Another example of tangible and intangible heritage is found in the region of Providencia, Rio de Janeiro's oldest favela. Savova (2009) writes that the municipality created the concept of the 'open-air/living museum' to celebrate history and to connect with large infrastructure development projects. Building off Cohen's (1979) concept of the 'centre', as the essence of morality and ideology with tourism mobilized as a catalyst for stimulating community cultural revival, Savova develops the concept of 'heritage kinaesthetics'. This concept is defined 'as the moving bodily practices that people imagine and enact to enliven the built environment's static aesthetic looks, or the immobile quality usually ascribed to historic sites' (Savova, 2009: 547). Savova's five main heritage kinaesthetics that are practised by local people and tourists are: visual (photographing, seeing versus looking); ambulatory (walking around versus exploration); performative (enacting intangible cultural heritage like the experience of a local guide, taking in a football game); oral (imaging history and telling stories); and acoustic (place-specific sounds). This type of thinking resonates with the work of Vidal González (2008) on intangible heritage tourism and identity using the example of Japanese flamenco tourism. One of the many effects of globalization is the unlinking of identity and local place attachment. In an effort to overcome this disunity, intangible heritage tourism can act as a source of identity that allows for deep integration into culture, which further provides for a more authentic experience.

GEOPARKS

The geopark concept is a programme developed by UNESCO in 1998 for the purpose of identifying and representing global geological heritage. Sustainability, economic development and cultural heritage and development are keystones to the geoparks concept, and so sustainable tourism (ST) and ecotourism have become important facets of the overall geoparks theme in balancing the use and preservation of heritage. As of 2019, there were 140 UNESCO Global Geoparks in 38 different countries (UNESCO, 2019c). The system is built on partnerships with communities, landowners, businesses and existing protected areas. In the case of the latter, geoparks 'piggyback' on various classifications of protected areas, including provincial and state parks, national parks, municipal parks, historical sites and other local, state, national or international jurisdictions.

As reported by the Canadian Federation of Earth Sciences (2017), there are several criteria for a region to be eligible as a geopark. These include areas that are:

- scientifically important, or especially striking, scenic, or unusual geologic phenomena;
- historically important sites where particular geologic features, rock types, landforms or type specimens of fossils were first recognized and described;
- outstanding examples of geologic features, structures, fossils, processes and landforms; and/or
- historical sites where cultural events were tied to an area's geologic features, such as those in the history of geology, mining and geology in early exploration and settlement.

For a region to become a geopark it first has to become an 'aspiring' geopark, which means the development of a committee and comprehensive management plan based on the identification of area, geological heritage, geo-conservation, economic activity and business plan, and supported by references from an inclusive list of stakeholders in the region. In the Niagara region of Ontario, Canada, the Ohnia:kara Aspiring Global Geopark (Ohnia:kara meaning 'neck between two bodies of water') has an application territory that includes the 12 municipalities of the Niagara region. The significance of the region is found not only in the world-famous Niagara Falls but also in history (this is where the British General Sir Isaac Brock defended the British interests (in what later would become Canada) against an American threat during the war of 1812), winemaking, education, festivals, major events and trails. The region also has a long history of Indigenous occupation of the region, and these Indigenous nations played an important role between British, American and French interests in the Niagara territory (Fennell, 2018b; see Chapter 6).

GOVERNANCE IN PROTECTED AREAS AND TOURISM

The innate challenges in the sustainable management of protected areas and tourism are readily apparent in consideration of what is lost and gained, i.e. what is at stake for people who rely on protected areas for their livelihoods or those who stand to gain economically from their relationship with protected areas – and other people. The literature on governance places these questions into sharp focus because it emphasizes the complex sets of actors and their relationships that influence, and have been influenced by, neoliberal programmes and policies that stimulate new forms of governance away from a single source of authority (Duffy, 2006). Duffy uses Madagascar as an example of how donor consortia are a form of global environmental governance. Such consortia often include several donor organizations, including the World Bank, several national governments, and NGOs such as

Conservation International and the World Wildlife Fund (WWF), connected by the aim of providing aid for Madagascar. Madagascar is attractive as a biodiversity hotspot and is therefore an important region for ecotourism investment and development. As such, and as Duffy observes, once the environment is saved through the efforts of donor funding, these places will be open for business. The neoliberal focus is evident in this calculus: there is much financial gain to be realized through destinations that offer so much resource potential.

'Good' governance is a concept that is being used liberally in the literature on resource management and community development, and the intersection of both, especially through social and ecological systems research. Saner and Wilson (2003) argue that good governance is built on 'noble ethical claims', suggesting that there is an ethical underlay to governance practised in the right way, while Lebel *et al.* (2006) have noted that 'good' includes the inclusion of '... participation, representation, deliberation, accountability, empowerment, social justice, and organizational features such as being multilayered and polycentric' (Lebel *et al.*, 2006). The value of ethics in good governance is the central part of the work of Fennell *et al.* (2008) on adaptive co-management, who argue that the three main traditions of ethical thinking all need to inform governance. These include deontology (universal principles, duties, norms and laws), teleology (to develop virtues, the greatest good for the greatest number) and existentialism (self-determination, freedom of choice and responsibility for action). Table 9.2 illustrates a set of principles for good governance that are directly applicable to the management of protected areas (Graham *et al.*, 2003) .

Table 9.2 Good governance principles for National Parks and protected areas management

The five good governance principles	***The UNEP principles on which they are based***
Legitimacy and voice	Participation Consensus orientation
Direction	Strategic vision, including human development and historical, cultural and social complexities
Performance	Responsiveness of institutions and processes to stakeholders Effectiveness and efficiency
Accountability	Accountability to the public and to institutional stakeholders Transparency
Fairness	Equity Rule of law

Source: Graham *et al.* (2003)

The problem with implementing good governance policies and actions in tourism (and in protected areas) is the multi-layered nature of issues and interactions. It is a difficult task indeed, as policy is difficult to institute at these various scales because of competing interests between actors and agencies (see Plummer & Fennell, 2009). Cooperation is therefore an essential goal to strive for (Milne & Ateljevic, 2001) in connecting actors vertically and horizontally through social learning approaches and regularized interactions focused on information sharing and problem articulation, among other strategies (Armitage *et al.*, 2008).

The breadth of governance issues in ST is large and diverse (Dangi & Jamal, 2016). While too extensive to list here in their entirety, governance issues according to these authors include: planning and strategic vision (e.g. development control); management and marketing (e.g. climate change and adaptation); power, rules and regulations (e.g. mobilizing resources); visitor safety and crisis management (e.g. safe working conditions); collaboration and coordination (e.g. multi-level integration); as well as participation, service delivery, accountability, transparency, equity, communication, leadership, the use of technology, and political support and participation. Underrepresented issues in governance according to Dangi and Jamal (2016) include equity, fairness and justice.

There are several cases in the literature that report on governance challenges in protected areas especially in consideration of how best to balance the interests of those who wish to benefit from tourism. Phong Nha-Ke Bang National Park in Vietnam is one such example of a region that has struggled to implement the right governance structures to balance stakeholder interests, tourism sustainability and conservation. Hūbner *et al.* (2014: 9) discovered that the model used to manage the park 'is significantly governed by opaque structures and processes and underlain by cultural values which define responsibilities, decision-making and the degree of involvement of a variety of tourism actors'. Politics and self-interest have emerged in the park, driven by economic concerns, which have inevitably led to a lack of participation and benefit sharing.

Box 9.1 The Tourism and Protected Areas Specialist Group (TAPAS)

The Tourism and Protected Areas Specialist Group (TAPAS) is a volunteer specialist group of the IUCN World Commission on Protected Areas. There are more than 500 professionals working within this network. The Vision, Mission and Objectives of this organization are as follows:

Vision

We advocate for a future where tourism enhances the conservation integrity of protected area systems, improves human wellbeing and provides benefits for the local population, and where there are accessible, inspiring, safe and educational opportunities for visitors through environmentally, socioculturally and economically sustainable products and experiences.

Mission

Our mission is to provide a platform for protected area practitioners and others, where expertise and knowledge is shared, sustainability awareness is enhanced, collaboration and dialogue are facilitated, leadership is developed and innovative solutions are fostered, in order to support the oversight of ST in protected area systems.

Objectives

- Strengthen the capacity, effectiveness and performance of protected area managers and policy makers and others in relation to ST, through learning, exchange and the development of information and guidance.
- Provide an interactive forum for individuals working on protected areas and tourism, enabling strategic networking, communication, collaboration, inspiration and partnerships.
- Provide strategic advice to protected area authorities, the tourism industry and other stakeholders on the optimum approaches to ST planning, development, management and monitoring in protected and natural area destinations.
- Build awareness and understanding of ST and protected areas by developing and disseminating knowledge, including case studies and best practice syntheses.
- Enhance the level of tourism's contribution to the goals of protected areas and protected area systems including biodiversity conservation, human wellbeing and the SDGs.
- Enhance the capacity of WCPA through cooperative ventures with other IUCN commissions and IUCN members as well as other networks and partners.
- Foster dynamic, innovative and sustainable solutions for financing and managing protected areas.

Source: IUCN (2019d)

TOURISM CONCESSIONS AND PARTNERSHIPS

We discuss public–private partnerships (PPPs) in detail in Chapter 8. PPPs have been widely used in the protected areas arena and are a signpost for a more intensive process of the commercialization of conservation in generating more revenue for the management and operation of parks. A PPP is defined as a 'long-term contract between a private party and a government entity, for providing a public asset or service, in which the private party bears significant risk and management responsibility and remuneration is linked to performance' (World Bank, 2017: 1).

Often these partnerships surround the management of concessions in protected areas. These services include, for example, lodging, transportation and restaurants, as well as recreational equipment rental for items such as kayaks, bicycles or climbing equipment. An example of a charismatic park that continues to offer concessions in the form of ecolodges (to boost revenue, provide a different experience for tourists and support conservation) is Kruger National Park in South Africa. Based on a study of 314 domestic visitors to the park, it was found that ecolodge concessions led to a shifting sense of place with the park that visitors have started to feel because of the commercialization process. Respondents claimed that the staff did not make them feel part of the park, and service quality and pricing generated several other negative sentiments (Coghlan & Castley, 2013).

It is not only visitors that are experiencing conflict – perhaps better stated as lack of agreement – with concessionaires. Dangi and Gribb (2018) report that park managers and concessionaires working in the Rocky Mountains National Park, USA were in agreement about the types of ecotourism activities that should take place in the park, such as hiking/backpacking, horseback riding, watching scenery by car, skiing, camping and climbing. However, they did not agree on carrying capacity, impacts from horseback riding and inter-group visitor conflicts, with park managers siding over the more cautious and protectionist stance on use, and concessionaires seeking to be more liberal.

Partnerships have also been discussed in protected areas for the purpose of responding to climate change and how climate change has an effect on biodiversity conservation. Monahan and Theobald (2018) argue that there is a potential for protected areas agencies to partner in efforts to conserve individual species challenged by climate change as well as biodiversity at regional scales. Partnerships were also said to hold value in improving biodiversity indicators and how these indicators need to change and adapt, with particular attention paid to invasive species and human land uses.

SUSTAINABILITY, PARKS AND TOURISM

There are several key documents that identify the important link between sustainability and protected areas. For example, by the early 1990s a comprehensive document was published by the Federation of Nature and National Parks of Europe on ST, asking whether we were loving these protected areas to death (FNNPE, 1993). The document stresses several disadvantages of non-sustainable tourism as well as the advantages of sustainable tourism (Table 9.3).

Table 9.3 Advantages and disadvantages of sustainable tourism

Disadvantages of non-sustainable tourism	***Advantages of sustainable tourism***
For conservation and protected areas: • Environmental damage • Visitor pressure • Pollution • Managing tourism consumes resources and diverts attention from other management priorities *For local people*: • Disturbance and damage to ways of life and social structure • Higher costs *For society*: • Pressure on resources	*For conservation and protected areas*: • Greater public and local awareness of protected areas and the environment • Political support which can help to attract funding and support the designation of new protected areas • Conservation of natural and cultural features through restoration projects and direct practical help • Additional finance from the tourism sector and from tourists *For the tourism sector*: • Support for businesses and employment • Development of new, high-quality, environmentally sound products based on nature and culture with a long-term future • Reduction of development costs through partnerships with protected areas • Improvement of company image • Attraction of customers looking for environmentally sound holidays *For local people and society*: • Improved income and living standards • Revitalization of local culture and traditional crafts and customs • Support to rural infrastructure • Improved economy • Improved physical and psychological health • Promotion of harmony between people from different areas

The recent IUCN-sponsored publication by Leung *et al.* (2018) is a comprehensive statement on state-of-the-art thinking around best practice management in the connection between sustainability, tourism and protected areas. There are chapters on the impacts of protected area tourism, aligning management objectives with tourism impacts, adaptive management for ST, capacity building for ST, managing tourism revenues, and the future of protected areas tourism, all with best practices sections. The best practices for impacts from protected area tourism are as follows (Leung *et al.*, 2018: 26):

- Encourage national tourism policies that fulfil the 'triple bottom line' by requiring protected area tourist activities to explicitly contribute to the conservation of nature, generate economic benefits to both protected area authorities and local communities, and account for and minimize negative social impacts.
- Support community-based delivery of tourism services that is market related. Consider partnerships between community enterprises and the private sector to improve the chances of commercial success.
- Build training in business development and management skills into community-based delivery of tourism services, and include community members, NGO representatives and protected area managers in the training.
- Reimagine recreational activities in protected areas as a way to meet community needs and address larger societal goals, such as those related to human health and wellbeing.

The UN's *Transforming our World: 2030 Agenda for Sustainable Development* (UN, 2015) is said to be the driving force behind the relationship between conservation and sustainable development for the foreseeable future (Dudley *et al.*, 2017a, 2017b). While the link between protected areas and SDGs 14 and 15 is obvious, i.e. these goals follow the Aichi Biodiversity Targets for marine and terrestrial protected areas (see Appendix 9.1: Aichi Biodiversity Targets), Table 9.4 shows how UN SDGs 1, 2, 3, 4, 6, 8, 11, 12 and 13 can be addressed through parks and protected areas.

There are also several key academic sources that attend to the topic of tourism, protected areas and sustainability. Often these publications, like the aforementioned new IUCN publication by Leung *et al.* (2018), come in the form of guidelines for the planning, development and management of ST in protected areas. Examples of this work include Eagles *et al.* (2002), and an older publication by Ceballos-Lascurain (1996) on tourism, ecotourism and protected areas, and guidelines for development. This book is based on studies presented at the IV World Congress on Parks and Protected Areas in Caracas, Venezuela, and highlights the many bad examples of how poor planning can lead to destruction within protected areas. Ecotourism is viewed as a solution through the implementation of good principles and guidelines which factor into the equation community development, the role of private organizations, and parks and protected areas agencies.

Table 9.4 Key links between SDGs and protected areas

SDG goals	*Approaches*
1.5: By 2030, build the resilience of the poor and those in vulnerable situations and reduce their exposure and vulnerability to climate-related extreme events and other economic, social and environmental shocks and disasters.	Highlighting the role of protected areas as tools for adaptation to climate change (Dudley *et al.*, 2009).
2.4: By 2030, ensure sustainable food production systems and implement resilient agricultural practices that increase productivity and production, that help maintain ecosystems, that strengthen capacity for adaptation to climate change, extreme weather, drought, flooding and other disasters and that progressively improve land and soil quality.	**1.** Basic supporting services such as soil production and stabilization of water supplies; **2.** Buffering against climate-related shocks; **3.** Promoting sustainable agriculture such as organic production within Category V protected landscapes (Phillips, 2002); **4.** Securing fish stocks in marine protected areas.
2.5: By 2020, maintain the genetic diversity of seeds, cultivated plants and farmed and domesticated animals and their related wild species, including through soundly managed and diversified seed and plant banks at the national, regional and international levels, and promote access to and fair and equitable sharing of benefits arising from the utilization of genetic resources and associated traditional knowledge, as internationally agreed.	Using protected areas to conserve crop wild relatives, and races and livestock wild relatives to help build agricultural resilience (Meilleur & Hodgkin, 2004; Stolton *et al.*, 2006).
3.4: By 2030, reduce by one-third premature mortality from non-communicable diseases through prevention and treatment and promote mental health and wellbeing.	Developing the Healthy Parks Healthy People concept in promoting the role of protected areas as green gyms and places for treatment of mental health and addiction issues (Stolton & Dudley, 2010).

(*Continued*)

Table 9.4 (*Continued*)

4.7: By 2030, ensure that all learners acquire the knowledge and skills needed to promote sustainable development including, among others, through education for sustainable development and sustainable lifestyles, human rights, gender equality, promotion of a culture of peace and non-violence, global citizenship and appreciation of cultural diversity and of culture's contribution to sustainable development.	Using protected areas near urban centres (Trzyna, 2014), to provide basic knowledge of ecosystem functioning, and to address nature-deficit problems in people of all ages
6.3: By 2030, improve water quality by reducing pollution, eliminating dumping and minimizing release of hazardous chemicals and materials, halving the proportion of untreated wastewater and substantially increasing recycling and safe reuse globally.	Promoting protected areas as water towers (Dudley & Stolton, 2003) in collaboration with major suppliers of municipal drinking water, by promoting these links particularly.
6.6: By 2020, protect and restore water-related ecosystems, including mountains, forests, wetlands, rivers, aquifers and lakes.	Expanding protected areas as a key tool for conservation of inland waters, some of the least protected habitats on Earth.
8.9: By 2030, devise and implement policies to promote ST that creates jobs and promotes local culture and products.	Providing important opportunities for nature tourism, the quickest growing tourism sector, in well-managed protected areas.
11.5: By 2030, significantly reduce the number of deaths and the number of people affected and substantially decrease the direct economic losses relative to global gross domestic product caused by disasters, including water-related disasters, with a focus on protecting the poor and people in vulnerable situations.	Recognizing and planning the role of protected areas as buffers for cities, both as important urban and peri-urban green space and for wetlands, coastal vegetation and mountain forests to provide an important disaster risk reduction function.

11.7: By 2030, provide universal access to safe, inclusive and accessible green and public spaces, in particular for women and children, older persons and persons with disabilities.	Arguing for more urban protected areas, particularly in rapidly growing cities (Trzyna, 2014).
11.b: By 2020, substantially increase the number of cities and human settlements adopting and implementing integrated policies and plans towards inclusion, resource efficiency, mitigation and adaptation to climate change, resilience to disasters, and develop and implement, in line with the Sendai Framework for Disaster Risk Reduction 2015–2030, holistic disaster risk management at all levels.	Using natural ecosystems in protected areas to provide mitigation of and adaptation to climate change, including urban nature reserves to provide cooling and absorption for flood water.
12.b: Develop and implement tools to monitor sustainable development impacts for ST that creates jobs and promotes local culture and products.	Providing a monitoring framework in collaboration with relevant UN agencies and as a contribution to the SDGs.
13.1: Strengthen resilience and adaptive capacity to climate-related hazards and natural disasters in all countries.	Using natural ecosystems in protected areas to provide mitigation of and adaptation to climate change (Gross *et al.*, 2016).

Source: Dudley *et al.* (2017b)

ECOTOURISM

Chapter 4 discussed a number of 'concepts' and 'tools' that are important in moving tourism along the road to sustainability. Ecotourism was highlighted as a concept because of its higher order focus on ecological and sociocultural factors that many argue should make it the greenest or most sustainable form of tourism in existence. Key tenets of ecotourism include the fact that it takes place in nature (i.e. there is a tangible connection to nature), it has a strong focus on environmental education, it is sustainable in the sense of being pro-community control and development while at the same time strong in the area of conservation and preservation, and it should be anchored by ethical planning, development and management (Fennell, 2014). Ecotourism, therefore, is different from the broader nature-based tourism which includes all the various types of tourism that take place in nature: swimming, going to the beach, off-road driving, mountaineering, surfing, hunting, fishing, etc. Ecotourism is also different from wildlife tourism chiefly on the basis of its consumptive versus non-consumptive orientation. Wildlife tourism also falls entirely within nature-based tourism and is defined as 'tourism based on encounters with non-domesticated (non-human) animals ... [that] can occur in either the animals' natural environment or in captivity' (Higginbottom, 2004: 2). While ecotourism is centred around being non-consumptive in its usage of the natural world, wildlife tourism can be both non-consumptive and consumptive (zoos, hunting, fishing). An article that graphically situates ecotourism, nature-based tourism and wildlife tourism is the work by Reynolds and Braithwaite (2001) (Figure 9.1). These three forms of tourists (ecotourists, nature-based tourists and wildlife tourists) are attracted to protected areas because of the natural attractions (e.g. birds, mammals, hiking trails, mountains, streams, lakes) that exist in these areas.

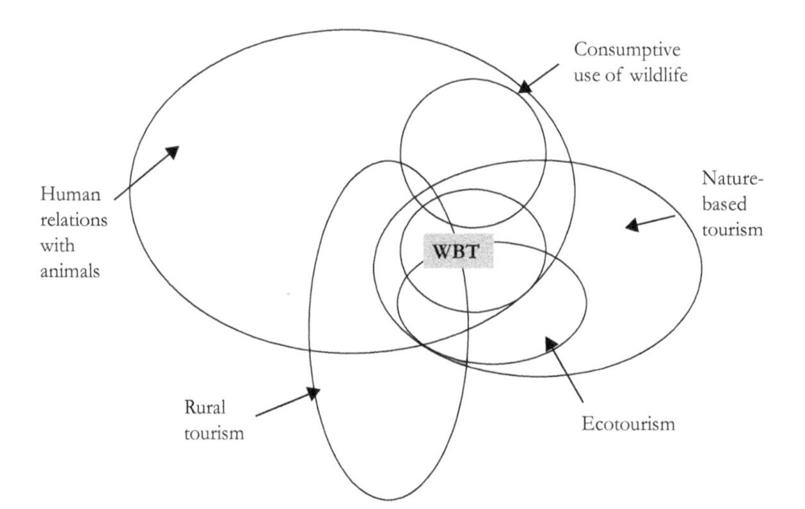

Figure 9.1 Wildlife-based tourism
Source: After Reynolds and Braithwaite (2001)

Ecotourism's 'big brother', nature-based tourism, is becoming big business. Several countries report on its increasing magnitude relative to conventional tourism, as well as how it must be differentiated from ecotourism through certification systems (Box 9.2). In Australia, for example, the nature-based tourism industry is so popular that it represents clearly one-third of all travel and tourism to the country (Buckley & Sommer, 2000).

Several scholars have investigated the connection between theory (ecotourism principles) and the practice of ecotourism to explore the legitimacy of the industry in various contexts. Donohoe and Needham (2008) explored the marketing practices of 25 Canadian ecotourism operators, and found that less than half of the sample adopted the main tenets of ecotourism in their operations. Ecotourism in name often does not

Box 9.2 Australia's ECO Certification Program

What is ECO Certification?

The ECO Certification Program assures travellers that certified products are backed by a strong, well-managed commitment to sustainable practices and provides high-quality nature-based tourism experiences. The ECO Certification Program is a world first and has been developed to address the need to identify genuine nature and ecotourism operators.

ECO-Certified nature tourism

Tourism in natural areas that leaves minimal impact on the environment.

ECO-Certified ecotourism

Tourism in a natural area that focuses on optimal resources use, leaves minimal impact on the environment and offers interesting ways to learn about the environment with operators that use resources wisely, contribute to conserving the environment and help local communities.

ECO-Certified advanced ecotourism

Australia's leading and most innovative ecotourism products that operate with minimal impact on the environment and provide opportunities to learn about the environment with operators who are committed to achieving best practice, using resources wisely, contributing to conserving the environment and helping local communities.

Source: ECO Tourism Australia (2018)

mean ecotourism in practice. Those providers that represented all of the tenets were referred to as 'genuine ecotourism', whereas those embracing fewer tenets were labelled as either 'ecotourism "lite"' or 'greenwashing'. The problem for Donohoe and Needhan and other scholars is that these less committed ecotourism operators and their programmes often begin to take on characteristics of nature-based tourism or even mass tourism. Kur and Hvenegaard (2012) found that whale-watching operators also marketed their programmes in a manner that was not inclusive of all the dimensions of ecotourism, based on an analysis of 62 brochures. While the operators emphasized the nature-based attraction and education dimensions of their programmes, the sustainability dimension was less often emphasized, including the conservation activities/involvement of these firms.

Other scholars have noted that whale watchers do not fit neatly on the specialization continuum from generalists to specialists like other forms of wildlife tourists such as divers (Bentz *et al.*, 2016; see also Lemelin *et al.*, 2008). Specialization in an activity often comes with a greater degree of skill development, better equipment and a stronger conservation ethic, which was documented by Bryan (1977) in the case of recreational fishing. Fly anglers (technique-setting specialists) have developed these characteristics over novices who are less invested in the activity. These characteristics have been investigated from the perspective of fly anglers who post their personal fly-fishing videos on Vimeo. Several key dispositions of this group were uncovered through the research such as type of water fished (rivers), type of fish caught (Salmonids), gratuitous lift of caught fish out of water, catch-and-release, treatment of fish once caught, clothing, and gender of anglers (Fennell, 2016). The ability to create these videos is a function of supply meeting the needs of demand. Stensland *et al.* (2017) have found that salmon angling in Norway is in a state of decline due to salmon abundance as a main leisure constraint (quality of fishing). This constraint was mitigated to a certain degree through various negotiation strategies such skills, knowledge and money. Their main conclusion is related to supply. Increase the supply of salmon and anglers will return to the rivers, but it is the way in which the resource base is managed that will be the marker of a sustained fishing industry over time.

A practical application of the tenets of ecotourism can be found in ecolodges which, as noted above in the section on concessions, are often affiliated with protected areas. For a property to be considered an ecolodge it must satisfy several main criteria: (1) it must be located within or near a natural area and provide benefits such as jobs to local communities; (2) it must actively work to conserve these lands; and (3) it must provide an educational component for both tourists and local people (Kwan *et al.*, 2010; Mehta *et al.*, 2002).

Linking with the discussion above on ecotourism lite and greenwashing, Lai and Shafer (2005) found that ecolodge owners only partially aligned with the principles of ST on the basis of their marketing strategies. The authors found that the environmental dimensions of sustainable ecotourism were more frequently marketed over the economic aspects, which were in turn more frequently marketed over the sociocultural dimensions. The educational dimension of sustainable ecotourism was largely ignored by ecolodge operators, which is inconsistent with the tenets of ecotourism as education ought to be one of the main driving forces of this type of tourism.

Osland and Mackoy (2004) report on a series of interviews conducted with owners and managers of 21 ecolodges in Mexico and Costa Rica on the performance goals used to gauge the success of their ecolodges. Sustainable economic development performance goals such as nature conservation, job creation, indirect economic benefits and community social/education benefits were the ones mentioned most often, but they were the ones most difficult to measure. These goals were often stated in an idealistic manner with the intent of capturing the spirit of ecotourism. So when operators said that their establishment was a social catalyst for the community, these operators were hard pressed to describe how to measure this type of impact or success. Results are different on performance in ecolodges when studies are conducted on patrons. Kwan *et al.* (2010) found that the performance of ecolodges far exceeded the importance that patrons placed on a number of different measures such as friendliness of staff, scenery, value for money, decent sanitary conditions and quality of environment/landscape. The authors observe that ecolodges have become adept at the provision of a collection of hospitality, personal service and environmental elements.

Nelson (2010) investigated the accommodation sector in Australia which has been eco-certified by ECO Tourism Australia (see Box 9.2). She found that although energy is a key issue contained within eco-certification guidelines in Australia, less than half of the 50 accommodation units studied provided any information on energy on their websites. Those that did discussed issues around reducing energy use, improving energy efficiency, using alternative forms of energy and offsetting emissions. Nelson concludes by suggesting that more emphasis is being placed on facilitating nature-based experiences in these enterprises than on minimizing environmental impacts (see also Lai & Shafter, 2005; Warnken *et al.*, 2004). This finding resonates with other inconsistencies that have been found on the ecological dimension of ecolodges. Food and wine operators in Australia were more likely to mention ethical food sourcing and sustainability on their websites than advanced ecotourism operators, as observed by Fennell and Markwell (2015; cf. Chapter 7).

Ecolodges have been very successful as part of private landholdings. These so-called private parks or private protected areas exist all over the world in places such as Costa Rica, Australia, Canada, Kenya, Tanzania and South Africa, and are excellent destinations for ecotourism. Several scholars have argued that not enough research has been conducted on private parks, and this includes arriving at an acceptable definition. Carter *et al.* (2008) argue that this designation ought to be defined according to several key factors. These include

> land of conservation importance that is directly under the ownership and/or management of a private sector conservation enterprise for the purpose of biodiversity conservation. This purpose may be singular (i.e. the entire mission of the organization is conservation), or it may be concurrent with other objectives (such as a business venture or other social imperative). (Carter *et al.*, 2008: 178)

These private enterprises, Carter *et al.* (2008) observe, include any type of non-state body or organization, including corporate interests, private individuals and trusts.

Langholz *et al.* (2000) argue that while many private parks or reserves are small in size, several of them are large – over 250,000 ha. In the 68 reserves that Langholz *et al.* (2000) studied in Costa Rica, reserves ranged in size from 20 ha to 22,000 ha, with a median size of 101 ha. The authors argue that bigger is not always better in privately owned parks, as the owners said it would be much more difficult to mange reserves that were too large. Furthermore, private reserves were found to be quite profitable, with one reserve in Costa Rica generating more revenue than all of Costa Rica's national parks combined because of the heightened interest in ecotourism (Langholz *et al.*, 2000). Grootbos Private Nature Reserve in Western Cape, South Africa is an example of a successful private park that has advanced tourism and sustainability on many different levels (Box 9.3).

COMMUNITY DEVELOPMENT

Getting community members onside in the development of ecotourism as well as protected areas and sustainability is a key feature of successful community development. Community-based tourism (CBT) has been conceived in many different ways. It may be as straightforward as an approach that engages local people in the planning and development of the tourism industry (Hall, 1996), and it often emphasizes a sense of shared purpose and common goals among constituents (Joppe, 1996). A more comprehensive definition that links community-based enterprises (CBEs), sustainability and tourism, is as follows: a CBE is a 'Sustainable, community-owned and community-based tourism initiative that enhances conservation and in which the local community is fully involved throughout its development and management and they are the main beneficiaries through community development' (Manyara & Jones, 2007: 637; see also Ellis & Sheridan, 2014).

Community development concepts have been operationalized in many different settings and in many different ways (McCool, 1995). For example, Zapata *et al.* (2011) tested whether CBT was better actualized using a top-down or a bottom-up approach in Nicaragua. Government and heavy reliance on external funding organizations define top down, whereas bottom up evolves more out of local initiatives. The authors discovered that top-down approaches were problematic because not only were they implanted in a relatively inorganic way, but they were also poorly adapted to local cultural conditions and time. These initiatives created the conditions for a greater degree of dependency if

Box 9.3 Grootbos Private Nature Reserve

Built on a hillside where a continent ends, two oceans collide and the smallest of the world's six floral kingdoms flourishes, the lodges of Grootbos Private Nature Reserve are uniquely positioned to capture the magic of South Africa's Western Cape. Below, Walker Bay stretches to the horizon, home to the 'marine big five': great white sharks, southern right whales, seals, penguins and dolphins; and vast coastal caves bear traces of a Stone Age civilization. Blanketing the hillside with bright, intricate flowers of every shade is the *fynbos*, or 'fine bush,' a hardy, complex plant family endemic to South Africa.

The reserve was named Grootbos – an Afrikaans word meaning 'big forest' – for the enchanting milkwood forests that cover the hillside. The lodges were built into natural clearings in these gnarled, ancient trees. Both acclaimed restaurants feature excellent cuisine with a focus on local seafood, and they, like the stylish suites and the elegant common areas, are all oriented towards an expansive, breath-taking vista of mountain, fynbos and sea.

Why we love this lodge

It's quite something to hear a lodge owner describe with glee the fire that charred his hotel and much of the surrounding land. But Michael Lutzeyer is not just any lodge owner, and as much as he loves the ecotourism retreat he has created, he also loves the fynbos – which rely on fire to germinate. In the aftermath of the 2006 fire, 70 new plant species were discovered on Grootbos Reserve, and an exquisite new Forest Lodge was built in record time. The fire has gone down in Grootbos history as a triumph of both nature and man.

Once the Lutzeyer family purchased this hillside in 1991, they spent years buying up adjacent plots, removing invasive species and returning farmland to fynbos. A recent survey revealed a remarkable 765 fynbos species at Grootbos, 100 of which are endangered and six that had never been discovered before. To conserve this unique biome and support the local community, Grootbos runs an extensive conservation and horticulture training programme for students from nearby villages; 80% of the staff also come from these communities. Close community connections are a hallmark of Grootbos, and art, music and local life are an integral part of the experience.

Spotlight on sustainability

From its earliest days, Grootbos has been dedicated to improving the livelihoods of local communities through ST practices. Their Green Futures College trains young adults from the Gansbaai area in conservation, giving them the skills and confidence necessary to become employable while contributing to the preservation and promotion of the region's unique biodiversity.

As part of the curriculum, students develop and maintain the environmentally friendly gardens at Grootbos. Their landscaping services and the sale of plants provide income to the college and help students pay for their tuition, thus creating a self-sustaining business model. On completion of their course, students are awarded a nationally accredited certificate in horticulture and assisted in work placement. Every year, three of the best students are given the opportunity to work at the Eden Project in Cornwall, England. The lodge proudly employs many of its graduates and has helped others find private employment elsewhere.

Source: National Geographic (2015)

communities could not move away from the funding model implemented throughout the life of the project. By contrast, bottom-up CBD projects tended to be more successful because: (a) organization emerged from the bottom of the community in regards to entrepreneurship and ownership; (b) there is a strong link between visitors and those places that were consumed; (c) cultural and social capital were built from the local context; (d) communities assume the risk of investing their own economic capital, building more of a sense of ownership; and (e) product development is based on local assets. (see Aquino *et al.*, 2018 for a model linking sustainable community development with tourism social entrepreneurship).

The concept of capital, as emphasized in the previous paragraph, is central to community development. Flora (2004) identified seven types of capital that are typically found in communities, and these have frequently been applied in a tourism context. Following McGehee *et al.* (2010) in the context of rural tourism development, these seven forms of capital include: financial capital (loans and credit, investment opportunities, tax credits, business-friendly structures); human capital (professional and educational growth and skill-building); built capital (physical structures in the community, including buildings, roads, mass transit and public facilities); cultural capital (preservation of local stories, history, art and so on); natural capital (plant and animal life, high-quality air and water); political capital (accessibility to power through local, regional, state and federal government); and social capital, which deals with trust, cooperation and reciprocity.

Of these various capital types, it is the latter form, social capital, that has received the most attention by theorists. Social capital is defined as 'features of social life – networks, norms and trust – that enable participants to act together more effectively to pursue shared objectives' (Putnam, 1995: 66). Putnam studied regional differences in the governance structures of northern, central and southern Italy. He argued that civic mindedness stems from trust and horizontal systems of governance rather than vertical ones. Putnam wrote that:

> These communities did not become civic simply because they were rich. The historical record strongly suggests precisely the opposite: They have become rich because they were civic. The social capital embodied in norms and networks of civic engagement seems to be a precondition for economic development, as well as for effective government. Development economists take note: Civic matters. (Putnam, 1993: 37)

The social capital concept has been applied to the sociocultural pillar of sustainability studies in the mountain resort community of Steamboat Springs, CO. Ooi *et al.* (2015) developed a conceptual framework to tie together a number of key components of social capital. Mountain resort tourism has an effect on social capital components (networks, norms and resources) that results in either positive or negative social capital outcomes, which indicate either socioculturally sustainable behaviour or unsustainable behaviour. An example of a positive social capital outcome is reinforcement of social capital norms, which indicates socioculturally sustainable behaviour in the form of improved quality of life and sense of community. An example of a negative social capital outcome is social exclusion, which indicates socioculturally unsustainable behaviour in the form of diminished quality of life, diminished sense of community and diminished community attachment.

The value of social capital to development theory and practice is germane. Many theorists, including Borlido (2016), argue that social capital is the 'missing link' in development (see also Macbeth *et al.*, 2004). The argument is that social networks and structures produce positive externalities beyond conventional economic thinking. In the context of tourism, social capital has had value in the discussions and debates on sustainability, poverty reduction and democratic decision making among other themes (Borlido, 2016). Specific to sustainability, Borlido argues that there have been challenges to attaining sustainability in Peneda-Gerês National Park in Portugal because:

1. an institutional void exists on the first administrative level;
2. community planning is scarce, mostly due to lack of education by the individuals that occupy administration posts in the territory;
3. communication flaws cause conflicts between the park's administration and the inhabitants;
4. the population feels they possess no influence in decision making, as elections are the only opportunity they have to participate in it;
5. the population is not interested in participating in the decision-making process. (Borlido, 2016: 40)

In a study of two communities on Jeju Island in South Korea, Hwang and Stewart (2017) found that the ability to participate in the tourism industry was a function of the quality of one's social networks. The closer one was to a community leader, the better the chance to be mobilized into the fabric of the tourism industry (see the theory of reciprocal altruism mentioned in Chapter 3). One of the primary conclusions of the study was the need for community leaders to reach out to other more marginalized residents of the community for the purpose of learning from these individuals. One of the most important resources for the development of any tourism industry is the human resources that are both manifest and latent. Tapping into this latent talent pool can only enrich the overall tourism product in a destination. Comprehensive tourism development depends on collective action. Park *et al.* (2012) have shown that the type of food product that a rural farmer in South Korea farms is a determinant in the level of social capital obtained. Local people who raise livestock and cultivate special crops have lower social capital than those farmers that cultivate fruit, rice and vegetables. This is because the latter group had a higher incentive to participate in cooperative agricultural activities, building social trust and compliance with social norms than the former group.

STAKEHOLDERS AND PARTICIPATION

The success or failure of community-based initiatives is often a function of the groups that have been identified as stakeholders – those allowed 'in' – and the degree of participation that these groups have in decision processes. Obviously potential stakeholders not being allowed to participate at all in decisions that affect them directly or even indirectly is problematic. Stakeholder theory was refined through Freeman's (1984) work on strategic management, where firms became more introspective on discerning the nature of their relationship with entities that were associated with or impacted by the activities of the firm. Freeman wrote that a stakeholder is 'any group or individual who can affect or is affected by the achievement of the organization's objective' (Freeman, 1984: 46). In the parks literature, stakeholders are 'individuals, groups or communities likely to affect or be affected by the management of the protected area' (Fonseca *et al.*, 2014: 462).

There are many potential stakeholders that should be involved in tourism and protected areas matters. These include the host community, tourists, government bodies, the tourism industry (operators), the not-for-profit sector including NGOs, and experts/academics. These stakeholders encompass a complex web of interested parties, which results in difficulty in reaching consensus on what ST means and how it can be achieved (Swarbrooke, 1999). Many of these stakeholders have different agendas, ultimately leading to conflict when trying to plan, develop and manage ST. For example, large multinational organizations are often less concerned about ST. Investments that show dwindling return over time are sold, with the organization developing other, newer properties in places that show more financial promise (Butler, 1980). Concern lies in the commitment and active participation among tour operators, especially in mass tourism, which is unlikely unless there are adequate incentives or penalties (Haywood, 1993): economic prosperity first, protection of the environment and local people second. As such, it is a well-established fact that local people may be the least engaged stakeholders of a range of interested parties in tourism (Choi & Murray, 2010). The implications of this finding are far reaching, especially because of the history and place attachment that local people have with the tourism destination. Furthermore, recognition of this local sense of stewardship is important, as are the values that developers and tourists alike impose on the environment. Local residents are cognizant of the intensity and quantity of tourism's impacts on the physical environment (Kousis, 2000; Lankford & Howard, 1994; Parker *et al.*, 2017; Tsartas, 1992).

Tosun (1999; see also Tosun, 2006) developed a typology of community participation which was adapted from work by Sherry Arnstein (1969) based on citizen participation and control within federal urban renewal, anti-poverty and Model Cities, as well as a typology developed by Pretty (1995). In essence, these typologies illustrate that participation exists on a ladder from 'being consulted' to playing an active role in the development process. As Cole (2006; see also Scheyvens, 2003) argues, even though all communities participate in tourism to a certain extent – sometimes only through menial jobs or sharing a despoiled environment – participation should be active and it should be about empowerment. Tosun's (1999) three main categories of community participation are as follows:

1. *Spontaneous participation bottom-up*: Active participation; direct participation; participation in decision making; authentic participation; self planning.
2. *Induced participation top-down*: Passive; formal; mostly indirect; degree of tokenism, manipulation; pseudo-participation; participation in implementation and sharing benefits; choice between proposed alternatives and feedback.
3. *Coercive participation top-down*: Passive; mostly indirect; formal; participation in implementation, but not necessarily sharing benefits; choice between proposed limited alternatives or no choice; paternalism; non-participation; high degree of tokenism and manipulation.

Surf tourism (a form of nature-based tourism) is a good example of community participation in the context of sustainability. In a study of dive tourism and local communities in three Malaysian islands on three different levels (backpackers, package tourism and upmarket dive tourism), Daldeniz and Hampton (2013) operationalized Tosun's (1999) typology and found that local participation in dive tourism did not deviate from the research on participation in tourism in general. The theorists found that local people had very little influence on the development of dive tourism or its management. Few local people managed or established their own dive shops, and few found direct employment in these enterprises. Most jobs were of the low-skilled variety.

In related work, Ponting *et al.* (2005) found that the conventional surfing tourism business model was flawed because it was founded on practices that are decidedly exploitative and inequitable with reference to the needs of local people. In order for surf tourism to be more sustainable (their work took place in the Mentawai Islands, Indonesia), three key elements need to be in place: (a) rejection of the neoliberal approach to development; (b) formalized planning that recognizes that there ought to be limits to growth; and (c) the implementation of cross-cultural understanding with local people which emphasizes the willingness to let local people define standards and symbols of representation, interpretation and integration. Towner and Milne (2017) also studied the dimensions of sustainability in the Menawai Islands and found that key impediments to sustainability were the negative influence of Western culture, the uncontrolled development of resorts, and conflicting opinions about the nature of sustainability between stakeholder groups. Whereas surfers wanted uncrowded and clean waves, local people wanted increased economic benefits from surfing tourism. This study has been pushed further through the addition of a fourth key element being added to building a sustainable surf tourism product in developing countries. Sport governing bodies have an important role to play in efforts to build a collaborative relationship between the sport (in this case surfing) and the utilization of local resources. As such, the International Surfing Association, the PNG Sports Federation and the Australian Surfing Association, for example, should take an active lead in balancing opportunities and costs for all involved (O'Brien & Ponting, 2013).

Another method of dealing with disparities in aspects of stakeholder participation is through the concept of empowerment. Scheyvens (2002) argued that success factors of empowerment in CBT include broad categories such as income and employment, community pride and self-esteem, community cohesion and shifts in power balance. Richards and Hall (2000) argue that empowerment needs to be considered from generative rather than distributive power vantage points. These authors argue that most current power structures are distributive, assuming a scarcity of resources which must be distributed, thus forcing different actors to compete with each other. In opposition, the generative view of power assumes that everyone has power, skills and capabilities, with the goal being to combine everyone's power in collective action for the common good. Although empowerment is a concept that is implicit in most views of ST, many models still assume a distributive form of empowerment to local communities from a higher level. Challenging these models requires an idea of how to generate local empowerment and link local communities at a global level. Even if consensus is reached within the community, its ability to control the scope and type of local tourism development may be limited by a number of factors. First, the power of the tourism industry is overwhelming since even if a destination area tries to control tourism, mass tourism operators may simply move on to another destination where constraints do not exist or are more lenient. One must also consider that externally based organizations may have a strong voice in the region because of their presence and large-scale ownership of the destination (e.g. transnational hotel chains). As well, government policies may overrule local concerns, such as in developing countries where a community may wish to limit tourism, yet the government may strive to maximize tourism in order to increase foreign revenues (Swarbrooke, 1999).

Given the breadth of interests that exist at the community level, there are several challenges involved in moving any type of agenda (conservation, ecotourism and so on) forward. For Blackstock (2005), CBT suffers from three major flaws. First, CBT often co-opts the community into thinking that there is power sharing when in fact the focus is more on economic profitability. Empowerment and social justice give way to an illusion of power sharing. Secondly, CBT tends to treat the community as homogeneous. Finally, there are serious structural constraints existing within the community that prevent local control from taking shape. There is the sense that too many powerful entities control and manipulate the industry and that the small, local 'player' is in a structurally weak position in the face of these external pressures.

When the conservation of protected areas is added to the equation, community development takes on a different complexion because of the amount of protection required at the expense of the community and vice versa. Consequently, the debate on pure protection of protected areas versus the CBD model is highly contentious. The community-based conservation (CBC) paradigm emphasizes community development, participation and conservation embraced by countless NGOs. On the other hand, the older imperialist fortress mentality forces local people to live outside the parks themselves instead of being able to harvest resources inside the park, including wildlife (Alcorn,1993; Ghimire & Pimbert, 1993). The main criticism of the fortress mentality position is that it places non-human interests over human interests, while the CBC model is criticized because it places too much emphasis on community and not nearly enough on conservation (Berkes, 2004), and it rarely eliminates the need for funding entirely (Kiss, 2004). Speaking in favour of the fortress paradigm, Spinage (1998) argues that CBC has become less about science and much more about money. In this capacity, Spinage (1998) observes that:

> One asks why are these resources [wildlife] assumed to be found only in the protected area? The answer must surely be because they have disappeared where people have freely exploited them ... so ... why are there greater densities of animals in national parks than outside them? [This is a problem that is not solved] by abandoning the small parks to those interests that created the unsustainable conditions around them. (Spinage, 1998: 271)

CONSERVATION

In the definition of ecotourism used at the outset of this section, community development (community development is given much more weight here because of the volume of material devoted to conservation in other parts of the book) and conservation are two essential prongs of sustainability. In fact, we often cannot separate conservation from community, as they are intricately interconnected. This is a theme that has emerged in other chapters of this book. The challenges in separating the two are illustrated above in reference to managing protected areas from the fortress standpoint versus the community-based model. Conservation is discussed below from two perspectives: biodiversity and culture.

Biodiversity

Several studies document the importance of conservation as an important element of ecotourism (and forms of nature-based tourism and wildlife tourism) because of human disturbance to specific species and taxa from a biological standpoint. A recent book by Blumstein *et al.* (2017) places these impacts into sharp focus. Geffroy *et al.* (2017) investigated physiological responses to human visitation (e.g. the effects of human visitation on basal stress hormone levels), and behavioural responses including avoidance, behavioural time budgets, responses to provisioning and animal personality, while other studies have investigated how ecotourism has impacts at population and community levels (Shannon, *et al.*, 2017). Topics include mortality, food provisioning, habitat degradation, biological invasions and disease. A chapter by Møller (2017) examines how animal behaviour changes across generations from ecotourism, which is said to be an area of research that is both scarce and scattered in the literature. Mechanisms of change are micro-evolutionary, epigenic, habituation and phenotypic sorting.

Trathan *et al.* (2014) studied the most important threats to the world's 18 species of penguins and what must be done to mitigate these threats. Sourcing the knowledge of 49 scientists and building a data set spanning over 250 years on all threats (harvesting adults for oil, skin and feathers and as bait for crab and rock lobster fisheries; harvesting of eggs; terrestrial habitat degradation; marine pollution; fisheries bycatch and resource competition; environmental variability and climate change; toxic algal poisoning and disease), habitat loss, fishing, pollution and climate change were deemed to be the most critical threats. Using the example of the Galapagos Islands, Trathan *et al.* (2015) drew on the research of Vargas (2009) and Boersma *et al.* (2005) to argue that although tourists themselves do not cause significant damage to penguin populations, the infrastructure built to develop penguin tourism (and other taxa) may have impacts. They argue that the impacts on breeding penguin colonies is unknown because of the wide variety of species studies, the locations of penguins and the levels and types of human activity to which penguins are exposed.

One of the areas of study that has received the most attention from a tourism and conservation perspective is the marine environment. Whale watching used to be a more specialized ecotourism pursuit. However, this quickly changed in the 1990s as bigger boats and more operators, based on increasing demand, turned the activity into a mass tourism pursuit (Muloin, 1998). Curtin (2003) illustrates that in Kaikoura Bay, New Zealand, one whale present in the bay generates approximately NZ$50,000 per day for one company when 500 people travel with that firm. This does not include indirect income. Given the magnitude of the whale-watching industry, government intervention involving regulation and licensing (and limiting supply) are essential in protecting this valuable resource (Curtin, 2003).

Because of the expanding magnitude of marine tourism, some jurisdictions have recommended scaling back on marine-based tourism activities because of adverse impacts to the resource base. At Shark Bay, Western Australia, Higham and Bejder (2008) write that studies have found that female dolphins exposed to high numbers of tour boats are less successful at reproduction than dolphins exposed to lower numbers of boats. In response, recommendations were made for a moratorium on the number of research permits and a reduction of tour vessel exposure by 50%. This, the authors, suggest, represents a paradigm shift in the management of the industry and an indication that the industry, at least in some contexts, is edging towards sustainability.

Culture

In support of a more robust type of conservation, McNelly (1993) argues that it is not just biodiversity (genes, species and ecosystems) that needs protection, but also traditional cultures. McNelly's point is that both need to be conserved together if either is to prosper. The knowledge that people have about their resources, and how they have been managed, is critical to all of humanity. His concept of 'biocultural diversity' connects well with the concept of traditional ecological knowledge (TEK), which refers to the intricate ways or systems by which traditional societies have managed resources. These systems are based on the knowledge of 'place' and have meanings tied to them that are very different from conventional scientific models. Those agencies that use a combination of TEK and conventional science are able to approach ecological cultural problems from a completely different vantage point, while at the same time building capacity for cooperation between Western and local perspectives. As Yi-fong (2012) writes, newer collaborative relationships between various stakeholders, including local people, have spawned a diverse network of ecotourism opportunities. For example, one programme called 'Follow the Footsteps of Indigenous Hunters' has ecotourists learning from Indigenous guides who have replaced hunting tools with cameras. 'Hunters' benefit from showing ecotourists rare and unique fauna. The rarer the find the higher the pay, according to Yi-fong. Tourism can also be an agent for the revitalization of language in Indigenous communities to emphasize the link between place and the individual and collective identity of people (Whitney-Squire, 2016).

Waitt (1999) provides a comprehensive evaluation of the marketing of Australia's Indigenous peoples via international advertisements by the Australian Tourist Commission. Waitt argues that Indigenous people are continuously marketed as either eco-angels (the angel within the ecosystem) or noble savages. This tends to perpetuate colonial power relations that trace back to the 19th century. The promise from these advertisements is that tourists will be able to escape civilization through contact with primitive hunter-gatherers. As such, 'Representations of Indigenous peoples are stereotyped: male, dark-skinned savages, loin-clothed primitives with boomerangs, stone axes and spears. The location is always in the outback' (Waitt, 1999: 155). One of the main problems, Waitt contends, is that advertisements downplay Indigenous people's modern-day lived experiences. The ATC's advertisements always project Indigenous people as eco-angels, Waitt argues, having lived peaceably and in harmony with their natural environment. Such discourse, however, fails to document the significant impact that Indigenous people had on the pre-historical environment through massive burning campaigns of plants and through the extinction of mega-fauna of the region (see also Fennell, 2008b).

A final example of an idea that attempts to bring together people, ecology, knowledge transfer (and thus has application to the next section on education) and conservation is the concept of the ecotourium (Fennell & Weaver, 2005). The intent of this initiative is to build an international network of ecotouria for the purpose of addressing the shortfalls of so-called soft or minimalist forms of ecotourism that venture away from the central tenets of the term. This erosion of values or willingness to operationalize ecotourism along mainstream trajectories has been addressed by Weaver (2005), who argued that such minimalist approaches to ecotourism are based more on elemental or nature-based attractions, learning is superficial and non-transformative, and sustainability is more about the status quo in site-specific situations. By contrast, comprehensive ecotourism is holistic, engenders deep understanding that is transformative, and works towards a type of sustainability that is based on enhancement, is global in its focus and is both environmental and sociocultural in nature. The essence of the ecotourium concept is articulated in the following mission statement introduced at the beginning of Fennell and Weaver's paper:

> This protected area is a special place. It has been recognized as an accredited Ecotourium under the auspices of the ________. Here your participation and financial contributions will be channelled into a number of essential programmes that will have a direct, positive impact on the protected area and its physical, ecological, and social environments. This area is one of a growing number of such entities situated around the world, and forms an international network of special places where ecotourism, research and education, conservation and biodiversity, community development, and partnerships are making a difference. (Fennell & Weaver, 2005: 373)

Although the role of visitors is central in the ecotourium context through the practice of pro-environmental behaviours (Kim *et al.*, 2018), its holistic agenda makes it an attractive option for further study.

EDUCATION

The discussion on pro-environmental attitudes and behaviours in Chapters 3 and 8 suggests that participants in ecotourism should have their pre-established notions of what is right and wrong in our treatment of the natural world tested according to new knowledge that stems from participation in ecotourism programmes (see Loyau & Schmeller, 2017). In short, ecotourism should change the way we view the world – for the better. But it is not just ecotourism but tourism in general where education has a real opportunity to affect some level of change in the individual. And because sustainability is relevant to all forms of tourism, education about how to be a better citizen or organization is essential.

Theorists argue that we are missing the opportunity to develop a sense of authentic education based on real purpose and engagement with and for the other – we are being educated to compete and consume rather than to care and conserve (Sterling, 2001). Education has the potential to make sustainability transformative through attempts to integrate instrumental and intrinsic views in building capacity in resilient learners who are able to navigate complex threats that test our social and ecological systems (Sterling, 2010). Key features of an educational platform that integrates sustainability and ethics in a deep and meaningful way are found in Table 9.5 (see also Fennell, 2018a; Warburton, 2003). Transformative education is characterized by actualizing the full potential of the human agent, which in turn, and collectively, will move us towards 'flexible, creative, adaptable, well-informed and inventive sustainable wellbeing communities' (Lehtonen

Table 9.5 Towards a sustainable education paradigm: Key characteristics

Ontology	*Realist/idealist (relationalist)*
Epistemology	Participatory
Theory of learning	Participative/systematic
Function of education	Remedial/developmental/transformative
Main emphasis	Towards transformative learning experiences
Focus	Meaning-making grounded in context
Seeks	Wholeness and sustainability in learning and living contexts
Reflects	Instrumental, and intrinsic and transformative values
Pedagogy	Transformative where possible and appropriate
Desired change	Contextually appropriate (i.e. healthy, sustainable relationships) in social–ecological systems at all levels

Source: Sterling (2010)

et al., 2019: 339). What is required of this radical change, according to Lehtonen *et al.* (2019), is a better understanding of the type of society we want in the pursuit of a good life on a finite planet that will soon support 9–10 billion people.

An example of deep, or authentic, or transformative thinking around the role that education might play in the reconfiguration of ST relates to the concept of time (Thaman, 2002). Colonialism and globalization have heavily influenced the discourse on sustainability, rendering it more in line with Western values and priorities. Thaman argues that time is thought of in a linear, scientific and financially driven manner, rather than as a circular perception which is more akin to Oceanic cultures. This latter perception amalgamates the past, present and future in an 'all embracing "now"' (Thaman, 2002: 234), linked in a presence that is envisioned as the future. Thaman's position is that sustainability must be rooted in people's cultural values. And beyond different notions of time, there are other important differences that contrast with Western views. Pacific Island cultures place more emphasis on trust, reciprocity, creativity, restraint, compassion and interdependence than Western culture. The bottom line for Thaman is that there must be enough elasticity in sustainability theory and practice to accommodate the special signs and symbols that make up different cultures.

There are several examples of Indigenous communities teaching Indigenous values through the platform of ecotourism, with the hope of fostering transformation in ecological consciousness (Higgins-Desboilles, 2009). For example, Zeppel and Muloin (2008) have shown that a combination of Western and Indigenous forms of interpretive messages can be beneficial to tourists in the context of Australian wildlife tourism. An Indigenous style of interpretation was said to be beneficial because of its focus on traditional uses of wildlife and personal stories, along with notions of Aboriginal Dreaming and animal species as totems. These sorts of attitudes, following the work of Stephen Kellert (1989), were judged to be moralistic (spiritual), utilitarian (food) and aesthetic (totems). By contrast, non-Indigenous styles of interpretation emphasized biological facts and species knowledge.

ETHICS

Our introduction to ethics as a defining characteristic of ecotourism is aided by a case study developed by Buckley (2005) on narwhal ecotourism in Canada's north (Pond Inlet), and the complications surrounding the Indigenous people who run these ecotours but who also hunt narwhal in close proximity to these tours and sell narwhal tusks to cruise line tourists who visit the region. Buckley observes that because of the hunt, which is frowned upon by the Canadian government as a result of low narwhal numbers, the narwhal are becoming increasingly difficult to spot. Hunting has made the animals warier than they have been in the past. And because the Inuit are provided with much of their lifestyle needs by the Canadian government, they should feel an obligation to observe

Canadian law – no hunting (poaching) of the endangered narwhal for the purpose of trading their tusks. Buckley asks the question as to whether it is ethical for tour companies to use the Inuit for ecotourism purposes, knowing full well that the poaching was taking pace. What should ecotourists do in finding out that such practices exist? Buckley argues that it is problematic to define ecotourism according to core ethical values, as some theorists are advocating, presumably because there are many circumstances that demand very different responses.

But are these cases not exactly why ethics are needed in tourism? Ethics, according to Saul (2001), allows us to move away from certainty. Organizations love certainty because it forces individuals to conform to established sets of practices and requirements. We do the same things over and over again. But when complicating circumstances enter the equation, we are not prepared to tackle such problems because of too much of a focus on efficiency, productivity and predictable behaviour. Ethics agitates against these rigid systems by being uncertain; ethics allows us to move beyond rigidity because it forces us to think laterally when confounding circumstances come to the fore. We are not stuck with preconceived notions of how to behave and think, but rather prepared to meet new challenges head on.

Scholars have discussed core values (core ethical theories) and codes of ethics widely in tourism and ecotourism. It is not the place to cover these theories in detail, but it is worthwhile revisiting the discussion on ethics earlier in this chapter. Deontological or means-based theories (i.e. what is ethically right is based on following rules or duties or pre-established principles) include theology and the Golden Rule (Heintzman, 1995), Kantian ethics and social contract ethics (codes of ethics). Teleological or ends-based theories are those that suggest that an act is good or bad based on their performance or non-performance: it is the consequences of the action that is the key feature of these theories. Types of teleological theories include utilitarianism (the greatest good for the greatest number), hedonism (the greatest good for the individual), and virtue theory which focuses not on what I ought to do, but rather on the type of person I ought to be. The third leg of the ethics platform, existentialism, underscores the importance of subjectivity in ethical thinking, i.e. moving away from normative ways of thinking about the world (deontology and teleology). Existentialism is characterized by the freedom and authenticity of the individual, but also taking responsibility for one's authentic actions.

Codes of ethics or codes of conduct are examples of sustainability concepts (Chapter 4) that have been used liberally in the tourism industry. A code of ethics is defined as 'a set of guiding principles which govern the behaviour of the target group in pursuing their activity of interest' (BC Ministry of Development, Industry and Trade, 1991: 21). These principles or standards are determined by moral values (Ray, 2000). In tourism, codes of ethics have been developed for a number of different entities or groups. These include types of tourism like ecotourism (Epler-Wood, 1993) and sustainable tourism (Blake & Becher, 2001; D'Amore, 1992), specific places like the Arctic (Mason, 1997), the environment (ATIA, 1992), hosts (Republique de Maurice, n.d.), airline pilots (Air Line Pilots Association International, 2017), travel agents (American Society of Travel Agents, 2013), and codes to protect children from sexual exploitation in tourism (ECPAT, 2012), and specific species or groups of species such as whales (Gjerdalen & Williams, 2000; Tobin, n.d.). In reference to this last category on specific species, Waayers *et al.* (2006) found that 77% of tourists ignored the voluntary code of conduct which is designed to minimize the effects of tourism on marine sea turtles in Ningaloo Marine Park in Western Australia: 51% of these tourist breaches included such behaviours as shining lights into turtles' eyes and using flash photography, making sudden movements, not staying behind turtles, and getting closer than 3m from turtles, which forced turtles to return to the sea without nesting. The voluntary nature of codes of ethics is perhaps the biggest drawback of these devices. They provide guidance, but there is little in the way of punitive action to actively dissuade tourists from depreciative behaviours.

Work by Fennell (2019b) addresses the concern that Buckley has about situations that demand different types of ethical thinking. Using Donaldson and Dunfree's (1994) research on integrated social contract theory, there are two different kinds of contracts that once combined can address all kinds of ethical dilemmas faced by tourism stakeholders. Macro social contracts (more broadly based sets of directives that are more dominant than micro contracts) are similar to the normative forms of ethics (deontology and teleology) mentioned above, and include questions and guidance around rights, justice, the greatest good for the greatest number, laws and other duties. The macro contract provides ground rules for the second type of contract. This latter category, micro contracts, is ethics for situations that are unique to small groups or communities, and includes four categories:

1. Local economic communities may specify ethical norms for their members through micro social contracts.
2. Norms-specifying micro social contracts must be grounded in informed consent buttressed by a right of exit.
3. In order to be obligatory, a micro social contract norm must be compatible with hypernorms.
4. In case of conflicts among norms that satisfy principles 1–3, priority must be established through the application of rules consistent with the spirit and letter of the macro social contract.

Hypernorms can be anything that the community values. In structuring a set of such norms for the future of tourism, Fennell argued that altruism, recognition, education, autonomy and rights, justice, respect and sustainability were essential in building a more ethical tourism industry. Since the actions of the Indigenous people in Buckley's account of ecotourism in Canada's north were contrary to the rights of the narwhal to exist, respect for the narwhal, as well as issues around sustainability, the act must be judged to be morally wrong in both local and cosmopolitan contexts. The ecotourism operator may also be criticized for putting self-interest ahead of a more altruistic attitude, which must expand to include both human and non-human entities, especially when it comes to ecotourism.

The lack of 'teeth' that codes of ethics have in addressing unethical behaviour on the part of stakeholder groups involved in tourism has necessitated a move more in the direction of regulation. Rodger *et al.* (2011) developed a generic assessment framework to aid in sustainable management of the marine tourism industry. Their framework is designed around ecological and environmental items for the first assessment framework, and social and operational items for the second. The framework is the culmination of experience, discussions with industry leaders and a thorough review of the literature. The authors conclude by suggesting that the framework has three principal applications. First, existing operations may be improved through a closer examination of sustainability criteria contained within the framework. Secondly, the framework acts as an auditing system as part of licensing provisions for this type of tourism. Thirdly, the framework allows decision makers to judge the sustainability of proposed ventures.

CONCLUSION

This chapter has sought to provide a general overview of many of the themes and topics relevant to protected areas and sustainability, and match these up with ecotourism as the form of tourism that fits best in protected areas. The focus was on the global protected area system and various types as identified by the IUCN, World Heritage Sites as an example of one of these types, governance, tourism concessions and partnerships, ecotourism and related forms of nature-based tourism, and the main characteristics that define ecotourism such as community development, conservation, education and ethics.

In the introduction to this chapter there was a positive story about the creation of a protected area in Tanzania for the highly endangered red colobus monkey. There were challenges in the creation of this protected area that took upwards of 40 years to rectify. Ecotourism was said to be the main advantage not only for local people to realize benefits from the protected area, but also for biodiversity that may be protected from indiscriminate human intrusion. The sustainable management of protected areas becomes a real priority in the face of so many different issues and concerns. In conclusion, it is prudent to identify what we continue to come up against in the protection of terrestrial and marine areas. The WWF (2017) argues that the main challenges include poor representation of habitats, lack of connectivity between protected areas (corridors), lack of funds, poor management, and the range of human activities that exist inside protected areas (e.g. poaching, logging, mining, agriculture) and outside these areas (e.g. invasive species, climate change). These are all firmly entrenched issues that will require significant resources and effort to overcome if protected areas are to become important vehicles for sustainable development.

END OF CHAPTER DISCUSSION QUESTIONS

1. What are the Aichi Biodiversity Targets? List some of the strategic goals built into these targets.

2. What percentage of the world's terrestrial ecosystems is designated as protected areas?

3. List some advantages and disadvantages of non-sustainable and sustainable tourism in parks and protected areas as observed by the Federation of Nature and National Parks in Europe.

4. List the range of protected areas categories as defined by the IUCN.

5. What is a World Heritage Site and what are the differences between cultural and natural sites?

6. What is the difference between tangible and intangible heritage?

7. How does ecotourism differ from wildlife tourism? How is it the same?

8. Identify the characteristics of an ecotourium.

9. How does ethics force us to live with uncertainty?

End of Chapter Case Study 9.1 The politics of managing a World Heritage Site: The complex case of Hadrian's Wall

This case study by Bell (2013) emphasizes the importance of planning and policy, and strong institutions as well as partnerships in generating good success in the management of a WHS. Working with the often-disparate needs of multiple stakeholders is a challenge, especially in complex environments. Bell writes that:

Managing sites that cover large areas or have multiple designations within their boundary can be challenging, particularly if several organizations are responsible for managing different aspects of the site. Hadrian's Wall WHS was designated for its archaeological importance, which is of universal significance. The management of Hadrian's Wall is formalized in the WHS Management Plan (2008–2014), which aims to cover different objectives with archaeology at the core of management. This article draws on management documents of the WHS, established visitor planning frameworks and qualitative interviews with managers along Hadrian's Wall to discuss the evolution of a values-based management approach. It is evident that managers have to balance different values, priorities and interests, and make trade-offs when they work in partnership on a multiple-use site so that conflicts can be resolved – and while this may look straightforward on paper, in practice, it can be much more difficult for institutions and individuals to compromise their own professional or personal values. Despite its challenges, this article advocates a values-based, or pluralistic, management approach as the most effective means of managing multiple-use sites, resolving management conflicts and working in partnership to agreed outcomes.

The evolution of management plan values can be seen in the changes that have taken place between the years of 2002–2007 and 2008–2014, when a pluralistic values-based approach was taken in pursuit of a shared vision in the latter time range.

2002–2007 management plan values	
Archaeological and historic values	The archaeological remains.
Natural values	The landscape setting, including habitats and species.
Contemporary values	Economic: tourism and industry (agriculture, forestry and quarrying); Recreational and educational: educational resource and recreational interest; Social and political: political interest in safeguarding the monument; WHS values: WHS inscription and how Hadrian's Wall fits WHS criteria and is of 'outstanding universal value'.

2008–2014 management plan values	
Evidential values	Concentrates on the archaeology and landscape and includes the complexity of the archaeology, the group value of the whole of the new WHS, the excavations; the landscape, including geology; the scale, rarity and international significance of the WHS.
Historic values	Includes documentation of historical records; direct association with Hadrian, boundaries of the Roman Empire and subsequent literature and other cultural links to the Wall.
Aesthetic values	The landscapes and designations such as the National Park and AONB, the views, agricultural landscape and tranquillity of the surroundings.
Communal values	Academic value: research resource, pre- and post-Roman history; Economic value: tourism and industries such as agriculture, forestry and quarrying; Educational value: formal and informal learning; Recreational value: activities; Social value: social identity of those living and working in the area; Natural values: habitats such as the Whin Sill, Solway estuary and the Cumbrian coast and designations such as SSSIs within the WHS.

Bell contends that three criteria need to be satisfied if the pluralistic and values-based approach is to work. First, the partnerships need to be formalized by taking into account the perspectives of a number of stakeholders with agreement based on compromise. Secondly, the plan must be assembled with flexibility in mind for the purpose of accommodating contingencies. Thirdly, partnerships need to be structured so that the agreements struck might be implemented. As more information and knowledge is developed for the site, the management plan must be reviewed, refined and adapted, keeping pace with these changes.

Source: Bell (2013)

Appendix 9.1 Aichi Biodiversity Targets

Strategic Goal A: Address the underlying causes of biodiversity loss by mainstreaming biodiversity across government and society

Target 1
By 2020, at the latest, people are aware of the values of biodiversity and the steps they can take to conserve and use it sustainably.

Target 2
By 2020, at the latest, biodiversity values have been integrated into national and local development and poverty reduction strategies and planning processes and are being incorporated into national accounting, as appropriate, and reporting systems.

Target 3
By 2020, at the latest, incentives, including subsidies, harmful to biodiversity are eliminated, phased out or reformed in order to minimize or avoid negative impacts, and positive incentives for the conservation and sustainable use of biodiversity are developed and applied, consistent and in harmony with the Convention and other relevant international obligations, taking into account national socioeconomic conditions.

Target 4
By 2020, at the latest, governments, business and stakeholders at all levels have taken steps to achieve or have implemented plans for sustainable production and consumption and have kept the impacts of use of natural resources well within safe ecological limits.

Strategic Goal B: Reduce the direct pressures on biodiversity and promote sustainable use
Target 5
By 2020, the rate of loss of all natural habitats, including forests, is at least halved and where feasible brought close to zero, and degradation and fragmentation is significantly reduced.

Target 6
By 2020 all fish and invertebrate stocks and aquatic plants are managed and harvested sustainably, legally and applying ecosystem-based approaches, so that

overfishing is avoided, recovery plans and measures are in place for all depleted species, fisheries have no significant adverse impacts on threatened species, and vulnerable ecosystems and the impacts of fisheries on stocks, species and ecosystems are within safe ecological limits.

Target 7
By 2020 areas under agriculture, aquaculture and forestry are managed sustainably, ensuring conservation of biodiversity.

Target 8
By 2020, pollution, including from excess nutrients, has been brought to levels that are not detrimental to ecosystem function and biodiversity.

Target 9
By 2020, invasive alien species and pathways are identified and prioritized, priority species are controlled or eradicated, and measures are in place to manage pathways to prevent their introduction and establishment.

Target 10
By 2015, the multiple anthropogenic pressures on coral reefs and other vulnerable ecosystems impacted by climate change or ocean acidification are minimized, so as to maintain their integrity and functioning.

Strategic Goal C: To improve the status of biodiversity by safeguarding ecosystems, species and genetic diversity

Target 11
By 2020, at least 17% of terrestrial and inland water and 10% of coastal and marine areas, especially areas of particular importance for biodiversity and ecosystem services, are conserved through effectively and equitably managed, ecologically representative and well-connected systems of protected areas and other effective area-based conservation measures, and integrated into the wider landscapes and seascapes.

Target 12
By 2020 the extinction of known threatened species has been prevented and their conservation status, particularly of those most in decline, has been improved and sustained.

Target 13
By 2020, the genetic diversity of cultivated plants and farmed and domesticated animals and of wild relatives, including other socioeconomically as well as culturally valuable species, is maintained, and strategies have been developed and implemented for minimizing genetic erosion and safeguarding their genetic diversity.

Strategic Goal D: Enhance the benefits to all from biodiversity and ecosystem services
Target 14
By 2020, ecosystems that provide essential services, including services related to water, and that contribute to health, livelihoods and wellbeing, are restored and safeguarded, taking into account the needs of women, Indigenous and local communities, and the poor and vulnerable.

Target 15
By 2020, ecosystem resilience and the contribution of biodiversity to carbon stocks has been enhanced through conservation and restoration, including restoration of at least 15% of degraded ecosystems, thereby contributing to climate change mitigation and adaptation and to combating desertification.

Target 16
By 2015, the Nagoya Protocol on Access to Genetic Resources and the Fair and Equitable Sharing of Benefits Arising from their Utilization is in force and operational, consistent with national legislation.

Strategic Goal E: Enhance implementation through participatory planning, knowledge management and capacity building
Target 17
By 2015 each Party has developed, has adopted as a policy instrument and has commenced implementing an effective, participatory and updated national biodiversity strategy and action plan.

Target 18
By 2020, the traditional knowledge, innovations and practices of Indigenous and local communities relevant for the conservation and sustainable use of biodiversity, and their customary use of biological resources, are respected, subject to

national legislation and relevant international obligations, and fully integrated and reflected in the implementation of the Convention with the full and effective participation of Indigenous and local communities, at all relevant levels.

Target 19
By 2020, knowledge, the science base and technologies relating to biodiversity, its values, functioning, status and trends, and the consequences of its loss, are improved, widely shared and transferred, and applied.

Target 20
By 2020, at the latest, the mobilization of financial resources for effectively implementing the Strategic Plan for Biodiversity 2011–2020 from all sources, and in accordance with the consolidated and agreed process in the Strategy for Resource Mobilization, should increase substantially from the current levels. This target will be subject to changes contingent to resource needs assessments to be developed and reported by Parties.

Source: CBD (2018)

CHAPTER 10
CROSS-CUTTING ISSUES IMPACTING ON SUSTAINABLE TOURISM

LEARNING OUTCOMES

This chapter focuses on a number of cross-cutting issues that impact on sustainable tourism. The chapter therefore deals with a number of the UN's Sustainable Development Goals (SDGs), including SDG 5 on gender equality and empowerment, SDG 8 on decent work and SDG 13 on climate action. The chapter begins with perhaps the most important issue of all, climate change and the challenge for tourism of how to transition to become a low-carbon sector. The chapter continues with a number of human-related issues – human resources, gender, mobilities and the less mobile and Indigenous rights. The chapter closes with the final cross-cutting but important issue, that of animals and animal rights and tourism. The chapter is designed to provide you with:

1. An awareness of climate change and the fact that tourism is both a vector and a victim of climate change.

2. An analysis of the issues related to tourism as 'decent work' and the growth of 'green jobs'.

3. An awareness of gender equity issues in sustainable tourism.

4. An understanding of the issues surrounding Indigenous rights and sustainable tourism.

5. An analysis of the concept of mobility and the challenges facing the less mobile.

6. An awareness of animal rights and tourism.

CLIMATE CHANGE

INTRODUCTION

We introduced climate change in Chapter 2 and revisit it in Chapter 11 when we discuss the future of sustainable tourism (ST). There is no doubt that climate change represents the single most important challenge to the world as well as to the tourism sector and as such has taken a political as well as a scientific dimension. It is represented as SDG 13 – take urgent action to combat climate change and its impacts. It is only in recent decades that the consequences of travel for climate change have been realized. As a result, tourism is now both 'a victim and a vector' of climate change. The additional activity of transportation due to tourism increases carbon emissions, with estimates suggesting that tourism accounts for around 5% of the world's total emissions, with transport generating 75% of these emissions. Aircraft are particularly damaging as their emissions take place at high altitude, but accommodation is also a major contributor to greenhouse gas (GHG) emissions as we saw in Chapter 6. Research is only now beginning to examine this issue – long-haul tourism, for example, has very high emissions, contributing 17% of global tourism-related carbon dioxide emissions, but only representing 2.7% of all tourist trips.

Climate change is a complex issue that we are only now beginning to understand and model. As a result, it has been realized that tourism development and activity as 'business as usual' is not sustainable for the environment, or for future economic development. Indeed, tourism decision making and decision makers are already impacted by climate change (see Buzinde *et al.*, 2009). This means that we are facing 'new realities' of tourism in a time of climate change, and these have begun to dominate international policy debates, fuelled by forecasts of economic and social disruption to lifestyles, health, social wellbeing and political stability – indeed destinations such as Venice, the Maldives and some Pacific islands are in danger for their very existence and will generate 'climate refugees'. Finally, central banks are increasingly concerned that the world's financial systems are too dependent on fossil fuels threatening the world's economic stability (see Schoenmaker, 2019).

Climate change is caused mainly by anthropogenic GHG emissions (Mooney *et al.*, 2009). There are two linked processes of climate change at work:

1. *Global warming*: With the accumulation of GHGs in the atmosphere generated by transportation, air conditioning and other processes, solar radiation is prevented from escaping from the Earth and therefore leads to a warming effect.
2. *Depletion of the ozone layer*: Release of certain gases into the atmosphere from devices such as air-conditioning units and fridges has reduced the effectiveness of the ozone layer to filter out harmful UVB sunrays. These rays can cause eye cataracts and skin cancer and have led to many beach resorts having to adjust their products to 'beach plus', where additional products such as sports and theme parks have been developed because visitors no longer want to spend all day in the sun.

Climate change has consequences for tourism destinations including:

- coral bleaching in iconic destinations such as the Australian Great Barrier Reef;
- retreat of the snowline for winter sports resorts in, say, the European Alps;
- excessive temperatures at beach resorts;
- rising sea level threatening destinations such as the Pacific islands;
- more extreme weather events such as hurricanes, heavy precipitation, typhoons and heatwaves;

- changing climate zones impacting on wetlands and deserts, creating water shortages;
- changing tourism consumer decision-making and demand patterns as transport costs and taxation increase; long-haul destinations in particular are vulnerable to these changes in demand patterns; this is an area that has yet to be researched in detail; and
- climate change is one of the principal causes of biodiversity loss globally (MEA, 2005), with effects across a broad spectrum of taxa (Thomas *et al.*, 2004). Climate change is having a global impact on forest biodiversity because of changes in temperature, rainfall, storm frequency and magnitude, the frequency of fires, and pest and disease outbreaks (Pawson *et al.*, 2013).

SUSTAINABLE TOURISM: ADDRESSING CLIMATE CHANGE

There are four key ways in which the tourism sector can address climate change:

1. Offset the carbon emissions created by tourism, for example by tree planting.
2. Mitigate the impact of tourism on climate change by changing industry practices and consumer behaviour. As Hopkins and Higham (2018) observe, mitigation can be achieved through behavioural (changing travel patterns), technological (the move towards electric vehicles) and policy approaches.
3. Adapt destinations and consumer behaviour to climate change as we showed in the final case study in Chapter 5. Adaptation recognizes that some degree of climate change is inevitable.
4. Work with climate scientists to increase our understanding of the linkages between tourism carbon emissions, climate change and societal needs and adaptability (Mooney *et al.*, 2009).

To address these pressing needs, the world's two leading agencies for tourism, the World Travel and Tourism Council (WTTC) and the UN World Tourism Organization (UNWTO), have both developed strategies that fast track the response of tourism to slow climate change.

The WTTC is committed to reducing GHG emissions and underlines the need for partnerships between the tourism industry, consumers, employees and government. The WTTC has partnered with UN Climate Change to work towards a carbon neutral world with the aim of:

- communicating the nature and importance of the inter-linkages between tourism and climate change;
- raising awareness of the positive contribution tourism can make to building climate resilience; and
- reducing the contribution of tourism to climate change and supporting quantitative targets and reductions.

The UNWTO has held a series of meetings to discuss the issue of tourism and climate change. These began in Djerba, Tunisia in 2003 and now include side events at international climate change conferences, such as the Copenhagen UN Climate Change Conference in 2012. One of the most significant meetings was in Davos, Switzerland in 2007. It led to the 'Davos Declaration' (UNWTO, 2009) which states that all tourism stakeholders – governments, the industry, destinations and research networks – need to harness their energies to address climate change, focusing in particular on the following points:

- Climate is a key resource for tourism and the sector is highly sensitive to climate change and global warming.
- Given the importance of tourism in the global challenges of climate change and poverty alleviation, there is an urgent need to consider policies that encourage truly sustainable tourism, reflecting a quadruple bottom line of environmental, economic, social and climate responsiveness. This recognizes that these global challenges are linked and reinforcing and must be addressed within the framework of the SDGs.
- The tourism sector must respond to climate change and progressively reduce its GHG emissions. This will need action to mitigate the GHG emissions of tourism, especially from transport and accommodation. Here, Juvan and Dolnicar (2014) discovered that even though there are a variety of calculators for tourists to use to better understand their carbon footprint, most were found to have limited functionality: they were difficult to use and were perceived to have poor credibility. Only four of 73 calculators reviewed were suitable in figuring the carbon footprint of a vacation.
- Tourism businesses and destinations must adapt to changing climate conditions.
- Existing and new technology must be applied to improve energy efficiency.
- Financial resources to help poor regions and countries in these actions must be secured.

The UNWTO has also been involved in global initiatives to reduce climate change, beginning with the 1997 Kyoto Protocol which determined binding emission reduction targets for 34 developed countries, but was generally agreed to have failed, and the 2015 Paris Agreement which determined self-defined, non-binding targets for 187 countries and included provisions for regular reporting towards a global stock-take every five years. For tourism, the key dimensions of the Paris Agreement are mitigation, adaptation and financing emission reductions. However, international aviation emissions are omitted from the agreement and, as a result, transport has become a major stumbling block to climate change reduction. This is partly due to insufficient action on the part of the UN's aviation agency – the International Civil Aviation Organization (ICAO) – despite the 2016 Montreal Agreement which attempted to get cooperation across the aviation sector to reduce GHG emissions (see also Chapter 6). This is a serious problem for tourism, which depends on high carbon transport (Hopkins & Higham, 2018). Nonetheless, tourism is a key part of the Paris Agreement (see ETC, 2018). For example, the UNWTO calculates that many countries see tourism either as a country priority, as part of their mitigation and adaptation strategies, or as a sector vulnerable to climate change.

It is clear that tourism has to tackle climate change in tandem with other tourism-related development issues, including poverty alleviation. If tourism continues 'business as usual' then carbon emissions from tourism will increase by 130% by 2035. In order not to threaten the economic and employment benefits that tourism can bring to the developed world, GHG targets will need to be applied differentially to different parts of the tourism industry, and to different destinations so as not to jeopardize tourism-related poverty alleviation projects. The tourism sector continues to need more evidence on essential climate-related information that is needed for better decision making. Box 10.1 shows how information can be given to small businesses to encourage them to adopt clean energy solutions.

Box 10.1 Hotel energy solutions

Introduction

The accommodation sector is dominated by small businesses. One of the challenges of ST is persuading these small and medium-sized enterprises (SMEs) that they should adopt environmentally friendly operational practices. The hotel energy solutions (HES) project is an ambitious attempt to import renewable energy and energy innovation into small accommodation businesses. HES provides an online mitigation toolkit to help hotels reduce their carbon footprint and operational costs, thus increasing business profits. The project was initiated by the UNWTO, working with a team of UN and EU agencies in tourism and energy. The project works by delivering information, technical support and training to SMEs in the tourism and accommodation sector across the EU to increase their energy efficiency and renewable energy usage.

SMEs and energy

There is no doubt that SME accommodation units rely on old and inefficient energy equipment and have failed to innovate with cleaner energy sources. One of the reasons for this is their lack of access to investment capital to change the way in which they operate. In addition, they have a limited awareness and knowledge of greener energy alternatives and their benefits.

HES project partners

One of the successes of the HES project has been the coming together of an impressive and influential team of partners to deliver the project. These partners are:

- Intelligent Energy – Europe;
- The UN World Tourism Organization;
- The UN Environment Programme;
- The International Hotel and Restaurant Association;
- The French Environment and Energy Management Agency; and
- The European Renewable Energy Council.

HES strategic objectives

The strategic objectives of the HES project are to enable and encourage accommodation SMEs to improve their sustainability and competitiveness. The strategic objectives are to:

- develop and disseminate tools and materials to change hotel management actions and investment decisions in their use of energy;
- promote exchanges of know-how and experience between hotels as energy users, and the suppliers and manufacturers of energy technologies and other key actors;
- raise awareness of hotel managers, decision makers, staff and consumers in relation to energy use and efficiency; and
- stimulate the establishment of networks with the commitment to disseminate and promote energy technologies to hotels.

The HES online toolkit

The HES online toolkit is a key output of the project and provides SMEs with an easy to use, free of charge set of resources. The toolkit allows SMES to assess their energy use and recommends appropriate renewable energy and energy efficiency technologies. The toolkit also recommends what savings on operating expenses hotels can expect from green investments through a 'return on investments calculator'.

Conclusion

The HES project is an important initiative for SMEs and energy professionals, facilitating exchange knowledge and joint work on identifying the best solutions in response to the energy needs of the SME accommodation sector. The HES project has created a regional network for distribution and adoption of the tools and materials developed by the project, allowing the initiative to live on long after the project itself has concluded.

Source: Cooper *et al.* (2018)

TOWARDS A LOW-CARBON ENERGY TOURISM SECTOR

A low-carbon sector is one that has a minimal output of GHG emissions. The energy that is used by the tourism sector is dominantly based on fossil fuels, with aviation delivering the majority of GHG emissions from tourism and motor transport almost one-third. There is no doubt, therefore, that a transition to a tourism sector based on low-carbon energy is needed (Eijgelaar *et al.*, 2018). This involves a shift in energy production away from fossil fuels towards cleaner forms of energy. For tourism the biggest challenge is the transport sector, particularly aviation which needs to move to a low emission mode and begin to use the most fuel-efficient technology possible. However, with increased demand for more travel and longer distances travelled, the 'business as usual scenario' for transport is not acceptable. Solutions include:

- use of bio fuels for aircraft and engineering solutions to make aircraft lighter;
- electric hybrid technology for motor transport;
- hydrogen power for cruise ships and ferries;
- use of renewable energy in accommodation and at attractions; renewables include solar, biomass, heat pumps, wind and wave power, hydro and geothermal; and
- tour operators demanding energy efficiency throughout their supply chains.

A more radical solution is presented by Eijgelaar *et al.* (2018), who state that a move away from geographical-based tourism towards experiential, technology-based tourism could reduce travel. Their 'green travel scenario' is based on reducing the amount of transportation taken by tourists. The key dimensions of the green travel scenario are:

- subsidies for bio fuel;
- decarbonisation of electricity production;
- additional efficiency measures for all transport modes;

- investment in public transport and light rail systems;
- carbon tax; and
- limiting airport slots.

A further challenge is persuading the tourism sector, particularly SMEs, that they should transition towards renewable energies, despite the fact that this can reduce their costs and meet the environmental expectations of government and their customers.

HUMAN RESOURCES FOR SUSTAINABLE TOURISM

One of the most important resources for ST is the human dimension, yet it is one that is often overlooked and its importance underestimated (Baum *et al.*, 2016). Human resources (HR) are represented in the SDGs as SDG 8 – promote sustained, inclusive and sustainable economic growth, full and productive employment and decent work for all. ST human resources can be defined as 'the people who make up the workforce of a sustainable tourism organization or business' (see Ladkin, 2018). These HR are an essential prerequisite to delivering the ST product and creating competitive sustainable destinations.

ST is labour intensive and supports a wide range of jobs in many different sectors, providing job opportunities for people of all ages, women and the less advantaged groups in society such as the disabled and the elderly (Ross & Pryce, 2010). These jobs in the future will demand the development of new skills as well as high levels of cultural awareness and adaptability, if the sector is to cope with issues of sustainability. Indeed, if tourism HR is to be truly sustainable, the sector will need to address issues of equal and fair pay, corporate social responsibility and ethics (Ladkin, 2018). Unfortunately, too, from the point of view of ST, investment in the physical product at the destination – such as accommodation – is often preferred to investment in workforce development. As a result, the tourism industry worldwide is facing an HR crisis (WTTC, 2015). Effectively, tourism is in a labour market trap: as more tourism is demanded so more jobs are needed, often with new skills in green jobs, technology and the digital economy (Baum, 2007; WTTC, 2015). The reasons for the HR crisis in tourism are complex and interlinked, but if we are to move to a truly sustainable form of tourism they need to be understood and addressed. The main causes are: (i) the shifting demographics and social attitudes of the 21st century (Riley *et al.*, 2002); and (ii) the type of work that tourism involves, the working conditions in the industry and the dominance of SMEs.

- **Demographics and attitudes**: In many countries of the world that are involved in tourism, birth rates are falling and their populations are ageing. Yet demand for tourism continues to grow, resulting in a 'demographic squeeze'. This is heightened by the fact that the new generations of workers, such as the millennials, have very different attitudes to work and careers, seeking out 'decent jobs' and often looking at the 'green' credentials of employers.

- **Tourism jobs can be characterized as follows**:
  - *Temporary*: Held by people with no career aspirations in tourism and no commitment to the industry.
  - *Young*: Dominated by younger people.
  - *Gendered*: Dominated by females; indeed, many tourism jobs and occupations demonstrate a clear gender bias – these include housekeepers (female) and chefs (male) (WTTC, 2015).
  - *Flexibility*: In terms of hours and working arrangements, but also uncertainty in terms of security with zero hours contracts and deregulated regimes common (see, for example, Terry, 2011, for HR arrangements in the cruise industry). Nonetheless these arrangements do suit specific groups in society such as working mothers or students.
  - *Accessibility*: With few or no barriers to entry in many jobs; indeed tourism is a highly 'under-qualified' industry, partly because many jobs are low skilled.
  - *Diversity*: They range from hospitality-related jobs, which tend to make up the bulk of employment in tourism but are also the lowest paid and the most seasonal, to the smaller number of jobs in air transport such as pilots, but also the highest paid.
  - *Low paid*: In many countries tourism jobs pay up to 20% below the average wage.
  - *Narrowly framed* in terms of occupations.
  - *Long and antisocial hours*: Often involving work during holidays, weekends and over meal times. This can impact on the balance between family and work life.
  - *Health and safety*: There are many accidents in kitchens, and workers face safety issues on the way home after late shifts.
  - *Sexual discrimination*: An issue in some tourism sectors.

TOWARDS SUSTAINABLE TOURISM HUMAN RESOURCES

There is no doubt that tourism human resources are problematic and that the current way of working is not sustainable. However, there are now a number of organizations working towards a solution to the crisis, notably the International Labour Organization (ILO) and the OECD. The solutions focus around:

- Greater understanding of the operation of the tourism labour market (Riley *et al.*, 2002). This also requires intervention by government and industry to promote jobs in tourism and to polish the image of working in the sector. One significant problem here is the lack of reliable labour market statistics but, fortunately, the role of the Tourism Satellite Account (TSA) is now helping us to understand and analyse labour markets.
- Recognition that tourism is in the knowledge economy and needs to invest in and develop its workforce rather than exploiting it. This will involve a shift in attitude and operation by the industry, particularly in terms of investing in training and education.

Education and training will support a more attractive labour market and business environment in tourism to maintain a skilled workforce and improve productivity. Upskilling people to become more proficient in their jobs, valuing and rewarding professional competence and supporting career development can improve the image of employment in the sector and create a more positive recruitment and retention cycle. In turn, this promotes enterprise and destination competitiveness and leads to better outcomes for workers (OECD, 2015). This involves embracing sustainability to provide employees for specialist fields such as guiding and special interest tourism, as well as providing a thorough grounding in business and peoples skills and technology. For example, the IUCN (2012) has partnered with the tour operator Kuoni to examine the skills needed for ecotourism operators, with a particular focus on importing business skills to the operators.

- SMEs dominate the tourism employment landscape, creating a problem of under-management in the sector and underutilization of the workforce, as many small entrepreneurs do not have grounding in good HR practice. There is also the problem of creating career pathways in SMEs. Solutions at the destination level are now being developed. These include destination management organizations (DMOs) encouraging SMEs to work together to create career paths across companies and to begin to think beyond managing people in organizations and make the jump to managing people in destinations.
- Government and the international agencies must encourage quality jobs in tourism and move away from simply treating tourism as a means to generate jobs with no eye for quality. The ILO's 'decent work' agenda is an important initiative here, with the mission of enhancing the quality of jobs and the working environment (www.ilo.org). A decent job involves opportunities for work that: is productive and delivers a fair income; provides security in the workplace and social protection for workers and their families; offers better prospects for personal development and encourages social integration; gives people the freedom to express their concerns, to organize and to participate in decisions that affect their lives; and guarantees equal opportunities and equal treatment for all (ILO, 2016).
- The decent work agenda forms the UN's SDG 8 – which aims to promote sustained inclusive and sustainable economic growth, full and productive employment and decent work for all (https//sustainabledevelpment.un.org/sdg8). It goes on to consider tourism in SDG Target 8.9 to 'devise and implement policies to promote sustainable tourism that creates jobs and promotes local culture and products'.
- The tourism experience is an amalgamation of services, be it hotel, restaurant, attraction or activity. Adopting a tourism value chain approach which takes into account the different skill levels in the sector and looks beyond specific occupations and branches can help to strengthen mobility and build the capacity of destinations to deliver quality tourism services.

GREEN JOBS IN TOURISM

For the future, the tourism sector will need to engage in the notion of 'green jobs'. As the world transitions from a fossil fuel economy to one of low carbon and renewable energy, tourism will need to play its part (UNIDO, 2011). According to the ILO (2016), 'green jobs' are decent jobs with an environmental and sustainability dimension such as renewable energy. In other words, a green job is any job or self-employment that genuinely contributes to a more sustainable world. The ILO (2016) states that green jobs aim to achieve a number of environmental goals, all of which are relevant to ST:

- improve energy and raw materials efficiency;
- limit GHG emissions;
- minimize waste and pollution;
- protect and restore ecosystems; and
- support adaptation to the effects of climate change.

The US Bureau of Labor Statistics classifies green jobs into two types:

1. jobs in businesses that produce goods or provide services that benefit the environment or conserve natural resources; and
2. jobs in which workers' duties involve making their establishment's production processes more environmentally friendly or use fewer natural resources.

These green jobs are found across all types of tourism organizations, including business, not-for-profit organizations or the public sector. For the private sector, green jobs are very much focused on SMEs, the self-employed and entrepreneurs. For tourism, these green jobs deliver products that benefit the environment, but they are not always based on green production processes and technologies. Therefore green jobs can also be distinguished by their contribution to more environmentally friendly processes. Green jobs see a partnership between the individual performing the role and the employer. Here, some green jobs in tourism require specific 'green' skills or education, such as a tour guide or environmental educator. Indeed, for many in green jobs it is part of their sustainable lifestyle.

The ILO's Green Jobs programme is an international agenda involving governments, employers and workers (ILO, 2016). The focus is on achieving sustainable development through the UN SDGs. The programme has three key sets of actions:

- *Society*: Cultivating the right mindsets, attitudes and values will foster sustainable and low-carbon economies. Increased awareness of and sensitization to the current pace of environmental degradation and the effects of climate change are decisive in driving consumer demand for green products and services and environmentally friendly production processes.

- *Governance*: Creating the right enabling environment in each country is essential to unleashing their green jobs potential and ensuring a just transition. This includes developing: macro-economic policies to redirect consumption and investment; sectoral policies for establishing environmental regulations; social and labour policies focusing on employment, social protection, training and skills development; and climate change mitigation and adaptation policies to promote employment in emerging new green activities, climate-resilient infrastructures and the rehabilitation of natural resources.
- *Enterprise and workplaces*: Any type of enterprise can be or become green. Green jobs can be created by green enterprises through start-ups by tapping into new markets or through the greening of existing enterprises by shifting towards more environmentally friendly production processes.

INCLUSIVITY, GENDER AND ECOFEMINISM

We introduced the concepts of gender and inequalities in Chapter 3. Gender equality is a core cross-cutting issue in ST represented as SDG 5 – achieve gender equality and empower all women and girls. Khoo-Lattimore and Ling Yang (2018) are clear that the 'tourist' has become more inclusive as women account for half the travel market. Yet, from an employment point of view, women make up the majority of the tourism workforce, they tend to also hold the lowest paid and lowest status jobs and they often perform unpaid work in entrepreneurial businesses and SMEs. Gender inequality is also a structural cause of poverty in terms of finances, skills and self-esteem and therefore fails to deliver 'decent jobs' to women. Here, SDG 8 focuses on decent work and requires a gender perspective in tourism employment policies, placing emphasis on salary gaps, sexual abuse and harassment by colleagues and tourists, and fostering female workers' participation and decision making.

In response, ST should stress the positive impact of tourism on women's lives and there are a number of initiatives working towards this aim, reflecting SDG 5. The UNWTO, for example, works to mainstream gender issues in the tourism sector and policies in a systematic and continuous manner. The UNWTO's work on gender began with their 2011 'Global Report on Women in Tourism 2010' (UNWTO, 2011a) and has continued with the development of internet portals stressing empowerment and tourism and gender (see http://ethics.unwto.org/en/content/tourism-and-gender-portal). The UNWTO is working to ensure that tourism can act as a vehicle for the empowerment of women in ST and local development. This will be achieved by using gender analysis and gender training to tackle inequality and gender-based discrimination in the tourism industry. The overall goal is to promote women's economic empowerment in tourism through partnerships with hotel chains and other stakeholders. Here it is important to recognize that to truly achieve ST, women's empowerment should be a core issue and not an 'add on'.

Despite these initiatives, Ferguson and Alarcon (2015) states that the impact on tourism has been minimal and that there is a resistance to incorporating gender as a core principle in

ST. This is confirmed by the organization Equality in Tourism – a non-profit international organization, created in 2012 to incorporate gender dimensions into tourism development (http://equalityintourism.org). Equality in Tourism specialize in promoting gender equality in tourism and raising the voices of women from all around the world about different issues in gender and tourism. They state that by promoting measures to reduce gender inequality in employment in tourism, sustainable development of the economy is encouraged, poverty reduced, and the policy frameworks and ethics of the sector are improved. In addition they ensure that gender equality is incorporated in the tourism value chain to ensure fair trade and equity in decision making. Equality in Tourism also promote responsible consumption. They stress that gender inequality is not simply economic but also links to environmental protection and empowerment: women do not have the same access and rights to power as men.

With regard to environmental protection, the ecofeminist movement, also known as ecological feminism, is notable. Ecofeminism is an activist movement that sees a connection between the domination of nature and the exploitation of women (Yudina & Fennell, 2013). It grew from the feminist, peace and ecology protest movements of the mid-1970s and it:

> takes from the green movement a concern about the impact of human activities on the non-human world and from feminism the view of humanity as gendered in ways that subordinate, exploit and oppress women. (Mellor, 1997: 1)

Ecofeminism focuses on the corporate world and its exploitation of natural resources, tackling the issue through ideas of respecting organic processes, holistic connections and the merits of intuition and collaboration. For example, women who live in subsistence societies and create livelihoods in partnership with nature have been experts in their own right of holistic and ecological knowledge of nature's processes. However, they claim, this type of knowledge is not recognized by the capitalist view of the world where women's work is 'invisible' and not appreciated.

Finally, the feminist movement – and indeed the ecofeminist movement – has been criticized for opening up and maintaining a binary divide between men and women when, in fact, the movements are trying to do exactly the opposite.

INDIGENOUS RIGHTS

Tourism has a long-standing, and often contentious, relationship with Indigenous people. As tourism has spread into increasingly remote regions it has had a sustained impact on many Indigenous groups, while Indigenous groups in the developed world – particularly in the USA, Canada, Australia and New Zealand – have long been associated with tourism. Indeed, there are now many Indigenous tourism businesses. Define Indigenous tourism as:

> Tourism activity in which indigenous people are directly involved either through control and/or by having their culture serve as the essence of the attraction. (Hinch & Butler, 1996: 9)

Here the two words 'control' and 'attraction' are key to sustainability. It is important that Indigenous people are respected and not seen simply as an attraction to be gazed at, and it is important that the locus of control of the relationship lies with the Indigenous community.

Weaver (2010) has developed a six-stage model of Indigenous tourism:

1. Pre-European *in situ* control is characterized by high local control and Indigenous themes where tourism may be represented, albeit on a small scale, by ceremonies of gift giving where no locals are invited.
2. *In situ* exposure to tourism occurs in the early stages of colonialism where explorers, scientists and anthropologists visit the communities.
3. *Ex situ* exhibitionism and exploitation occur as native artefacts are displayed in museums and exhibitions.
4. *In situ* exhibitionism and exploitation are now evident in the relationship as remnant Indigenous spaces are opened up to tourist visits. Hinch and Butler (1996) term this 'culture dispossessed tourism' where, despite the fact that cultural themes are present, control by the Indigenous groups is low.
5. Strategies of resistance result in *in situ* empowerment as Indigenous tourism businesses are established. Here we move to Hinch and Butler's (1996) stage of Indigenous 'culture-controlled tourism'.
6. *Ex situ* quasi empowerment and the presence of 'shadow Indigenous tourism' where Indigenous peoples re-establish the ownership of their traditional lands.

Clearly, in the early stages of Weaver's model tourism numbers are small, but as we progress through the stages tourism numbers increase and initially Indigenous control decreases until Stages 5 and 6 where control is regained. From a sustainability point of view, these latter stages allow the Indigenous groups to regain their culture and control over tourism as an activity.

In Weaver's model, the notion of control can also be thought of as 'power'. For example, Richards and Hall (2000) argue that while most models of sustainability focus on distributive power, which typically comes from above, the notion of generative power focuses on empowerment distilled from within the community. Even though external forces often weigh heavily against localities, Richards and Hall argue that generative empowerment holds potential in 'leveling the playing field'. A manner in which to operationalize this type of thinking is through resource management systems that are based on traditional ecological knowledge (TEK). The systems that are based on the intricate knowledge of 'place' have meaning tied to them that is very different from conventional scientific models. Those agencies that use a combination of TEK and conventional science are able to approach ecological cultural problems from a completely different vantage point, while at the same time building capacity for cooperation between Western and local perspectives. As Yi-fong (2012) writes, newer collaborative relationships between various stakeholders, including local people, have spawned a diverse network of ecotourism opportunities. For example, one programme called 'Follow the Footsteps of Indigenous Hunters' has ecotourists learning from Indigenous guides who have replaced hunting tools with cameras. 'Hunters' benefit from showing ecotourists rare and unique fauna. The rarer the find, the higher the pay, according to Yi-fong.

As we move into Stage 4 of the model, it is clear that the Indigenous group has lost control of tourism and the process is unsustainable. Here, issues around respect are at the forefront of the dilemma between Aboriginal people and the cruise industry in the Kimberley region of North West Australia (Smith *et al.*, 2009). The authors note that non-Aboriginal cruise operators continue to capitalize on a range of cultural resources but with little to no contact with traditional owners (TOs), who have inherent rights and obligations specific to taking care of the country. Documented impacts include lack of contact with TOs, permission to gain access to sites not being requested and cultural protocols not being followed. In reference to proper protocols, the authors observe, based on conversations with TOs of the region, that non-Aboriginal people do not respect the dead in the same way as Aboriginal people do. There is evidence, the TOs note, that some burial sites (the deceased are covered with stones) are being disturbed by cruise tourists. The end of chapter case study outlines an innovative organization and its work with Indigenous communities in Africa.

Much of the notion of respect is linked to understanding the Indigenous worldview. Thaman (2002), for example, offers a persuasive interpretation of sustainability packaged through the call for a more holistic approach based on valuing Indigenous worldviews. Colonialism and globalization have heavily influenced the discourse on sustainability, rendering it more in line with Western values and priorities. Using the example of time, Thaman argues that time is thought of in a linear, scientific and financially driven manner, rather than as a circular perception which is more akin to Oceanic cultures. This latter perception amalgamates the past, present and future in an 'all embracing "now"' (Thaman, 2002: 234) and linked in a presence that is envisioned as the future. Thaman's position is that sustainability must be rooted in people's cultural values. Moreover, beyond different notions of time, other important differences contrast with Western views. Pacific Island cultures appear to place more emphasis on trust, reciprocity, creativity, restraint, compassion and interdependence than Western culture. The bottom line for Thaman is that there must be enough elasticity in sustainability theory and practice to accommodate the special signs and symbols that make up different cultures.

In a similar argument relating to Indigenous world views, Zeppel and Muloin (2008) have shown that a combination of Western and Indigenous forms of interpretive messages can be beneficial to tourists in the context of Australian wildlife tourism. Indigenous styles of interpretation were said to be beneficial because of their focus on traditional uses of wildlife and personal stories, along with notions of Aboriginal Dreaming and animal species as totems. These sorts of attitudes, following the work of Kellert (1989), were judged to be moralistic (spiritual), utilitarian (food) and aesthetic (totems). By contrast, non-Indigenous styles of interpretation emphasized biological facts and species knowledge. This is but one of countless examples of Indigenous communities teaching Indigenous values through the platform of ecotourism, with the hope of fostering transformation in ecological consciousness (Higgins-Desboilles, 2009).

Similarly, McNelly (1993) argues that it is not just biodiversity (genes, species and ecosystems) that needs protection, but also traditional cultures. McNelly's point is that both need to be conserved together if either is to prosper. This refers to the knowledge that people have about their resources, and how they have been managed, being critical to all of humanity. His concept of 'biocultural diversity' connects well with the concept of TEK. This links back to our earlier discussion on carrying capacity, where scholars quickly recognized that we could not separate ecological and social carrying capacity: the two are intricately intertwined. Of course, language is a key agent of culture. Here, researchers have investigated the global erosion of Indigenous language which poses a serious threat to cultural heritage. In a study of the Haida Gwaii of British Columbia, Canada, Whitney-Squire (2016) developed a framework on why language should be used, how it should be used and what processes need to take place in order to share language appropriately. Whitney-Squire argues that tourism can be an agent of the revitalization of language in Indigenous communities to emphasize the link between place and the individual and collective identity of people.

MOBILITIES

The mobilities paradigm provides a useful lens through which to view ST. The paradigm combines social and spatial approaches encompassing,

> both the large scale movement of people, objects and capital ... as well as the more local processes of daily transportation, movement ..., and the travel of material things. (Hannam *et al.*, 2006: 1)

The paradigm challenges the traditional view of tourism, conceiving instead of tourism as a form of 'voluntary temporary mobility in relation to home' (Hall, 2005). In other words, tourism is seen as but one of a range of other activities in a mobile society. Larsen *et al.* (2006) justify this approach by arguing that the world in the 21st century is a highly mobile one and, because tourism is relatively inexpensive and convenient, it blends with other forms of mobility and connections.

From a sustainability perspective, the mobilities paradigm highlights the consequences of the patterns and processes of movement and the consequences for sustainability and communities at the nodes. The key question, therefore, is to understand and explain what influences these movements, from micro-personal resources to macro geopolitics and the restructuring of economies and territories, as well as the very sustainability of movement with its consequences for energy and climate change (Cooper, 2017).

However, not everyone can be part of this world of movement. The concept of differentiated mobilities reflects structures and hierarchies of power, recognizing that there are many in the world whose mobility is highly limited (Cohen & Cohen, 2015). This notion of differentiated mobilities highlights the fact that ST must consider those tourists with limited mobility and attempt to address the issues if tourism is to be truly inclusive. The ability to travel involves mustering the personal resources needed, and yet some argue that the mobilities approach ignores these inequalities. The less mobile are an important and valuable research area for ST but are largely ignored (Hannam *et al.*, 2014). This was first recognized in the 1980 Manila Declaration on World Tourism, which stated that the right to have holidays and travel is a basic human right. However, despite this statement, in many countries tourism is not recognized as a right and in many only as a moral right, not a basic human right.

There is much more statistical coverage of those that do travel rather than those that do not. However, there is some research on non-travellers. In fact, the majority of the world's population do not travel internationally and those that do are mainly in the developed world, although there are exceptions such as Brazil, China, India, Malaysia and Indonesia showing rapid growth in air travel. Peeters *et al.* (2006) estimate that only 2–3% of the world's population participate in international air travel and the pattern is not evenly distributed across countries. And of those that do travel some are 'hypermobile', a kinetic elite engaging in many trips in any one year. These highly mobile individuals take frequent trips, often over great distances. For example, Dubois *et al.* (2011) estimated that at Gothenburg Airport, Sweden, 3.8% of hypermobile air travellers accounted for more than a quarter of total trips taken. Across the EU's 27 countries it is estimated that almost two-thirds of those at risk of poverty did not have the financial ability to take a holiday (Hall, 2010b). Resources are clearly influential in facilitating mobility. In the USA, for example, the trip generation rate almost triples in the transition from the very low-income group to a very high-income group (Hall, 2005).

Social tourism is a long-standing movement that attempts to redress the immobility of both individuals and particular groups in society. It is defined as:

> all activities, relationships and phenomena in the field of tourism resulting from the inclusion of otherwise disadvantaged and excluded groups in participation in tourism. The inclusion of these groups in tourism is made possible through financial or other interventions of a well defined and social nature. (Minnaert *et al.*, 2012: 29)

Social tourism supports those with limited mobility through provision of finance, infrastructure, respite support or transport. It makes an important contribution to inclusion, as estimates suggest that at any one point in time 40–45% of the world's population do not engage in travel (Diekmann *et al.*, 2018). Of course, limited resources mean that not all groups can be supported by social tourism. A particular group that has received considerable attention both in the sector and in research is those with disabilities, giving rise to the term 'accessible tourism' (see Buhalis & Darcy, 2010). Around 15% of the world's population have some form of disability and not being engaged fully in travel is an infringement of their human rights. Infrastructure and superstructure at destinations has tended to be designed for able-bodied tourists and this immediately disadvantages participation by the disabled. The European Network for Accessible Tourism (ENAT), for example, is a non-profit association for organizations that study, promote and practise accessible tourism; they have developed a code of conduct for destinations (https://www.accessibletourism.org/). The UNWTO (2013) define accessible tourism as:

> A form of tourism that involves a collaborative process among stakeholders that enables people with access requirements, including mobility, vision, hearing and cognitive dimensions of access, to function independently and with equity and dignity through the delivery of universally designed tourism products, services and environments. (UNWTO, 2013: 13)

Taking this argument one step further, Lee and Jamal (2008) have developed a distributive and procedural justice framework that reflects the inequalities of mobilities in terms of the global framework of destinations and their costs and benefits (see Figure 10.1). They argue that ST may only be truly achievable if industry is able to bridge the gap between the benefits that go primarily to industrialized countries and the costs that go to the lesser developed countries. Distributive justice includes elements such as access to natural and tourism resources, distribution of economic benefits from the use of environmental and cultural goods, and measures to avoid the displacement of local people because of tourism development. Specific examples often cited in tourism include access to natural capital, economic benefits, water consumption, groundwater use, water quality and air quality. Lee and Jamal (2008: 59) state that procedural justice 'focuses on the process through which environmental decisions are made'. Local people must be given the opportunity to participate in and have some measure of control over the types of decisions that are made that have a direct impact on their lives and livelihoods. Policies, norms and guidelines must be developed, upheld and constantly changed so that collaborative approaches reflect dynamic situations and circumstances (see also Fennell, 2018a).

ANIMALS

There is an emerging link in both scholarship and practice between ST and animal ethics, reflecting the 'animal turn' in the social sciences more generally and as we introduced in the tiger case study in Chapter 3 (Fennell, 2012, 2013). For example, Dietz *et al.* (2017) surveyed the literature on environmental values, attitudes and decision making and found that an animal-focused value orientation comprises a motivation that is distinct from other values used in this research. More specifically, they argue that an animal-focused value orientation is different from biospheric altruism, humanistic altruism, self-interest, traditionalism, openness to change and hedonism. Even though the biospheric value is closely related to the animal focus, both represent different domains, as above, in terms of

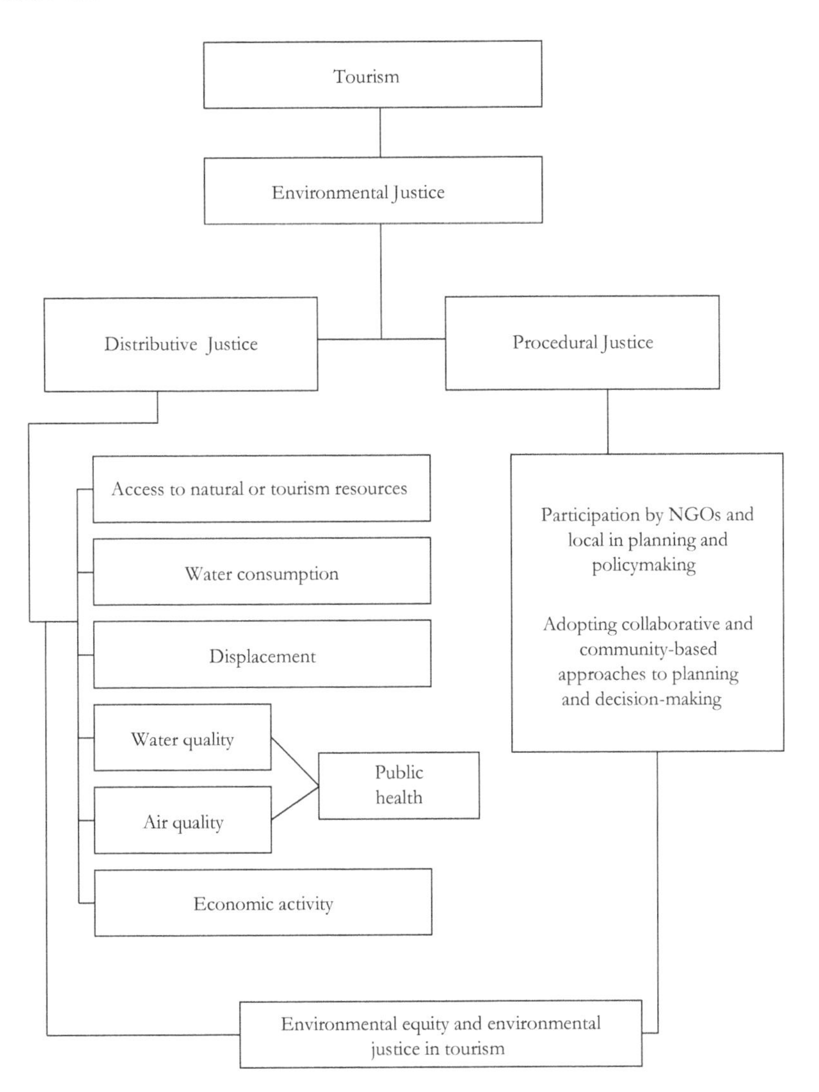

Figure 10.1 Environmental justice and environmental equity in tourism: Examples of distributive and procedural elements
Source: After Lee and Jamal (2008)

environmentalism. The former adheres more closely to the environmental movement, the latter to the animal liberation movement.

ANIMAL WELFARE AND TOURISM

The World Society for the Protection of Animals (WSPA), now known as World Animal Protection, created a comprehensive document on why animal welfare is important if we are to have any chance of being sustainable (WSPA, 2012). The WSPA, like many other organizations that support animal welfare in tourism, adheres to the Five Freedoms of Animal Welfare as developed by the UK Farm Animal Welfare Council, now known as the Animal Welfare Council (AWC; https://www.gov.uk/government/groups/farm-animal-welfare-committee-fawc#assessment-of-farm-animal-welfare---five-freedoms-and-a-life-worth-living):

1. freedom from hunger;
2. freedom from discomfort;
3. freedom from pain, injury and disease;
4. freedom to express normal behaviour; and
5. freedom from fear and distress.

The clearest and most direct line between tourism, animal welfare and sustainability to the WSPA beyond the use of the Five Freedoms is to advocate and implement measures that move towards certification and regulation, education and philanthropy to fund animal welfare initiatives, creating shared value in the production and consumption of tourism, and to be sustainable in our resource use.

Other organizations have also embraced the animal welfare platform in animal ethics for a better tourism industry. They include:

- The Brooke;
- SPANA;
- AnimalsAsia; and
- RIGHT-Tourism.

The Brooke is an international animal welfare organization based in the UK which is dedicated to improving the lives of working equines, including horses, donkeys and mules (https://www.thebrooke.org/). The focus of the Brooke is on many of the world's poorest nations. There are hundreds of thousands of equines working in the service of the tourism industry, often severely overworked under adverse conditions, and these animals are often provided with few resources to guarantee their health and wellbeing. The Brooke has developed its own code of ethics – the Happy Horses Holiday Code (Brooke, 2019b) – to better inform tourists and the tourism industry about these welfare issues.

The London-based organization, *SPANA*, is another not-for-profit organization geared towards the welfare of working animals in tourism. This organization, as noted in their *Holiday Hooves Guide*, focuses on supporting those operators that treat their working animals with respect (SPANA, 2018).

AnimalsAsia, founded in 1998, takes a broader view of the many different types of animals used for entertainment in the tourism industry. There is a particular focus on animals used in captive environments such as circuses and zoos, which include apes, big cats,

Table 10.1 AnimalAsia's vision and values

Vision
• To end cruelty and restore respect for all animals throughout Asia.
Values
• The welfare of all animals is our first priority. • We are responsible guardians of the animals under our care. • We are committed to responsible stewardship of donor funds. • We strive to find compassionate solutions that benefit people as well as animals. • We act ethically, responsibly and fairly in all our dealings.

Source: AnimalsAsia (2019)

elephants and other types of charismatic mammals, but especially farmed bears. This organization actively educates the industry around issues related to safety, where these animals come from and how they are kept, the educational value of using animals in such ways, and conservation. AnimalsAsia's vision and values are found in Table 10.1.

The organization responsibletravel.com has also taken an active role with regard to several issues in tourism that have implications and consequences for animals that are abused. This organization has taken a stand on animal welfare issues at rodeos and stampedes (https://www.responsibletravel.com/copy/animal-welfare-issues-at-rodeos-and-stampedes), the purchase of animal souvenirs, some of which are derived from endangered species (https://www.responsibletravel.com/copy/about-to-go-on-holiday-think-twice-about-your-souvenirs), and a general statement on animal welfare issue in tourism with a focus on poor conditions in zoos, animal circuses, and dancing bears (https://www.responsibletravel.com/copy/animal-welfare-issues-in-tourism) and canned hunting (https://www.responsibletravel.com/copy/canned-hunting). Canned hunting is an especially heinous practice where charismatic animals, usually lions, are raised on farms and released at about the age of four in a confined, caged space, to be easy prey for hunters.

Other more generalist ethics organizations are also coming to the table in reference to the treatment of animals in tourism. For example, *RIGHT-Tourism* has developed its own set of policies on the use of animals in tourism. Their new campaign to stop the practice of taking 'selfies' with animals was launched in early 2018.

TOUR OPERATORS AND ANIMAL WELFARE

Welford *et al.* (1999) argue that the major tourism operators likely to have the most power in the tourism industry ought to be more demanding of the suppliers in tourism to offer more sustainable options for animals in tourism. Since their call there has been evidence that there are important changes taking place with regard to the welfare considerations of animals as tour operators are becoming active in the area of standards for animal welfare.

The Association of British Travel Agents (ABTA) animal welfare vision is that.

> all animals encountered through tourism are treated humanely, with respect and in accordance with transparent and robust animal welfare standards which adhere to the Five Freedoms.

ABTA members and destinations strive to satisfy the following six areas (ABTA, 2015):

1. to understand the scope of tourism's impact on animals;
2. to raise awareness of animal welfare best practice with customers, suppliers and governments;
3. to assess and improve performance within the tourism supply chain;
4. to review and report on actions;
5. to set targets for improvements; and
6. to influence and encourage governmental policy around animal welfare and tourism.

The British trade association, ABTA (2013), developed the Global Animal Welfare Guidance for Animals in Tourism, which is a set of manuals that provide best practice and minimum requirements for the proper treatment of animals used in tourism. Specific manuals have been developed for animals in captive environments, dolphins in captive environments, elephants in captive environments, wildlife viewing, working animals and unacceptable and discouraged practices.

The world's largest tour operator, TUI (2018), collaborated with ABTA in the development of the Global Animal Welfare Guidance for Animals in Tourism, and their media release under the category of 'Sustainability' identifies the degree to which they are committed to animal welfare. Following the lead of the World Animal Protection and Born Free Foundation, TUI put into practice the following policy as of July 2015:

> to withdraw from elephant riding and elephant shows across all TUI operations. Our full withdrawal from such activities was achieved in 2016 and we now offer elephant friendly excursions where our customers can see and learn about elephants in a way that avoids unnatural behaviours.

Thomas Cook, too, was active in the area of animal ethics, launching its animal welfare policy in 2016, and following the audit structure of the Global Animal Welfare Guidance for Animals in Tourism lead (Thomas Cook Group, 2018). Their state-of-the-art policy contains the following procedures (as noted in the media release by the Born Free Foundation (Born Free, 2016):

- all animal attractions must be fully compliant with the ABTA Global Welfare Guidance for Animals in Tourism;
- a decision to no longer sell or promote any new attractions or hotels keeping wild animals in captivity that do not comply with industry minimum welfare requirements;

- ensuring that existing facilities keeping captive wild animals, contracted by Thomas Cook, meet the highest animal welfare standards;
- agreeing to phase out practices that are known to severely compromise the welfare and survival of animals; and
- supporting attractions and excursions that safeguard the welfare and protection of animals.

Finally, TripAdvisor has joined forces by stopping selling tickets to many of the world's cruellest tourism activities, after being criticized for profiting from these types of attractions by World Animal Protection (Cision, 2016).

The number of large tourism firms participating in the struggle to enhance the welfare of animals used in tourism and to reduce animal suffering is therefore impressive. It serves notice that a moral turn has taken place in the industry in the last few years – one that is no doubt here to stay.

Zoos, circuses and aquaria

Despite the fact that, traditionally, establishments such as zoos have been significant tourism attractions, there are increasing moral issues tied to the keeping of animals in such captive environments. These sentiments have intensified consistently over the years, stemming from a growing environmental consciousness that began in the early 1960s (Catibog-Sinha, 2008). Proponents of zoos argue that such environments are morally acceptable because they focus on education and conservation. However, research indicates that these more lofty goals are secondary to entertainment. Ryan and Saward (2004), for example, found that while visitors to a New Zealand zoo were motivated to view animals in the zoo, they were far less likely to obtain information on them.

What is also vague in tourism studies and practice is the extent to which zoos and aquaria should be considered as ecotourism attractions. While some argue that because of their conservation and education focus these institutions should rightfully be viewed as ecotourism (Mason, 2000), others feel that the issue is far more complicated. Fennell (2013) contends that because animals are not free and are unable to live normal lives because of confinement in zoos and aquaria, such attractions should never be classified as ecotourism. The following 'first principle' might act as a guiding force in decisions about what is ecotourism and what is not:

> Reject as ecotourism all practices that are based on or support animal capture and confinement, or other forms of animal use that cause suffering, for human pleasure and entertainment. Embrace as ecotourism interactions that place the interests of animals over the interests of humans. This would include encounters with free-living animals that would have the liberty to engage or terminate interactions independent of human influence. (Fennell, 2013: 10)

SUSTAINABLE WILDLIFE TOURISM

One example of the relationship between tourism and animals is safari tourism. Here, non-consumptive wildlife tourism is focused on the wildlife-watching market segment. Wildlife watching takes place in non-captive, natural environments, commonly in protected areas, often savannah grasslands where viewing is unobstructed (UNWTO, 2015). In Africa, for example, viewing the 'big five' is the central pillar of the safari product. The quality of the commercial viewing experience for tourists depends on how close they can get to the animals and how reliable the sightings are, specifically to see the big five of African tourism – the African lion, leopard, elephant, buffalo and black rhino. For tourists, the result of the wildlife-watching experience can lead to greater awareness, appreciation and connection to the environment and nature, and a greater sense of their stewardship of the environment.

However, despite its seemingly green credentials, wildlife watching has generated a number of issues for the environments and societies within which it operates. These include:

- poaching and crime – poaching is a real issue in Africa, threatening the sustainability of safari tourism as it targets the 'big five' animals.
- human pressure on protected areas;
- invasive species and habitat loss;
- conflict between animals and local communities, such as destruction of crops;
- lack of equity in the distribution of wildlife-watching and national park revenues to local communities;
- negative social impact of tourists on local village communities; and
- the big five animals are particularly vulnerable to negative impacts of wildlife watching: not only do they become accustomed to humans at close range, but their hunting patterns and reproductive patterns are disturbed. There are also disruptions to their habitats to allow for better viewing such as thinning woodland and constructing viewing towers.

TOWARDS SUSTAINABLE WILDLIFE TOURISM

There is no doubt that wildlife watching is a highly significant economic contributor to national economies, prompting governments to take action to develop and protect it (see UNWTO, 2015). It also has the potential to: (i) redistribute the benefits of tourism to local communities in terms of income from accommodation, guiding and restaurants as well as opportunities for local entrepreneurs; and (ii) change the attitudes of tourists towards the environment through interpretation, education and use of technological innovations such as virtual and augmented reality. The UNWTO (2014) World Bank (2013) has outlined a clear pathway to encourage the transition towards sustainable wildlife watching. This includes:

- taking difficult strategic decisions on the scale of wildlife watching at particular destinations and deciding whether to focus development in one area or disperse across a wider

area. Here scale is linked to judgements of both carrying capacity and the absorptive capacity of the industry;
- putting in place political support, policies and a stable investment environment to support the financial sustainability of operators and entrepreneurs;
- protecting the natural asset base from degradation through sound environmental management;
- extending the benefits of wildlife tourism to local communities in order to make the positive impacts of tourism mutually beneficial with the industry; here, sustainability is a central concern for corporate management and they are partnering with local communities;
- ensuring the nuts and bolts of tourism are in place – product development, access and connectivity and appropriate promotion;
- sustaining and deepening wildlife tourism by looking at spatial and evening up seasonality; and
- ensuring high-quality standards through tools such as certification, as has been done in Costa Rica.

CONCLUSION

This chapter has provided an analysis of the major cross-cutting issues that impact on sustainable tourism and in so doing considers UN SDGs 5, 8 and 13. Clearly, one of the most important issues is climate change, and tourism can be considered as both a victim and a vector of climate change. Climate change has the potential to change the nature and distribution of tourism significantly. As a result there is an imperative for tourism to transition to become a low-carbon sector but the challenges of air transport in particular are a stumbling block here. The chapter then went on to analyse human resources in tourism and the fact that many jobs in tourism cannot be classified as 'decent work'. In terms of sustainable tourism, the transition towards 'green jobs' in tourism is an important one. Closely linked to human resources is the issue of gender equity in tourism. Women represent a majority of employees in tourism but they tend to be less well paid and appreciated. A number of initiatives are taking place to address this issue as well as the development of the ecofeminism movement which is concerned with women's role in environmental stewardship. The rights of Indigenous communities and tourism is an important concern for sustainable tourism, particularly the locus of control of the relationship. The chapter then went on to consider the mobilities concept in tourism and how it identifies the less mobile such as the disabled. This has given rise to the 'accessible tourism' movement. Finally the chapter examined the role of animals in tourism and the issue of animal rights and sustainable wildlife tourism.

END OF CHAPTER DISCUSSION QUESTIONS

1. In class, debate the fact that climate action is urgent, but as climate change is invisible the tourism sector will be slow to act.
2. Taking a tourism organization of your choice, assess their attempts to instil gender equity into their operations.
3. Assess the future of zoos as tourist attractions.
4. Taking a destination that you know well, assess the level of accessibility for the disabled.
5. Choose a particular 'green job' in tourism. What skills and competencies will be needed for that job?

End of Chapter Case Study 10.1 Isoitok Camp Manyara and the African Roots Foundation

This end of chapter case study outlines an innovative organization and its work with Indigenous communities in Africa.

The camp

The Isoitok Camp Manyara is a tented camp located on the Maasai Steppe in Tanzania. It offers 'a grass roots cultural experience and a true look behind the scenes into Maasai culture' (http://isoitok.com/). The camp is set in a rich wild savannah bush location and was founded in 2007. The camp has a close partnership with the local tribes people, the Maasai, and aims to have a low impact on the area through the use of initiatives such as solar energy.

The camp's unique partnership with the local Maasai people provides the opportunity to offer a menu of cultural and lifestyle activities that allow tourists to understand the tribespeople's way of life. Local English-speaking Maasai guide scouts take tourists on walks and to various real (not staged) Maasai activities and ceremonies. These include:

- Maasai medicine walk – covering root and plant extracts still used by the Maasai.
- Maasai boma (dwelling) visit – meetings with the Elders, warriors, women and their children.
- Traditional goat sacrifice.
- Viewpoint walk with bush refreshments.

The local Maasai community not only has the unique partnership with the camp but also has its own foundation, the 'African Roots Foundation' (ARF).

The Foundation

The ARF is a not-for-profit community development organization, created in 2007, based on small-scale, sustainable projects which aim to improve the daily life of the communities surrounding the camp. To quote the website, the ARF is:

> an initiative of Bush2Beach Safaris to create a platform for inter-cultural experiences through which Western and African cultures combine their strengths to increase the self support opportunities of local rural communities through the initiation of integrated, community based development projects that are actively and physically supported by eco-tourism. (http://africanrootsfoundation.org)

The ARF is involved in a range of projects which include:

- education on the conservation of the environment and wildlife, health and HIV awareness;
- promotion of cultural traditions and practices;
- fair trade of locally manufactured cultural crafts;
- health awareness and improvement programmes;
- water management and sustainable use of natural resources; and
- natural environment and wildlife, eco- and low-impact tourism.

These projects are designed to meet the agreed immediate needs of the communities and 'must provide additional income and employment through the promotion of fair trade and sustainable tourism' (http://africanrootsfoundation.org). The ARF continually monitors the success and outcome of the projects and is in constant touch with the elders and representatives of the communities.

Discussion questions

1. How comfortable are you with the tourists of the Isoitok camp visiting intimate parts of the lives of the tribespeople? Is it voyeurism?

2. Visit the two websites below. Who really controls the activity with the Maasai – the ARF or the tour operators?

3. Ecotourism is the economic basis for the community development initiatives – is this a sustainable model?

Sources: ARF (2019; http://africanrootsfoundation.org/); Isoitok Camp Manyara (2017; http://isoitok.com/)

CHAPTER 11

SUSTAINABLE TOURISM FUTURES

LEARNING OUTCOMES

This chapter focuses on sustainable tourism futures, outlining the key drivers of these futures and possible future scenarios. The chapter is designed to provide you with:

1. A disciplined approach to analysing and anticipating sustainable tourism futures.

2. An understanding of the key drivers of sustainable tourism futures and their relationship to the UN's Sustainable Development Goals (SDGs).

3. An awareness of the ways in which sustainable tourism destinations will evolve in the future.

4. An approach to understanding how the tourism sector will be shaped in the future in terms of sustainability.

5. An awareness of the various sustainable tourism scenarios.

INTRODUCTION

In this final chapter we turn to sustainable tourism (ST) futures. Here we move into uncharted territory where few authors dare to tread: the future is unknown, which means that anything is possible in terms of ST futures. That is why we use the term *sustainable tourism futures* rather than a *sustainable tourism future*, because there are an infinite number of futures for us to consider, some more likely than others. There is no doubt that the future for ST will be challenging. Tourism will grow to approach 2 billion international trips by 2030 with domestic travel far exceeding that volume. Set against this growth is the fact that the earth is becoming more crowded and available resources scarce and we have a real challenge to create an ST sector and destinations that are resilient to change. In particular, this level of travel is likely to contribute to growth in emissions of CO_2 by a factor of 2.5 in the period 2005–2035, mostly as a result of increasing air travel as we have already seen in previous chapters (Gössling *et al.*, 2010; UNWTO, 2008). Here we can see the influence of the UN SDGs on tourism futures, certainly into the medium term.

In predicting the future it is easy to get carried away in terms of discussing the future of ST and provide a sensationalist account of time travel and virtual reality theme parks. Of course the future is an exciting one – one where we can envisage hydrogen-powered aircraft, carbon sinks and transformational travel to move people seamlessly and rapidly between continents. Indeed, despite pessimistic accounts of the human race being wiped out by disease, meteorites or aliens, it is likely that we will be around for some time to come and the near future may not be too different from the present day. But predicting ST futures is not an easy task, simply because tourism is inextricably integrated with society, political systems and other economic sectors. Here, the European Tourism Futures Institute (ETFI, 2014) observes that the future, especially the long-term future, is difficult to grasp for a number of reasons:

- We only have partial control over the future because human influence is constrained by previous decisions and behaviour in the past (Postma, 2014).
- The complex nature of tourism renders it difficult both to understand and to influence change completely. Tourism is not an isolated sector and is linked to other demographic, economic, social, technological, environmental and political and institutional drivers of the future in a highly globalized and hyperconnected society. A further complicating issue is the fact that tourism services and facilities are owned, managed or influenced by a diversity of public, semi-public and private parties (Postma & Jenkins, 1997).
- Tourism is also characterized by contrasting styles of change. Butler (2009), for example, characterizes tourism, first, as prone to a high level of dynamism, and secondly, in contrast, destinations and organizations demonstrate a slow level of evolutionary change, building on their original resources and features.
- The role of technology, allied to an increasingly environmentally conscious consumer, is changing tourism behaviours and creating uncertainty in terms of ST futures.

These bullet points recognize that ST futures are exposed to many changes and influences. As a result, this chapter has taken a disciplined and structured approach, analysing the drivers of ST futures and presenting a series of possible future scenarios.

THE DRIVERS OF SUSTAINABLE TOURISM FUTURES

ST futures will be fundamentally shaped by a number of driving forces which map onto the UN SDGs. Some drivers are already known, some are on the horizon, while others are yet to emerge. It is this level of uncertainty that means that the tourism sector and destinations need to build resilience into their systems in order to be prepared for an unpredictable future. It is a mistake to treat each driver in isolation as they are interlinked and reinforcing. For example, social and economic drivers will encourage the growth of tourism, but they will also determine social attitudes to processes such as climate change and as a consequence we will see the very nature of tourism operations begin to change. However, at certain points in time, some trends will tip and become significant – and irreversible. The adoption of the internet in the 1990s is an obvious example here, as is the changing attitude to single-use plastics and their impact on the oceans. Finally, underpinning these drivers will be cross-cutting variables such as technology which will pervade every aspect of ST in the future.

DEMOGRAPHIC AND SOCIAL DRIVERS OF SUSTAINABLE TOURISM FUTURES

Demographic drivers

We looked at population growth in Chapter 2. Despite the fact that the world will be home to over 9 billion people in 2050, for most of the traditional generators of domestic and international tourism population growth is either static or even negative, with populations ageing as people live longer in North America, Europe and Japan (UNWTO, 2010a, 2010b). By 2020 one in eight people will be aged 60 years or more as average life expectancy lengthens, and we can see that this links to SDG 3 on good health and wellbeing. Ageing populations tend to be associated with urbanization and conservative politics, and markets for their goods and services have clear implications for the ST sector: the ageing baby boomer generations of the developed world are currently one of the most influential market segments – fitter, healthier and more demanding than previous generations – but in the future their influence will wane. However, in the future there is uncertainty about just how active this generation will be in terms of travel, and whether a shifting set of economic models and social security changes will allow them the resources to travel. This is because there is increasing evidence that 'a lack of personal savings, a ballooning crisis over national pensions and costly medical bills in old age mean there's little cash left for holidaying' (Forum for the Future, 2009: 9).

Yet it is the younger generations who will shape ST futures. Generations X, Y and Z – the Millennials – will remain in the youth market longer as they marry later and continue with their youth lifestyle, so changing the nature of the traditional nuclear family household. Generation Y, roughly speaking those born between 1978 and 2000, is the largest population bulge since the baby boomers and will therefore influence future consumer behaviour as technologically adept and more savvy consumers, sceptical about marketing messages, seeking networked and virtual communities and caring of the environment. Here SDG 12 on responsible consumption and production maps onto the consumer behaviour of future generations.

So what does this mean for ST futures? There is no doubt that the importance of a sustainable approach to tourism will grow in the future, with more and more consumers and businesses acting and travelling responsibly. However, we are still in the early stages of a tourism market that is wholly committed to responsible travel, with the vast majority of the market unlikely to change their holiday plans to reduce the environmental impact of their trip. The Northern Ireland Tourist Board (NITB, n.d.), for example, states that responsible travellers are still in a well-meaning minority with the mainstream market believing that it is the responsibility of industry and government to act rather than the traveller.

Social drivers

The knowledge economy will place a premium on education, creating a 'creative class' of well-educated, networked and self-motivated individuals where English is the dominant language. This maps onto SDG 4 on quality education. With education comes an increasing concern for ethical consumption and environmental concerns which will change attitudes to mobility in general, and a conflicting emphasis on pleasure-seeking conspicuous consumption. Here the Future Foundation (2015) has created six future traveller tribes for 2030 based on personality and the psychology of travel, influenced by 'social media, ethical concerns and a desire for wellbeing' (Future Foundation, 2015: 3). The tribes are fluid and consumers may identify with more than one. They are:

1. simplicity searchers based on ease and transparency in travel;
2. cultural purists immersing themselves in the destination;
3. social capital seekers seeking social reward from travel;
4. reward hunters seeking a return on their travel investment;
5. obligation meeters such as business travellers; and
6. ethical travellers shaped by their conscience.

From the point of view of ST futures it is the final category that is of most interest – ethical travellers. They will differ from earlier travellers in their embrace of ethical travel and their concern for the destinations and the communities who live there. We will see an increase in philanthropic, conscientious travel and the growth of fair trade tourism as they recognize their responsibilities to the destination (Hall & Brown, 2008). In response to this trend, new forms of tourism such as volunteering (volun-tourism) will grow and ethical travellers will demand greater behind-the-scenes access and accountability from big business, while suppliers will embrace the idea of corporate social responsibility (CSR) particularly with respect to local communities. Operators will diversify their products to achieve market advantage and government agencies and destination management organizations (DMOs) will facilitate a move towards more sustainable forms of tourism through 'greener' policies and codes of conduct (Williams & Ponsford, 2009).

The genesis of ethical travellers is rooted in increased public awareness of the impacts of all forms of travel as they are reported on in social media (Forum for the Future, 2009). While air travel is the sector most identified with climate change, in the future we can expect that all sectors will come under scrutiny. Here, the consumer psychology of tourism will be critical in determining whether tourists really care about climate change and environmental impacts enough not to fly or to shun visits to fragile destinations. At the moment few travellers draw a strong link between their personal travel behaviour and climate change, but if this changes in the future it will have an impact beyond just the choice of transport and could affect where people want to visit, where they stay, the activities they undertake and how they behave on holiday. Dolnicar (2010), for example, identifies two principal indicators pointing to environmentally friendly behaviour – high income and moral obligation. This view is supported by Hughes (1995), who argues that such behaviour is ethical in nature and based on the moral drive that individuals have in operationalizing sustainability.

Yet, while research can tell us about people's attitudes, it is much more difficult to discern future travel behaviour. This is neatly summed up by Yeoman *et al.* (2014), who portray New Zealand as a paradise in a world of scarcity of resources. Their idea is the realization that consumers have to make sacrifices for the common good. Consumers realize that in order to save the planet, behaviours have to change. So, in altruistic fashion, collective responsibility overrides individual desires.

POLITICAL DRIVERS OF SUSTAINABLE TOURISM FUTURES

ST futures are intricately linked to politics and economies at all levels, mapping onto SDG 10 on reduced inequalities between countries. Initiatives at different geographical scales are changing the world order as we witness the rising economic power of newly emerging economies such as Brazil, Russia, India and China (BRIC economies) and the new group of strong economies of Mexico, Indonesia, Nigeria and Turkey (MINT), all reducing the influence of the USA on the world stage. At the same time, political and economic disrupters such as the election of US president Trump and the UK's proposed departure from the EU will also influence travel and attitudes to sustainability. We can identify three political drivers of ST futures.

1. Trade blocs versus regionalism

Opportunities for tourism will be enhanced by the formation of a number of trading blocs across the globe as country groupings come together in deregulated economic alliances. Notable here are the North American Free Trade Agreement (NAFTA), the Association of South East Asian Nations (ASEAN) and the creation and expansion of the EU. But set against this is the contradictory trend of the rise of regionalism and a search for cultural identity, evidenced by Brexit and independence campaigns in Scotland and Catalonia. This trend is supportive of sustainability and local-level initiatives for sustainable destinations.

2. Globalization

Underlying the changing world order is the globalization of the tourism sector. Globalization is a powerful force shaping national and regional economies which are increasingly linked and interdependent. But it is the larger, international companies that can take advantage of these consequences as globalization encourages increased concentration in the tourism industry with major companies gaining market share and market influence. At the same time, we are seeing the concentration of capital in a few tourism companies – a trend that also drives tourism towards the key performance indicators (KPIs), return on investment (ROI) measures and business practices demanded by the finance industry. This reflects Macbeth's (1994) view that the move towards sustainability is more reactionary than progressive. This smacks of enlightened self-interest, according to Macbeth, especially on the part of big business which continues to be highly exploitative – a 'ruling capitalistic hegemony that is antithetical to sustainability' (Macbeth, 1994: 44). Sustainable development will not come to fruition until or unless this capitalistic system is radically altered.

A particular sustainability problem associated with globalization is that most of the larger corporations do not have a relationship with a specific destination. They may therefore be less sensitive to the impact of their operations on that destination and its community. In addition, small and medium-sized enterprises (SMEs) and local destinations fear the 'neo-colonial' relationship which can emerge from dealing with large companies. This is an important consideration for tourism where, at the end of the day, the product is delivered locally – hence the conundrum of balancing global forces on an essentially 'local' product. Here, Duffy (2006) uses the example of Madagascar to analyse how 'donor consortia' have become a form of global environmental governance. Such consortia often include several donor organizations, including the World Bank, several national governments, and NGOs such as Conservation International and the World Wildlife Fund, all connected by the aim of providing aid for Madagascar. Madagascar is attractive as a biodiversity hotspot and is therefore an important region for ecotourism investment and development. As such, and as Duffy observes, once the environment is saved through the efforts of donor funding, these places will be open for business. The neoliberal focus is evident in this calculus: there is much financial gain to be realized through destinations that offer so much resource potential.

One further consequence of globalization for ST futures is the fact that environmental problems such as pollution do not respect political boundaries. Here, Potts and Harrill (1998) propose a travel ecology framework based on a political ecology platform and critique of the global economy. Political ecology, they argue, investigates the root causes behind environmental change – often pinpointing how broad social and political institutions are the main drivers of these transformations – and what systems or processes need to be implemented at these levels that move us in the direction towards sustainability. This approach echoes the work of Stonich (1998), who articulated how national and international forces play a vital role in affecting action at local levels in reference to the natural world and the behaviour of community stakeholders. Travel ecology, Potts and Harrill (1998) contend, should push tourism planning in the direction of creating communities that are resilient enough to withstand the pressures that these exogenous forces place on localities.

3. Environmental legislation and carbon tax

In the past, attitudes towards tourism by many governments have generally been favourable with support for tourism development. However, Butler (2009) notes that attitudes may be changing, particularly at local levels where the impacts and costs of tourism are at their most visible. Equally, at national and international levels there is no doubt that controls over emissions of greenhouse gases (GHGs) across the economy will steadily increase in the future. The UK, for example, has a legally binding commitment to reduce GHG emissions, currently by 34% on 1990 levels by 2020 (Forum for the Future, 2009). However, we do not know the level of severity of legislation and taxation and the potential impact on both short- and long-haul travel, which may in fact have a greater impact on the cost of travel than consumer attitudes.

SCIENCE AND THE ENVIRONMENTAL DRIVERS OF SUSTAINABLE TOURISM FUTURES

Science helps us to understand environmental change and its potential impact on tourism futures as the sector continues its imperative for sustainable development (Fayos Sola & Cooper, 2018). As Dwyer *et al.* (2007) so clearly state, environmental change will be with us for the remainder of the century, whereas most of the other drivers of the futures identified in this chapter will have a shorter time horizon. Tourism is by its very nature resource intensive, particularly in terms of energy, water, soil and raw materials (see Robaina & Madaleno, 2018). It is therefore important that the future of tourism becomes *eco-efficient* to assure the optimal use of natural resources and to minimize impact on the natural environment (Fayos Sola & Cooper, 2018). The World Business Council for Sustainable Development (WBCSD) defines eco-efficiency as the delivery of:

> ... competitively priced goods and services that satisfy human needs and bring quality of life, while progressively reducing ecological impacts and resource intensity throughout the life-cycle to a level at least in line with the Earth's estimated carrying capacity. (WBCSD, 2006: 4)

To achieve eco-efficiency will involve 'eco-innovation in tourism', as a path to sustainable technologies ensuring clean production processes in areas such as construction, integrated water cycles, energy conservation and waste management relating to tourism (Robaina & Madaleno, 2018).

Climate change

We introduced climate change in Chapters 2 and 10 and provided a case study of destination adaptation in Chapter 5. There is no doubt that climate change will be the major driver of ST futures, both positively and negatively, with estimates suggesting that the tourism sector contributes around 5% of global GHG emissions. This is because the scientific evidence is compelling, and also because tourism is a climate-sensitive sector, mapping onto SDG 13 on climate action. Climate determines tourism seasons, destination choice and impacts on tourism product development. Concern for the impact of human activity on the climate is altering consumer behaviour and has increasingly become a focus for tourism policy and management initiatives.

Our understanding of the science of climate change is increasing, as is the world's awareness of the seriousness of the issue. As we have seen in this book, global climate change includes long-term factors such as global warming and the erosion of the ozone layer. In order to analyse the impact of climate change on tourism we need to consider the total tourism system, including transport. On the supply side, there is no doubt that the raising of the earth's temperature and the consequent rise in sea level will affect tourism destinations such as wetlands, deserts, islands, mountains and coastal areas. Much of tourism investment is found in locations that fringe the coast, and global warming will irrevocably alter vital tourism resources. This includes the disappearance of iconic destinations such as the Maldives, the deterioration of coral reefs through coral bleaching, and the loss of snow in winter resorts. On the demand side, fear of skin cancer and eye cataracts may reduce the demand for products such as beach tourism, which in turn will impact on destination and product development. Finally, we have to recognize that some transport modes used for tourism contribute to climate change and will need to change.

Resource use and biodiversity

Consumption of water across the globe is growing at double the rate of population growth and tourism is a voracious consumer of water, mapping onto SDG 6 on clean water and sanitation. In the future this will have to be carefully managed as water becomes a precious asset in some regions of the world: indeed, estimates suggest that by 2025, 1.8 billion people will live in areas of absolute water shortage (see Gössling *et al.*, 2015). Biodiversity too is being impacted on by development in many regions of the world, and by climate change. For an industry like tourism that depends on natural resources as part of its attraction this remains a serious issue.

Energy

We introduced energy issues in Chapter 2. By its very definition, tourism is a transport-intensive activity and is therefore exposed to changes in energy and fluctuating oil prices, mapping onto SDG 7 on affordable and clean energy. As we reach the years of 'peak oil' when the maximum amount of oil is being extracted, there is a growing dialogue supporting the transition to a low-carbon economy. This will depend on development of renewable energy technologies for tourism operations and 'fossil fuel free' efficient and smart transport systems (Becken, 2015). While the cost of energy has concerned tourism forecasters, it is alternative forms of energy that will have the most impact on ST futures. Forum for the

Future (2009: 6) reports that new transport and fuel technologies have the potential to 'offer steady improvements (such as in energy efficiencies) as well as dramatic breakthroughs (for example with algae-based fuels'. The success of these new technologies will depend on their ability to deliver low-emission travel without other unintended environmental consequences, and indeed whether the pace of their development can match consumer demand for travel. Nonetheless, we are already witnessing the fact that use of renewable energy is not only technically feasible, but can even save costs in many tourism-related facilities.

Transportation

Environmental factors will be an increasing concern for all transport modes in the future, particularly because in the case of air transport, emissions are unlikely to be reduced in the medium term. Indeed, the Intergovernmental Panel on Climate Change (IPCC) estimates that air transport accounted for 2% of total carbon emissions in the early years of the century and will rise to 3% by 2050 (www.ipcc.ch/). Environmental factors may influence ST futures in two ways:

- first, the consumer has an increasing concern for energy consumption and this may lead to a gradual modal shift in transport away from air and towards surface modes; and
- secondly, this trend may be reinforced by the imposition of environmental taxes on both air and car travel which will prompt further changes in choice of transport mode as tourism is asked to pay its way.

In response, the industry is developing more environmentally efficient, high-capacity, high-speed passenger vehicles. As a result, competition between transport modes will increase in the future, characterized by improved rail services and products, the realization of the environmental advantages of rail and continued technological developments in the area of high-speed train networks. A magnetically levitated (Maglev) fast train service is already operating in Shanghai (smtdc.com/en/) and commentators suggest that by 2030 Maglev technology may be replacing more traditional forms of traction. Road transport has not escaped this revolution, with driverless cars and traffic prediction enabled by WiFi and sensors and of course hybrid and electric cars increasing in popularity as battery technology improves. Forecasts of international transport predict that technological developments, use of aviation biofuels, increased airline efficiency and labour productivity savings will offset any rises in aviation fuel prices. This is supported by the fact that on short-haul routes the low-cost carriers are gaining market share from the traditional 'scheduled' carriers. This maps onto SDG 9 on industry, innovation and infrastructure.

Food futures

We introduced the importance of sustainable food supplies in Chapter 7. Food security will be a major concern for ST futures and is reflected in SDG 2 on zero hunger. With more than

9 billion people predicted to be living on the planet by 2050, demand for food will increase by 70%. But at the same time three factors will complicate the future for food. First, climate change will impact on food production, boosting yields in some parts of the world but also triggering drought, loss of biodiversity, soil degradation and the decline of pollinating bees in others. Secondly, food is already political and we are seeing shifting patterns of global trade – partly as a result of changing global politics, but also due to disruptive climate events such as hurricanes. Finally, the contemporary food system is built for a previous era based on non-renewable energy sources. Meat production, for example, has a high carbon footprint and livestock account for 18% of GHG emissions and consume 35% of our food production. Yet by 2040 meat production will grow by up to 70%. Clearly this is an unsustainable scenario and ST futures will need to move into a 'post-cow' era. This will involve alternative food sources such as insects and bugs, algae and seaweed, powdered food such as Queal (https://queal.com/), plant-based meat substitutes and lab-grown meat. The future of food supplies for ST futures will therefore depend on alternative sources of protein, genetically engineered food and a rethinking of our taste and cultural acceptability for food.

TECHNOLOGICAL DRIVERS OF SUSTAINABLE TOURISM FUTURES

The UNWTO has stated that the world in the year 2020 will be characterized by the penetration of technology into all aspects of life. ST futures will be determined and facilitated by technology, where – to paraphrase the well-known commentator on technology trends, Donald Norman (2007) – the most successful technology will be invisible to the user and, where it does interface with the user, the experience will be an organic one.

Fayos Sola and Cooper (2018) identify a series of technological drivers that will impact on ST futures. These include:

- the use of *big data* and *artificial intelligence* to enhance the visitor experience;
- the blending of the physical and digital worlds in mixed *augmented reality* (AR), made possible by advances in computer vision and graphical processing power among other innovations;
- the *processing of natural language* to improve search engines in regard to tourism destinations and services;
- the use of *virtual reality* (VR) as both a promotional and an immersive tool; and
- the customization of *artificial intelligence* (AI) to personalize tourism experiences.

These and other innovations will condition the ability of tourism to adapt to (and co-shape) the future, while the ability of destinations and organizations to survive will be challenged by developing technological paradigms. The fusion of information and communication technologies (ICT) will allow tourism enterprises to network with their consumers to communicate CSR and sustainability initiatives. Indeed, everything and everyone will be connected in the future. ICT and Big Data will increasingly allow customization of products and real-time access to demand, allowing segmentation almost down to the individual as we saw with the work of the Future Foundation (2015) on Traveller Tribes. In the market we are also seeing the growth of online tourism communities: here, the fusion of internet technologies with social activities such as blogging will facilitate content-rich sites and enable networked online tourism communities who share similar sustainability values and ethical principles. Effectively, technology will increase the connections between the actors in the ST system.

Artificial intelligence and robotics

The sustainability of tourism human resources has long been challenged by organizations such as the International Labour Organization (ILO), claiming that jobs in tourism are not quality jobs, characterized rather by low-skilled, part-time and seasonal work. This is reflected in SDG 8 on decent work and economic growth. Indeed, some say that the tourism labour market is not sustainable: as more tourism is demanded, so more jobs are needed (WTTC, 2015). Here we will see the rise of artificial intelligence (AI) and robotics to fill this gap in the future. This is known as the fourth industrial revolution and asks questions about the shape and composition of the future tourism workforce. The McKinsey Global Institute (MGI, 2017) suggests that up to 30% of current jobs will be obsolete by the early 2030s and that robots, particularly in the hospitality sector, will be the norm in five years' time. This is because tourism sectors such as hospitality are especially prone to automation as they are highly structured and predictable working environments.

As a result, the hospitality sector has begun adopting robots in their properties as technology costs have fallen and guests have become less fearful of robots and see them as both entertaining and a novelty with families (Murphy *et al.*, 2017). Robots will play two key roles in the hospitality system: of course they are ideal for routine and dirty jobs, but also they are increasingly being used to interact with guests. Additionally, they can free up human staff for other roles and they can constantly collect data as they operate. However, as Cooper and Hall (2019) note,

> while robots are intended to replace more fallible humans – who are expensive, get sick, make mistakes and leave, robots are not perfect. They still cannot make beds and have to be monitored 24/7, meaning that the hotel environment has many security cameras, and needs real people to monitor to both the safety of guests and the security of the expensive robots. (Cooper & Hall, 2019: 334)

Nonetheless, there are quite a few examples of robots working in hospitality environments – of these one of the most well known is the 'Weird Hotel' in Sasebo, southwest Japan. The hotel is 'manned' almost totally by robots to save labour costs (www.h-n-h.jp/en/).

So will AI and robotics contribute to ST futures? There is no doubt that the use of AI and robots will grow in tourism – they can replace fallible humans who get sick, emotional and take maternity leave and they will tend to replace female and blue-collar jobs. So the jury is still out on their benefits, but one school of thought counters the negative view of job losses and suggests that AI and robots can increase occupancy levels, which in turn will generate new jobs.

THE RESPONSE – SUSTAINABLE TOURISM FUTURES

Given these drivers of change identified above, we now turn to examine how they will shape ST futures. Here we identify two key themes: (i) tourism destination futures and (ii) innovation for ST futures:

1. TOURISM DESTINATION FUTURES

To create ST futures will demand that the tourism sector, communities and governments work together, rather than as single organizations working in isolation, mapping onto SDG 17 on partnerships for goals (Forum for the Future, 2009). There are three key responses required here. First, sustainable and smart destinations will need to be managed carefully and ensure that the many actors involved in tourism – in the public as well as the private sectors – assume their respective roles and responsibilities to work towards a common goal of sustainability. Currently there is a concern that environmental initiatives are uncoordinated and piecemeal. Communities need to be listened to and shown the economic benefits of tourism while increasing the destination's value and appeal to the market. In this way communities can be more involved in the future development and visioning of their destination. Williams and Ponsford (2009) observe that destination stakeholders are making a slow transition towards more sustainable practices, particularly noting the response of the private sector as their consumers become more aware of the adverse environmental effects of tourism on destinations (see, for example, the work of Tourism Concern, 2018; Williams and Ponsford (2009) observe:

> an emerging combination of science-based and industry approved codes of conduct, along with more eco-efficient transportation, communication and service production technologies are helping some tourism operators transition into a more environmentally sustainable approach to their businesses. Built on principles associated with 'softer' forms of sustainable development, a range of government policy, planning and management initiatives are encouraging more industry and destination based 'green' sensitivities. (Williams & Ponsford, 2009: 3)

However, the urgency of responding to issues such as climate change means that we must question whether such 'gradual' transitions are enough.

Managing the sustainable destination of the future

To deliver ST futures, there is no doubt that destinations of the future will need to be better planned and managed and show more concern and respect for their environment and host community. In the future, the focus of tourism will be on the destination as new intelligent management systems and techniques are adopted and the attention to volume will give way to concepts of visitor experience and value. These concerns will be addressed by enhanced tourism planning and visitor management techniques and a clear agenda to involve local communities in the futures of their destinations. In this way, the imperative will be for the sustainable management of tourism destinations and the conservation of their unique characteristics. Examples here include the management of UNESCO (2019a) World Heritage Sites such as Machu Picchu in Peru or Mount Everest.

The central issue is the gradual shift from short-term to longer term thinking and planning in tourism which maps onto SDG15 on life on land. It is no longer acceptable for the industry to exploit and use up destinations and then move on; indeed, we are already seeing the results of this in the demise of some of the mass tourism resorts built in the 1960s and 1970s – Acapulco in Mexico (acapulco.com/en/), or Benidorm in Spain (benidorm.com/). The concepts of the tourism area life cycle and strategic planning provide a much-needed long-term perspective in this respect. This means that destinations can decide to remain at a particular point on the life cycle by using marketing and planning approaches, rather than being inexorably driven to grow – or decline. On the demand side there are also drivers of sustainability as consumers place pressure on the industry and destination managers to behave in a responsible manner; if they do not, then their destination may be shunned as environmentally unacceptable to visit.

Destinations will respond to these demands in a variety of ways. Resource-based destinations are adopting sophisticated planning, management and interpretive techniques to provide both a welcome and a rich experience for the tourist while at the same time ensuring protection of the resource itself. Enterprises at the destination are also responding to the drive for sustainable destinations in three ways:

1. The tourism industry is anxious to demonstrate that it is responsible and acting to curb some of the excesses of past development. Increasingly, ST practices are being adopted as guidelines and manuals; certification and eco-labelling are examples here, as we have seen in this book. They encourage tourism enterprises to 'raise their game' in terms of sustainability and allow the consumer to discern those enterprises that are attempting to be sustainable in their practices. Indeed, it could be argued that sustainability has become a driver of innovation in the tourism sector with the larger companies adopting three guiding principles for their actions:

a. sustainability is driven by business imperatives such as competitiveness;
b. all the organization's stakeholders are involved – including customers and suppliers; and
c. sustainability is underpinned by detailed research to guide decisions. An example of this trend is the hotel company Accor which has commissioned deep research into the sustainability of its operations through its 'earth guest' programme, as we saw in Chapter 6 (https://www.accorhotels.com/gb/sustainable-development/index.shtml).

2. Destinations will benefit from future trends in the tourism supply chain. In the past, this was a combat zone with each member feeling that they had to compete. For destinations, this resulted in exploitation by tour operators who failed to recognize that the destination was, in fact, their product. In the future, tourism businesses will begin to recognize the importance of working with other members of the chain and this will include tour operators investing in the destination.
3. Networks or alliances of businesses and consumers along value chains will increase business efficiencies and improve communication. This trend is critical for the tourism sector and is leading to a shift in thinking away from management of individual sectors of the industry to the concept of integrated management.

Destinations, too, are embracing the 'smart' concept by using technology. With the emergence of the 'Internet of Things', there will be a growth of embedded technology in destinations allowing, for example, hand-held devices for guides and navigation (such as nodeexplorer.com). When combined with Big Data, mobile computing and the growing participatory media culture, this means that managers will be able to track a tourist's digital and contextual footprint in order to communicate market opportunities, facilitating dynamic destination management which can influence movement and so protect more fragile resources. It will also allow for individually curated visits to the destination.

Addressing climate change

As we have learned in this book, the tourism sector is a climate-sensitive activity and destinations in the future will recognize the sustainability challenge of adaptation to climate change. Climate is an important resource for destinations and a critical part of the destination's economic base. Changes in climatic elements will trigger human responses in terms of demand and will therefore pose a threat to a destination's competitiveness, sustainability and economic viability (Cavlek *et al.*, 2019). The vulnerability of a destination to climate change will depend on its exposure to climate change, the sensitivity of the tourism system and the adaptive capacity of the destination to cope with change. In the future we will see increasing attention given to destination adaptation and a mainstreaming of climate change into destination policies and strategies (Simpson *et al.*, 2008). Destination adaptation to climate change involves adjustments in practices, processes or structures to take account of changing climate conditions in order to moderate potential damages, or to benefit from the opportunities associated with climate change (McCarthy *et al.*, 2001).

Growing recognition of the seriousness of climate change has prompted tourism policy and management intervention. For ST futures, the solutions are challenging. One of the ways in which tourism contributes to climate change is through carbon emissions from tourism-related activity such as transportation and accommodation. While approaches such as green taxes on carbon emitters, or carbon trading schemes, will undoubtedly target tourism in the future, carbon offsetting has emerged as a popular but controversial way to neutralize the GHGs emitted by travel. For the future, solutions will coalesce around four areas:

- consumer education with a view to adapting behaviour;
- destination adaptation to climate change;
- reduction of carbon emissions and adoption of carbon trading schemes; and
- scientific advances in the use of alternative energy forms and the sequestering of carbon through natural 'sinks' such as forests.

All of these solutions combine to demand mainstreaming policies and technologies that change the way tourists and the industry operate, adopting a quadruple bottom line where carbon is added as the fourth element, and accounting for the carbon footprint of tourism using contemporary approaches such as the Tourism Satellite Account (TSA).

Sustainable destination futures

So what of the future for ST destinations? We can discern two clearly different and divergent trends:

- The first is the trend towards the use of artificially, technologically enhanced destinations such as theme parks, cruises and resorts. Examples here include Las Vegas (www.visitlasvegas.com/), the Disney theme parks and Carnival Cruise Line (www.carnival.com). The product is unashamedly artificial – creating a fantasy world that will be increasingly part of the 'experience' economy. Technology plays a major role here with theme parks re-engineering their parks to become computer interfaces to communicate with visitors' devices such as wristbands.
- The second trend is for authentic, well-managed contact with nature and Indigenous communities. Here eco-tourism and heritage tourism are the obvious examples, such as sympathetic encounters with wildlife (gorilla encounters in Rwanda) or native peoples (meeting with Maori in New Zealand). This type of destination demands a different type of management from the artificial fantasy destination as here it is the resource that is paramount in delivering the experience.

2. INNOVATION FOR SUSTAINABLE TOURISM FUTURES

ST futures will not be achieved without technological innovation to create a low-carbon, low-impact sector as reflected in SDG 9 on industry, innovation and infrastructure. As the Forum for the Future (2009) states, this will involve trialling new technologies, increasing energy efficiency and exploring the use of non-renewables. Here there has long been the opinion on the part of some researchers that all environmental problems are manageable, and technical innovations will provide the needed solutions to the many of these serious broad-scale problems (Smil, 1984). In tourism, however, Hjalager (1997) suggests that innovations are launched by the tourism industry primarily as defensive strategies. She identifies five types of innovation in ST:

- product innovations like the incorporation of different design standards and emission limits;
- classical process innovations that raise the performance of operations already functioning;
- process innovations in information handling such as the use of information technology for the monitoring of environmental issues;
- management innovations which include the training of staff to have better environmental practices; and
- institutional innovations, which might include, for example, the decision to externalize environmental burdens or risks to third-party organizations in efforts to avoid costs.

Innovation in ST comes as a result of creativity, employing a problem-solving approach and developing new ways of thinking (Moscardo, 2008). This is demonstrated by Hjalager's (2010) work on innovation systems in tourism based on angling tourism in Denmark. This illustrates the challenges in moving away from environmental degradation to a more sustainable future, which could only be based on the right institutional frameworks that enable several stakeholders to work together for human benefit (recreational enjoyment and commercial success) and the rehabilitation of ecosystems for the health of sea trout populations. Innovation was based on:

1. a systematic removal of watercourse obstructions, which hinder the free migration of trout to and from the hatching, breeding and feeding places;
2. breeding and release of large numbers of sea trout for the benefit of the environment as well as a resource for anglers; and
3. development of services related to recreational fishery and subsequent marketing of the products.

Proponents of sea trout angling needed to be successful in convincing politicians that the cost of making these quite significant changes could be offset by the development of a robust tourism industry around the resource, which included the cost of permits and the heavy regulation of anglers.

Bramwell and Lane (2012) argue that there are few efforts stemming from ST research that are truly innovative in the tourism industry and indeed that most innovations are 'incremental' rather than 'breakthrough' moments (Carlsen *et al.*, 2008). Most innovations come from outside the field and are applied in a tourism or sustainable tourism context (Fennell, 2018a). For example, it was the computer industry that devised the concept of e-intermediaries, a concept that has revolutionized how tourism is sold. Another example of the use of technology from outside the field is the creative use of databases and Big

Data to understand patterns from social media posts and to develop sophisticated customer relationship management.

While innovations in science and technology will play their part in delivering ST futures, we believe that it is the softer skills that will be needed in order to succeed. In particular, contemporary insights in knowledge management and adaptive organizational learning will be essential in realizing the benefits of generating and managing knowledge for innovation in ST (Cooper, 2006, 2015; ETFI, 2014). In this approach, we can think of the tourism system as a complex networked system of actors, organizations and destinations that have to cope with the new imperative of ST futures through adaptive learning. If the tourism sector is to successfully navigate these difficult waters it will require strong leadership and direction.

SCENARIOS OF SUSTAINABLE TOURISM FUTURES

As we saw at the beginning of this chapter, futurists say that the future cannot be predicted but alternative futures can, and that in order to be useful, futures studies should be linked to strategic planning and policy. Scenario planning is an approach that is being increasingly used in tourism to predict alternative futures and to assist agencies in planning and it is a useful tool to analyse ST futures. The approach was developed by the oil company Shell in the 1960s and has grown out of futures studies and strategic analysis. It is a tool that develops a range of pictures and stories of multiple, plausible futures, constructed using future-shaping drivers and trends. In tourism the approach has been pioneered by the 2025 Scenario Planning Group based at VisitScotland (VisitScotland, 2005).

Scenario planning uses a systems approach that recognizes the interconnectedness of tourism. For example, in Australia the Tourism Futures Simulator has been created which views tourism as a complex system (Walker *et al.*, 1998). By applying systems thinking to tourism, scenarios of the future can be built utilizing all the elements of the system and their relationships to see how destinations and their markets might evolve. The advantage of this approach is that it prevents agencies from focusing on a single issue such as technology; the individual issues can be located within the wider system and the policy response can be more effective and less 'reactive' to single issues.

Scenario planning is a process rather than an answer and has to be deeply rooted in research. The scenarios are not forecasts, nor are they one-off exercises, but they develop and evolve. Organizations can then gain a competitive edge by utilizing the scenarios to think and act in a particular way and to develop a strategic 'conversation' with stakeholders. This allows the organization to understand the future and to develop plans and policies accordingly in order to align themselves with the future scenarios. The next section outlines a number of examples where tourism has used scenario planning to show how ST futures may develop (Yeoman & McMahon-Beattie, 2005).

FORUM FOR THE FUTURE SCENARIOS TO 2023

Using expert opinion across the sector, the Forum for the Future (2009) has assembled four scenarios for tourism to 2023 which strongly feature ST and resource use. The scenarios are:

1. *Boom and burst*: Growth in the economy creates growth in tourism globally with travellers going further, more often and faster than before. Political stability, growth in alternative fuels and forms of propulsion have facilitated growth while the transport sector has kept pace with legislation to control carbon emissions. However, this has involved the transport sector effectively financing the decarbonization of other sectors of the economy. As a result, the economics of tourism are tight and there has been consolidation across the sector. The market is strong with the elderly travelling frequently, often for health tourism, while flexible, part-time and semi-retirement working has also fuelled demand. The resultant scenario is one of overcrowded destinations with wilderness being the scarcest resource as tourism has reached into the furthest corners of the world.
2. *Divided disquiet*: In this scenario, extreme climate change, clashes over resources and social unrest create an uncertain and fearful world. Travel is therefore unpopular and has caught many destinations unawares. Tourism is reduced to an 'adventurous' activity in some of the worst hit parts of the world such as the Middle East. Protectionism is strong and border crossings difficult with tight security disrupting travel. Safer destinations are popular and overcrowded and the practice of doomsday tourism – visiting threatened places – is on the rise. Other forms of 'virtual' travel are replacing tourism.
3. *Price and privilege*: Here oil process have soared and the travel and aviation sector have been hit hard. Cost becomes the key concern for the mass market, although a small elite still travels regularly. Affordable travel becomes a distant memory and surface transport becomes popular again.
4. *Carbon clampdown*: In this scenario, governments have developed tradable carbon allowances to help combat climate change. Allowances are seen as the fairest way of allocating the 'right to pollute' equally and are used in a discriminating way to encourage certain types of ST through carbon rebates. Public backing for regulation is high as there have been a number of natural disasters. Pensions are in crisis with companies not honouring their obligations, leading to a decline in travel among the older population. The experience becomes more important than the destination.

EUROPEAN TOURISM FUTURES INSTITUTE SCENARIOS TO 2040

The European Tourism Futures Institute (ETFI, 2014) has developed a set of four scenarios for tourism in 2040. These are based on future states where the commercial interests of the tourism sector are balanced and moderated by concerns for sustainability to create competitive advantage. For each scenario, the types of business involved and the implications of these were mapped together with strategic questions (Postma, 2014). The four scenarios are:

1. *Back to the seventies*: In this scenario, a rapidly growing European economy leads to an increase in energy demand leading to a revival of the growth paradigm at the cost of sustainability. There is support for non-renewable energy such as shale gas while others fear for the environment. The tourism sector favours the management of concentrated mass tourism at the expense of small-scale localized tourism.
2. *Captured in fear*: Here the European economy is in depression and the BRIC countries and Northern Africa are booming. Europe takes a protectionist position, slowing technological development and inhibiting innovation in alternative energy sources. This threatens environmental quality as well as social welfare and prosperity, while control of the media hampers knowledge exchange and impacts on creativity and innovation in tourism.
3. *Shoulders to the wheel*: This scenario sees a resurgent Europe with revival of international trade and clear benefits to southern Europe. With increased prosperity, tourism becomes inventive, working cooperatively towards a sustainable future and boosting technological innovation.
4. *Unique in the world*: Europe's economy is slowly rising and a protectionist policy benefits tourism. There is a strong belief in sustainability and sustainable resources to give competitive advantage, which in turn supports small-scale tourism and helps protect the environment.

NORTHERN IRELAND'S DRIVERS SCENARIOS TO 2030

The Northern Ireland Tourist Board (NITB, n.d.) has predicted what tourism will look like in 2030. From a number of drivers and results, two scenarios focus particularly on ST futures:

1. *Zero impact tourism*: Here the NITB uses the example of the United Arab Emirates project to create the world's first zero carbon city at Masdar, powered by solar, wind and other renewable energy sources (https://masdar.ae/). This is a utopian vision of a future destination but the investment required ($22 billion in the case of Masdar) makes it difficult to achieve.
2. *Changing trends, preferences and policies in sustainability*: Here the NITB identifies a growing demographic known as Lifestyles of Health and Sustainability (LOHAS). This group is driven by eco-conscious, well-educated consumers.

CONCLUSION

Sustainable tourism futures are exciting. This chapter has examined the key approaches to ST, which confirm the fact that tourism is a difficult sector to predict. Each of the drivers of the future that we identify in the chapter is influential in its own right and reflected in the UN SDGs, but when combined they deliver a set of powerful forces shaping the futures of ST. Futurists counsel against predictions of a single 'future' of tourism, but we can begin to see some of the possible ST scenarios. We discussed the type of tourists we may expect to find in the future and how they will differ from the tourists of today. At the same time, the destination will come under increasing pressure and will need careful and appropriate management, while transportation will be greener and more efficient. Of course, ST cannot control all the forces that impact on it and there is no doubt that change is happening more quickly and in unexpected ways. The future for ST is therefore about managing that change and building resilience into tourism systems (Wall, 2018). The message of this book has been that, whatever tourism futures bring, tourism will only be successful if we innovate and take a scientific and disciplined approach.

END OF CHAPTER DISCUSSION QUESTIONS

1. If the 'new tourist' were truly concerned about climate change they would not travel at all. Debate this contentious statement in class.
2. Critically review the key drivers of sustainable tourism futures and show how they are related.
3. How can destinations build in resilience to climate change?
4. Write a report on how you see artificial intelligence and robotics impacting on human resources in tourism in the future. How can companies prepare for this new scenario?
5. In a sustainable future, ethical consumption of tourism and CSR concerns will be increasingly important. Discuss this statement in the light of the need for companies to make a profit and return on investment.

End of Chapter Case Study 11.1 Virtual tourism

In the realm of science fiction we find virtual reality (VR) has long been talked about – a completely immersive digital world that is neurologically indistinguishable from the outside world. There is an ongoing debate as to whether VR may one day replace the authentic travel experience altogether; indeed, the very nature of tourism as an 'experience' lends itself perfectly to VR – virtual reality is simply a further step along the road of engineering tourist experiences and one that reinforces the trend of leisure activity based on the home with, say, gaming. It is, for example, already possible to take a virtual tour of Tutankhamun's tomb or Stonehenge – surely a key tool of sustainability as it takes away pressure from the resource itself. With VR the impacts are minimal: no environmental degradation, carbon emissions, cultural demonstration effect or risk of disease. However, opponents contend that VR will simply whet the appetite for more travel through enhanced exposure to, and awareness of, the product, as VR is used simply as an advanced form of tourist brochure.

There is no doubt that truly immersive VR is some years away, but the development of products such as Oculus Rift is bringing the reality closer. Oculus Rift delivers a 3D immersive experience for the user and such is the promise of the technology that it has been purchased by Facebook. VR is already being used in a variety of ways for tourism – planning and management, marketing, entertainment, education, accessibility and heritage preservation (Guttentag, 2010; Guttentag & Griffin, 2018). But of course the key question is just how easily will the market accept VR substitutes for the real thing – if, for example, VR is to be used to protect heritage by diverting visitation from authentic heritage?

Discussion questions

1. Expand on the above list of the 'pros' and 'cons' of virtual tourism.

2. Debate in class: Virtual reality is the most sustainable form of tourism – visitation without impact?

3. Do you think that alternative realities will ever replace authentic travel?

Sources: Guttentag (2010); Guttentag and Griffin (2018)

CHAPTER 12
CONCLUSION

Because of your country, my country's reef is dying

INTRODUCTION

We have seen in this book that 'tourism is and can be a tool for sustainable development' (Font *et al.*, 2019: 6) and there is a very broad base of engagement of tourism with sustainability challenges. We have contextualized these challenges within the framework of the UN's Sustainable Development Goals (SDGs), partly because these are all-encompassing in terms of sustainability challenges, but also because it allows us to locate tourism within a broader global research agenda. Yet, in trying to find silver bullet solutions to the world through sustainability it is wise to heed the words of Zolli (2012) who argues that politics and marketing have not aided the sustainability agenda because there is the view that the world will likely return to some state of equilibrium. This will never happen, Zolli adds, because, by nature, the world is constantly in a state of flux or disequilibrium through cycles of trying, failing, adapting, learning and evolution. The failures are the opportunities for growth and learning, if these perturbations are correctly understood. Here, a theme of this book is that of resilience, transitions and transformation, at both organizational and destination levels, creating the capacity to deal with change.

Who will lead the necessary changes required for a sustainable tourism (ST) now and in the future is a question of some weight. Skirting the issue is futile, as we continue along a path of continued population growth, poverty extremes, resource over-utilization, climate change, inequality and so on, all fuelled by dire need, as well as individual and organizational self-interest. These firmly entrenched issues have prompted scholars to refine definitions of sustainable development (SD) with more of a focus on safeguarding the Earth's life-support systems. The definition of SD that we adopt in Chapter 1 by Griggs *et al.* (2013) is a reflection of the need to place sustainability alongside, or even over, the needs of development, as difficult as this may be in theory and especially in practice.

In this final chapter we conclude with what we view as central themes around ST research, as well as ST from an applied standpoint based on recent articles on state-of-the-art practices that were published around the time of the book's submission. The chapter then moves towards a brief discussion on sustainability ethics in keeping with our conceptual framework illustrated in Chapter 1. We believe that if we are to achieve a move towards securing the Earth's life support systems, much would need to be done as a moral imperative. Given the breadth of challenges that we are up against globally, there is little evidence that we can achieve SD or ST without a massive change in the way we think and act both individually and collectively.

SUSTAINABLE TOURISM RESEARCH

We have drawn on a wide range of sources in each of the book's chapters. Of course, this not only reflects the broad nature of the subject area, but also the necessity for ST researchers to look beyond the tourism literature and reach out to other disciplines, sciences and themes. Buckley (2012), for example, has examined the relevance of ST researchers for different communities of practice (Table 12.1). In this respect we are seeing a transformation in the nature of ST research in terms of both coverage and impact, as reflected in the 2019 editorial of the *Journal of Sustainable Tourism* (JOST) (Font *et al.*, 2019). The editorial identifies a range of future directions for research into the subject, including examining:

- individual, organizational and societal change in sustainability;
- how behaviour change can be created and measured;
- the role of entrepreneurs and innovation for sustainability; and
- empowerment of stakeholders to take charge of their destinations.

Table 12.1 Sustainability significance, industry influence and research effort

	Parks, biodiversity, conservation	***Pollution, climate change***	***Prosperity, poverty alleviation***	***Peace, security, safety***	***Population, stabilization and reduction***
Significance for sustainability	*****	*****	*****	*****	*****
Influence of tourism sector	*****	****	***	**	*
Attention by tourism industry	**	***	*****	**	–
Effort by tourism researchers	*	**	*****	*	*
Effort by science researchers	***	*****	*	*	–

Key: Number of stars indicates the scale or importance of factors in each row, for the components in each column: ***** most; * least; – none or negligible.
Source: Buckley (2012)

We would add to this the following research themes:

- A focus on the development of more refined tools in order to enable communities to monitor the effects of the implementation for ST – more monitoring tools such as many of the indicators of ST mentioned earlier in the book are required.
- Similarly, one area of study that requires greater attention is the evaluation of the actual implementation of the community-based tourism development processes. Longitudinal research could provide information on actual successes, changes in power structures, and the influence of community members on the decision-making processes (Joppe, 1996). Murphy (1992) argues that a major void exists for individual destination communities in attaining locally relevant information that increases their understanding of the local tourism product.
- Investigating all of the SDGs in the context of tourism, as well as how tourism interfaces with other industries in achieving sustainability within regions.
- Pushing for a greater degree of interdisciplinarity or transdisciplinarity in tourism research on sustainability.
- Finally, more research is needed on human resources and ST and the fact that many jobs in tourism cannot be classified as 'decent work'. In terms of ST, the transition towards 'green jobs' and the 'sustainable professional' in tourism is an important one.

In their JOST editorial, Font *et al.* (2019) also echo a major theme of this book in calling for greater collaboration and co-creation of research between academics, industry and government. This book recognizes the wide range of organizations involved in ST, from government, through the private sector, to charities, lobby groups and individuals. This reflects SDG 16, which is about building accountable and inclusive institutions, and SDG 17 on the need for partnerships. What is clear is that the private sector now plays a greater and more influential role than in the past and that here ST is a source of innovation. In Chapters 6 and 7 we saw the degree of innovation taking place across the tourism industry in terms of reducing carbon emissions, waste, water and energy, but this has to be tempered by the fact that transport remains a high-carbon sector, outside the Paris Climate Change Agreement, and one that therefore taints tourism as a significant environmental polluter. Here, Macbeth (1994) argues that, in terms of the private sector, the move towards sustainability is more reactionary than progressive. The problem with this is that the history of capitalism is marked by numerous examples of how reactionary tendencies have been co-opted by the capitalist system to reinforce and justify its own existence. This smacks of enlightened self-interest, according to Macbeth, especially on the part of big business, which continues to be highly exploitative – a 'ruling capitalistic hegemony that is antithetical to sustainability' (Macbeth, 1994: 44). SD will not come to fruition until or unless this capitalistic system is radically altered.

This implies that countries such as the UK, which have committed themselves to environmental stewardship through various means such as greenhouse gas (GHG) emissions being subject to legally binding reduction strategies (Ball & Bebbington, 2008), should start to be very choosy about who they do business with. This means that there ought to be normative structures put into place which are designed to apply pressure on those trading partners that are not doing their fair share when it comes to environmental (and social) improvement. Otherwise greener countries become guilty by association with those other regions that are not playing by the 'rules', informal as these may be. This should not be restricted to the public sphere alone, as leadership can come from the corporate world too.

This raises the issue of public sector policy for tourism and how the sector ought to be regulated. Despite the move to integrated 'whole of destination' policy and governance, some commentators still doubt the effectiveness of ST policies given the continued negative impact of tourism, particularly the transport sector. Here, attempts to address climate change are central to ST. Tourism can be considered as both a victim and a vector of climate change. Climate change has the potential to change the nature and distribution of tourism significantly. As a result, there is an imperative for tourism to transition to become a low-carbon sector and we have reviewed the key initiatives in this direction taken by both government and the private sector.

SUSTAINABLE TOURISM PRACTICE

While researchers have been calling attention to the impacts of tourism for many years, there have been signs that industry is now on board with their practices. Trade publications often extol the virtues of companies or case studies where sustainability is the overriding criterion as to how best to do business. We wish to briefly share a few of these success stories that were in the news during the final stages of the writing of this book. These few stories that we highlight have become newsworthy because of their recognition around international standards and awards. As such, what passes as ethical in the new age of tourism is very much tied to recognition via these sorts of awards.

1. THE HALIBURTON FOREST & WILD LIFE RESERVE, SOUTH-CENTRAL ONTARIO, CANADA

The Haliburton Forest & Wild Life Reserve (https://www.haliburtonforest.com/) has a 100-year business plan to return its 100,000-acre forest to a healthy, viable ecosystem. They do this through a variety of ecotourism and nature-based tourism programmes such as a walk in the clouds forest canopy, snowmobiling, hiking, observatory, wolf centre, sled-dog tours and cross-country skiing. They also build log homes using a stand of hemlock blown down by a hurricane several years ago, and they sustainably harvest genetically poor or redundant trees by removing them using traditional methods (powered by horses and people). The Haliburton Forest & Wild Life Reserve is Canada's first sustainable forest, as recognized by the Forest Stewardship Council (2019), which is an international non-profit organization which, through standards, promotes the responsible use and management of the Earth's forests.

2. MACHU PICCHU, PERU

Travel Wire News (2019) reported that Machu Picchu was recognized as Latin America's first 100% sustainable city through the management of its waste. The city uses a process referred to as pyrolysis where waste is decomposed at high temperatures in the absence of oxygen. The waste that is generated from this process is a type of bio-coal, which may be used as fertilizer to enhance agricultural productivity. Other tools used by the city include a plastic compactor to recycle the vast amount of trash generated by locals and tourists, which processes upwards of 14 tons of plastic per day. Travel Wire News also reports that in 2018 a biodiesel and glycerine plant was implemented in the Machu Picchu Pueblo Hotel, which produces biodiesel from used vegetable oil from homes, lodges, hotels and restaurants. These initiatives were enough to award Machu Picchu the German Die Goldene Palme award in the category of responsible tourism.

3. COPENHAGEN, DENMARK

This European Green Capital City in 2014 has taken dead aim at being the world's first carbon-neutral city by 2025. The VisitCopenhagen (2019) website states that:

> Danes are often said to be the happiest people in the world, and Copenhagen is among the world's most liveable cities. Some of the reasons are the large number of green oases and open spaces with fresh air, some of the cleanest water in the world – for drinking and swimming – as well as a city overflowing with bicycles, and a high availability and consumption of organic produce. Denmark is also one of the safest countries in the world, and has an excellent and efficient infrastructure. When you are in Copenhagen, there are plenty of ways to be an ecotourist. Finding a sustainable place to stay in Copenhagen is actually easier than finding a non-sustainable one, as over 70% of all the city's hotel rooms hold an official eco-certification. More than half of the hotels in Copenhagen have an environmental plan regarding water, laundry, house cleaning, waste, energy consumption, food, smoking, indoor climate, and administration. While biking is a way of life in Copenhagen (The International Cycling Union, UCI, appointed Copenhagen the first official Bike City in the world from 2008–2011), eating organic is just as natural. The New Nordic kitchen, which has become so popular in recent years, is also good news for the eco conscious. With its focus on seasonal ingredients and a largely organic use of produce you will find many climate-friendly and organic restaurants in all price ranges in Copenhagen. Furthermore, in Copenhagen, buying organic produce is not considered a luxury, but merely logical. Thus organic food makes up 24% of the total food sale in Copenhagen, which is the highest in Denmark. Even better, 88% of the food consumption in the City of Copenhagen's public institutions, such as day-care centres, nursing homes and schools, is organic. (VisitCopenhagen, 2019)

4. BIOSPHERE TOURISM AND SKÅL INTERNATIONAL

Skål International is a professional organization with the goal of uniting all branches of the travel and tourism industry in the promotion of global tourism and friendship (Skål International, 2019), and launched an ST award in 2002. Skål has recently teamed up with Biosphere Tourism in the development of a new award called the 'Special Skål Biosphere Award'. Biosphere Tourism is an organization that certifies industry players who are able to balance sociocultural, economic and ecological factors within a tourism destination – following the principles of the Biosphere Reserve programme introduced in Chapter 9. The benefits of Biosphere Tourism certification include increased efficiency around minimizing ecological impacts and improving profitability, improved position within the industry through competitiveness, and help with ST strategies around the 17 SDGs (Biosphere Tourism, 2019). Biosphere Tourism certifications are awarded in areas such as: destinations and tourist routes; accommodation; tourist sites and centres; parks; and other certifications in areas such as tour operation, mobility, shopping and events (Travel Wire News, 2019).

5. THE TREADRIGHT FOUNDATION

In 2019 the TreadRight Foundation (www.treadright.org) was recognized by the UNWTO's annual awards for its pioneering work in sustainability. It is a not-for-profit organization created as a joint initiative across The Travel Corporation's (TTC) family of brands, which include Contiki, Trafalgar and Insight Vacations. The Foundation's mission is to ensure that travel has a positive impact on people and communities, to protect wildlife and marine life and to care for the planet. By 2019 the Foundation had supported over 50 ST projects worldwide, structured into three themes – people, wildlife and planet (TreadRight Foundation, 2019).

CONTRADICTIONS

There is an interesting tension that exists around the motivations for behaving sustainably that deserves more critical investigation. On the one hand, organizations may choose to be sustainable as a good unto itself (i.e. we choose to be sustainable on the basis of intrinsic motivation). For example, the natural world does not simply have instrumental value, but rather has a good of its own which must be recognized through polices and procedures. By contrast, there is a worry that organizations might only engage in sustainable practices for competitive advantage or if these practices amount to win-win situations. Companies will argue that they are sustainable if they encourage patrons to hang up towels to use the next day, or if they install spring taps because these are good tools by which to 'save the environment'. It is hard to get past the reality that these measures are simply saving the organization money by not having to employ more people and use more resources (water, electricity, detergent) in order to wash more towels. If we choose to employ measures on the basis of extrinsic motivation (because we will save money, or because of the competitive advantage that we generate from our actions, or because of the awards bestowed on us), there is something lost in the act by operating on the basis of bad faith. We argue through the myriad examples situated throughout this book that we have a duty to make a 'just' effort to perform sustainable actions based on honesty, integrity, transparency and many other virtues and characteristics (see Fennell, 2019b) because there are bigger issues at hand. As such, award programmes such as the ones mentioned above, although important, cannot be the gold standard of what it means to be sustainable in the new age of ST (see Hunt & Durham, 2012).

On 3 September 2019, around the time that this book was going off to press, the British Prince Harry announced that his new charitable foundation, Sussex Royal, was partnering with several large travel firms, including Booking.com, Ctrip, Skyscanner, TripAdvisor and Visa. The purpose of the partnership, called Travalyst, is to raise awareness about the importance of sustainable and responsible travel. Prince Harry explained that one of the catalysts behind the decision to tackle ST stemmed from his brief conversation with a seven-year-old boy from the Caribbean, who made the statement quoted at the outset of this chapter. But while the Prince was extolling the virtues of being more sustainable, others were quick to accuse him of hypocrisy on the basis of the fact that he continues to fly around the world, frequently, on private jets (The Guardian, 2019).

This is but one example of the contradictions that continue to plague tourism industry practices. We think that we should curtail our travel in the interests of creating fewer impacts, but travel is just too savoury. We tell ourselves that we have earned time away, or that the meeting simply will grind to a halt if 'I'm not there'. In the end, it is our human nature, our self-interest, that bubbles to the surface and drives us to do things that we know we should not do. This weakness of will, termed *akrasia* (Fennell, 2015), offers an explanation as to why we often act irrationally in tourism. In countless other cases, however, people simply do not know. It becomes the role of education, therefore, to get the word out regarding what is acceptable and what is not when it comes to the travel and tourism industry. Perhaps in the future, those who travel frequently in the absence of a sustainability ethic will endure social costs? At present, travel and tourism is something that is celebrated and envied, not something to be criticized. As tourism continues to grow at phenomenal rates (5–6% annually), the job of becoming sustainable becomes that much more important – and ominous.

ETHICS

We end this chapter with an extension of the previous section, which also happens to be a core tenet of this book, and indeed of sustainability itself: ethics. At various places throughout the text, ethics has been an important topic of discussion in reference to our views – and the views of many others – on sustainability: in Chapter 1, ethics and human nature are fundamental components of the conceptual framework for this text; ethical trading in shops and restaurants was discussed in Chapter 6, as well as ethical food systems; the UNWTO Global Code of Ethics for Tourism was a key topic in Chapter 8; ethics and good governance in Chapter 9; moral issues tied to the use of animals in Chapter 10; and the importance of education for ethical consumption in Chapter 11.

Yet, as Fennell (2019a) has noted, the focus on ethics in sustainability is surprisingly fleeting given the importance of attending to sociocultural, ecological and economic elements and the challenges inherent in each. This is especially the case in tourism, where there are few efforts to connect moral theory and sustainability in a direct manner. One of the first papers to underscore the importance of ethics in sustainability comes from Hughes (1995), who argued that there are two broad domains in which ST has been considered, bridging from the environmental movement discourse. The first of these, and the most prominent, is based on environmental concern, biodiversity loss and systems theory. The second is ethical in nature and is based on the moral drive that individuals and organizations ought to have in operationalizing sustainability. We could not agree more.

The evolution of Hughes' (1995) manner of thinking has been advanced on the back of several decades of environmental discourse, which have been vital in laying the groundwork for change in the way in which humans interact with the environment. In the latter part of the 1990s, however, disciplines started to build a more cohesive body of knowledge around the direct synthesis of ethics and sustainability: the field or study of *sustainability ethics* merged, first stemming from engineering (Duffell, 1998), then biology (Cairns, 2003) as well as ecology (Minteer & Collins, 2005). The most comprehensive statement on the topic of sustainability ethics, however, comes from Becker (2012; as reported in Fennell, 2019a, in reference to tourism). In building his thesis around sustainability ethics, Becker defines it as 'the ability to establish continuance as a means of orienting human actions and life toward the threefold relatedness of human existence to contemporaries, future generations, and nature' (Becker, 2012: 14).

'Continuance' makes reference to the importance of time, as in systems (ecosystems), entities (e.g. a historic building) or processes (e.g. protected areas). 'Orientation' refers to the notion that sustainability has normative and evaluative meaning, and that we should work to achieve it as a major long-term goal. Sustainability is also about the relationships that exist between humans and ecology. The ethical dimension of sustainability, Becker (2012) adds, must be built around three sets of relationships, including people of the present day, future generations and nature. The best way to improve these relationships, Becker observes, is by moving away from a reliance on normative ethics (e.g. utilitarianism) to solve our problems, towards a focus on virtue ethics and ethics of care. Virtue ethics focuses on the type of person we ought to be in living the good life (Foot, 2003; see also Jamal, 2004, in a tourism context). Care ethics places better balance on the often-unbalanced relationships between men, women and the natural world based on power, which have been stabilized over time (Held, 2004; see also Stark, 2002, in a tourism context).

What is important in Becker's broader message on sustainability ethics is the need to change 'meta-structures', which are the dominant social and global systems such as technology, science and economy that dictate how people relate to each other and the natural world at all scales. For example, neoliberalism, the principle of non-satiation, selfishness and pareto efficiency (no-one can be made better off without making other worse off) are entrenched principles and practices that have created great disparities in the world. In reference to changing meta-structures, Becker argues that:

1. we need to move beyond reliance on narrow bands of knowledge based on economic and scientific thought in achieving a more holistic ethos; this demands the development of people who are rational, creative, communicative, emotional and relational instead of selfish utility maximizers;
2. a new paradigm of sustainability is needed based on evolving social contracts, which provides the foundation for an ethical orientation in all that we do, including science, technology, education, the economy and so on; and
3. a move away from the overriding influence and complexity of the dominant meta-structures. Although we will always need to develop, Becker (2012) contends that such development should be built around stability and simplicity. The current paradigm centred on automatic, omnidirectional and uncontrolled growth must be replaced with an explicit, deliberate and orientated dynamic (Becker, 2012: 102).

Although it is beyond the scope of this chapter to fully engage with Becker's treatise on sustainability ethics, it must be viewed as a step in the right direction in the need to derail the contemporary orthodox manner of placing individual and organizational interests above all else. It will be a tall order, Becker claims, to design a good life for *all* of the people of the planet, now and in the future, while securing the Earth's functions and services, given the tremendous power imbalances that exist between different cultures and industries.

Here, at the end, we argue that there is no doubt that sustainability in tourism is now an expectation and not an ambition. Tourism is inextricably linked to social, economic, political and technological innovations and, as we move towards a sustainability transformation in tourism, we have to recognize these interconnections (Budeanu *et al.*, 2016). As a result, tourism is a difficult sector to predict, influenced by a range of external influences and drivers that will combine to create sustainable futures. For example, there is no doubt that the sustainable tourist of the future will be more knowledgeable and aware of sustainability issues; at the time of writing, for example, the climate change protests in London combined with the strikes by schoolchildren in protest at climate inaction have been influential in changing public opinion. Technology, too, through the 'Internet of Things' and 'alternate realities' will be influential in creating the sustainable tourism of the future. Here the future for sustainable tourism will therefore be about managing these changes and building resilience into tourism systems, particularly destinations. The message of this book has been that, whatever tourism futures bring, tourism will only be successful if we innovate and take a scientific and disciplined approach.

REFERENCES

Aas, C., Ladkin, A. and Fletcher, J. (2005) Stakeholder collaboration and heritage management. *Annals of Tourism Research* 32 (1), 28–48.

Abbasi, D.R. (2006) *Americans and Climate Change: Closing the Gap between Science and Action.* New Haven, CT: Yale School of Forestry and Environmental Studies.

Abram, M. and Jarząbek, J. (2016) Corporate social responsibility in hotel industry – Environmental implications. *Ecocycles* 2 (1), 9–16.

Abson, D.J, Fischer, J., Leventon, J., Newig, J., Schomerus, T., Vilsmaier, U., von Wehrden, H., Abernethy, P., Ives, C.D., Jager, N.W., *et al.* (2017) Leverage points for sustainability transformation. *Ambio* 46, 30–39.

ABTA (2013) *Global Welfare Guidance for Animals in Tourism*, See https://www.abta.com/sites/default/files/media/document/uploads/ABTA%20Global%20Welfare%20Guidance%20for%20Animals%20in%20Tourism.pdf.

ABTA (2015) *Animals in Tourism*, See https://abta.com/working-with-the-industry/animal-welfare.

Acapulco.com (2017) Website, www.acapulco.com/en/.

Accor (2011) *Sustainable Hospitality: Ready to Check in.* Paris: Accor.

Accor (2013a) *2013 Business Review*, See http://www.accorhotels-group.com/fileadmin/user_upload/Contenus_Accor/Finance/pdf/2014/UK/2013_business_review_accor.pdf.

Accor (2013b) *The Planet 21 Programme.* Paris: Accor.

Accor (2015) *What Are the Links between Business and CSR Performance for AccorHotels's Strategic and Key B2B Accounts.* Paris: Planet 21 Research.

Accor (2016a) *Our Socioeconomic Footprint (2016). Paris: Accor*, https://group. accor.com/en/commitment/sharing-our-knowledge/our-footprint.

Accor (2016b) *Accorhotels' Socio-economic Footprint.* Paris: Accor.

Accor (2018) *Sustainable Development*, See http://www.accorhotels.com/gb/sustainable-development/index.shtml.

Adams, C.J. (1993) Introduction. In C.J. Adams (ed.) *Ecofeminism and the Sacred* (pp. 1–9). New York: Continuum.

Adetiba, A.S. (2017) Higher education and sustainable development in Africa. In B.U. Anaemene and O.J. Bolarinwa (eds) *Agenda 2030 and Africa's Development in the 21st Century* (pp. 235–244). Kuala Lumpur: UN University International Institute for Global Health.

Adler, J. (2002) Akratic believing?. *Philosophical Studies* 110, 1–27.

Adventure World (2018) *Responsible Tourism*, See https://www.adventureworld.com.au/responsibility/.

Agarwal, A. (2005) Environmentality: Community, intimate government, and the making of environmental subjects in Kumaon, India. *Current Anthropology* 46 (2), 161–190.

Agarwal, S. (2012) Relational spatiality and destination restructuring. *Annals of Tourism Research* 39 (1), 134–154.

Ahmed, F., Draz, M.U., Su, L., Ozturk, I. and Rauf, A. (2018) Tourism and environmental pollution: Evidence for the one belt one road provinces of Western China. *Sustainability* 10, 3520.

Aiga, H. (2015) Hunger measurement complexity: Is the global hunger index reliable?. *Public Health* 129 (9), 1288–1290.

Air Line Pilots Association International (2017) *Code of Ethics*, See http://www.alpa.org/en/about-alpa/what-we-do/code-of-ethics

Ajaps, S. and McLellan, R. (2015) "We don't know enough": Environmental education and pro-environmental behaviour perceptions. *Cogent Education* 2 (1), 1124490.

Ajuntament de Barcelona (2010) *City of Barcelona Strategic Tourism Plan: Diagnosis and Strategic Proposal. Executive Summary*, See http://ajuntament.barcelona.cat/turisme/en/documents.

Ajuntament de Barcelona (2015) *Barcelona, City and Tourism: Basics for a Local Agreement*, See http://ajuntament.barcelona.cat/turisme/en/documents.

Ajuntament de Barcelona (2016) *Barcelona Strategic Tourism Plan for 2020: Strategic Diagnosis*, See http://ajuntament.barcelona.cat/turisme/en/documents.

Ali, V., Cullen, R. and Toland, J. (2015) ICTs and tourism in small island developing states: The case of the Maldives. *Journal of Global Information Technology Management* 18, 250–270.

Allen, C.M. and Edwards, S.R. (1995) The sustainable-use debate: Observations from IUCN. *Oryx* 29 (2), 92–98.

Allen, T. and Prosperi, P. (2016) Modeling sustainable food systems. *Environmental Management* 57, 956–975.

Alonso-Almeida, M., Llach, J. and Marimon, F. (2014) A closer look at the "Global Reporting Initiative" sustainability reporting as a tool to implement environmental and social policies: A worldwide sector analysis. *Corporate Social Responsibility and Environmental Management* 21, 318–335.

American Society of Travel Agents (2013) *Code of Ethics*, See http://www.asta.org/about/content.cfm?ItemNumber=745

Andersson, T.D. and Lundberg, E. (2013) Commensuratability and sustainability: Triple impact assessment of a tourism event. *Tourism Management* 37, 99–109.

AnimalsAsia (2019) *Vision and Values*, See https://www.animalsasia.org/uk/about-us/vision-and-values.html.

Anker, R., Chernyshev, I., Egger, P., Mehran, F. and Ritter, J.A. (2003) Measuring decent work with statistical indicators. *International Labour Review* 142 (2), 147–177.

Anywhere Inc (2018) *Costa Rica – Certification of Sustainable Tourism (CST)*. See www.anywhere.com/costa-rica/sustainable/cst-sustainable-tourism.

APEC (2002) *Public/Private Partnerships for Sustainable Tourism: Delivering a Sustainability Strategy for Tourism Destinations*. Singapore: Asia-Pacific Economic Cooperation.

Aquino, R.S., Lück, M. and Schänzel, H.A. (2018) A conceptual framework of tourism and social entrepreneurship for sustainable community development. *Journal of Hospitality and Tourism Management* 37, 23–32.

Archer, B. (1996) Sustainable tourism – Do economists really care?. *Progress in Tourism and Hospitality Research* 2 (3–4), 217–222.

Arden, Rt. Hon. Lady Justice (2016) Water for all? Developing a human right to water in National and International Law. *International and Comparative Law Quarterly* 65, 771–789.

ARF (African Roots Foundation) (2019) Website, http://africanrootsfoundation.org.

Ariño, O.B. (2002) Sustainable tourism and Taxes: An insight into the Balearic Eco-tax Law. *European Environmental Law Review* June, 169–174.

Armitage, D., Plummer, R., Berkes, F., Arthur, R., Davidson-Hunt, I., Diduck, A., Doubleday, N.C., Johnson, D.S., Marschke, M., McConney, P., Pinkerton, E.W. and Wollenberg, E.K. (2008) Adaptive co-management for social–ecological complexity. *Frontiers in Ecology* 7 (2), 95–102.

Armstrong, E.K. and Kern, C.L. (2011) Demarketing manages visitor demand in the Blue Mountains National Park. *Journal of Ecotourism* 10 (1), 21–37.

Arnstein, S.R. (1969) A ladder of citizen participation. *Journal of the American Institute of Planners* 35 (4), 216–224.

Arp, A.-K., Kiehne, J., Dünnweber, M., Marquardt, K. and Murswieck, R.G.D. (2018) Waste management in aviation – Recycling is not enough. *Ecoforum* 7 (2), 1–5.

ASQ (2018) *What is Auditing?*, See http://asq.org/learn-about-quality/auditing/.

ATIA (1990) *Environmental Guidelines for Tourist Developments*. Canberra: Australian Tourism Industry Association.

Atilgan, B. and Azapagic, A. (2015) Life cycle environmental impacts of electricity from fossil fuels in Turkey. *Journal of Cleaner Production* 106, 555–564.

Australian Academy of Science (2018) *What is Climate Change?*, See https://www.science.org.au/learning/general-audience/science-booklets-0/science-climate-change/1-what-climate-change.

AWC (Animal Welfare Committee) (2019) *Five Freedoms and a Life Worth Living*, See https://www.gov.uk/government/groups/farm-animal-welfare-committee-fawc#assessment-of-farm-animal-welfare---five-freedoms-and-a-life-worth-living.

Axelrod, R. (1984) *The Evolution of Cooperation.* New York: Basic Books.

Badia, F. and Donato, F. (2013) Performance measurement at world heritage sites: Per Aspera ad Astra. *International Journal of Arts Management* 16 (1), 20–34.

Baggio, R. and Cooper, C. (2010) Knowledge transfer in a tourism destination: The effects of network structure. *Service Industries Journal* 30 (10), 1–15.

Baker, C. (2009) Marriott thinks green for its supply Chain. *Journal of Hospitality Marketing & Management* 1, 8.

Baker, J.E. (1997) Trophy hunting as a sustainable use of wildlife resources in Southern and Eastern Africa. *Journal of Sustainable Tourism* 5 (4), 306–321.

Bakker, K. (2012) Water security: Research challenges and opportunities. *Science* 337, 914–915.

Ball, A. (2005) *Advancing Sustainability Reporting for Public Service Organizations.* London: CIPFA.

Ball, A. and Bebbington, J. (2008) Accounting and reporting for sustainable development in Public Service Organizations. *Public Money & Management* 28 (6), 323–326.

Ballantyne, R. and Packer, J. (2011) Using tourism free-choice learning experiences to promote environmentally sustainable behaviour: The role of post-visit "Action Resources". *Environmental Education Research* 17 (2), 201–215.

Ballantyne, R., Packer, J. and Falk, J. (2011) 'Visitors' learning for environmental sustainability: Testing short- and long-term impacts of wildlife tourism experiences using structural equation modelling. *Tourism Management* 32 (6), 1243–1252.

Ballouard, J.M., Brischoux, F. and Bonnet, X. (2011) Children prioritize virtual exotic biodiversity over local biodiversity. *PLoS ONE* 6, e23152.

Balmford, A., Green, J.M.H., Anderson, M., Beresford, J., Huang, C., Naidoo, R., Walpole, M. and Manica, A. (2015) Walk on the wild side: Estimating the global magnitude of visits to protected areas. *PLoS Biology* 13 (2), 1–6.

Băltescu, C.A. (2017) Green marketing strategies with romanian tourism enterprises. *Annals of the Constantin Brâncuşi* 4, 83–89.

Barber, C.P., Cochrane, M.A., Souza, C.M. and Laurance, W.F. (2014) Roads, deforestation, and the mitigating effect of protected areas in the Amazon. *Biological Conservation* 177, 203–209.

Barbier, E.B. (1987) The concept of sustainable economic development. *Environmental Conservation* 14 (2), 101–110.

Barbieri, C. (2013) Assessing the sustainability of agritourism in the US: A comparison between agritourism and other farm entrepreneurial ventures. *Journal of Sustainable Tourism* 21 (2), 252–270.

Barbiroli, G. (2009) *Principles of Sustainable Development, Volume 1.* Paris: EOLSS Publications.

Baroni, L., Cenci, L., Tettamanti, M. and Berati, M. (2006) Evaluating the environmental impact of various dietary patterns combined with different food production systems. *European Journal of Clinical Nutrition* 61, 279–286.

Barton, J. and Pretty, J. (2010) What is the best dose of nature and green exercise for improving health? A multi-study analysis. *Environmental Science & Technology* 44, 3947–3955.

Bauer, I. (2017) Improving global health – Is tourism's role in poverty elimination perpetuating poverty, powerlessness and 'Ill-being'?. *Global Public Health* 12 (1), 45–64.

Baum, T. (2007) Human resources in tourism: Still waiting for Change?. *Tourism Management* 28 (6), 1383–1399.

Baum, T., Kralj, A., Robinson, R.N.S. and Solnet, D.J. (2016) Tourism workforce research: A review, Taxonomy and agenda. *Annals of Tourism Research* 60, 1–22.

Bava, L., Zucali, M., Sandrucci, A. and Tamburini, A. (2017) Environmental impact of the typical heavy pig production in Italy. *Journal of Cleaner Production* 140, 685–691.

BC Ferries (2019) *SeaForward*, See https://www.bcferries.com/about/seaforward/.

BC Ministry of Development, Industry and Trade (1991) *Developing Code of Ethics: British Columbia's Tourism Industry*. Victoria, BC: Ministry of Development, Trade and Tourism.

B Corp (2018) *The B Corp Declaration of Interdependence*, See https://bcorporation.net/about-b-corps.

Beatley, T. (2011) *Biophilic Cities: Integrating Nature Into Urban Design and Planning*. Washington, DC: Island Press.

Beaumont, N. (2001) Ecotourism and the conservation ethic: Recruiting the uninitiated or preaching to the converted?. *Journal of Sustainable Tourism* 9 (4), 317–338.

Bebbington, J., Higgins, C. and Frame, B. (2009) Initiating sustainable development reporting: Evidence from New Zealand. *Accounting, Auditing & Accountability Journal* 22 (4), 588–625.

Becken, S. (2001) Tourism and Transport in New Zealand: Implications for Energy Use. Tourism Recreation Research and Education Centre (TRREC) Report No. 54, Lincoln University, New Zealand.

Becken, S. (2002) Analysing international tourist flows to estimate energy use associated with air travel. *Journal of Sustainable Tourism* 10 (2), 114–131.

Becken, S. (2014) Water equity – Contrasting tourism water use with that of the local community. *Water Resources and Industry* 7–8, 9–22.

Becken, S. (2015) *Tourism and Oil*. Clevedon: Channel View.

Becken, S. and McLennan, C. (2017) Evidence of the water–energy nexus in tourist accommodation. *Journal of Cleaner Production* 144, 415–425.

Becken, S. and Patterson, M. (2006) Measuring national carbon dioxide emissions from tourism as a key step towards achieving sustainable tourism. *Journal of Sustainable Tourism* 14 (4), 323–338.

Becken, S., Simmons, D.G. and Frampton, C. (2003) Energy associated with different travel choices. *Tourism Management* 24, 267–277.

Becker, C.U. (2012) *Sustainability Ethics and Sustainability Research.* New York: Springer.

Beckford, F.B. (2019) *Poverty and Climate Change: Restoring a Global Biogeochemical Equilibrium.* New York: Routledge.

Beesley, L. (2005) The management of emotion in collaborative tourism research settings. *Tourism Management* 26, 261–275.

Belboom, S., Renzoni, R., Deleu, X., Digneffe, J.-M. and Leonard, A. (2011) Electrical Waste Management Effects on Environment using Life Cycle Assessment Methodology: The Fridge Case Study'. In SETAC Europe 17th LCA Case Study Symposium: Sustainable Lifestyles. Budapest, Hungary, Brussels: SETAC, 2.

Belinskij, A. and Kotzé, L.J. (2016) Obligations arising from the right to Water in Finland and South Africa. *Aquatic Procedia* 6, 30–38.

Bell, V.R. (2013) The politics of managing a world heritage site: The complex case of Hadrian's wall. *Leisure Studies* 32 (2), 115–132.

Benidorm.com (2014) Website, www.benidorm.com/.

Benton, T. and Redfearn, S. (1996) The politics of animal rights – Where is the left?. *New Left Review* 215, 43–58.

Bentz, J., Lopes, F., Calado, H. and Dearden, P.P. (2016) Managing marine wildlife tourism activities: Analysis of motivations and specialization levels of divers and whale watchers. *Tourism Management Perspectives* 18, 74–83.

Berenguer, J. (2007) The effect of empathy in proenvironmental attitudes and behaviors. *Environment and Behavior* 39 (2), 269–283.

Berkes, F. (1984) Competition between commercial and sport fisherman: An ecological analysis. *Human Ecology* 12 (4), 413–429.

Berkes, F. (2004) Rethinking community-based conservation. *Conservation Biology* 18 (3), 621–630.

Berkes, F., Colding, J. and Folke, C. (2001) *Linking Social–Ecological Systems.* Cambridge: Cambridge University Press.

Bermudez, G.M.A. and Lindemann-Matthies, P. (2018) "What matters is species richness" – High school students' understanding of the components of biodiversity. *Research in Science Education* doi: 10.1007/s11165-018-9767-y.

Berno, T. and Bricker, K. (2001) Sustainable tourism development: The long road from theory to practice. *International Journal of Economic Development* 3 (3), 1–18.

BEST EN (Building Excellence for Sustainable Tourism – an Education Network) (2019) Website, https://www.besteducationnetwork.org/.

Biggs, R.H., Simons, H., Bakkenes, M., Scholes, R.J., Eickhout, B., van Vuuren, D. and Alkemade, R. (2008) Scenarios of biodiversity loss in Southern Africa in the 21st Century. *Global Environmental Change* 18, 296–309.

Biosphere Tourism (2019) *The Sustainability Certification with International Recognition*, See https://www.biospheretourism.com/en.

Biro, A. (2002) Towards a denaturalized ecological politics. *Polity* 35 (2), 195–212.

Blake, B. and Becher, A. (2001) *The New Key to Costa Rica* (15th edn). Berkeley, CA: Ulysses Press.

Blake, A., Sinclair, M.T. and Sugiyarto, G. (2001) The Economy-wide Effects of Foot and Mouth Disease in the UK Economy. TTRI Discussion Paper No. 2001/3, University of Nottingham.

Blackman, A. and Rivera, J. (2011) Producer-level benefits of sustainability certification. *Conservation Biology* 25 (6), 1176–1185.

Blackman, A., Naranjo, M.A., Robalino, J., Alpizar, F. and Rivera, J. (2012) Does Tourism Eco-certification Pay? Costa Rica's Blue Flag Program. Discussion Paper No. RFF DP 12–50, Resources for the Future, Washington, DC.

Blackstock, K. (2005) A critical look at community based tourism. *Community Development Journal* 40 (1), 39–49.

Blancas, F.J., Lozano-Oyola, M., Gonzálex, M., Guerrero, F.M. and Caballero, R. (2011) How to use sustainability indicators for tourism planning: The case of rural tourism in Andalusia (Spain). *Science of the Total Environment* 412–413, 28–45.

Blay-Palmer, A., Sonnino, R. and Custot, J. (2016) A food politics of the possible? Growing sustainable food systems through networks of knowledge. *Agriculture and Human Values* 33, 27–43.

Blue, F. (2019) *Our Programme*, See https://www.blueflag.global/our-programme

Blumstein, T.D., Geffroy, B., Samia, D.S.M. and Bessa, E. (eds) (2017) *Ecotourism's Promise and Peril: A Biological Evaluation*. Cham: Springer.

Bodansky, D. (1994) the precautionary principle in US environmental law in T. O'Riordan, and J. Cameron (eds) *Interpreting the Precautionary Principle* (pp. 203–28), London: Earthscan.

Boehmer-Christiansen, S. (1994) The precautionary principle in Germany – Enabling government. In T. O'Riordan and J. Cameron (eds) *Interpreting the Precautionary Principle* (pp. 31–60). London: Earthscan.

Boersma, P.D., Vargas, H. and Merlen, G. (2005) Living laboratory in peril. *Science* 308, 925.

Boes, K., Buhalis, D. and Inversini, A. (2015) Conceptualising smart tourism destination dimensions. In I. Tussyadiah and A. Inversini (eds) *Information and Communication Technologies in Tourism 2015* (pp. 391–403). Cham: Springer.

Bohdanowicz, P., Zientara, P. and Novotna, E. (2011) International hotel chains and environmental protection: An analysis of Hilton's WE CARE! programme (Europe, 2006–2008). *Journal of Sustainable Tourism* 19 (7), 797–816.

Böhler, A., Grischkat, S., Haustein, S. and Hunecke, M. (2006) Encouraging environmentally sustainable holiday travel. *Transportation Research Part A* 40, 652–670.

Bolwell, D. and Weinz, W. (2008) *Reducing Poverty through Tourism.* Geneva: International Labour Organization.

Boniface, B., Cooper, R. and Cooper, C. (2012) *Worldwide Destinations: The Geography of Travel and Tourism* (6th edn). Oxford: Elsevier Butterworth-Heinemann.

Boorstin, D.J. (1987) *The Image: A Guide to Pseudo-events in America.* New York: Atheneum.

Borden, D.S., Coles, T. and Shaw, G. (2017) Social marketing, sustainable tourism, and small/medium size tourism enterprises: Challenges and opportunities for changing guest behaviour. *Journal of Sustainable Tourism* 25 (7), 903–920.

Borlido, T. (2016) Law, social capital and tourism at Peneda-Gerês: An exploratory analysis of the leadership and decision-making processes. *Dos Algarves: A Multidisciplinary e-Journal* 28, 29–44.

Born Free (2016) *Thomas Cook Calls on the Travel Industry to Better Protect Animals in Tourism,* See https://twitter.com/BornFreeFDN/status/808623388478636032.

Boulding, K.E. (1966) The economics of the coming spaceship earth. In H. Jarrett (ed.) *Environmental Quality in a Growing Economy: Essays from the Sixth Resources for the Future Forum* (pp. 3–14). Baltimore, MD: Johns Hopkins Press.

Bourguignon, F. and Platteau, J.-P. (2017) Does aid availability affect effectiveness in reducing poverty? A review article. *World Development* 90, 6–16.

Bradley, R. (2009) *Capitalism at Work: Business, Government, and Energy.* Salem, MA: M&M Scrivener Press.

Bramwell, B. and Lane, B. (2008) Priorities in sustainable tourism research. *Journal of Sustainable Tourism* 16 (1), 1–4.

Bramwell, B. and Lane, B. (2012) Towards innovation in sustainable tourism research?. *Journal of Sustainable Tourism* 20 (1), 1–7.

Bramwell, B., Henry, I., Jackson, G., Prat, A., Richards, G. and Van der Straaten, J. (1996) (eds) *Sustainable Tourism Management: Principles and Practice.* Tilburg: Tilburg University Press.

Brand, F.S., Berger, V., Hetze, K., Schmidt, J.E.U., Weber, M.-C., Winistörfer, H. and Dauget, C.-H. (2018) Overcoming current practical challenges in sustainability and integrated reporting: Insights from a Swiss Field Study. *Nachhaltigkeits Management Forum/Sustainability Management Forum* 26 (1), 35–46.

Brandt, W. (1980) *North–South Report of the Independent Commission on International Development Issues.* Pan Books, London.

Brau, R. (2008) Demand-driven sustainable tourism? A choice modelling analysis. *Tourism Economics* 14 (4), 691–708.

Braun, P. (2004) Regional tourism networks: The Nexus between ICT diffusion and change in Australia. *Information Technology & Tourism* 6, 231–243.

Brenner, L. and Guillermo Aguilar, A. (2002) Luxury tourism and regional economic development in Mexico. *The Professional Geographer* 54 (4), 500–520.

Briassoulis, H. (2002) Sustainable tourism and the question of the commons. *Annals of Tourism Research* 29 (4), 1065–1085.

Bricker, K. (2018) Positioning sustainable tourism: Humble placement of a complex enterprise. *Journal of Park and Recreation Administration* 36, 205–211.

Broccardo, L., Culasso, F. and Truant, E. (2017) Unlocking value creation using an Agritourism business model. *Sustainability* 9, 1618.

Brockhagen, D., Höller, C. and Tschopp, J. (1997) *Flugverkehr – Wachstum auf Kosten der Umwelt.* Wien: Verkehrsclub Österreich.

Bronfenbrenner, U. (1995) The bioecological model from a life course perspective: Reflections of a participant observer. In P. Moen, G.H. Elder, Jr. and K. Lüscher (eds) *Examining Lives in Context: Perspectives on the Ecology of Human Development* (pp. 599–618). Washington, DC: American Psychological Association.

Brooke (2019a) *Animals, People and Communities.* See thebrooke.org/.

Brooke (2019b) *Happy Horses Holiday Code*, See https://www.thebrooke.org/get-involved/responsible-use-animals-tourism/happy-horses-holiday-code.

Brooks, T.M., Mittermeier, R.A., Mittermeier, C.G., Da Fonesca, G.A.B., Rylands, A.B., Konstant, W.R., Flick, P., Pilgrim, H., Oldfield, S., Magin, G. and Hilton-Taylor, C. (2002) Habitat loss and extinction in the hotspots of biodiversity. *Conservation Biology* 16 (4), 909–923.

Brunner, P. and Rechberger, H. (2004) Practical handbook of material flow analysis. *International Journal of Life Cycle Assessment* 9, 337–338.

Bruser, D. and Poisson, J. (2016) Star investigation: A poisoned people. *The Star*, 24 July, https://www.thestar.com/news/canada/2016/07/24/disability-board-approves-mercury-poisoning-claims-from-grassy-narrows-first-nation.html.

Bryan, H. (1977) Leisure value systems and recreational specialization: The case of trout fishermen. *Journal of Leisure Research* 9, 174–187.

Buckley, R. (2005) In search of the Narwhal: Ethical dilemmas in ecotourism. *Journal of Ecotourism* 4 (2), 129–134.

Buckley, R. (2012) Sustainable tourism: Research and reality. *Annals of Tourism Research* 39 (2), 528–546.

Buckley, R. and Sommer, M. (2000) *Tourism and Protected Areas: Partnerships in Principle and Practice.* Gold Coast, Qld: Cooperative Research Centre for Sustainable Tourism.

Buckley, R., Gretzel, U., Scott, D., Weaver, D. and Becken, S. (2015) Tourism megatrends. *Tourism Recreation Research* 40 (1), 59–70.

Budeanu, A. (2007) Sustainable tourist behavior – A discussion of opportunities for change. *International Journal of Consumer Studies* 31, 499–508.

Budeanu, A., Miller, G., Moscardo, G. and Ooi, C.-S. (2016) Sustainable tourism, progress, challenges and opportunities: Introduction to this special volume. *Journal of Cleaner Production* 111, 285–294.

Budowski, G. (1976) Tourism and environmental conservation: Conflict, coexistence, or symbiosis. *Environmental Conservation* 3 (1), 27–31.

Buhalis, D. and Darcy, S. (2010) *Accessible Tourism: Concepts and Issues.* Bristol: Channel View.

Bulgarian State Tourism Agency (2006) *Strategic Plan for the Strategy for the Development of Bulgarian Tourism for the Period 2006–2009.* Sofia: BSTA.

Bures, F. (2006) Open Road, Clear Conscience. *Mother Jones*, July/August, See http://www.motherjones.com/politics/2006/07/open-road-clear-conscience/.

Burkhard, B., de Groot, R., Costanza, R., Seppelt, R., Jørgensen, S.E. and Potschin, M. (2012a) Solutions for sustaining natural capital and ecosystem services. *Ecological Indicators* 21, 1–6.

Burkhard, B., Kroll, F., Nedkov, S. and Muller, F. (2012b) Mapping ecosystem service supply, demand and budgets. *Ecological Indicators* 21, 17–29.

Burr, S.W. (1995) Sustainable tourism development and use: Follies, foibles, and practical approaches. In S.F. McCool and A.E. Watson (eds) *Linking Tourism, the Environment, and Sustainability* (pp. 8–13), Papers compiled from a session of the National Recreation and Park Association, Minneapolis, MN, 12–14 October, General Technical Report INTGTR-323. Ogden, UT: USDA, Forest Service.

Bushell, R. (2018) Refocusing sustainable tourism: Poverty alleviation in iconic world heritage destinations in Southeast Asia. In Y. Wang, A. Shakeela, A. Kwek and C. Khoo-Lattimore (eds) *Managing Asian Destinations: Perspectives on Asian Tourism* (pp. 159–176). Singapore: Springer.

Butcher, J. (1997) Sustainable development or development?. In M. Stabler (ed.) *Tourism and Sustainability: Principles to Practice* (pp. 27–38). Wallingford: CABI.

Butler, R.W. (1980) The concept of a tourist area cycle of evolution. *Canadian Geographer* 24, 5–12.

Butler, R.W. (1990) Alternative tourism: Pious hope or trojan horse?. *Journal of Travel Research* 28, 40–45.

Butler, R.W. (1991) Tourism, environment, and sustainable development. *Environmental Conservation* 18 (3), 201–209.

Butler, R.W. (1993) Tourism – An evolutionary perspective. In J.G. Nelson, R.W. Butler, and G. Wall (eds) *Tourism and Sustainable Development: Monitoring, Planning, Managing* (pp. 27–44), Department of Geography Publication No. 37, Waterloo, Ont.: University of Waterloo.

Butler, R.W. (1996) Sustainable tourism: A state-of-the-art review. *Tourism Geographies* 1 (1), 7–25.

Butler, R.W. (1997) Modelling tourism development. In S. Wahab and J.J. Pigram (eds) *Tourism Development and Growth: The Challenge of Sustainability* (pp. 109–125). London: Routledge.

Butler, R.W. (1998) Sustainable tourism – Looking backwards in order to progress? In C.M. Hall and A.A. Hew (eds) *Sustainable Tourism: A Geographical Perspective* (pp. 25–34). New York: Addison Wesley Longman.

Butler, R. (2009) Tourism in the future: Cycles, waves or wheels?. *Futures* 41 (6), 346–352.

Butler, R.W. (2015) Tourism area life cycle. In C. Cooper (ed.) *Contemporary Tourism Reviews, Volume 1* (pp. 183–226). Oxford: Goodfellow.

Butler, R.W. (ed.) (2017) *Tourism and Resilience.* Wallingford: CABI.

Butler, R. (2018) Sustainability and resilience: Two sides of the same coin? In C. Cooper, S. Volo, W.C. Gartner and N. Scott (eds) *The Sage Handbook of Tourism Management, Volume 1* (pp. 407–421). London: Sage.

Buzinde, C.N., Manuel-Navarrete, D., Yoo, E.E. and Morais, D. (2009) "Tourists" perceptions in a climate of change eroding destinations. *Annals of Tourism Research* 37 (2), 333–354.

CAAWS (2018) *What is Gender Equity?. Toronto: Canadian Association for the Advancement of Women and Sport and Physical Activity*, https://www.caaws.ca/gender-equity-101/what-is-gender-equity/.

Cairns, J., Jr. (2000) World peace and global sustainability. *International Journal of Sustainable Development & World Ecology* 7 (1), 1–11.

Cairns, J., Jr. (2003) A preliminary declaration of sustainability ethics: Making peace with the ultimate bioexecutioner. *Ethics in Science and Environmental Politics*, May, 34–48.

CalMac (2018) *CalMac Environmental Strategy 2018–2020*, See https://www.calmac.co.uk/article/6875/CalMac-Environmental-Strategy-2018-2020.

Calzada, I. (2019) Seeing tourism transformations in Europe through algorithmic, techno-political, and city-regional lenses. In A. Alzua (ed.) *Transforming Tourism from a Regional Perspective.* Brussels: Coppieters Foundation.

Canadian Federation of Earth Sciences (2017) *Guidelines and Criteria for Canadian Areas seeking UNESCO Global Geopark Designation.* Calgary: Canadian National Committee for Geoparks.

Carey, S., Gountas, Y. and Gilbert, D. (1997) Tour operators and destination sustainability. *Tourism Management* 18 (7), 425–431.

Carlsen, J., Liburd, J., Edwards, D. and Forde, P. (eds) (2008) *Innovation for Sustainable Tourism: International Case Studies.* Esbjerg: Best Education Network, University of Southern Denmark.

Carnival Cruise Line (2019) Website, www.carnival.com.

Carrera, P.M. and Bridges, J.F.P. (2006) Globalization and healthcare: Understanding health and medical tourism. *Expert Review of Pharmacoeconomics Outcomes Research* 6 (4), 447–454.

Carr-Harris, A. and Lang, C. (2018) Sustainability and tourism: A case study of the United States' first offshore wind farm. *Resource and Energy Economics* 57, 51–67.

Carrington, D. (2018) What is biodiversity and why does it matter to us?. *The Guardian*, 12 March, https://www.theguardian.com/news/2018/mar/12/what-is-biodiversity-and-why-does-it-matter-to-us.

Carroll, A.B. (1979) A three-dimensional conceptual model of corporate performance. *Academy of Management Review* 4 (4), 497–505.

Carroll, A.B. (1991) The pyramid of corporate social responsibility: Toward the moral management of organizational stakeholders. *Business Horizons* 34 (4), 39–48.

Carroll, A.B. (2009) A history of corporate social responsibility: Concepts and practice. In A. Crane, D. Matten, A. McWilliams, J. Moon and D.S. Siegel (eds) *The Oxford Handbook of Corporate Social Responsibility* (pp. 19–46). Oxford: Oxford University Press.

Carrots & Sticks (2016) *Global Trends in Sustainability Reporting Regulation and Policy*, See https://www.carrotsandsticks.net/wp-content/uploads/2016/05/Carrots-Sticks-2016.pdf.

Carson, R. (1962) *Silent Spring*. Boston, MA: Houghton Mifflin.

Carter, E., Adams, W. and Hutton, J. (2008) Private protected areas: Management regimes, tenure arrangements and protected area categorization in East Africa. *Oryx* 42 (2), 177–186.

Catibog-Sinha, C. (2008) Zoo tourism: Biodiversity conservation through tourism. *Journal of Ecotourism* 7 (2&3), 160–178.

Čavlek, N., Cooper, C., Krajinović, V., Srnec, L. and Zaninović, K. (2019) Destination climate adaptation. *Journal of Hospitality & Tourism Research* 43 (2), 314–322.

Cazcarro, I., Hoekstra, A.Y. and Sánchez Chóliz, J. (2014) The water footprint of tourism in Spain. *Tourism Management* 40, 90–101.

Cazcarro, I., Duarte, R. and Sánchez Chóliz, J. (2016) Tracking water footprints at the Micro and Meso scale: An application to Spanish tourism by regions and municipalities. *Journal of Industrial Ecology* 20 (3), 446–461.

CBD (Convention on Biological Diversity) (2004) *Guidelines on Biodiversity and Tourism Development*. Montreal: Secretariat of the Convention on Biological Diversity. See https://www.cbd.int/doc/publications/tou-gdl-en.pdf.

CBD (Convention on Biological Diversity) (2018) *Aichi Biodiversity Targets*. See www.cbd.int/sp/targets/.

Ceballos-Lascurain, H. (1996) *Tourism, Ecotourism, and Protected Areas: The State of Nature-based Tourism around the World and Guidelines for its Development*. Cambridge: International Union for Conservation of Nature (IUCN).

Cel, W., Kujawska, J. and Wasąg, H. (2017) Impact of hydraulic fracturing on the quality of natural water. *Journal of Ecological Engineering* 18 (2), 63–68.

Cerruti, J. (1964) The two Acapulcos. *National Geographic Magazine* 126 (6), 848–878.

Chafe, Z. and Honey, M. (eds) (2005) Consumer Demand and Operator Support for Socially and Environmentally Responsible Tourism. CESD/TIES Working Paper No. 104, Center on Ecotourism and Sustainable Development (CESD) and

The International Ecotourism Society (TIES), Washington, DC, http://www.travelersphilanthropy.org/resources/documents/ Consumer_Demand_Report.pdf.

Chan, W.W., Mak, L.M., Chen, Y.M., Wang, Y.H., Xie, H.R., Hou, G.Q. and Li, D. (2008) Energy saving and tourism sustainability: Solar control window film in hotel rooms. *Journal of Sustainable Tourism* 16 (5), 563–574.

Cheer, J.M. and Lew, A.A. (eds) (2019) *Tourism, Resilience and Sustainability: Adapting to Social, Political and Economic Change*. London: Routledge.

Chen, Y.-H., Wen, X.-W. and Luo, M.-Z. (2016) Corporate social responsibility spillover and competition effects on the food industry'. *Australian Economic Papers* 55 (1), 1–13.

Chi, C.G., Cai, R. and Li, Y. (2017) 'Factors influencing residents' subjective well-being at world heritage sites'. *Tourism Management* 63, 209–222.

Chiarelli-Helminiak, C.M. and Lewis, T.O. (2018) Sustainable development goal 4: When access to education is not enough. *Journal of Access, Retention & Inclusion in Higher Education* 1, 17–27.

Choi, H.C. and Murray, I. (2010) Resident attitudes toward sustainable community tourism. *Journal of Sustainable Tourism* 18 (4), 575–594.

Choi, H.C. and Sirakaya, E. (2006) Sustainability indicators for managing community tourism. *Tourism Management* 27 (6), 1274–1289.

Choi, H.C. and Turk, E.S. (2006) Sustainability indicators for managing community tourism. *Tourism Management* 27, 1274–1289.

Choo, H. and Jamal, T. (2009) Tourism on organic farms in South Korea: A new form of ecotourism. *Journal of Sustainable Tourism* 14 (4), 431–454.

Cision (2016) Tripadvisor ends sale of cruel wildlife tourism attractions following world animal protection campaign. *Cision*, 12 October, https://www.newswire.ca/news-releases/tripadvisor-ends-sale-of-cruel-wildlife-tourist-attractions-following-world-animal-protection-campaign-596827901.html.

Clark, C.W. (1973) Profit maximization and the extinction of animal species. *Journal of Political Economy* 81, 950–961.

Clark, W.C. (1985) Scales of climate impacts. *Climatic Change* 7, 5–27.

Clarke, J. (1997) A Framework of approaches to sustainable tourism. *Journal of Sustainable Tourism* 5 (3), 224–233.

Cleverdon, R. and Kalisch, A. (2000) Fair trade in tourism. *International Journal of Tourism Research* 2, 171–187.

Clowney, D. (2013) Biophilia as an environmental virtue. *Journal of Agricultural and Environmental Ethics* 26, 999–1014.

CN (2016) *Delivering Responsibly*. Montreal: Canadian National.

CN (2018) *Working with our Neighbours for a Brighter Future*. Montreal: Canadian National.

CN (2019) Canadian National Railway website, www.cn.ca.

Cobbinah, P.B., Erdiaw-Kwasie, M.O. and Anoateng, P. (2015) Rethinking sustainable development within the framework of poverty and urbanisation in developing countries. *Environmental Development* 13, 18–32.

Cócola, A. (2014) The invention of the barcelona gothic quarter. *Journal of Heritage Tourism* 9 (1), 18–34.

Coghlan, A. and Castley, J.G. (2013) 'A Matter of perspective: Residents', regulars' and locals' perceptions of private tourism Ecolodge Concessions in Kruger National Park, South Africa', *Current Issues in Tourism* 16 (7), 682–699.

Coghlan, A. and Fennell, D.A. (2009) Myth or substance: An examination of altruism as the basis of volunteer tourism. *Annals of Leisure Research* 12 (3&4), 377–402.

Coglianese, C. and Nash, J. (eds) (2001) *Regulating from the Inside: Can Environmental Management Systems Achieve Policy Goals?*. Washington, DC: Resources for the Future.

Cohen, E. (1978) The impact of tourism on the physical environment. *Annals of Tourism Research* 5 (2), 215–237.

Cohen, E. (1979) A phenomenology of tourist experiences. *Sociology* 13 (2), 179–201.

Cohen, E. (1987) "Alternative tourism" – A critique. *Tourism Recreation Research* 12 (2), 13–18.

Cohen, E. (2013) "Buddhist Compassion" and "Animal Abuse" in Thailand's Tiger Temple. *Society & Animals* 21 (3), 266–283.

Cohen, E. (2019) The fall and reincarnation of thailand's tiger temple. *International Journal of Tourism Anthropology* 7 (2), 115–131.

Cohen, E. and Cohen, S.A. (2015) A mobilities approach to tourism from emerging world regions. *Current Issues in Tourism* 18 (1), 11–43.

Cole, S. (2006) Information and Empowerment: The Keys to Achieving Sustainable Tourism. *Journal of Sustainable Tourism* 14 (6), 629–644.

Cole, S. (2014) Tourism and water: From stakeholders to rights holders, and what tourism businesses need to do. *Journal of Sustainable Tourism* 22 (1), 89–106.

Commoner, B. (1972) *The Closing Circle: Confronting the Environmental Crisis*. London: Jonathan Cape.

Connell, J. (2000) The role of tourism in the socially responsible university. *Current Issues in Tourism* 3 (1), 1–19.

Considerate Group (2018) Website, www.consideratehoteliers.com.

Consumer Goods Forum (2011) *Top of Mind Survey 2011*. Levallois-Perret: Consumer Goods Forum.

Convention on Biological Diversity (2019) *Use of Terms*, See https://www.cbd.int/convention/articles/default.shtml?a=cbd-02

Conway, F.J. (2014) Local and public heritage at a world heritage site. *Annals of Tourism Research* 44, 143–155.

Cook, J., Nuccitelli, D., Green, S., Richardson, M., Winkler, B., Painting, R., Way, R., Jacobs, P. and Skuce, A. (2013) Quantifying the consensus on anthropogenic global warming in the scientific literature. *Environmental Research Letters* 8 (2), 1–7.

Cooper, C. (1995) Strategic planning for sustainable tourism: The case of the offshore Islands of the UK. *Journal of Sustainable Tourism* 3 (4), 1–19.
Cooper, C. (2006) Knowledge management and tourism. *Annals of Tourism Research* 33 (1), 47–64.
Cooper, C. (2015) Managing tourism knowledge. *Tourism Recreation Research* 40 (1), 106–119.
Cooper, C. (2017) Commentary on tourism and mobilities. *The End of Tourism as We Know it? e-Review of Tourism Research* 14 (3.4), 82.
Cooper, C. and Hall, M. (2019) *Contemporary Tourism*. Oxford: Goodfellow.
Cooper, C., Fayos Sola, E., Jafari, J., Lisboa, C., Maron, C., Perdomnao, Y. and Urosevic, Z. (2018) Case studies in technological innovation. In E. Fayos Spal and C. Cooper (eds) *The Future of Tourism: Innovation and Sustainability* (pp. 111–127). Cham: Springer.
Coppock, J.T. (1982) 'Tourism and conservation'. *Tourism Management* 3 (4), 270–276.
Cornwall Destination Management Organisation (2007) *Destination Audit*. Exeter: Roger Tym and Partners.
Costanza, R. and Daly, H.E. (1992) Natural capital and sustainable development. *Conservation Biology* 6, 37–46.
Costanza, R., d'Arge, R., de Groot, R., Farber, S., Grasso, M., Hannon, B., Limburg, K., Naeem, S., O'Neill, R.V., Paruelo, J., Raskin, R.G., Sutton, P. and van den Belt, M. (1997) The value of the world's ecosystem services and natural capital. *Nature* 387, 253–260.
Cotswolds Tourism Partnership (2009) *Cotswolds and Forest of Dean Destination Management Organization Sustainability Plan*. Cirencester: Cotswolds Tourism Partnership.
Crabbé, P. (1998) Sustainable development: Passing fancy or pragmatism?. *Micro* 5 (1), 3–4.
Crabtree, A., O'Reilly, P. and Worboys, G. (2002) Setting a Worldwide Standard for Ecotourism. Paper presented at the World Ecotourism Summit. Quebec, Canada.
Craig, R.K. (2012a) *Comparative Ocean Governance: Place-based Protections in a Climate Change Era*. Cheltenham: Edward Elgar.
Craig, R.K. (2012b) Marine biodiversity, climate change, and governance of the oceans. *Diversity* 4, 224–238.
Crane, A., Matten, D. and Moon, J. (2004) Stakeholders as citizens? Rethinking rights, participation, and democracy. *Journal of Business Ethics* 53 (1–2), 107–122.
Crane, A., Matten, D. and Moon, J. (2008) *Corporations and Citizenship*. Cambridge: Cambridge University Press.
Crane, A., Matten, D., McWilliams, A., Moon, J. and Siegel, D.S. (eds) (2009) *The Oxford Handbook of Corporate Social Responsibility*. Oxford: Oxford University Press.
Creo, C. and Fraboni, C. (2011) Awards for the sustainable management of coastal tourism destinations: The example of the blue flag program. *Journal of Coastal Research* 61, 378–381.
Cronin, L. (1990) A strategy for tourism and sustainable developments. *World Leisure and Recreation Journal* 32 (3), 12–18.

Crossett, J. (1961) More and Seneca. *Philosophical Quarterly* 15 (4), 577–580.

Crossman, N.D. and Bryan, B.A. (2009) Identifying cost-effective hotspots for restoring natural capital and enhancing landscape multifunctionality. *Ecological Economics* 68, 654–668.

Crossman, N.D., Burkhard, B., Nedkov, S., Willemen, L., Petz, K., Palomo, I., Drakou, E.G., Martin-Lopez, B., McPhearson, T., Boyanova, K., Alkemade, R., Egoh, B., Dunbar, M.B. and Maes, J. (2013) A blueprint for mapping and modelling ecosystem services. *Ecosystem Services* 4, 4–14.

Crouch, D. (2000) Places around us: Embodied lay geographies in leisure and tourism. *Leisure Studies* 19 (2), 63–76.

Crouch, G. (2013) *Homo Sapiens* on vacation: What can we learn from Darwin?. *Journal of Travel Research* 52 (5), 575–590.

Crouch, G. and Ritchie, J.R.B. (2012) *Competitiveness and Tourism.* London: Edward Elgar.

Čuček, L., Klemeš, J.J. and Kravanja, Z. (2012) A review of footprint analysis tools for monitoring impacts on sustainability. *Journal of Cleaner Production* 34, 9–20.

Culiberg, B. and Elgaaied-Gambier, L. (2016) Going green to fit in – Understanding the impact of social norms on pro-environmental behaviour, a cross-cultural approach. *International Journal of Consumer Studies* 40, 179–185.

Curcovic, S., Melnyk, S., Calantone, R. and Handfield, R. (2000) Investigating the linkage between total quality management and environmentally responsible manufacturing. *IEEE Transactions in Engineering Management* 47, 444–464.

Curry, P. (2011) *Ecological Ethics: An Introduction* (2nd edn). Cambridge: Polity Press.

Curtin, S. (2003) Whale-watching in Kaikoura: Sustainable destination development?. *Journal of Ecotourism* 2 (3), 172–193.

Curtin, S. (2009) Wildlife tourism: The intangible, psychological benefits of human–wildlife encounters. *Current Issues in Tourism* 12 (5–6), 451–474.

Curtis, K. and Slocum, S. (2016) The potential impacts of green certification programs focused on food waste reduction on the tourism industry. *Journal of Food Distribution Research* 47 (1), 6–11.

Czarnezki, J.J. (2014) Greenwashing and self-declared seafood ecolabels. *Tulane Environmental Law Journal* 28 (1), 37–52.

Czech, B. (2006) If Rome is burning, why are we fiddling?. *Conservation Biology* 26 (6), 1563–1565.

Dahlberg, K.A. (1993) Regenerative food systems: Broadening the scope and agenda of sustainability. In P. Allen (ed.) *Food for the Future: Conditions and Contradictions of Sustainability* (pp. 75–102). New York: John Wiley.

Daily, G.C., Alexander, S., Ehrlich, P.R., Goulder, L., Lubchenco, J., Matson, P.A., Mooney, H.A., Postel, S., Schneider, S.H., Tilman, D. and Woodwell, G.M. (1997) Ecosystem services: Benefits supplied to human societies by natural ecosystems. *Issues in Ecology* 2, 1–16.

Daldeniz, B. and Hampton, M.P. (2013) Dive tourism and local communities: Active participation or subject to impacts? Case studies from Malaysia. *International Journal of Tourism Research* 15, 507–520.

Daly, H.E. (2008) *A Steady-State Economy*. London: Sustainable Development Commission.

Dangi, T.B. and Gribb, W.J. (2018) Sustainable ecotourism management and visitor experiences: Managing conflicting perspectives in Rocky Mountain National Park, USA. *Journal of Ecotourism* 17 (3), 338–358.

Dangi, T.B. and Jamal, T. (2016) An integrated approach to "Sustainable Community-based Tourism". *Sustainability* 8, 475.

Dann, G.M.S. (2015) Understanding tourism: Once more Greek Philosophy to the rescue. *Tourism Recreation Research* 40 (2), 262–264.

Dans, E. (2017) Cities, tourism and the tragedy of the commons. *Medium*, 21 August, https://medium.com/enrique-dans/cities-tourism-and-the-tragedy-of-the-commons-62c9c665f58b.

Darnell, N., Jolley, J. and Handfield, R.B. (2008) Environmental management systems and green supply chain management: complements for sustainability? *Business Strategy and the Environment* 18, 30–45.

Dasgupta, S. (2019) After 40 years of effort, Tanzania creates new protected area for endangered monkeys. *PacificStandard*, 22 January, https://psmag.com/environment/tanzania-creates-new-protected-lands-for-rare-species.

Davidson, M.D. (2017) Equity and the conservation of global ecosystem services. *Sustainability* 9, 339.

Daw, T., Brown, K., Rosendo, S. and Pomeroy, R. (2011) Applying the ecosystem services concept to poverty alleviation: The need to disaggregate human well-being. *Environmental Conservation* 38 (4), 370–379.

Dawkins, R. (1989) *The Selfish Gene*. New York: Oxford.

De Beer, A., Rogerson, C.M. and Rogerson, J.M. (2014) Decent work in the South African tourism industry: Evidence from tourist guides. *Urban Forum* 25 (1), 89–103.

de Bloom, J., Nawijn, J., Geurts, S., Kinnunen, U. and Korpela, K. (2017) Holiday travel, staycations, and subjective well-being. *Journal of Sustainable Tourism* 25 (4), 573–588.

de Castro, F. and Begossi, A. (1996) Fishing at Rio Grande (Brazil): Ecological Niche and competition. *Human Ecology* 24 (3), 401–411.

Defra (2006) *Pesticides: Code of Practice for Using Plant Protection Products*. London: Department of Environment, Food and Rural Affairs, Crown Copyright.

Defra (2009) *Protecting our Water, Soil and Air: A Code of Good Agricultural Practice for Farmers, Growers and Land Managers*. Belfast: TSO.

Defra (2010) *2010 Guidelines to Defra/DECC's GHG Conversion Factors for Company Reporting, Version 1.2.1, Final*, See http://www.sthc.co.uk/documents/DEFRA-guidelines-ghg-conversion-factors_2010.pdf.

de França Doria, M. (2010) Factors influencing public perception of drinking water quality. *Water Policy* 12, 1–19.

DeFur, P. and Kaszuba, M. (2002) Implementing the precautionary principle. *Science of the Total Environment* 288, 155–165.

de Groot, R.S., Wilson, M.A. and Boumans, R.M.J. (2002) A typology of the classification, description and valuation of ecosystem functions, goods and services. *Ecological Economics* 41, 393–408.

de Groot, R.S., Alkemade, R., Braat, L., Hein, L. and Willemen, L. (2010) Challenges in integrating the concept of ecosystem services and values in landscape planning, management and decision making. *Ecological Complexity* 7, 260–272.

de Grosbois, D. (2011) Carbon footprint of the hotel companies: Comparison of methodologies and results. *Tourism Recreation Research* 36 (3), 231–245.

de Grosbois, D. (2016) Corporate social responsibility reporting in the cruise tourism industry: A performance evaluation using a new institutional theory based model. *Journal of Sustainable Tourism* 24 (2), 245–269.

de Grosbois, D. and Fennell, D.A (2011) Carbon footprint of the global hotel companies: Comparisons of methodologies and results. *Tourism Recreation Research* 36 (3), 231–246.

de Haan, G. (2006) The BLK "21" programme in Germany: A "Gestaltungskompetenz"-based model for education for sustainable development. *Environmental Education Research* 12 (1), 19–32.

De Kadt, E. (1992) Making the alternative sustainable: Lessons from development for tourism. In V.L. Smith and W.R. Eadington (eds) *Tourism Alternatives* (pp. 47–75). Philadelphia, PA: University of Pennsylvania Press.

Delgado, C., Rosegrant, M., Steinfeld, H., Ehui, S. and Courbois, C. (2001) 'Livestock to 2020: The next food revolution. *Outlook on Agriculture* 30 (1), 27–29.

Delmas, M.A. and Lessem, N. (2017) Eco-premium or Eco-penalty? Ecolabels and quality in the organic wine market. *Business & Society* 56 (2), 318–356.

Delpla, I., Jung, A.-V., Baures, E., Clement, M. and Thomas, O. (2009) Impacts of climate change on surface water quality in relation to drinking water production. *Environment International* 35, 1225–1233.

Dewey, J. (1915) *The School and Society*. Chicago, IL: University of Chicago Press.

Diamantis, D. (1998) Environmental auditing: A tool in ecotourism development. *Eco-Management and Auditing* 5, 15–21.

Diamond, J. (2005) *Collapse: How Societies Choose to Fail or Succeed*, Toronto: Penguin Books.

Díaz, S., Demissew, S., Joly, C., Lonsdale, W.M. and Larigauderie, A. (2015) A Rosetta stone for nature's benefits to people. *PLoS Biology* 13 (1), e1002040.

Díaz Pérez, F.J., Chinarro, D., Mouhaffel, M., Díaz Martín, R., and Pino Otín, M.R. (2019) Comparative study of carbon footprint of energy and water in hotels of canary

Islands Regarding Mainland Spain. *Environment, Development and Sustainability* 21 (4), 1763–1780.

Dickinson, J.E. and Peeters, P. (2014) Time, tourism consumption and sustainable development. *International Journal of Tourism Research* 16, 11–21.

Dickson, B., Blaney, R., Miles, L., Regan, E., van Soesbergen, A., Vaananen, E., Blyth, S., Harfoot, M., Martin, C., McOwen, C., Newbold, T. and van Bochove, J.-W. (2014) *Toward a Global Map of Natural Capital: Key Ecosystem Assets*, Nairobi: UN Environmental Programme.

Dieckmann, A, McCabe, S. and Kakadoukakis, K.I (2018) Human rights, disabilities and social tourism: Management issues and challenges. In C. Cooper, S. Volo, W.C. Gartner and N.N. Scott (eds) *The Sage Handbook of Tourism Management. Applications of Theories and Concepts to to Tourism* (pp. 61–74). Sage, London.

Dierksmeier, C. and Pirson, M. (2009) 'Oikonomia versus Chrematistike: Learning from Aristotle about the future orientation of business management. *Journal of Business Ethics* 88, 417–430.

Dietz, T., Stern, P.C. and Guagnano, G.A. (1998) Social structural and social psychological bases of environmental concern. *Environment and Behaviour* 30 (4), 450–471.

Dietz, T., Allen, S. and McCright, A.M. (2017) Integrating concern for animals into personal values. *Anthrozoos* 30 (1), 109–122.

Dilek, S.E. and Fennell, D.A. (2018) Discovering the hotel selection factors of vegetarians: The case of Turkey. *Tourism Review* 73 (4), 492–506.

Di Minin, E., Leader-Williams, N. and Bradshaw, C.J.A. (2016) Banning trophy hunting will exacerbate biodiversity loss. *Trends in Ecology & Evolution* 31 (2), 99–102.

Dimitrov, D., Slocum, S. and Webb, K. (2018) Meeting the 2030 Sustainable Development Goals in Higher Education Tourism Programs. TAKE 2018: Theory and Applications in the Knowledge Economy, Conference Book of Abstracts. New York: E4C, 73–76.

Dinarès, M. and Saurí, D. (2015) Water consumption patterns of hotels and their response to droughts and public concerns regarding water conservation: The case of the barcelona hotel industry during the 2007–2008 episode. *Documents d'Anàlisi Geogràfica* 61 (3), 623–649.

Dirnböck, T., Djukic, I., Kitzler, B., Kobler, J., Mol-Dijkstra, J.P., Posch, M., Reinds, G.J., Schlutow, A., Starlinger, F. and Wamelink, W.G.W. (2017) Climate and air pollution impacts on habitat suitability of Austrian forest ecosystems. *PLoS ONE* 12 (9), e0184194.

Discover Corps (2018) *The World's Best Responsible Travel Organizations*, See https://discovercorps.com/blog/best-responsible-travel-organizations/.

Dixit, K. (2018) Contributions of Education for Sustainable Development (ESD) to quality education. *Journal of Industrial Relationship, Corporate Governance & Management Explorer* 2 (1), 14–17.

Dobson, A. (1996) Environmental sustainabilities: An analysis and a typology. *Environmental Politics* 5 (3), 401–428.

Dodds, R. and Walsh, P.R. (2019) Assessing the factors that influence waste generation and diversion at Canadian festivals. *Current Issues in Tourism* 22 (19), 2348–2352.

Dolnicar, S. (2010) Identifying tourists with smaller environmental footprints. *Journal of Sustainable Tourism* 18, 717–734.

Dolnicar, S., Crouch, G.I. and Long, P. (2008) Environment-friendly tourists: What do we really know about them?. *Journal of Sustainable Tourism* 16 (2), 197–210.

Dolnicar, S., Laesser, C. and Mattus, K. (2010) Short-haul city travel is truly environmentally sustainable. *Tourism Management* 31 (4), 505–512.

Donohoe, H.M. and Needham, R. (2008) Internet-based ecotourism marketing: Evaluating Canadian sensitivity to ecotourism tenets. *Journal of Ecotourism* 7 (1), 15–40.

Doria, M.F. (2010) Factors influencing public perception of drinking water quality. *Water Policy* 12 (1), 1–19.

Doughty, M.R.C. and Hammond, G.P. (2003) Cities and sustainability. In A. Winnett and A. Warhurst (eds) *Towards an Environment Research Agenda* (pp. 81–105). London: Palgrave Macmillan.

Dovers, S.R. and Handmer, J. (1995) Ignorance, the precautionary principle and sustainability. *Ambio* 24 (2), 92–97.

Dowling, R.K. (ed.) (2006) *Cruise Tourism: Issues, Impacts, Cases.* Wallingford, CABI.

Downton, P., Jones, D. and Zeunert, J. (2016) Biophilia in urban design: Patterns and principles for smart Australian cities. Paper presented at the 9th International Urban Design Conference. Canberra, ACT, 7–9 November.

D'Sa, E. (1999) Wanted: Tourists with a social conscience. *International Journal of Contemporary Hospitality Management* 11 (2/3), 64–68.

Dubois, G., Peeters, P., Ceron, J.P. and Gössling, S. (2011) The future tourism mobility of the world population: Emission growth versus climate policy. *Transportation Research Part A: Policy and Practice* 45 (10), 1031–1042.

du Cros, H. (2001) A new model to assist in planning for sustainable cultural heritage tourism. *International Journal of Tourism Research* 3, 165–170.

Dudley, N. and Stolton, S. (eds) (2003) Running pure: The importance of forest protected areas to drinking water. Gland and Washington, DC: WWF and The World Bank.

Dudley, N., Stolton, S., Belokurov, A., Krueger, L., Lopoukhine, N., MacKinnon, K., Sandwith, T. and Sekhran, N. (eds) (2009) *Natural Solutions: Protected Areas Helping People Cope with Climate Change*, Gland, Washington, DC and New York: IUCN-WCPA, TNC, UNDP, WCS, World Bank and WWF.

Dudley, N., Ali, N. and MacKinnon, K. (2017a) *Natural Solutions: Protected Areas Helping to Meet the Sustainable Development Goals*, Briefing. Gland, Switzerland: IUCN World Commission on Protected Areas.

Dudley, N., Ali, N., Kettunen, M. and MacKinnon, K. (2017b) Editorial essay: Protected areas and the sustainable development goals. *Parks* 23 (2), 9–12.

Duffell, R. (1998) Toward the environment and sustainability ethic in engineering education and practice. *Journal of Professional Issues in Engineering Education and Practice* 124 (3), 78–90.

Duffy, R. (2006) Global environmental governance and the politics of ecotourism in Madagascar. *Journal of Ecotourism* 5 (1&2), 128–143.

Duffy, S. (1991) "Silent Spring" and "A Sand County Almanac": The two most significant environmental books of the 20th century. *Nature Study* 44 (2–3), 6–8.

Duran, P. (2002) The impact of Olympic games on tourism. In M. Moragas and M. Botella (eds) *Barcelona: l'Herència dels Jocs, 1992–2002*. Barcelona: Centre d'Estudis Olímpics UAB.

DW (2018) *Venice Installs Turnstiles to Limit Massive Tourist Flow*, See https://www.dw.com/en/venice-installs-turnstiles-to-limit-massive-tourist-flow/a-43578659.

Dwyer, L., Edwards, D., Mistilis, N., Scott, N., Cooper, C. and Roman, C. (2007) *Trends Underpinning Tourism to 2020: An Analysis of Key Drivers for Change*. Gold Coast, Qld: STCRC.

Dwyer, L., Forsyth, P., Spurr, R. and Hoque, S. (2010) Estimating the carbon footprint of Australian tourism. *Journal of Sustainable Tourism* 18 (3), 355–376.

Dziuba, R. (2016) Sustainable development of tourism: EU Ecolabel standards illustrated using the example of Poland. *Comparative Economic Research* 19 (2), 111–127.

Eagles, P.F.J., McCool, S.F. and Haynes, C.D. (2002) *Sustainable Tourism in Protected Areas: Guidelines for Planning and Management*, Gland: IUCN.

Eber, S. (1992) *Beyond the Green Horizon: Principles for Sustainable Tourism*, London: WWF.

EC (European Commission) (2010) *Being Wise to Waste: The EU's Approach to Waste Management*, See https://ec.europa.eu/environment/waste/pdf/WASTE%20BROCHURE.pdf.

EC (2014) *Communication from the Commission to the European Parliament, the Council, the European Economic and Social Committee and the Committee of the Regions: A European Strategy for more Growth and Jobs in Coastal and Maritime Tourism*. Brussels: European Commission.

EC (European Commission) (2019) *EU Ecolabel*, See https://ec.europa.eu/environment/ecolabel/.

ECO Tourism Australia (2018) *ECO Certification*, See https://www.ecotourism.org.au/our-certification-programs/eco-certification/.

ECPAT (2012) *The Code*, See http://www.thecode.org

Edgell, D.L. (2006) *Managing Sustainable Tourism: A Legacy for the Future*. London: Routledge.

Edgell, D. (2016) *Managing Sustainable Tourism: A Legacy for the Future* (2nd edn). London: Routledge.

Edwards, P.J. and Abivardi, C. (1998) The value of biodiversity: Where ecology and economy blend. *Biological Conservation* 83 (3), 239–246.

EEA (2016) *European Aviation Report.* Copenhagen: European Environment Agency.

EEA (2019) *About Sustainability Transitions.* Copenhagen: European Environment Agency, https://www.eea.europa.eu/themes/sustainability-transitions/intro.

Egler, H.P. and Frazao, R. (2016) Sustainable Infrastructure and Finance: How to Contribute to a Sustainable Future. Report for Designing of a Sustainable Financial System, Working Paper 16/09, UNEP, Nairobi.

Ehrlich, P.R. (1968). *The Population Bomb.* New York: Ballantine Books.

Ehrlich, P.R. (2000) *Human Natures: Genes, Cultures, and the Human Prospect.* New York: Penguin.

Eijgelaar, W., Amelung, B., Filimonau, V., Guiver, J. and Peeters, P. (2018) Tourism in a low carbon energy future. In C. Cooper, S. Volo, W.C. Gartner and N. Scott (eds) *The Sage Handbook of Tourism Management, Volume 2* (pp. 552–566). London: Sage.

Ellabban, O., Abu-Rib, H. and Blaabjerg, F. (2014) Renewable energy sources: Current status, future prospects and their enabling technology. *Renewable and Sustainable Energy Reviews* 39, 748–764.

Elliott, M. (2003) Biological pollutants and biological pollution – an increasing cause for concern. *Marine Pollution Bulletin* 46, 275–280.

Ellis, S. and Sheridan, L.M. (2014) The legacy of war for community-based tourism development: learnings from Cambodia. *Community Development Journal* 49, 129–142.

ENAT (European Network for Accessible Tourism) (2019) Website, https://www.accessibletourism.org/.

Enriquez, J., Tipping, D.C., Lee, J.-J., Vijay, A., Kenny, L., Chen, S., Mainas, N., Holst-Warhaft, G. and Steenhuis, T.S. (2017) Sustainable water management in the tourism economy: Linking the Mediterranean's traditional rainwater cisterns to modern needs. *Water* 9, 868.

Enz, C.A. and Siguaw, J.A. (1999) Best hotel environmental practices. *Cornell Hotel and Restaurant Administration Quarterly* 40 (5), 72–77.

Epler Wood, M. (2017) *Sustainable Tourism on a Finite Planet: Environmental, Business and Policy Solutions.* London: Routledge.

Epuran, G., Tescaşiu, B., Todor, R.D., Sasu, K.-A. and Cristache, N. (2017) Responsible consumption – Source of competitive advantages and solution for tourist protection. *Amfiteatru Economic* 19 (45), 447–462.

Equality in Tourism (2019) Website, http://equalityintourism.org.

Erfurt-Cooper, P. and Cooper, M. (2009) *Health and Wellness Tourism: Spas and Hot Springs,* Toronto: Channel View.

Esteves, M. (2018) Gender equality in education: A challenge for policy makers. *PEOPLE: International Journal of Social Sciences* 4 (2), 893–905.

ETFI (2014) *The Future of European Tourism.* Leeuwarden: European Tourism Futures Institute.

Etzioni, A. (2003) 'Introduction: Voluntary simplicity – Psychological implications, societal consequences. In D. Doherty and A. Etzioni (eds) *Voluntary Simplicity: Responding to Consumer Culture* (pp. 29–41). Oxford: Rowan & Littlefield.

EU DG XXIII (1998) *Agenda 2010 for Small Businesses in the "World's Largest Industry"*, Final communiqué. Llandudno: EU Directorate General.

EUROPARC Federation (2019) *Become a Sustainable Destination*, See https://www.europarc.org/sustainable-tourism/become-a-sustainable-destination/.

European Travel Commission (2018) *Tourism and Climate Change Mitigation Embracing the Paris Agreement: Pathways to Decarbonisation*. Brussels: ETC.

Eurostar (2019) *Tread Lightly*, See https://www.eurostar.com/uk-en/about-eurostar/community-environment/treadlightly.

EU Urban Access (2019) *Madrid – ZEZ – Traffic Limited Zone*, See https://urbanaccess-regulations.eu/countries-mainmenu-147/spain/madrid-access-restriction.

Everett, S. (2009) Beyond the visual Gaze? The pursuit of an embodied experience through food tourism. *Tourist Studies* 8 (3), 337–358.

Fahrig, L. (2003) Effects of habitat fragmentation on biodiversity. *Annual Review of Ecology, Evolution, and Systematics* 34, 487–515.

Fairtrade Canada (2016) *Fair Trade Town Action Guide*, See https://promo.fairtrade.ca/materials-resources/fair-trade-town-action-guide/.

Falchi, F. (2019) Light pollution. In S.M. Charlesworth and C.A. Booth (eds) *Urban Pollution* (pp. 147–152). Oxford: John Wiley.

Falkenmark, M. (1989) The massive water scarcity now threatening Africa: Why isn't it being addressed?. *Ambio* 18 (2), 112–118.

FAO (Food and Agriculture Organization) (2006) Livestock Impacts on the Environment: Remedies Urgently Needed'. *FAO Newsroom*, 29 November, http://www.fao.org/newsroom/en/news/2006/1000448/index.html.

FAO, International Fund for Agricultural Development and World Food Program (2015) *The State of Food Insecurity in the World 2015: Strengthening the Enabling Environment for Food Security and Nutrition*. Rome: Food and Agriculture Organization, http://www.fao.org/hunger/en/.

Farrell, B.H. (1999) Conventional or sustainable tourism? No room for choice. *Tourism Management* 20, 189–191.

Farsari, I. (2012) The development of a conceptual model to support sustainable tourism policy in North Mediterranean destinations. *Journal of Hospitality Marketing & Management* 21 (7), 710–738.

Fayos Sola, E. and Cooper, C. (2018) Conclusion: The future of tourism: innovation for inclusive sustainable development. In E. Fayos Sola and C. Cooper (eds) *The Future of Tourism: Innovation and Sustainability* (pp. 325–337). Berlin: Springer.

Fennell, D.A. (2004) Towards interdisciplinarity in tourism: Making a case through complexity and shared knowledge. *Recent Advances and Research Updates* 5 (1), 95–110.

Fennell, D.A. (2006) Evolution in tourism: The theory of reciprocal altruism and tourist-host interactions. *Current Issues in Tourism* 9 (2), 105–124.

Fennell, D.A. (2008a) Responsible tourism: A Kierkegaardian interpretation. *Tourism Recreation Research* 33 (1), 3–12.

Fennell, D.A. (2008b) Ecotourism and the myth of indigenous stewardship. *Journal of Sustainable Tourism* 16 (2), 129–149.

Fennell, D.A. (2012) Tourism and animal rights. *Tourism Recreation Research* 37 (2), 157–166.

Fennell, D.A. (2013) Ecotourism, animals and ecocentrism: A re-examination of the billfish debate. *Tourism Recreation Research* 38 (2), 189–202.

Fennell, D.A. (2014) *Ecotourism* (4th edn). London: Routledge.

Fennell, D.A. (2015) *Ecotourism* (4th edn). London: Routledge.

Fennell, D.A. (2016) Fly fishing on Vimeo: Cas(t)ing the channel. *Human Dimensions of Wildlife* 22 (1), 18–29.

Fennell, D.A. (2018a) *Tourism Ethics* (2nd edn). Toronto: Channel View.

Fennell, D.A. (2018b) *Ohnia:kara Management Plan: St. Catharines.* Ontario: Brock University.

Fennell, D.A. (2018c) On tourism, pleasure and the Summum Bonum. *Journal of Ecotourism* 17 (4), 383–400.

Fennell, D.A. (2019a) Sustainability ethics in tourism: the imperative next imperative. *Tourism Recreation Research* 44 (1), 117–130.

Fennell, D.A. (2019b) The future of ethics in tourism. In C. Cooper and E. Fayos Sola (eds) *The Future of Tourism: Innovation and Sustainability* (pp. 155–177). London: Springer.

Fennell, D.A. and Butler, R.W. (2003) A human ecological approach to tourism interaction. *International Journal of Tourism Research* 5, 1–14.

Fennell, D.A. and Ebert, K. (2004) Tourism and the precautionary principle. *Journal of Sustainable Tourism* 12 (6), 461–479.

Fennell, D.A. and Malloy, D.C. (2007) *Tourism Codes of Ethics: Practice, Theory, Synthesis.* Clevedon, UK: Channel View Publications.

Fennell, D.A. and Markwell, K. (2015) Ethical and sustainability dimensions of foodservice in Australian Ecotourism Businesses. *Journal of Ecotourism* 14 (1), 48–63.

Fennell, D.A. and Weaver, D.B. (2005) The ecotourium concept and tourism symbiosis. *Journal of Sustainable Tourism* 13 (2), 373–390.

Fennell, D.A., Plummer, R. and Marschke, M. (2008) Is adaptive co-management ethical?. *Journal of Environmental Management* 88 (1), 62–75.

Ferguson, L. (2011) Promoting gender equality and empowering women? Tourism and the third millennium development goal. *Current Issues in Tourism* 14 (3), 235–249.

Ferguson, L. and Alarcon, D.M. (2015) Gender and sustainable tourism: Reflections on theory and practice. *Journal of Sustainable Tourism* 23 (3), 401–416.

Ferraro, P.J. and Kiss, A. (2002) Direct payments to conserve biodiversity. *Science* 298, 1718–1719.

Ferraz, J., Ferreira, C.M. and Serpa, S. (2018) Special issue: Tourism, human rights, social responsibility and sustainability. *Societies*, https://www.mdpi.com/journal/societies/special_issues/tourism_human_rights.

Ferus-Comelo, A. (2014) CSR as corporate self-reporting in India's tourism industry. *Social Responsibility Journal* 10 (1), 53–67.

Filimonau, V., Dickinson, J., Robbins, D. and Huijbregts, M.A.J. (2011) Reviewing the carbon footprint analysis of hotels: Life Cycle Energy Analysis (LCEA) as a holistic method for carbon impact appraisal of tourist accommodation. *Journal of Cleaner Production* 19, 1917–1930.

Filimonau, V., Dickinson, J. and Robbins, D. (2014) The carbon impact of short-haul tourism: A case study of UK travel to Southern France using life cycle analysis. *Journal of Cleaner Production* 64, 628–638.

Finkelstein, J. (1989) *Dining Out: A Sociology of Modern Manners.* Cambridge: Polity.

Fiore, M., Silvestri, R., Conto, F. and Pellegrini, G. (2017) Understanding the relationship between green approach and marketing innovations tools in the wine sector. *Journal of Cleaner Production* 142, 4085–4091.

FirstCarbon Solutions (2018) *The Role of Technology in Sustainability*, See https://firstcarbon-solutions.com/resources/newsletters/may-2015-the-role-of-technology-in-sustainability/the-role-of-technology-in-sustainability.

Fisher, B., Turner, R.K. and Morling, P. (2009) Defining and classifying ecosystem services for decision making. *Ecological Economics* 68, 643–653.

Flora, C.B. (2004) Community dynamics and social capital. In D. Rickerl and C. Francis (eds) *Agroecosystems Analysis* (pp. 93–107). Madison, WI: American Society of Agronomy, Crop Science Society of America and Social Science Society of America.

FNNPE (1993) *Loving them to Death? Sustainable Tourism in Europe's Nature and National Parks.* Grafenau: Federation of Nature and National Parks of Europe.

FoEI (Friends of the Earth International) (2019) Website, https://foe.org/.

Foodtank (2019) Meat's Large Water Footprint: Why Raising Livestock and Poultry for Meat is so Resource-intensive, See https://foodtank.com/news/2013/12/why-meat-eats-resources/

Folke, C., Carpenter, S.R., Walker, B., Scheffer, M., Chapin, T. and Rockström, J. (2010) Resilience thinking: Integrating resilience, adaptability and transformability. *Ecology & Society* 15 (4), 20.

Fonseca, C., Pereira da Silva, C., Calado, H., Moniz, F., Bragagnolo, C., Gil, A., Phillips, M., Pereira, M. and Moreira, M. (2014) Coastal and marine protected areas as key elements for tourism in small Islands. *Journal of Coastal Research* 70, 461–466.

Font, X. and Sallows, M. (2002) Setting global sustainability standards: The sustainable tourism stewardship council. *Tourism Recreation Research* 27 (1), 21–32.

Font, X., Sanabria, R. and Skinner, E. (2003) Sustainable tourism and ecotourism certification: Raising standards and benefits. *Journal of Ecotourism* 2 (3), 213–218.

Font, X., Tapper, R., Schwartz, K. and Kornilaki, M. (2008) Sustainable supply chain management in tourism. *Business Strategy and the Environment* 17, 260–271.

Font, X., Higham, J., Miller, G. and Pourfakhimi, S. (2019) Research engagement, impact and sustainable tourism. *Journal of Sustainable Tourism* 27 (1), 1–11.

Foot, P. (1978) *Virtues and Vices and Other Essays in Moral Philosophy*, Berkeley, CA: University of California Press.

Foot, P. (2003) *Natural Goodness*, Oxford: Clarendon Press.

Forest Stewardship Council (2019) Website, https://www.fsc-uk.org/en-uk.

Forestry England (2019) *We Are Forestry England*, See https://www.forestryengland.uk/.

Forman, R.T.T. (1990) 'Ecologically Sustainable Landscapes: The Role of Spatial Configuration'. In I.S. Zonneveld and R.T.T. Forman (eds) *Changing Landscapes: An Ecological Perspective*. New York: Springer.

Forrester, J.W. (1971) *World Dynamics*. Portland, OR: Productivity Press.

Forsyth, T. (1995) Business attitudes to sustainable tourism: Self-regulation in the UK outgoing tourism industry. *Journal of Sustainable Tourism* 3 (4), 210–231.

Forum for the Future (2009) *Tourism 2023*. London: Forum for the Future.

Fox, D. and Xu, F. (2017) Evolutionary and socio-cultural influences on the feelings and attitudes towards nature: A cross-cultural study. *Asia Pacific Journal of Tourism Research* 22 (2), 187–199.

Fox, M.A. (2000) Vegetarianism and planetary health. *Ethics and the Environment* 5 (2), 163–174.

Frankena, W.K. (1963) *Ethics*. Englewood Cliffs, NJ: Prentice-Hall.

Frankl, V. (1985) *Man's Search for Meaning*. New York: Washington Square Press.

Fransson, N. and Garling, T. (1999) Environmental concern: Conceptual definitions, measurement methods, and research findings. *Journal of Environmental Psychology* 19, 369–382.

Frazier, J. (2007) Sustainable use of wildlife: The view from archaeozoology. *Journal of Nature Conservation* 15, 163–173.

Frearson, A. (2015) Designs unveiled for giant rotterdam wind turbine you could live inside. *Dezeen*, 20 April, https://www.dezeen.com/2015/04/20/designs-unveiled-for-giant-rotterdam-wind-turbine-you-could-live-inside/.

Freeman, R.E. (1984) *Strategic Management: A Stakeholder Approach*. Boston, MA: Pitman.

Friends of the Earth UK (2019) Website, https://friendsoftheearth.uk/.

Frischknecht, R., Althaus, H.J., Bauer, C., Doka, G., Heck, T., Jungbluth, N., Kellenberger, D. and Nemecek, T. (2007) The environmental relevance of capital goods in life cycle assessments of products and services. *International Journal of Life Cycle Assessment* 12, 7–17.

Fromm, E. (1964) *The Heart of Man*. New York: Harper & Row.

Fromm, E. (1968) *The Revolution of Hope: Toward a Humanized Technology*. New York: Bantam Books.

Future Foundation (2015) *Future Traveller Tribes, 2030*. London: Future Foundation.

Fyall, A. and Garrod, B. (1997) Sustainable tourism: Towards a methodology for implementing the concept. In M. Stabler (ed.) *Tourism and Sustainability: Principles to Practice* (pp. 51–68). Wallingford: CABI.

Gabarda-Mallorquí, A., Garcia, X. and Ribas, A. (2017) Mass tourism and water efficiency in the hotel industry: A case study. *International Journal of Hospitality Management* 61, 82–93.

García-Hernández, M., de la Calle-Vaquero, M. and Yubero, C. (2017) Cultural heritage and urban tourism: Historic city centres under pressure. *Sustainability* 9 (8), 1346.

Garfì, M., Tondelli, S. and Bonoli, A. (2009) Multi-criteria decision analysis for waste management in Saharawi refugee camps. *Waste Management* 29, 2729–2739.

Garrod, B. and Fyall, A. (1998) Beyond the rhetoric of sustainable tourism?. *Tourism Management* 19 (3), 199–212.

Gaughran, A. (2012) Business and human rights and the right to water. *Proceedings of the Annual Meeting (American Society of International Law)* 106, 52–55.

Gawel, E. and Bretschneider, W. (2017) Specification of a human right to water: A sustainability assessment of access hurdles. *Water International* 42 (5), 505–526.

Geels, F.W. (2004) Understanding system innovations: A critical literature review and a conceptual synthesis. In B. Elzen, F.W. Geels and K. Green (eds) *System Innovation and the Transition to Sustainability: Theory, Evidence and Policy* (pp. 19–47). Cheltenham: Edward Elgar.

Geels, F.W. (2005) The dynamics of transitions in socio-technical systems: A multi-level analysis of the transition pathway from horse-drawn carriages to automobiles (1860–1930). *Technology Analysis & Strategic Management* 17, 445–476.

Geels, F.W. (2006) The hygienic transition from cesspools to Sewer Systems (1840–1930): The dynamics of regime transformation. *Research Policy* 35, 1069–1082.

Geerts, W. (2014) Environmental certification schemes: Hotel managers' views and perceptions. *International Journal of Hospitality Management* 39, 87–96.

Geffroy, B., Sadoul, B. and Ellenberg, U. (2017) Physiological and behavioural consequences of human visitation. In D.T. Blumstein, B. Geffroy, D.S.M. Samia and E. Bessa (eds) *Ecotourism's Promise and Peril: A Biological Evaluation* (pp. 9–28). Cham: Springer.

Geopark (2019) *Welcome to the Burren and Cliffs of Moher UNESCO Global Geopark Official Website*, http://www.burrengeopark.ie/.

Getz, D. (1986) Models in tourism planning: Towards integration of theory and practice. *Tourism Management* 7, 21–32.

Getz, D. and Page, S. (2016) Progress and prospects for event tourism research. *Progress in Tourism Management* 52, 593–631.

Gheribi, E. (2017) The activities of foodservice companies in the area of corporate social responsibility – on the Example of International Fast Food Chain. *Journal of Positive Management* 8 (1), 64–77.

Ghilarov, A.M. (1995) Vernadsky's biosphere concept: An historical perspective. *Quarterly Review of Biology* 70 (2), 193–203.

Giannetti, B.F., Agostinho, F., Almeida, C.M.V.B. and Huisingh, D. (2015) A review of limitations of GDP and alternative indices to monitor human wellbeing and to manage eco-system functionality. *Journal of Cleaner Production* 87, 11–25.

Giddens, A. (2002) *Runaway World*. London: Profile Books.

Gignac, J. (2017) Local animal rights advocates protest indigenous hunt near St. Catharines. *The Star*, 14 November, See https://www.thestar.com/news/gta/2017/11/14/local-animal-rights-advocates-protest-indigenous-hunt-near-st-catharines.html.

GIZ (Deutsche Gesellschaft für Internationale Zusammenarbeit) (2019) Website, https://www.giz.de/en/html/index.html.

Gjerdalen, G. and Williams, P.W. (2000) An evaluation of the utility of a whale watching code of conduct. *Tourism Recreation Research* 25 (2), 27–37.

Gligor-Cimpoieru, D.C., Munteanu, V.P., Nițu-Antonie, R.D., Schneider, A. and Preda, G. (2017) Perceptions of Future Employees toward CSR Environmental Practices in Tourism. *Sustainability* 9, 1631.

Globe '90 (1991) *Tourism Stream Conference: Action Strategy for Sustainable Tourism Development.* Ottawa: Tourism Canada.

Goetz, J., Keltner, D. and Simon-Thoms, E. (2010) Compassion: An evolutionary analysis and empirical review. *Psychological Bulletin* 136, 351–374.

Goggins, G. and Rau, H. (2016) Beyond calorie counting: Assessing the sustainability of food provided for public consumption. *Journal of Cleaner Production* 112, 257–266.

Goodall, B. (1995) Environmental auditing: A tool for assessing the environmental performance of tourism firms. *Geographical Journal* 161 (1), 29–37.

Goodall, B. and Cater, E. (1996) Self-regulation for Sustainable Tourism? *Ecodecision* 20 (Spring), 43–45.

Goodman, A. (2000) Implementing sustainability in service operations at Scandic Hotels. *Interfaces* 30 (3), 202–214.

Goodwin, H. (2005) Responsible tourism and the market. Occasional Paper No. 4, International Centre for Responsible Tourism, Faversham.

Goodwin, H., Spenceley, A. and Maynard, B. (2002) Development of responsible tourism guidelines for South Africa. NRI Report No. 2692, Project Code V0149, DFID, London.

Gordon, H.S. (1954) The economic theory of a common property resource: The fishery. *Journal of Political Economy* 62, 124–142.

Gössling, S. (1999a) Sustainable tourism development in developing countries: Some aspects of energy use. *Journal of Sustainable Tourism* 8 (5), 410–425.

Gössling, S. (1999b) Ecotourism: A means to safeguard biodiversity and ecosystem functions? *Ecological Economics* 29, 303–320.

Gössling, S. (2001a) The consequences of tourism for sustainable water use on a Tropical Island: Zanzibar, Tanzania. *Journal of Environmental Management* 61, 179–191.

Gössling, S. (2001b) Tourism, economic transition and ecosystem degradation: Interacting processes in a Tanzanian coastal community. *Tourism Geographies* 3 (4), 430–453.

Gössling, S. (2002) Global environmental consequences of tourism. *Global Environmental Change* 12, 283–302.

Gössling, S. (2015) New performance indicators for water management in tourism. *Tourism Management* 46, 233–244.

Gössling, S. and Hall, C.M. (eds) (2006) *Tourism and Global Environmental Change*. London: Routledge.

Gössling, S., Hall, C.M. and Weaver, D.B. (eds) (2008) *Sustainable Tourism Futures: Perspectives on Systems, Restructuring and Innovations*. London: Routledge.

Gössling, S., Hall, C.M., Peeters, P. and Scott, D. (2010) The future of tourism: Can tourism growth and climate policy be reconciled? A mitigation perspective. *Tourism Recreation Research* 35 (2), 119–130.

Gössling, S., Peeters, P., Hall, C.M., Ceron, J.P., Dubois, G., Lehmann, L.V. and Scott, D. (2012) Tourism and water use: Supply, demand, and security. An international review. *Tourism Management* 33, 1–15.

Gössling, S., Hall, C.M. and Scott, D. (2015) *Tourism and Water*. Clevedon: Channel View.

Government of PEI Environmental Advisory Council (2013) *Principles of Sustainable Development 2013*. Charlottetown: EAC, See https://www.princeedwardisland.ca/sites/default/files/publications/principles_of_sustainable_development.pdf.

Government of Quebec (2018) *The Principles of Sustainable Development: A Guide for Action*, See http://www.environnement.gouv.qc.ca/developpement/principes_en.htm.

Graci, S. and Dodds, R. (2010) *Sustainable Tourism in Island Destinations*. London: Routledge.

Graham, J., Amos, B. and Plumptre, T. (2003) *Governance Principles for Protected Areas in the 21st Century: A Discussion Paper*. Ottawa: Institute on Governance.

Green Marine (2018) Website, www.green-marine.org.

Greenpeace (2019) Website, https://www.greenpeace.org.uk/.

Gretzel, U. and Jamal, T. (2009) Conceptualizing the creative tourist class: Technology, mobility, and tourism experiences. *Tourism Analysis* 14 (4), 471–481.

Gretzel, U., Sigala, M., Xiang, Z. and Koo, C. (2015) Smart tourism: Foundations and developments. *Electronic Markets* 25 (3), 179–188.

Grey, D. and Sadoff, C.W. (2007) Sink or swim? Water security for growth and development. *Water Policy* 9, 545–571.

Greyhound Bus Company (2019) *Going Green*, See https://www.greyhound.com/en/discover-greyhound/going-green.

GRI (Global Reporting Initiative) (2018) *About GRI*, See https://www.globalreporting.org/Information/about-gri/Pages/default.aspx.

Griggs, D., Stafford-Smith, M., Gaffney, O., Rockstrom, J., Ohman, M.C., Shyamsundar, P., Steffen, W., Glaser, G., Kanie, N. and Noble, I. (2013) Sustainable development goals for people and planet. *Nature* 495, 305–307.

Gronau, W. and Kagermeier, A. (2007) Key factors for successful leisure and tourism public transport provision. *Journal of Transport Geography* 15, 127–135.

Grønhøj, A. and Thøgersen, J. (2017) Why young people do things for the environment: The role of parenting for adolescents' motivation to engage in pro-environmental behaviour. *Journal of Environmental Psychology* 54, 11–19.

Gross, J.E., Woodley, S., Welling, L.A. and Watson, J.E.M. (eds) (2016) *Adapting to Climate Change: Guidance for Protected Area Managers and Planners*. Best Practice Protected Area Guidelines Series No. 24. Gland, Switzerland: IUCN.

Grosser, K. (2009) Corporate social responsibility and gender equality: Women as stakeholders and the european union sustainability strategy. *Business Ethics: A European Review* 18 (3), 290–307.

Grunert, K.G. (2011) Sustainability in the food sector: A consumer behaviour perspective. *International Journal of Food System Dynamics* 2 (3), 207–218.

GSTC (2016) *Hotel Criteria, Version 3, with Suggested Performance Indicators*. Washington, DC: Global Sustainable Tourism Council, See https://www.gstcouncil.org/wp-content/uploads/2015/11/GSTC-Hotel_Industry_Criteria_with_hotel_indicators_21-Dec-2016_Final.pdf.

GSTC (Global Sustainable Tourism Council) (2019a) *What is the GSTC?*, See https://www.gstcouncil.org/about/about-us/.

GSTC (Global Sustainable Tourism Council) (2019b) Website, https://www.gstcouncil.org.

GSTC (2019c) *Roadmap for Sustainable Destinations*. Washington, DC: Global Sustainable Tourism Council.

Guerrero, L.A., Maas, G. and Hogland, W. (2013) Solid waste management challenges for cities in developing countries. *Waste Management* 33, 220–232.

Gulenko, M. (2018) Mandatory CSR reporting – Literature review and future developments in Germany. *Nachhaltigkeits Management Forum* 26 (1–4), 3–17.

Gullino, P., Beccaro, G.L. and Larcher, F. (2015) Assessing and monitoring the sustainability in rural world heritage sites. *Sustainability* 7, 14186–14210.

Gullone, E. (2000) The biophilia hypothesis and life in the 21st century: Increasing mental health or increasing pathology? *Journal of Happiness Studies* 1, 293–321.

Gunderson, R. (2014) Social barriers to biophilia: Merging structural and ideational explanations for environmental degradation. *Social Science Journal* 51, 681–685.

Guttentag, D.A. (2010) Virtual reality: Applications and implications for tourism. *Tourism Management* 31 (5) 637–651.

Guttentag, D.A. and Griffin, T. (2018) Virtual tourism/augmented reality. In C. Cooper, W. Gartner, N. Scott and S. Volo (eds) *The Sage Handbook of Tourism Management.* London: Sage.

Habibullah, N.S., Din, B.H., Chong, C.W. and Radam, A. (2016) Tourism and biodiversity loss: Implications for business sustainability. *Procedia Economics and Finance* 35, 166–172.

Hadjikakou, M., Chenoweth, J. and Miller, G. (2013) Estimating the direct and indirect water use of tourism in the eastern Mediterranean. *Journal of Environmental Management* 114, 548–556.

Haliburton Forest & Wild Life Reserve (2019) Website, https://www.haliburtonforest.com/.

Halkos, G. and Matsiori, S. (2017) Environmental attitude, motivations and values for marine biodiversity protection. *Journal of Behavioral and Experimental Economics* 69, 61–70.

Hall, C.M. (1996) *Introduction to Tourism in Australia: Impacts, Planning and Development.* Melbourne: Addison Wesley Longman.

Hall, C.M (2005) *Rethinking the Social Science of Mobility Prentice Hall.* Harlow.

Hall, C.M. (2009) Degrowing tourism: Décroissance, sustainable consumption and steady-state tourism. *Anatolia: An International Journal of Tourism and Hospitality Research* 20 (1), 46–61.

Hall, C.M. (2010a) Tourism and biodiversity: More significant than climate change? *Journal of Heritage Tourism* 5 (4), 253–266.

Hall, C.M. (2010b) Changing paradigms and global change: From sustainable to steady-state tourism. *Tourism Recreation Research* 35 (2), 131–143.

Hall, C.M. (2011a) Consumerism, tourism and voluntary simplicity: We all have to consume, but do we really have to travel so much to be happy? *Tourism Recreation Research* 36 (3), 298–303.

Hall, C.M. (2011b) Policy learning and policy failure in sustainable tourism governance: From first- and second-order to third-order change? *Journal of Sustainable Tourism* 19 (4–5), 649–671.

Hall, C.M. and Lew, A.A. (eds) (1998) *Sustainable Tourism: A Geographical Perspective.* New York: Addison Wesley Longman.

Hall, D. and Brown, F. (2008) The tourism industry's welfare responsibilities: An adequate response? *Tourism Recreation Research* 33 (2), 213–218.

Hall, C.M., Sharples, L., Mitchell, R., Macionis, N. and Cambourne, B. (2003) *Food Tourism around the World: Development, Management and Markets.* New York: Routledge.

Hall, C.M., Gössling, S. and Scott, D. (2015) *The Routledge Handbook of Tourism and Sustainability.* London: Routledge.

Hall, C.M., Prayag, G. and Amore, A. (2017) *Tourism and Resilience: Individual, Organisational and Destination Perspectives.* Bristol: Channel View.

Hamilton, W.D. (1964) The genetical evolution of social behaviour (I and II). *Journal of Theoretical Biology* 7, 1–52.

Hamilton, M., Fischer, A.P., Guikema, S.D. and Keppel-Aleks, G. (2018) Behavioral Adaptation to Climate Change in Wildfire-prone Forests. *Wiley Interdisciplinary Reviews: Climate Change* doi: 10.1002/wcc.553.

Han, Z., Song, W., Deng, X. and Xu, X. (2017) Trade-offs and synergies in ecosystem service within the three-rivers headwater region, China. *Water* 9, 588.

Hannam, K., Sheller, M and Urry, J. (2006) Editorial: Mobilities, immobilities and moorings. *Mobilities* 1 (1), 1–22.

Hannam, K., Butler, G. and Paris, C.M. (2014) Developments and key issues in tourism mobilities. *Annals of Tourism Research* 44, 171–185.

Hanski, I. (2011) Habitat loss, the dynamics of biodiversity, and a perspective on conservation. *Ambio* 40, 248–255.

Hardesty, D.L. (1975) The niche concept: Suggestions for its use in human ecology. *Human Ecology* 3, 71–85.

Hardin, G. (1968) The tragedy of the commons. *Science* 162, 1243–1248.

Hardy, A.L. and Beeton, R.J. (2001) Sustainable tourism or maintainable tourism: Managing resources for more than average outcomes. *Journal of Sustainable Tourism* 9 (3), 168–179.

Harrington, I. (1971) The trouble with tourism unlimited. *New Statesman* 82, 176.

Harris, R., Griffin, T. and Williams, P. (2012) *Sustainable Tourism* (2nd edn). London: Routledge.

Harrison, L.C., Jayawardena, C. and Clayton, A. (2003) Sustainable tourism development in the Caribbean: Practical challenges. *International Journal of Contemporary Hospitality Management* 15 (5), 294–298.

Hartmann, M. (2011) Corporate social responsibility in the food sector. *European Review of Agricultural Economics* 38 (3), 297–324.

Hawkes, S. and Williams, P. (1993) *The Greening of Tourism: From Principles to Practice.* Burnaby, British Columbia: Centre for Tourism Policy and Research, Simon Fraser University.

Hawkins, D.E. and Mann, S. (2007) The world Bank's role in tourism development. *Annals of Tourism Research* 34 (2), 348–363.

Hawkins, R. and Bohdanowicz, P. (2011) *Responsible Hospitality.* Oxford: Goodfellow.

Hayward, T. (2016) A global right of water. *Midwest Studies on Philosophy* xl, 217–233.

Haywood, K.M. (1988) Responsible and responsive tourism planning in the community. *Tourism Management* June, 105–118.

Haywood, K.M. (1993) Sustainable development for tourism: A commentary with an organizational perspective. In J.G. Nelson, R.W. Butler and G. Wall (eds) *Tourism and Sustainable Development: Monitoring, Planning, Managing* (pp. 233–241). Waterloo, ON: Heritage Resources Centre.

He, C., Liu, Z., Tian, J. and Ma, Q. (2014) Urban expansion dynamics and natural habitat loss in China: A multiscale landscape perspective. *Global Change Biology* 20, 2886–2902.

Healy, R.G. (1992) The role of tourism in sustainable development. Paper presented at the 4th World Congress on Parks and Protected Areas, Caracas, Venezuela, 10–21 February.

Hebb, T. (2019) Investing in sustainable infrastructure. In S. Arvidsson (ed.) *Challenges in Managing Sustainable Business* (pp. 251–273). London: Palgrave Macmillan.

Heintzman, P. (1995) Leisure, ethics, and the golden rule. *Journal of Applied Recreation Research* 20 (3), 203–222.

Held, V. (2004) Care and justice in the global context. *Ratio Juris* 17 (2), 141–155.

Henry, I.P. and Jackson, G.A.M. (1996) Sustainability of management processes and tourism products and contexts. *Journal of Sustainable Tourism* 4 (1), 17–28.

Herr, M.L., Muzira, T.J. and ILO (2009) *Value Chain Development for Decent Work: A Guide for Development Practitioners, Government and Private Sector Initiatives.* Geneva: International Labour Office.

Herskovitz, J. (2014) Permit to hunt endangered rhino sells for $350,000 despite protests. *Scientific American*, See https://www.scientificamerican.com/article/permit-to-hunt-endangered-rhino-sel/.

Hervani, A.A., Helms, M.M. and Sarkis, J. (2005) Performance measurement for green supply chain management. *Benchmarking: An International Journal* 12 (4), 330–353.

HES (Hotel Energy Solutions) (2019) Website, www.hotelenergysolutions.net.

Higginbottom, K. (2004) Wildlife tourism: An introduction. In K. Higginbottom (ed.) *Wildlife Tourism: Impacts, Management and Planning* (pp. 1–14). Gold Coast, Australia: Common Ground Publishing in association with the Cooperative Research Centre for Sustainable Tourism.

Higgins-Desboilles, F. (2008) Justice tourism and alternative globalisation. *Journal of Sustainable Tourism* 16 (3), 345–364.

Higgins-Desboilles, F. (2009) Indigenous ecotourism's role in transforming ecological consciousness. *Journal of Ecotourism* 8 (2), 144–160.

Higgins-Desboilles, F. (2010) The elusiveness of sustainability in tourism: The culture-ideology of consumerism and its implications. *Tourism and Hospitality Research* 10 (2), 116–129.

Higham, J.E.S. and Bejder, L. (2008) Managing wildlife-based tourism: Edging slowly towards sustainability? *Current Issues in Tourism* 11 (1), 75–83.

Hillary, R. (ed.) (2000) *Small and Medium Sized Enterprises and the Environment.* Sheffield: Greenleaf.

Hilton (2019) *Environmental Impact*, See www.cr.hilton/environment.

Hinch, T. and Butler, R. (1996) *Indigenous Tourism: A Common Ground for Discussion.* Wallingford: CABI.

Hindrichson, D. (1994) Putting the bite on planet earth: Rapid human population growth is devouring global natural resources. *International Wildlife*, September/October, See http://dieoff.org/page120.htm.

HIS Hotel Group (2019) Website, www.h-n-h.jp/en/.

Hjalager, A.M. (1997) Innovation patterns in sustainable tourism. *Tourism Management* 18 (1), 35–41.

Hjalager, A.M. (1999) Consumerism and sustainable tourism. *Journal of Travel & Tourism Marketing* 8 (3), 1–20.

Hjalager, A.M. (2010) Regional innovation systems: The case of angling tourism. *Tourism Geographies* 12, 192–216.

Hjalager, A.M. and Richards, G. (eds) (2002) *Tourism and Gastronomy*. New York: Routledge.

Hoekstra, A.Y. (2008) Water Neutral: Reducing and Offsetting the Impacts of Water Footprints. Value of Water Research Report Series No. 28, UNESCO-IHW, Delft.

Hoekstra, J.M., Boucher, T.M., Ricketts, T.H. and Roberts, C. (2005) Confronting a biome crisis: Global disparities of habitat loss and protection. *Ecology Letters* 8, 23–29.

Hof, A. and Schmitt, T. (2011) Urban and tourist land use patterns and water consumption: Evidence from Mallorca, Balearic Islands. *Land Use Policy* 28 (4), 792–804.

Hofferth, S.L. (2009) Changes in American children's time – 1997 to 2003. *Electronic International Journal of Time Use Research* 6, 26–47.

Holcomb, J.L., Upchurch, R.S. and Okumus, F. (2007) Corporate social responsibility: What are top hotel companies reporting? *International Journal of Contemporary Hospitality Management* 19 (6), 461–475.

Holden, A. (1999) High impact tourism: A suitable component of sustainable policy? The case of downhill skiing development at Cairngorm, Scotland. *Journal of Sustainable Tourism* 7 (2), 97–107.

Holden, A. (2005) Achieving a sustainable relationship between common pool resources and tourism: The role of environmental ethics. *Journal of Sustainable Tourism* 13 (4), 339–352.

Holland America Line (2019) *About Us – Sustainability*, See https://www.hollandamerica.com/en_US/our-company/sustainability.html.

Holland America Line (n.d.) *Health, Environment, Safety, Security and Sustainability Policy*. Seattle: Holland America Line, See http://www.hollandamerica.com.

Holling, C.S. (1973) Resilience and stability of ecological systems. *Annual Review of Ecology, Evolution, and Systematics* 4, 1–23.

Holt, A.R., Alix, A., Thompson, A. and Maltby, L. (2016) Food production, ecosystem services and biodiversity: We can't have it all everywhere. *Science of the Total Environment* 573, 1422–1429.

Hopcraft, J.G.C., Bigurube, G., Lembeli, J.D. and Borner, M. (2015) Balancing conservation with national development: A socio-economic case study of the alternatives to the Serengeti Road. *PLoS ONE* 10 (7), e0130577.

Hopkins, R. (2008) *The Transition Handbook: From Oil Dependency to Local Resilience.* Totnes: Green Books.

Hopkins, D. and Higham, J. (2018) Climate change and tourism: Mitigation and global climate agreements. In C. Cooper, S. Volo, W.C. Gartner and N. Scott (eds) *The Sage Handbook of Tourism Management*, Volume 2 (pp. 422–436). London: Sage.

Hřebíček, J., Faldík, O., Kasem, E. and Trenz, O. (2015) Determinants of sustainability reporting in food and agriculture sectors. *ACTA Universitatis Agriculturae et Silviculturae Mendelianae Brunensis* 63 (2), 539–552.

Huang, G., Zhou, W. and Ali, S. (2011) Spatial patterns and economic contributions of mining and tourism in biodiversity hotspots: A case study in China. *Ecological Economics* 70, 1492–1498.

Hübner, A., Phong, L.T. and Châu, T.S.H. (2014) Good governance and tourism development in protected areas: The case of Phong Nha-Ke Bang National Park, Central Vietnam. *Koedoe* 56 (2), Art. 1146.

Hudson, S. and Miller, G. (2005) The responsible marketing of tourism: The case of Canadian Mountain Holidays. *Tourism Management* 26, 133–142.

Huggett, R.J. (1999) Ecosphere, biosphere, or gaia? What to call the global ecosystem. *Global Ecology and Biogeography* 8, 425–431.

Hughes, G. (1995) The cultural construction of sustainable tourism. *Tourism Management* 16 (1), 49–59.

Hughes, E. and Scheyvens, R. (2016) Corporate social responsibility in tourism post-2015: A development first approach. *Tourism Geographies* 18 (5), 469–482.

Hughes, T.P., Baird, A.H., Bellwood, D.R., Card, M., Connolly, S.R., Folke, C., Grosberg, R., Hoegh-Guldberg, O., Jackson, J.B.C., Kleypas, J., Lough, J.M., Marshall, P., Nyström, M., Palumbi, S.R., Pandolfi, J.M., Rosen, B. and Roughgarden, J. (2003) Climate change, human impacts, and the resilience of coral reefs. *Science* 301, 929–933.

Hughes, M., Weaver, D.B. and Pforr, C. (ed.) (2015) *The Practice of Sustainable Tourism: Resolving the Paradox.* London: Routledge.

Humphrey, M. (2000). "Nature" in deep ecology and social ecology: Contesting the core. *Journal of Political Ideologies* 5 (2), 247–268.

Humphreys, A. and Grayson, K. (2008) The intersecting roles of consumer and producer: A critical perspective on co-production, co-creation and prosumption. *Sociology Compass* 2, 1–18.

Hunt, C.A. and Durham, W.H. (2012) Shrouded in fetishistic mist: Commoditisation of sustainability in tourism. *International Journal of Tourism Anthropology* 2 (4), 330–347.

Hunt, C. and Stronza, A. (2014) Stage-based tourism models and resident attitudes towards tourism in an emerging destination in the developing world. *Journal of Sustainable Tourism* 22 (2), 279–298.

Hunter, C. (1995) On the need to re-conceptualise sustainable tourism development. *Journal of Sustainable Tourism* 3 (3), 155–165.

Hunter, C. (1997) Sustainable tourism as an adaptive paradigm. *Annals of Tourism Research* 24 (4), 850–867.

Hunter, C. (2002) Sustainable tourism and the touristic ecological footprint. *Environment, Development and Sustainability* 4, 7–20.

Hunter, C. and Green, H. (1995) *Tourism and the Environment: A Sustainable Relationship?* London: Routledge.

Huntsman, A.G. (1944) Fishery depletion. *Science* XCIX, 534.

Husbands, W. and Harrison, L.V. (1996) Practicing responsible tourism: Understanding tourism today to prepare for tomorrow. In W. Husbands and L.C. Harrison (eds) *Practicing Responsible Tourism: International Case Studies in Tourism Planning, Policy and Development* (pp. 1–15). New York: John Wiley & Sons.

Hwang, D. and Stewart, W.P. (2017) Social capital and collective action in rural tourism. *Journal of Travel Research* 56 (1), 81–93.

IAATO (International Association of Antarctica Tour Operators) (2019) *Guidance for Visitors to the Antarctic*, See https://iaato.org/en_GB/visitor-guidelines.

IATA (2018) *IATA Factsheet: Climate Change & CORSIA*. Montreal: International Air Transport Association.

Ibitayo, D.O. (2017) International obligations and African norms: Sustainable development goals and the reality of gender equality in Africa. In B.U. Anaemene and O.J. Bolarinwa (eds) *Agenda 2030 and Africa's Development in the 21st Century* (pp. 197–210). Kuapa Lumpur: UN University International Institute for Global Health.

IEA (International Energy Agency) (2006) *Key World Energy Statistics*. Paris.

IEA (International Energy Agency) (2017a) *World Energy Outlook 2017*, See https://www.iea.org/weo2017/#section-1-5.

IEA (International Energy Agency) (2017b) *Key World Energy Statistics 2017*, See https://www.iea.org/publications/freepublications/publication/keyworld2017.pdf.

iEduNote (2018) *Difference Between Values and Attitudes*, See https://iedunote.com/values-attitudes-difference.

IFPRI (2013) *Global Hunger Index. The Challenge of Hunger: Building Resilience to Achieve Food and Nutrition Security*. Washington, DC: International Food Policy Research Institute.

IHG (2018a) *2017 Responsible Business Executive Summary*. Denham: Intercontinental Hotels Group.

IHG (Intercontinental Hotels Group) (2018b) *IHG Green Engage System*, See https://www.ihg.com/content/us/en/about/green-engage.

IHRA (2014) *Managing the Future of Hospitality, Today*. Geneva: International Hotel & Restaurant Association.

IHRA (International Hotel & Restaurant Association) (2019) Website, www.ih-ra.org.

IISD (1992) *Indicators for Sustainable Management of Tourism: Report of the International Working Group on Indicators of Sustainable Tourism to the Environment Committee World Tourism Organisation*. Winnipeg: International Institute for Sustainable Development.

ILO (International Labour Organization) (2016) *ILO Implementation Plan 2030: Agenda for Sustainable Development*, See https://www.ilo.org/wcmsp5/groups/public/---dgreports/---dcomm/---webdev/documents/publication/wcms_510122.pdf.

ILO (International Labour Organization) (2018) *Decent Work*, See https://www.ilo.org/global/topics/decent-work/lang--en/index.htm.

IMO (International Maritime Organization) (2019) Website, www.imo.org.

Indian Express (2015) In "Smart City" move, NDMC plans solar panels in schools. *Indian Express* 21 April, See https://indianexpress.com/article/cities/delhi/in-smart-city-move-ndmc-plans-solar-panels-in-schools/.

Indriatmoko, Y. (2010) Rapid human population growth and its impacts on Danau Sentarum. *Borneo Research Bulletin* 41,101–108.

Intrepid Travel (2019) Website, intrepidtravel.com.

Intergovernmental Panel on Climate Change (IPCC) (2019) Website, www.ipcc.ch/.

Ireland (2019) *Ireland Leave No Trace: Promoting Responsible Use of the Outdoors*, See http://www.leavenotraceireland.org/.

Irish Department of Justice and Equality (2018) *Gender Equality in Ireland*, See http://genderequality.ie/en/GE/Pages/WP13000092.

Isaac, A.-Y., Nielsen, M. and Nielsen, R. (2017) *Comparison of Seafood and Agricultural Ecological Premiums*, See http://orgprints.org/31426/1/Premium%20Ecolabels.pdf.

Ishwaran, N., Persic, A. and Tri, N.H. (2008) Concept and practice: The case of UNESCO biosphere reserves. *International Journal of Environment and Sustainable Development* 7 (2), 118–131.

ISO (2010) *ISO 26000:2010 Guidance on Social Responsibility*. Geneva: International Organization for Standardization.

ISO (2012) *Sustainable Events with ISO*. Geneva: International Organization for Standardization.

ISO (2018) *ISO 14000 Family – Environmental Management*. Geneva: International Organization for Standardization, See https://www.iso.org/iso-14001-environmental-management.html.

Isoitok Camp Manyara (2017) Website, http://isoitok.com/.

ISSG (2004) *100 of the World's Worst Invasive Alien Species*. Auckland: Invasive Species Specialist Group, See http://www.issg.org/pdf/publications/worst_100/english_100_worst.pdf.

ISSG (2015) *About the Global Invasive Species Database*. Auckland: Invasive Species Specialist Group, See http://www.iucngisd.org/gisd/about.php.

ISSG (2018) *Invasive Species Specialist Group*, See http://www.issg.org/publications.htm.

ITC (International Training Centre) (2018) *Decent Work and Sustainable Development*, See https://www.itcilo.org/en/areas-of-expertise/decent-work-and-sustainable-development.

ITP (International Tourism Partnership) (2019) Website, www.tourismpartnership.org.

IUCN (2008) *Biodiversity: My Hotel in Action – A Guide to Sustainable Use of Biological Resources.* Gland: International Union for the Conservation of Nature.

IUCN (2012) *Integrating Business Skills into Ecotourism Operations.* Vienna: IUCN

IUCN (International Union for the Conservation of Nature) (2019a) *Protected Areas,* See https://www.iucn.org/theme/protected-areas/about.

IUCN (International Union for the Conservation of Nature) (2019b) *Protected Area Categories,* See https://www.iucn.org/theme/protected-areas/about/protected-area-categories.

IUCN (International Union for the Conservation of Nature) (2019c) Website, https://www.iucn.org/.

IUCN (International Union for the Conservation of Nature) (2019d) *Tourism – TAPAS,* See https://www.iucn.org/commissions/world-commission-protected-areas/our-work/tourism-tapas.

Ives, C.D. and Kendal, D. (2013) values and attitudes of the urban public towards peri-urban agricultural land. *Land Use Policy* 34, 80–90.

Jafari, J. (1990) The basis of tourism education. *Journal of Tourism Studies* 1, 33–41.

Jain, S.C. (1996) *Marketing Planning and Strategy* (5th edn). Cincinnati, OH: Southwestern College.

Jain, P. and Jibril, L. (2018) Achieving sustainable development through libraries: Some preliminary observations from botswana public libraries. Paper presented at IFLA WLIC 2018, Kuala Lumpur.

Jakes (2019) Hotel website, www.jakeshotel.com.

Jamal, T. (2004) Virtue ethics and sustainable tourism pedagogy: Phronesis, principles and practice. *Journal of Sustainable Tourism* 12 (6), 530–545.

Jamal, T. and Getz, D. (1995) Collaboration theory and community tourism planning. *Annals of Tourism Research* 22 (1), 186–204.

Jang, Y.C., Hong, S., Lee, J., Lee, M.J. and Shim, W.J. (2014) Estimation of lost tourism revenue in Geoje Island from the 2011 marine debris pollution event in South Korea. *Marine Pollution Bulletin* 81, 49–54.

Jarman-Walsh, J. (2018) 2030 sustainable development goals (SDGs) and global hotel chain in leadership models in Japan. *Journal of Yasuda Women's University* 46, 199–206.

Jarvis, N., Weeden, C. and Simcock, N. (2010) The benefits and challenges of sustainable tourism certification: A case study of the green tourism business scheme in the west of England. *Journal of Hospitality and Tourism Management* 17 (1), 83–93.

Javorić Barić, D. (2016) Water agreements in central Asia and their impact on human rights. *Pravnik* 50 (1), 123–132.

Jenner, S. and Lamadrid, A.J. (2013) Shale Gas vs. Coal: Policy implications from environmental impact comparisons of shale gas, conventional gas, and coal on air, water, and land in the United States. *Energy Policy* 53, 442–453.

Jennifer, S.M., Anu, C. and Arun, D.O.P. (2018) Probing the hosts' and tourists' responsibilities and participation towards solid waste management in destination Wayanad. *Journal of Hospitality Application & Research* 13 (1), 57–72.

Jensen, M.C. (1994) Self-interest, altruism, incentives and agency theory. *Journal of Applied Corporate Finance* 7 (2), 40–45.

Jensen, M.C. and Meckling, W.H. (1994) The nature of man. *Journal of Applied Corporate Finance* 7 (2), 4–19.

Johnson, D. (2002) Environmentally sustainable cruise tourism: A reality check. *Marine Policy* 26, 261–270.

Johnsson-Latham, G. (2007) A Study on Gender Equality as a Prerequisite for Sustainable Development. Report to the Environment Advisory Council No. 2007:2, Stockholm.

Jones, H. (1972) Gozo – The living showpiece. *Geographical Magazine* 45 (1), 53–57.

Jonkute, G. and Staniškis, J.K. (2016) Realising sustainable consumption and production in companies: The sustainable and responsible company (SURESCOM) model. *Journal of Cleaner Production* 138, 170–180.

Jopp, R., DeLacy, T. and Mair, J. (2010) Developing a framework for regional destination adaptation to climate change. *Current Issues in Tourism* 13 (6), 591–605, doi: 10.1080/13683501003653379

Joppe, M. (1996) Sustainable community tourism development revisited. *Tourism Management* 17 (7), 475–479.

Jovicic, D.Z. (2014) Key issues in the implementation of sustainable tourism. *Current Issues in Tourism* 17 (4), 297–302.

Joye, Y. and De Block, A. (2011) "Nature and I are two": A critical examination of the Biophilia hypothesis. *Environmental Values* 20, 189–215.

Juvan, E. and Dolnicar, S. (2014) Can tourists easily choose a low carbon footprint vacation? *Journal of Sustainable Tourism* 22 (2), 175–194.

Kaennel, M. (1998) Biodiversity: A diversity in definition. In P. Bachmann, M. Köhl and R. Päivinen (eds) *Assessment of Biodiversity for Improved Forest Planning* (pp. 71–81). Dordrecht: Kluwer Academic.

Kaiser, M. (1997) The precautionary principle and its implications for science. *Foundations of Science* 2, 201–205.

Kalisch, A. (2002) *Corporate Futures: Social Responsibility in the Tourism Industry; Consultation on Good Practice.* London: Tourism Concern.

Kaltenborn, B.P. (1996) Keeping tourism under ecological limits. *Ecodecision* 20, 25–28.

Kangas, H. (2007a) European Charter for Sustainable Tourism in Protected Areas: Ecolabel as a Tool for Area Management and Tourists' Response to Labels. Study Visit Report, Europarc Federation, Alfred Toepher Foundation.

Karatzoglou, B. and Spilanis, I. (2010) Sustainable tourism in Greek Islands: The integration of activity-based environmental management with a destination environmental scorecard based on the adaptive resource management paradigm. *Business Strategy and the Environment* 19, 26–38.

Kasim, A., Gursoy, D., Okumus, F. and Wong, A. (2014) The importance of water management in hotels: A framework for sustainability through innovation. *Journal of Sustainable Tourism* 22 (7), 1090–1107.

Kassinis, G.I. and Soteriou, A.C. (2003) Greening the service profit chain: The impact of environmental management practices. *Production and Operations Management* 12 (3), 386–403.

Kellert, S.R. (1984) American attitudes toward and knowledge of animals: An update. In M.W. Fox and L.D. Mickley (eds) *Advances in Animal Welfare Science 1984/85* (pp. 177–213). Washington, DC: The Humane Society of the United States.

Kellert, S.R. (1989) Perceptions of animals in America. In R.J. Hoage (ed.) *Perceptions of Animals in American Culture* (pp. 129–151). Washington, DC: Smithsonian Institution Press.

Kellert, S.R. (2008) Dimensions, elements, attributes of Biophilic design. In S.F. Kellert, J.H. Heerwagen and M.L. Mador (eds) *Biophilic Design* (pp. 3–19). Hoboken, NJ: Wiley.

Kelly, J. and Williams, P. (2007) Modelling tourism destination energy consumption and greenhouse gas emissions: Whistler, British Columbia, Canada. *Journal of Sustainable Tourism* 15 (1), 67–90.

Kharas, H. and Zhang, C. (2014) New agenda, new narrative: What happens after 2015? *SAIS Review of International Affairs* 34 (2), 25–35.

Khazaei, A., Elliot, S. and Joppe, M. (2015) An application of stakeholder theory to advance community participation in tourism planning: The case for engaging immigrants as fringe stakeholders. *Journal of Sustainable Tourism* 23 (7), 1049–1062.

Khoo-Lattimore, C. and Ling Yang, E.C. (2018) Tourism gender studies. In C. Cooper, S. Volo, W.C. Gartner and N. Scott (eds) *The Sage Handbook of Tourism Management. Applications of Theories and Concepts to to Tourism* (pp. 256–268). London: Sage.

Khoshnood, Z. (2017) Effect of environmental pollution on fish: A short review. *Transylvanian Review of Systematical and Ecological Research* 19 (1), 29–61.

Kiddee, P., Naidu, R. and Wong, M.H. (2013) Electronic waste management approaches: An overview. *Waste Management* 33, 1237–1250.

Kiley, H.M., Ainsworth, G.B., Van Dongen, W.F.D. and Weston, M.A. (2017) Variation in public perceptions and attitudes towards terrestrial ecosystems. *Science of the Total Environment* 590–591, 440–451.

Kim, M.S., Ki, J. and Thapa, B. (2018) Influence of environmental knowledge on affect, nature affiliation and pro-environmental behaviors among tourists. *Sustainability* 10 (9), 3109.

Kinnier, R.T., Kernes, J.L. and Dautheribes, T.M. (2000) A short list of universal moral values. *Counseling and Values* 45 (1), 4–17.

Kirstges, T. (1995) *Sanfter Tourismus* (2nd edn). Munich: Oldenbourg.

Kiss, A. (2004) Is community-based ecotourism a good use of biodiversity conservation funds? *Trends in Ecology & Evolution* 19 (5), 232–237.

Kitcher, P. (1993) The evolution of human altruism. *Journal of Philosophy* 90 (10), 497–516.

Klein, R.A. (2011) Responsible cruise tourism: Issues of cruise tourism and sustainability. *Journal of Hospitality and Tourism Management* 18 (1), 107–116.

Kleindorfer, P.R., Singhal, K. and Wassenhove, L.N.V. (2005) Sustainable operations management. *Production and Operations Management* 14 (4), 482–492.

Kleinman, G., Kuei, C. and Lee, P. (2017) Using formal concept analysis to examine water disclosure in corporate social responsibility reports. *Corporate Social Responsibility and Environmental Management* 24, 341–356.

Kline, C. (ed.) (2018) *Animals, Food, and Tourism.* New York: Routledge.

Klinsky, S. and Golub, A. (2016) Justice and sustainability. In H. Heinrichs, P. Martens, G. Michelsen and A. Wiek (eds) *Sustainability Science: An Introduction* (pp. 161–173). Dordrecht: Springer.

Knapp, M., Saska, P., Knappová, J., Vonička, P., Moravec, P., Kurka, A. and Anděl, P. (2013) The habitat-specific effects of highway proximity on ground-dwelling arthropods: Implications for biodiversity conservation. *Biological Conservation* 164, 22–29.

Konan, D.E. and Chan, H.L. (2010) Greenhouse gas emissions in hawaii: household and visitor expenditure analysis. *Energy Economics* 32, 210–219.

Kotler, P. and Zaltman, G. (1971) Social marketing: An approach to planned social change. *Journal of Marketing* 35 (3), 3–12.

Kourgialas, N.N., Karatzas, G.P., Dokou, Z. and Kokorogiannis, A. (2018) Groundwater footprint methodology as policy tool for balancing water needs (agriculture and tourism) in water Scarce Islands – The Case of Crete, Greece. *Science of the Total Environment* 615, 381–389.

Kousis, M. (2000) Tourism and the environment: A social movements perspective. *Annals of Tourism Research* 27 (2), 468–489.

KPMG (2017) *The Road Ahead: The KPMG Survey of Corporate Responsibility Reporting 2017*, See https://home.kpmg.com/uk/en/home/insights/2017/11/kpmg-international-survey-of-corporate-responsibility-reporting-2017.html.

Krelling, A.P., Williams, A.T. and Turra, A. (2017) Differences in perception and reaction of tourist groups to beach marine debris that can influence a loss of tourism revenue in coastal areas. *Marine Policy* 85, 87–99.

Krippendorf, J. (1977) *Les Devoreurs des Paysages.* Lausanne: 24 Heures.

Krueger, R. (2017) Sustainable development. In D. Richardson (ed.) *The International Encyclopedia of Geography: People, the Earth, Environment, and Technology.* Chichester: Wiley (online resource and 15-volume set).

Kucukusta, D., Mak, A. and Chan, X. (2013) Corporate social responsibility practices in four- and five-star hotels: Perspectives from hong kong visitors. *International Journal of Hospitality Management* 34, 19–30.

Kuo, N.-W. and Chen, P.-H. (2009) Quantifying energy use, carbon dioxide emission, and other environmental loads from island tourism based on a life cycle assessment approach. *Journal of Cleaner Production* 17, 1324–1330.

Kur, N.T. and Hvenegaard, G.T. (2012) Promotion of ecotourism principles by whale-watching companies' marketing efforts. *Tourism in Marine Environments* 8 (3), 145–151.

Kwan, P., Eagles, P.F.J. and Gebhardt, A. (2010) Ecolodge patrons' characteristics and motivations: A study of Belize. *Journal of Ecotourism* 9 (1), 1–20.

Laderchi, C.R., Saith, R. and Stewart, F. (2003) Does it matter that we do not agree on the definition of poverty? A comparison of four approaches. *Oxford Development Studies* 31 (3), 243–274.

Lai, P. and Shafter, S. (2005) Marketing ecotourism through the internet: An evaluation of selected ecolodges in Latin America and the Caribbean. *Journal of Ecotourism* 4 (3), 143–160.

Ladkin, A. (2018) Tourism human resources. In C. Cooper, S. Volo, W.C. Gartner and N. Scott (eds) *The Sage Handbook of Tourism Management. Theories, Concepts and Disciplinary Approaches to Tourism* (pp. 256–268). London: Sage.

Lamers, M., Eijgelaar, E. and Amelung, B. (2015) The environmental challenges of cruise tourism. In C.M. Hall, S. Gossling and D. Scott (eds) *The Routledge Handbook of Tourism and Sustainability* (p. 430). London: Routledge.

Lamsfus, C., Martin, D., Alzua-Sorzabal, A. and Torres-Manzanera, E. (2015) Smart tourism destinations: An extended conception of smart cities focusing on human mobility. In I. Tussyadiah and A. Inversini (eds) *Information and Communication Technologies in Tourism 2015* (pp. 363–375). Heidelberg: Springer.

Landorf, C. (2009) Managing for sustainable tourism: A review of six cultural world heritage sites. *Journal of Sustainable Tourism* 17 (1), 53–70.

Langholz, J.A., Lassoie, J.P., Lee, D. and Chapman, D. (2000) Economic considerations of privately owned parks. *Ecological Economics* 33, 173–183.

Lankford, S.V. and Howard, D.R. (1994) Developing a tourism impact attitude scale. *Annals of Tourism Research* 21, 121–139.

Larsen, J., Axhausen, K. and Urry, J. (2006) Geographies of social networks: meetings, travel and communications. *Mobilities* 1 (2), 261–283.

Las Vegas (2019) UK website, https://www.visitlasvegas.com/uk/.

Lash, S. and Urry, J. (1994) *Economies of Signs and Space*. London: Sage.

LaVanchy, G.T. (2017) When wells run dry: Water and tourism in Nicaragua. *Annals of Tourism Research* 64, 37–50.

Leach, M., Rockström, J., Raskin, P., Scoones, I., Stirling, A.C., Smith, A., Thompson, J., Millstone, E., Ely, A., Around, E., Folke, C. and Olsson, P. (2012) Transforming innovation for sustainability. *Ecology and Society* 17 (2), 11.

Lebel, L., Anderies, J., Campbell, B., Folke, C., Hatfield-Dodds, S., Hughes, T.P. and Wilson, J. (2006) Governance and the capacity to manage resilience in regional social-ecological systems. *Ecology and Society* 11 (1), 19.

Lee, K.F. (2001) Sustainable tourism destinations: The importance of cleaner production. *Journal of Cleaner Production* 9 (4), 313–323

Lee, S. and Jamal, T. (2008) Environmental justice and environmental equity in tourism: Missing links to sustainability. *Journal of Ecotourism* 7 (1), 44–67.

Lee, C.-Y. and Heo, H. (2016) Estimating willingness to pay for renewable energy in South Korea using contingent valuation method. *Energy Policy* 94, 150–156.

Lee-Ross, D. and Pryce, J. (2010) *Human Resources and Tourism: Skills, Culture and Industry.* Clevedon: Channel View Publications.

Legrand, W., Sloan, P. and Chen, J.S. (2017) *Sustainability in the Hospitality Industry* (3rd edn). London: Routledge.

Lehtonen, A., Salonen, A.O. and Cantell, H. (2019) Climate change education: A new approach for a world of wicked problems. In J.W. Cook (ed.) *Sustainability, Human Well-being, and the Future of Education* (pp. 339–376). Cham: Palgrave Macmillan.

Leighly, J. (1987) Ecology as metaphor: Carl sauer and human ecology. *Professional Geographer* 39 (4), 405–412.

Leiper, N. (1979) The framework of tourism: towards a definition of tourism, tourist, and the tourist industry. *Annals of Tourism Research* 6 (4), 390–407.

Le Klahn, D.T. and Hall, C.M. (2015) Tourist use of public transport at destinations: A review. *Current Issues in Tourism* 18 (8), 785–803.

Lemelin, R.H., Fennell, D.A. and Smale, B. (2008) Polar bear viewers as deep ecotourists: How specialised are they? *Journal of Sustainable Tourism* 16 (1), 42–62.

Leopold, A. (1949/1989) *A Sand County Almanac and Sketches Here and There.* New York: Oxford University Press.

Le Roy, E. (1928) *Les origines humaines et l'évolution de l'intelligence, III.* Paris: La noosphère et l'hominisation, pp. 37–57.

Leslie, D. (1994) Sustainable tourism or developing sustainable approaches to lifestyle? *World Leisure and Recreation* 36 (3), 30–36.

Leung, Y.-F., Spenceley, A., Hvenegaard, G. and Buckley, R. (eds) (2018) *Tourism and Visitor Management in Protected Areas: Guidelines for Sustainability.* Best Practice Protected Area Guidelines Series No. 27. Gland: IUCN.

Levy, S.E. and Park, S.-Y. (2011) An analysis of CSR activities in the lodging industry. *Journal of Hospitality and Tourism Management* 18, 147–154.

Lewis, S.L. and Masin, M.A. (2015) Defining the Anthropocene. *Nature* 519, 171–180.

Likhotal, A. (2007) Building a global culture of peace and sustainability. *Social Alternatives* 26 (3), 31–33.

Lim, C.C. and Cooper, C. (2009) Beyond sustainability: Optimising Island tourism development. *International Journal of Tourism Research* 11, 89–103.

Lin, T., Xie, L.H. and Liu, X.P. (2005) Social economic benefits and countermeasure of building energy-conserving. *Construction Economy* 7, 91–94.

Linden, E. (1993) Sustainable follies. *Time Magazine*, 24 May, 56–57.

Little, M.E. (2017) Innovative recycling solutions to waste management challenges in Costa Rica tourism communities. *Journal of Environmental and Tourism Analyses* 5 (1), 33–52.

Liu, Z. (2003) Sustainable tourism development: A critique. *Journal of Sustainable Tourism* 11 (6), 459–475.

Liu, Q.Q., Yu, M. and Wang, X.L. (2015) Poverty reduction within the framework of SDGs and Post-2015 development agenda. *Advances in Climate Change Research* 6, 67–73.

Livingstone, S.W., Cadotte, M.W. and Isaac, M.E. (2018) Ecological engagement determines ecosystem service evaluation: A case study from rouge national urban park in Toronto, Canada. *Ecosystem Services* 30, 86–97.

Loftus, A. (2015) Water (In)security: Securing the right to water. *Geographical Journal* 181 (4), 350–356.

Lombardi, A., Caracciolo, F., Cembalo, L., Lerro, M. and Lombardi, P. (2015) How does corporate social responsibility in the food industry matter. *New Mediterranean* 3, 2–9.

Loorbach, D. (2010) Transition management for sustainable development: A prescriptive, complexity-based governance framework. *Governance* 23 (1), 161–183.

Loorbach, D. and Rotmans, J. (2006) Managing transitions for sustainable development. In X. Olshoorn and A.J. Wieczorek (eds) *Understanding Industrial Transformation: Views from Different Disciplines* (pp. 187–206). New York: Springer.

Lopez De Avila, A. (2015) Smart destinations: XXI century tourism. Paper presented at the ENTER2015 Conference on Information and Communication Technologies in Tourism, Lugano, Switzerland, 4–6 February.

Lorek, S. and Fuchs, D. (2013) Strong sustainable consumption governance – Precondition for a Degrowth Path? *Journal of Cleaner Production* 38, 36–43.

Lovelock, J.E. (1979) *Gaia: A New Look at Life on Earth*. Oxford: Oxford University Press.

Loyau, A. and Schmeller, D.S. (2017) Positive sentiment and knowledge increase tolerance towards conservation actions. *Biodiversity and Conservation* 26 (2), 461–478.

Lucas, R.C. (1964) Wilderness perception and use: The example of the boundary waters canoe area. *Natural Resources Journal* 3 (3), 394–411.

Lück, M., Maher, P.T. and Stewart, E.J. (eds) (2010) *Cruise Tourism in Polar Regions: Promoting Environmental and Social Sustainability?* New York: Earthscan.

Luhmann, H. and Theuvsen, L. (2017) Corporate social responsibility: Exploring a framework for the agribusiness sector. *Journal of Agricultural and Environmental Ethics* 30, 241–253.

Lund, H. (2007) Renewable energy strategies for sustainable development. *Energy* 32 (6), 912–919.

Macbeth, J. (1994) To sustain is to nurture, to nourish, to tolerate and to carry on: Can tourism? Trends in sustainable rural tourism development. *Parks and Recreation Magazine* 31 (1), 42–45.

Macbeth, J., Carson, D. and Northcote, J. (2004) Social capital, tourism and regional development: SPCC as a basis for innovation and sustainability. *Current Issues in Tourism* 7 (6), 502–522.

MacCannell, D. (2013) *The Tourist: A New Theory of the Leisure Class.* Berkeley, CA: University of California Press.

MacGillivray, A. and Zadek, S. (1995) *Accounting for Change: Indicators for Sustainable Development Indicators.* London: New Economics Foundation.

Mair, J. and Jago, L. (2010) The development of a conceptual model of greening in the business events sector. *Journal of Sustainable Tourism* 18 (1), 77–94.

Maisels, F., Strinberg, S., Blake, S., *et al.* (2013) Devastating decline of forest elephants in Central Africa. *PLoS ONE* 8 (3), 1–13.

Maloni, M.J. and Brown, M.E. (2006) Corporate social responsibility in the supply chain: An application in the food industry. *Journal of Business Ethics* 68, 35–52.

Manning, T. (1996) Tourism: Where are the limits? *Ecodecision* 20 (Spring), 35–39.

Manyara, G. and Jones, E. (2007) Community-based tourism enterprises development in Kenya: An exploration of their potential as avenues of poverty reduction. *Journal of Sustainable Tourism* 15, 628–644.

Marchese, D., Reynolds, E., Bates, M.E., Morgan, H., Spierre Clark, S. and Linkov, I. (2018) Resilience and sustainability: Similarities and differences in environmental management applications. *Science of the Total Environment* 613–614, 1275–1283.

Margoni, F. and Surian, L. (2017) The emergence of sensitivity to biocentirc intentions in preschool children. *Journal of Environmental Psychology* 52, 37–42.

Markides, C. and Geroski, P. (2005) *Fast Second: How Smart Companies Bypass Radical Innovation to Enter and Dominate New Markets.* San Francisco, CA: Jossey-Bass.

Markwell, K. (1998) Taming the chaos of nature: cultural construction and lived experience in nature-based tourism. PhD Dissertation, University of Newcastle, Newcastle, Australia.

Marriott International (2017a) *Sustainability and Social Impact Goals*, Bethesda, MD: Marriott International.

Marriott International (2017b) *Marriott Social Sustainability and Social Impact Report.* Bethesda, MD: Marriott International.

Marshall, A. (2016) Antarctica's tourism industry is designed to prevent damage, but can it last? *The Guardian* 26 June, See https://www.theguardian.com/world/2016/jun/26/antarctica-tourism-regulations-cruises-field-trips.

Martinez-Ibarra, E. (2015) Climate, water and tourism: Causes and effects of droughts associated with urban development and tourism in Benidorm (Spain). *International Journal of Biometeorology* 59, 487–501.

Masdar (2019) *Deploying Clean Energy Worldwide*, See https://masdar.ae/.

Maseyk, F.J.F., Mackay, A.D., Possingham, H.P., Dominati, E.J. and Buckley, Y.M. (2017) Managing natural capital stocks for the provision of ecosystem services. *Conservation Letters* 10 (2), 211–220.

Masset, E. (2011) A review of hunger indices and methods to monitor country commitment to fighting hunger. *Food Policy* 36 (1), S102–S108.

Mason, P. (2000) Zoo tourism: The need for more research. *Journal of Sustainable Tourism* 8 (4), 333–339.

May, V. (1991) Tourism, environment and development: Values, sustainability and stewardship. *Tourism Management* 21 (1), 112–118.

Mayr, E. (1988) *Toward a New Philosophy of Biology: Observations of an Evolutionist.* Cambridge, MA: Belknap Press.

McAllister, R.R.J. and Taylor, B.M. (2015) Partnerships for sustainability governance: A synthesis of key themes. *Current Opinion in Environmental Sustainability* 12, 86–90.

McCabe, S. and Johnson, S. (2013) The happiness factor in tourism: Subjective well-being and social tourism. *Annals of Tourism Research* 41, 42–65.

McCabe, S., Minnaert, L. and Diekmann, A. (eds) (2011) *Social Tourism in Europe: Theory and Practice.* Bristol: Channel View.

McCarthy, J.J., Canziani, O.F., Leary, N.A., Dokken, D.J. and White, K.S. (2001) *Climate Change 2001: Impacts, Adaptation, and Vulnerability. Third Assessment Report of the Intergovernmental Panel on Climate Change (IPCC) Working Group II.* Cambridge: Cambridge University Press.

McClanahan, T.R. and Kaunda-Arara, B. (1996) Fishery recovery in a coral-reef marine park and its effect on the adjacent fishery. *Conservation Biology* 10, 1187–1199.

McCool, S.F. (1995) Linking tourism, the environment, and concepts of sustainability: Setting the stage. In S.F. McCool and A.E. Watson (eds) *Linking Tourism, the Environment and Sustainability* (pp. 3–7). General Technical Report INT-GTR-323. Ogden, UT: USDA.

McCool, S.F. and Bosak, K. (2015) *Reframing Sustainable Tourism.* New York: Springer.

McDermott, J. (2017) *The History of Airline (De)regulation in the United States*, See https://aeronauticsonline.com/the-history-of-airline-deregulation-in-the-united-states/.

McGehee, N.G., Lee, S., O'Bannon, T.L. and Perdue, R.R. (2010) Tourism-related social capital and its relationship with other forms of capital: An exploratory study. *Journal of Travel Research* 49 (4), 486–500.

McGeoch, M.A., Butchart, S.H.M., Spear, D., Marais, E., Kleynhans, E.J., Symes, A., Chanson, J. and Hoffman, M. (2010) Global indicators of biological invasion: Species numbers, biodiversity impact and policy responses. *Diversity and Distributions* 16, 95–108.

McKee, J.K., Sciulli, P.W., Fooce, C.D. and Waite, T.A. (2003) Forecasting global biodiversity threats associated with human population growth. *Biological Conservation* 115, 161–164.

McKeown, R. (2000) *Education for Sustainable Development Toolkit, Version 1.* Knoxville, TN: Centre for Geography and Environmental Education.

McKeown, R., Hopkins, C.A., Rizzi, R. and Chrystalbride, M. (2002) *Education for Sustainable Development Toolkit, Version 2.* Knoxville, TN: Centre for Geography and Environmental Education.

McKercher, B. (1993a) Some fundamental truths about tourism: Understanding tourism's social and environmental impacts. *Journal of Sustainable Tourism* 1 (1), 6–16.

McKercher, B. (1993b) The unrecognised threat to tourism: Can tourism survive sustainability? *Tourism Management* 14 (2), 131–136.

McKercher, B. (2016) Towards a taxonomy of tourism products. *Tourism Management* 54, 196–208.

McLaren, D. (2006) The responsible travel movement. In D. Beardsley, D. Clemmons and M. DelliPriscoli (eds) *Responsible Travel Handbook 2006* (pp. 10–13). Bennington, VT: Transition Abroad.

McMinn, S. (1997) The challenge of sustainable tourism. *The Environmentalist* 17 (2), 135–141.

McMurtry, J.J. (2010) *Living Economics: Canadian Perspectives on the Social Economy, Co-operatives, and Community Economic Development.* Toronto: Emond Montgomery.

McNelly, J. (1993) Nature and culture: Conservation needs them both. *World Leisure & Recreation* 35 (2), 34–38.

McPadden, R. and Margerum, R.D. (2014) Improving national park service and non-profit partnerships – lessons from the national trail system. *Society and Natural Resources* 27, 1321–1330.

McPherson, T. (2014) *The Rise of Resilience: Linking Resilience and Sustainability in City Planning*, See https://www.thenatureofcities.com/2014/06/08/the-rise-of-resilience-linking-resilience-and-sustainability-in-city-planning/.

MEA (Millennium Ecosystem Assessment) (2003) *Ecosystems and Human Well-being: A Framework for Assessment.* Washington, DC: Island Press.

MEA (Millennium Ecosystem Assessment) (2005) *Ecosystems and Human Well-being: Synthesis.* Washington, DC: Island Press.

Meadows, D. (1999) *Leverage Points: Places to Intervene in a System.* Hartland, VT: Sustainability Institute.

Meadows, D.H., Meadows, D.L., Randers, J. and Behrens, W.W. III (1972) *The Limits to Growth: A Report for the Club of Rome's Project on the Predicament of Mankind.* New York:

Universe Books, See http://www.donellameadows.org/wp-content/userfiles/Limits-to-Growth-digital-scan-version.pdf.

Medrado, L. and Jackson, L.A. (2016) Corporate nonfinancial disclosures: An illuminating look at the corporate social responsibility and sustainability reporting practices of hospitality and tourism firms. *Tourism and Hospitality Research* 16 (2), 116–132.

Mehra, R. and Shebi, K. (2018) *Economic Programs in India: What Works for the Empowerment of Girls and Women*. Washington, DC: 3D Program.

Mehta, B., Baez, A. and O'Loughlin, P. (2002) *International Ecolodge Guidelines*. North Bennington, VT: International Ecotourism Society.

Mele, A.R. (1994) Self-control and belief. *Philosophical Psychology* 7 (4), 419–436.

Meilleur, B.A. and Hodgkin, T. (2004) In situ conservation of crop wild relatives: Status and trends. *Biodiversity and Conservation* 13, 663–684.

Mellor, M. (1997) New woman, new earth–setting the agenda. *Organization & Environment* 10 (3), 296–308.

Méndez, V.E., Bacon, C.M., Cohen, R. and Gliessman, S.R. (eds) (2015) *Agroecology: A Transdisciplinary, Participatory and Actions-oriented Approach*. Boca Raton, FL: CRC Press.

Meylan, G., Lai, A., Hensley, J., Stauffacher, M. and Krütli, P. (2018) Solid waste management of small island developing states – the case of the seychelles: A systematic and collaborative study of Swiss and Seychellois students to support policy. *Environmental Science and Pollution Research* 25 (36), 35791–35804.

MGI (2017) *A Future that Works: Automation, Employment and Productivity*. New York: McKinsey Global Institute.

Mi, T., Qingwen, M., Fei, L., Zheng, Y., Fuller, A.M., Lun, Y., Yongxun, Z. and Jie, Z. (2015) Evaluation of tourism water capacity in agricultural heritage sites. *Sustainability* 7, 15548–15569.

Michalena, E., Hills, J. and Amat, J.P. (2009) Developing sustainable tourism, using a multicriteria analysis on renewable energy in Mediterranean islands. *Energy for Sustainable Development* 13, 129–136.

Mihalič, T. and Fennell, D. (2015) In pursuit of a more just international tourism: The concept of trading tourism rights. *Journal of Sustainable Tourism* 23 (2), 188–206.

Miles, A., Delonge, M.S. and Carlisle, L. (2017) Triggering a positive research and policy feedback cycle to support a transition to agorecology and sustainable food systems. *Agroecology and Sustainable Food Systems* 41 (7), 855–879.

Milgrath, L. (1989) An inquiry into values for a sustainable society: A personal statement. In L. Milgraph (ed.) *Envisioning a Sustainable Society* (pp. 58–87). Albany, NY: SUNY Press.

Miller, K.R. (1976) Global Dimensions of Wildlife Management in Relation to Development and Environmental Conservation in Latin America. *Proceedings of Regional Expert Consultation on Environment and Development, Bogota, 5–10 July*, Santiago: Food and Agriculture Organization.

Miller, K.R. (1978) *Planning National Parks for Ecodevelopment: Methods and Cases from Latin America.* Ann Arbor, MI: Center for Strategic Wildland Management Studies, School of Natural Resources, University of Michigan.

Miller, G. (2001) Corporate responsibility in the UK tourism industry. *Tourism Management* 22, 589–598.

Miller, G. (2003) Consumerism in sustainable tourism: A survey of UK consumers. *Journal of Sustainable Tourism* 11 (1), 17–39.

Miller, J.R. (2005) Biodiversity conservation and the extinction of experience. *Trends in Ecology & Evolution* 20 (8), 430–434.

Miller, G. and Twinning Ward, L. (2005) *Monitoring for a Sustainable Tourism Transition: The Challenge of Developing and Using Indicators.* Wallingford: CABI.

Miller, D., Merrilees, B. and Coghlan, A. (2015) Sustainable urban tourism: Understanding and developing visitor pro-environmental behaviours. *Journal of Sustainable Tourism* 23 (1), 26–46.

Millstone, E. and Lang, T. (2003) *The Penguin Atlas of Food: Who Eats What, Where and Why.* Toronto: Penguin Books.

Milne, S. and Ateljevic, I. (2001) Tourism, economic development and the global–local nexus: Theory embracing complexity. *Tourism Geographies* 3 (4), 369–393.

Milne, M.J. and Gray, R. (2007) Future prospects for corporate sustainability reporting. In J. Unerman, J. Bebbington and B. O'Dwyer (eds) *Sustainability Accounting and Accountability.* London: Routledge.

Milne, M.J., Tregidga, H. and Walton, S. (2009) Words not actions! The ideological role of sustainable development reporting. *Accounting, Auditing & Accountability Journal* 22 (8), 1211–1257.

Minnaert, L., Diekmann, A. and McCabe, S. (2012) Defining social tourism and its historical context. *Social Tourism in Europe: Theory and Practice* 18–30.

Ministry of Natural Resources and Forestry (2017) Short Hills Provincial Park to Host First Nation Deer Hunt. *Ontario*, 6 November, See https://news.ontario.ca/mnr/en/2017/11/short-hills-provincial-park-to-host-first-nation-deer-hunt.html.

Ministry of Tourism of the Republic of Bulgaria (2016) Website, http://www.tourism.government.bg/en.

Minteer, B.A. and Collins, J.P. (2005) Why we need an "Ecological Ethics". *Frontiers in Ecology and the Environment* 3 (6), 332–337.

Miranda-Ackerman, M.A. and Azzaro-Pantel, C. (2017) Extending the scope of eco-labelling in the food industry to drive change beyond sustainable agriculture practices. *Journal of Environmental Management* 204, 814–824.

Mitchell, B. (1994) Sustainable development at the village level in Bali, Indonesia. *Human Ecology* 22 (2), 189–211.

Møller, A.P. (2017) Transgenerational consequences of human visitation. In D.T. Blumstein, B. Geffroy, D.S.M. Samia and E. Bessa (eds) *Ecotourism's Promise and Peril: A Biological Evaluation* (pp. 47–58). Cham: Springer.

Molnar, J.L., Gamboa, R.L., Revenga, C. and Spalding, M.D. (2008) Assessing the global threat of invasive species to marine biodiversity. *Frontiers in Ecology and the Environment* 6, 485–492.

Molz, J.G. (2009) Representing pace in tourism mobilities: Staycations, slow travel and *the amazing race*. *Journal of Tourism and Cultural Change* 7 (4), 270–286.

Monahan, W.B. and Theobald, D.M. (2018) Climate change adaptation benefits of potential conservation partnerships. *PLoS ONE* 13 (2): e0191468.

Montoya, I.D. and Richard, A.J. (1994) A comparative study of codes of ethics in health care facilities and energy companies. *Journal of Business Ethics* 13, 713–717.

Mooney, H., Larigauderie, A., Cesario, M., Elmquist, T., Hoegh-Guldberg, O., Lavorel, S., Mace, G.M., Palmer, M., Scholes, R. and Yahara, T. (2009) Biodiversity, climate change, and ecosystem services. *Current Opinion in Environmental Sustainability* 1, 46–54.

Moore, H.L. (2015) Global prosperity and sustainable development goals. *Journal of International Development* 27 (6), 801–815.

Morinville, C. and Rodina, L. (2013) Rethinking the human right to water: Water access and dispossession in Botswana's Central Kalahari Game Reserve. *Geoforum* 49, 150–159.

Moscardo, G. (2008) Sustainable tourism innovation: Challenging basic assumptions. *Tourism and Hospitality Research* 8 (1), 4–13.

Moscardo, G. (2011) Exploring social representations of tourism planning: Issues for governance. *Journal of Sustainable Tourism* 19, 423–436.

Moscardo, G. (2016) Building excellence in sustainable tourism: 15 years of building excellence in sustainable tourism education network (BEST EN) practice. *Journal of Cleaner Production* 11, 538–539.

Moscardo, G. and Murphy, L. (2014) There is no such thing as sustainable tourism: Re-conceptualizing tourism as a tool for sustainability. *Sustainability* 6 (5), 2538–2561.

Moscardo, G., Morrison, A.M. and Pearce, P.L. (1996) Specialist accommodation and ecologically-sustainable tourism. *Journal of Sustainable Tourism* 4 (1), 29–52.

Motto, A.L. and Clark, J.R. (1968) Paradoxum senecae: The epicurean stoic. *The Classical World* 62 (2), 37–41.

Moula, M.M.E., Maula, J., Hamdy, M., Fang, T., Jung, N. and Lahdelma, R. (2013) Researching social acceptability of renewable energy technologies in Finland. *International Journal of Sustainable Built Environment* 2, 89–98.

Muller, H. (1994) The thorny path to sustainable tourism development. *Journal of Sustainable Tourism* 2 (3), 131–136.

Muloin, S. (1998) Wildlife tourism: The psychological benefits of whale watching. *Pacific Tourism Review* 2 (3/4), 199–213.

Munar, A.M., Munar, A.M., Biran, A., Budeanu, A., Caton, K., Chambers, D., Dredge, D., Gyimóthy, S., Jamal, T., Larson, M., Nilsson Lindström, K., Nygaard, L. and Ram, Y. (2015) *The Gender Gap in the Tourism Academy: Statistics and Indicators of Gender Equality. While Waiting for the Dawn.* Copenhagen.

Munasinghe, M. (1994) Economic and policy issues in natural habitats and protected areas. In M. Munasinghe and J. McNeely (eds) *Protected Area Economics and Policy: Linking Conservation and Sustainable Development.* Washington, DC: World Bank, 263–269.

Munt, I. (1992) A great escape? *Town and Country Planning* July/August, 212–214.

Murphy, P.E. (1983) Tourism as a community industry: Anecological model of tourism development. *Tourism Management* 4 (3), 181.

Murphy, J., Hofacker, C. and Gretzel, U. (2017) Dawning of the age of robots in hospitality and tourism: Challenges for teaching and research. *European Journal of Tourism Research* 15, 104–111.

Murphy, P.E. (1992) Data gathering for community-oriented tourism planning: Case study of Vancouver Island, British Columbia. *Leisure Studies* 11, 65–79.

Myers, N. (1993) Biodiversity and the precautionary principle. *Ambio* 22 (2–3), 74–79.

Myers, D.G. (2003) Wealth and happiness: A limited relationship. In D. Doherty and A. Etzioni (eds) *Voluntary Simplicity: Responding to Consumer Culture* (pp. 41–52). Oxford: Rowan & Littlefield.

Myers, N., Mittermeier, R.A., Mittermeier, C.R., De Fonseca, G.A.B. and Kent, J. (2000) Biodiversity hotspots for conservation priorities. *Nature* 403, 853–858.

Naess, A. (1984) A defence of the deep ecology movement. *Environmental Ethics* 6 (3), 265–270.

Nah, S.-L. and Chau, C.-F. (2010) Issues and challenges in defeating world hunger. *Trends in Food Science & Technology* 21, 544–557.

Narayana, G.J. (2013) Water resources and tourism promotion: A case study of Hyderabad. *International Journal of Research in Commerce and Management* 4 (7), 110–112.

Nash, R.F. (1989) *The Rights of Nature: A History of Environmental Ethics.* Madison, WI: University of Wisconsin Press.

National Geographic (2015) *Unique Lodges of the World: Grootbos Private Nature Reserve*, See https://www.nationalgeographiclodges.com/lodges/africa/grootbos/#.XFH-2i57mWh.

Nava, M. (1991) Consumerism reconsidered: Buying and power. *Cultural Studies* 5 (2), 157–173.

Naylon, J. (1967) Tourism – Spain's most important industry. *Geography* 52, 23–40.

Nelson, G. (1992) Sustainable development: A heritage and human ecological perspective. *The Operational Geographer* 10 (1), 6–8.

Nelson, V. (2010) Promoting energy strategies on eco-certified accommodation websites. *Journal of Ecotourism* 9 (3), 187–200.

Netinger Grubeša, I. and Barišić, I. (2016) Environmental impact analysis of heavy metal concentrations in waste materials used in road construction. *e-GFOS* 13, 23–29.

Neto, F. (2003) A new approach to sustainable tourism development: Moving beyond environmental protection. *Natural Resources Forum* 27, 212–222.

Neuhofer, B., Buhalis, D. and Ladkin, A. (2014) A typology of technology-enhanced tourism experiences. *International Journal of Tourism Research* 16, 340–350.

New Forest National Park (2019) Website, www.newforestnpa.gov.uk.

Newman, A. (1990) *Tropical Rainforest.* New York: Facts on File.

Newswander, C.B., Matson, A. and Newswander, L.K. (2017) The recovery of self-interest well understood as a regime value: What is at stake/why this is important. *Administration & Society* 49 (4), 552–574.

NFDC (1994) *Living with the Enemy.* Lyndhurst: New Forest District Council.

NFDC (1996) *Making New Friends.* Lyndhurst: New Forest District Council.

NFDC (2003) *Our Future Together.* Lyndhurst: New Forest District Council.

NFDC (New Forest District Council) (2019) Website, www.nfdc.gov.uk.

NFTA (New Forest Tourism Association) (2019) Website, www.nfta.co.uk/.

Nguyen, N.C. and Bosch, O.J.H. (2013) A systems thinking approach to identify leverage points for sustainability: A case study in the Cat Ba biosphere reserve, Vietnam. *Systems Research and Behavioural Science* 30, 104–115.

Nibert, D. (2002) *Animal Rights and Human Rights: Entanglements of Oppression and Liberation.* New York: Rowman & Littlefield.

Nidumolu, R., Prahalad, C.K. and Rangaswami, M.R. (2009) Why sustainability is now the key driver of innovation. *Harvard Business Review* 87 (9), 57–64.

Niezgoda, A. (2004) Problems of implementing sustainable tourism in Poland. *Economics and Business Review* 4 (1), 30–42.

NITB (n.d.) *The Future of Sustainable Tourism.* Belfast: Northern Ireland Tourist Board.

Node Explorer (2019) Website, https://nodeexplorer.com/.

Nodoushani, O., Stewart, C. and Kaur, M. (2016) Recycling and its effects on the environment. *Competition Forum* 14 (1), 65–69.

Norman, D.A. (2007) *The Design of Future Things.* New York: Basic Books.

Noronha, L., Siqueira, A., Sreekesh, S., Qureshy, L. and Kazi, S. (2002) Goa: Tourism, migrations, and ecosystem transformations. *Royal Swedish Academy of Sciences* 31 (4), 295–302.

Norwegian (2018) In 10 Years, Norwegian has Reduced Emissions per Passenger by 30%, See http://nmagazine.ink-live.com/html5/reader/production/default.aspx?pubname &edida16a4935-f172-4345-bd9a-af262984c68a.

Novelli, M. and Hellwig, A. (2011) The UN millennium development goals, tourism and development: The tour operators' perspective. *Current Issues in Tourism* 14 (3), 205–220.

Nowaczek, A. and Fennell, D.A. (2002) Ecotourism in postcommunist Poland: An examination of tourists, sustainability and institutions. *Tourism Geographies* 4 (4), 372–395.

NRC (National Research Council) (2006) *Food Insecurity and Hunger in the United States: An Assessment of the Measure.* Washington, DC: National Academies Press.

NRI (2002) *What are Criteria, Indicators and Verifiers?* NRET Theme Papers on Codes of Practice in the Fresh Produce Sector No. 3. London: Natural Resources Institute.

O'Brien, D. and Ponting, J. (2013) Sustainable surf tourism: A community centered approach in Papua New Guinea. *Journal of Sport Management* 27, 158–172.

OECD (2001) *Extended Producer Responsibility: A Guidance Manual for Governments.* Paris: Organisation for Economic Cooperation and Development.

OECD (2007) *Fostering SME and Entrepreneurship Development in the Tourism Sector in Bulgaria.* Paris: Organisation for Economic Cooperation and Development.

OECD (2013) *Women and Financial Education: Evidence, Policy Responses and Guidance.* Paris: Organisation for Economic Cooperation and Development.

OECD (2015) Supporting Quality Jobs in Tourism. OECD Tourism Papers No. 2015/02, Organisation for Economic Cooperation and Development, Paris.

OECD (2017) *Policy Statement – Tourism Policies for Sustainable and Inclusive Growth.* Paris: Organisation for Economic Cooperation and Development.

OECD (2019) Website, www.oecd.org.

Offorma, G.C. and Obiefuna, C.A. (2017) Teacher educators' use of active learning strategies for attainment of sustainable development goal 4. *African Journal of Sustainable Development* 7 (1), 121–133.

Ogino, A., Sonmart, K., Subepang, S., Mitsumori, M., Hayashi, K., Yamashita, T. and Tanaka, Y. (2016) Environmental impacts of extensive and intensive beef production systems in Thailand evaluated by life cycle assessment. *Journal of Cleaner Production* 112, 22–31.

Omidiani, A. and HashemiHazaveh, S. (2016) Waste management in hotel industry in India: A review. *International Journal of Scientific and Research Publications* 6 (9), 670–680.

Ooi, N., Laing, J. and Mair, J. (2015) Social capital as a heuristic device to explore sociocultural sustainability: A case study of mountain resort tourism in the community of steamboat springs, Colorado, USA. *Journal of Sustainable Tourism* 23 (3), 417–436.

Orams, M. (1997) The effectiveness of environmental education: Can we turn tourists into "Greenies"? *Progress in Tourism and Hospitality Research* 3 (4), 295–306.

Orban, A. (2018) Norwegian Named Most Fuel-efficient Airline on Transatlantic Routes for a Second Time (and British Airways the Least Efficient). *Aviation24*, 12 September, https://www.aviation24.be/airlines/norwegian-air-shuttle/norwegian-named-most-fuel-efficient-airline-on-transatlantic-routes-for-a-second-time/.

Oreskes, N. (2004) The scientific consensus on climate change. *Science* 306, 1686.

O'Riordan, T. (1981) *Environmentalism.* London: Pion.

O'Riordan, T. and Cameron, J. (1994) The history and contemporary significance of the precautionary principle. In T. O'Riordan and J. Cameron (eds) *Interpreting the Precautionary Principle* (pp. 12–30). London: Earthscan.

Osano, P.M., Said, M.Y., De Leeuw, J., Ndiwa, N., Kaelo, D., Schomers, S., Birner, R. and Ogutu, J.O. (2013) Why keep lions instead of livestock? Assessing wildlife tourism-based payment for ecosystem services involving herders in the Maasai Mara, Kenya. *Natural Resources Forum* 37, 242–256.

Osland, G.E. and Mackoy, R. (2004) Ecolodge performance goals and evaluations. *Journal of Ecotourism* 3 (2), 109–128.

Ostrom, E. (1990) *Governing the Commons: The Evolution of Institutions for Collective Action.* New York: Cambridge University Press.

Ostrom, E., Walker, J. and Gardner, R. (1992). Covenants with and without a sword: Self-governance is possible. *The American Political Science Review* 86 (2) (June, 1992), 404–417.

Ottman, J. (1993) *Green Marketing: Challenges and Opportunities.* Lincolnwood, IL: NTC Business Books.

Oxford Living Dictionaries (2018) *Infrastructure*, See https://www.lexico.com/en/definition/infrastructure.

Padurean, L. and Maggi, R. (2011) TEFI values in tourism education: A comparative analysis. *Journal of Teaching in Travel & Tourism* 11 (1), 24–37.

Page, S.J. and Thorn, K. (2002) Towards sustainable tourism development and planning in New Zealand: The public sector response revisited. *Journal of Sustainable Tourism* 10 (3), 222–238.

Pandy, W.R. and Rogerson, C.M. (2018) Tourism and climate change: Stakeholder perceptions of at risk tourism segments in South Africa. *EuroEconomica* 1 (37), 104–118.

Paraskevaidis, P. and Andriotis, K. (2017) Altruism in tourism: Social exchange theory vs altruistic surplus phenomenon in host volunteering. *Annals of Tourism Research* 62, 26–37.

Park, D.B., Lee, K.W., Choi, H.S. and Yoon, Y. (2012) Factors influencing social capital in rural tourism communities in South Korea. *Tourism Management* 33, 1511–1520.

Park, J., Seager, T.P., Rao, P.S.C., Convertino, M. and Linkov, I. (2013) Integrating risk and resilience approaches to catastrophe management in engineering systems. *Risk Analysis* 33 (3), 356–367.

Parker, P., Rollins, R., Murray, G., Chafey, A. and Cannessa, R. (2017) Community perceptions of the contributions of parks to sustainability in Canada. *Leisure/Loisir* 41 (3), 365–389.

Parks Canada (2018) *Rouge National Urban Park*, See https://www.pc.gc.ca/en/pn-np/on/rouge.

Parsons, M. (2018) Extreme floods and river values: A social–ecological perspective. *River Research and Applications* 10.1002/rra.3355.

Parven, A. and Hasan, M.S. (2018) Trans-boundary water conflicts between Bangladesh and India: Water governance practice for conflict resolution. *International Journal of Agricultural Innovation & Technology* 8 (1), 79–84.

PATA (Pacific Asia Travel Association) (2015) *APEC/PATA Code*, See http://sustain.pata.org/about/apec-pata-code/.

PATA (Pacific Asia Travel Association) (2019) Website, See https://www.pata.org/.

Pawson, S.M., Brin, A., Brockerhoff, E.G., Lamb, D., Payn, T.W., Paquette, A. and Parrotta, J.A. (2013) Plantation forests, climate change and biodiversity. *Biodiversity Conservation* 22, 1203–1227.

Pearce, D.W. (1992) Towards Sustainable Development through Environmental Assessment. CSERGE Working Paper No. PA 92–11, Norwich: Centre for Social and Economic Research on the Global Environment.

Peeters, P., Gossling, S. and Becken, S. (2006) Innovation towards tourism sustainability: Climate change and aviation. *International Journal of Innovation and Sustainable Development* 1 (3), 184–200.

Peeters, P.M., Williams, V. and Gössling, S. (2007) Air transport greenhouse gas emissions. In P.M. Peeters (ed.) *Tourism and Climate Change Mitigation: Methods, Greenhouse Gas Reductions and Policies* (pp. 29–50). Breda: NHTV.

Peeters, P., Higham, J., Cohen, S., Eijgelaar, E. and Gössling, S. (2019) Desirable tourism transport futures. *Journal of Sustainable Tourism* 27 (2), 173–188.

Peglau, R. (2005) *ISO 14001 Certification of the World*. Berlin: Federal Environmental Agency.

Peisley, T. (2006) *The Future of Cruising*. Harlow: Pearson.

Perch-Nielsen, S., Sesartic, A. and Stucki, M. (2010) The greenhouse gas intensity of the tourism sector: The case of Switzerland. *Environmental Science & Policy* 13, 131–140.

Peters, M.A. and Reveley, J. (2015) Noosphere rising: Internet-based collective intelligence, creative labour, and social production. *Thesis Eleven* 130 (1), 3–21.

Petrevska, B., Serafimova, M. and Cingoski, V. (2018) Managing Sustainable Tourism and Hotel Industry in Macedonia: Energy Resources Strategic Approach in *Proceedings of the 26th International Conference on Ecological Truth and Environmental Research, 12–15 June, Bor Lake, Serbia* (pp. 401–406), Belgrade: University of Belgrade.

Pfattheicher, S., Sassenrath, C. and Schindler, S. (2016) Feelings for the suffering of others and the environment: Compassion fosters proenvironmental tendencies. *Environment and Behavior* 48 (7), 929–945.

Pforr, C. (2004) Policy-making for sustainable tourism. *WIT Transactions on Ecology and the Environment* 76.

Pham Phu, S.T.P., Hoang, M.G. and Fujiwara, T. (2018) Analyzing solid waste management practices for the hotel industry. *Global Journal of Environmental Science and Management* 20 (2), 19–30.

Phillip, S., Hunter, C. and Blackstock, K. (2010) A typology for defining agritourism. *Tourism Management* 32, 754–758.

Phillips, A. (2002) *Management Guidelines for IUCN Category V Protected Areas: Protected Landscapes/Seascapes*. Gland and Cambridge: IUCN.

Piemontese, L., Jaramillo, F., Fetzer, I. and Rockström, J. (2018) Future hydroclimatic challenges for Africa: Beyond the Paris agreement. *Geophysical Research Abstracts* 20, EGU2018-13855.

Pigram, J.J. (1990) Sustainable tourism – Policy considerations. *Journal of Tourism Studies* 1 (2), 2–9.

Pigram, J.J. (1996) Best practice environmental management and the tourism industry. *Progress in Tourism and Hospitality Research* 2 (3–4), 261–271.

Pimentel, D. (2003) Ethanol fuels: Energy balance, economics, and environmental impacts are negative. *Natural Resources Research* 12 (2), 127–134.

Pinker, S. (2002) *The Blank Slate: The Modern Denial of Human Nature*. New York: Viking.

Piper, L.A. and Yeo, M. (2011) Ecolabels, Ecocertification and Ecotourism. Paper presented at 1st International Scientific Conference, Tourism in South East Europe 2011, Sustainable Tourism: Socio-cultural, Environmental and Economic Impact, Opatija, Croatia, 4–7 May (pp. 279–294), Opatija: Faculty of Tourism and Hospitality Management.

Pirani, S.I. and Arafat, H.A. (2014) Solid waste management in the hospitality industry: A review. *Journal of Environmental Management* 146, 320–336.

Plummer, R. and Fennell, D.A. (2009) Managing protected areas for sustainable tourism: Prospects for adaptive co-management. *Journal of Sustainable Tourism* 17 (2), 149–168.

Ponting, J., McDonald, M.G. and Wearing, S.L. (2005) Deconstructing Wonderland: Surfing tourism in the Mentawai Islands, Indonesia. *Society and Leisure* 28 (1), 141–162.

Ponton, D.M. and Asero, V. (2018) Representing cruise tourism: A paradox of sustainability. *Critical Approaches to Discourse Analysis across Disciplines* 10 (1), 45–62.

Porter, P.W. (1978) Geography as human ecology. *American Behavioral Scientist* 22 (1), 15–39.

Postma, A. (2014) Anticipating the future of European tourism. In I. Yeoman, A. Postma and J. Oskam (eds) *The Future of European Tourism* (pp. 290–305). Leeuwarden: European Tourism Futures Institute, Stenden University of Applied Sciences.

Postma, A. and Jenkins, A.K. (1997) Improving the tourists' experience: Quality management applied to tourist destinations. In P.E. Murphy (ed.) *Quality Management in Urban Tourism* (pp. 183–198). New York: John Wiley.

Potts, T.D. and Harrill, R. (1998) Enhancing communities for sustainability: A travel ecology approach. *Tourism Analysis* 3, 133–142.

Powell, R.B. and Ham, S.H. (2008) Can ecotourism interpretation really lead to pro-conservation knowledge, attitudes and behaviour? evidence from the galapagos Islands. *Journal of Sustainable Tourism* 16 (4), 467–489.

Pretty, J. (1995) The many interpretations of participation. *In Focus* 16, 4–5.

Prideaux, B. (2018) Tourism and surface transport. In C. Cooper, S. Volo, N. Scott and W. Gartner (eds) *The Sage Handbook of Tourism Management: Theories, Concepts and Disciplinary Approaches to Tourism* (pp. 297–313). London: Sage.

Prigogene, I., Nicolis, G. and Babloyantz, A. (1972) Thermodynamics of evolution. *Physics Today* 25 (11), 23–28.

Principato, L, Pratesi, C.A. and Secondi, L. (2018) Towards zero waste: An exploratory study on restaurant managers. *International Journal of Hospitality Management* 74, 130–137.

Prosperi, P., Allen, T., Cogill, P. and Peri, I. (2016) Towards metrics of sustainable food systems: A review of the resilience and vulnerability literature. *Environment Systems and Decisions* DOI: 10.1007/s10669-016-9584-7

Protected Planet (2019a) *Explore the World's Marine Protected Areas*, See https://www.protectedplanet.net/marine.

Protected Planet (2019b) *World Database on Protected Areas*, See https://www.protectedplanet.net.

Pryor, A., Carpenter, C. and Townsend, M. (2005) Outdoor education and bush adventure therapy: A social-ecological approach to health and wellbeing. *Australian Journal of Outdoor Education* 9 (1), 3–13.

Pulido-Fernández, J.I. and López-Sánchez, Y. (2016) Are tourists really willing to pay more for sustainable destinations? *Sustainability* 8, 1240.

Puskar (2011) *Sustainable Tourism: A Case Study of Nainital, Utterakhand, India*, See https://greencleanguide.com/sustainable-tourism-a-case-study-of-nainitaluttarakhand/

Putman, R. (1993) *Making Democracy Work: Civic Traditions in Modern Italy*. Princeton, NJ: Princeton University Press.

Putman, R. (1995) Bowling alone: America's declining social capital. *Journal of Democracy* 6 (1), 65–78.

Pyke, S., Hartwell, H., Blake, A. and Hemingway, A. (2016) Exploring well-being as a tourism product resource. *Tourism Management* 55, 94–105.

Pyle, R.M. (1978) The extinction of experience. *Horticulture* 56, 64–67.

Pyle, R.M. (1993) *The Thunder Tree: Lessons from an Urban Wildland*. Boston, MA: Houghton Mifflin.

Queal (2019) *Complete Meals to Simplify your Food*, See https://queal.com/.

Radwan, H.R.I., Jones, E. and Minoli, D. (2012) Solid waste management in small hotels: A comparison of green and non-green small hotels in wales. *Journal of Sustainable Tourism* 20, 533–550.

Raj, R. and Musgrave, J. (2009) *Event Management and Sustainability*. Wallingford: CABI.

Rana, P., Platts, J. and Gregory, M. (2009) Exploration of Corporate Social Responsibility (CSR) in Multinational Companies within the Food Industry. Paper Presented at Corporate Responsibility Research Conference, September 2008, Queen's University, Belfast, See https://www.researchgate.net/publication/237457674_Exploration_of_corporate_social_responsibility_CSR_in_multinational_companies_within_the_food_industry.

Raudsepp-Hearne, C., Peterson, G.D. and Bennett, E.M. (2010) Ecosystem service bundles for analyzing tradeoffs in diverse landscapes. *Proceedings of the National Academy of Sciences of the USA* 107, 5242–5247.

Rawls, J. (1972) *A Theory of Justice*. Oxford: Clarendon Press.

Rawls, J. (1999) *The Law of Peoples*. Cambridge, MA: Harvard University Press.

Ray, R. (2000) *Management Strategies in Athletic Training* (2nd edn). Champaign, IL: Human Kinetics.

Ray, C. and Jain, R. (2011) *Drinking Water Treatment: Focusing on Appropriate Technology and Sustainability*. New York: Springer.

Rayner, G., Barling, D. and Lang, T. (2008) Sustainable food systems in Europe: Policies, realities and futures. *Journal of Hunger & Environmental Nutrition* 3 (2/3), 145–168.

Reagans, R. and McEvily, B. (2003) Network structure and knowledge transfer: The effects of cohesion and range. *Administrative Science Quarterly* 48 (2), 240–267.

Redclift, M. (1987) *Sustainable Development: Exploring the Contradictions*. London: Methuen.

Reed, M. and Harvey, D.L. (1992) The new science and the old: Complexity and realism in the social sciences. *Journal for the Theory of Social Behaviour* 22 (4), 353–380.

Rees, W.E. (1992) Ecological footprint and appropriated carrying capacity: What urban economics leaves out. *Environment and Urbanization* 4 (2), 121–130.

Regan, T. (2004) *The Case for Animal Rights*. Berkeley, University of California Press.

Renner, S. (2015) Perth's Carnegie Wave Energy Project Produces Power AND Fresh Water from the Motion of the Ocean. *Inhabitat*, 22 March, See https://inhabitat.com/perths-carnegie-wave-energy-project-produces-clean-power-and-potable-water-from-the-motion-of-the-ocean/.

Republique de Maurice (nd). *Mauritian Code of Ethics for Tourism*, See http://www.dodolidays.com/page_content-86-172

Responsible Travel (2018) *Responsible Tourism*, See https://www.responsiblevacation.com/copy/responsible-tourism.

Responsibletravel.com (2018a) *About to Go on Holiday? Think Twice About Your Souvenirs*, See https://www.responsibletravel.com/copy/about-to-go-on-holiday-think-twice-about-your-souvenirs.

Responsibletravel.com (2018b) *Animal Welfare Issues at Rodeos and Stampedes*, See https://www.responsibletravel.com/copy/animal-welfare-issues-at-rodeos-and-stampedes.

Responsibletravel.com (2018c) *Hunting, Canned Hunting and Volunteering at Wildlife Reserves*, See https://www.responsibletravel.com/copy/canned-hunting.

Responsibletravel.com (2018d) *Our Stance on Captive Animals, Animal Welfare & Tourism*, See https://www.responsibletravel.com/copy/animal-welfare-issues-in-tourism.

Reyers, B., O'Farrell, P.J.O., Cowling, R.M., Egoh, B.N., Le Maitre, D.C. and Vlok, J.H.J. (2009) Ecosystem services, land-cover change, and stakeholders: Finding a sustainable foothold for a semiarid biodiversity hotspot. *Ecology and Society* 14 (1), 38.

Reynolds, P.C. and Braithwaite, D. (2001) Towards a conceptual framework for wildlife tourism. *Tourism Management* 22, 31–42.

Richards, G. and Hall, D. (2000) The community: A sustainable concept in tourism development? In G. Richards and D. Hall (eds) *Tourism and Sustainable Community Development* (pp. 1–13). New York: Routledge.

Riddell, R. (1981) *Ecodevelopment*. New York: St Martin's Press.

Ridley, M. (2003) *Nature Via Nurture*. Toronto, ON: HarperCollins.

Riley, M., Ladkin, A. and Szivas, E. (2002) *Tourism Employment, Analysis and Planning*. Clevedon: Channel View.

RINA (Royal Institution of Naval Architects) (2019) Website, See https://www.rina.org/en.

Ripple, W.J., Newsome, T.M. and Kerley, G.I.H. (2016) Does trophy hunting support biodiversity? A response to Di Minin *et al. Trends in Ecology & Evolution* 31 (7), 495–496.

Risse, M. (2014) The human right to water and common ownership of the earth. *Journal of Political Philosophy* 22 (2), 178–203.

Rivera, J. (2002) Assessing a voluntary environmental initiative in the developing world: The costa rican certification for sustainable tourism. *Policy Sciences* 35, 333–360.

Rivera, J. and De Leon, P. (2004) Is greener whiter? The sustainable slopes program and the voluntary environmental performance of Western Ski areas. *Policy Studies Journal* 32, 417–437.

Rivera Huerta, A., Güereca, L.P. and Lozano, M.R. (2016) Environmental impact of beef production in Mexico through life cycle assessment. *Resources, Conservation and Recycling* 109, 44–53.

Robaina, M. and Madeleno, M. (2018) Resources: Eco-efficiency, sustainability and innovation in tourism. In E. Fayos Sola and C. Cooper (eds) *The Future of Tourism: Innovation and Sustainability* (pp. 19–41). Berlin: Springer.

Robbins, J. (1987) *Diet for a New America*. Walpole, NH: Stillpoint.

Roberts, S. and Tribe, J. (2008) Sustainability indicators for small tourism enterprises – an exploratory perspective. *Journal of Sustainable Tourism* 16 (5), 575–594.

Robinson, R., Martins, A., Solnet, D. and Baum, T. (2019) Sustaining precarity: Critically examining tourism and employment. *Journal of Sustainable Tourism* 27 (7), 1008–1025.

Rodger, K., Smith, A., Newsome, D. and Moore, S.A. (2011) Developing and testing an assessment framework to guide the sustainability of the marine wildlife tourism industry. *Journal of Ecotourism* 10 (2), 149–164.

Roe, D. and Urquhart, P. (2002) *Pro-poor Tourism: Harnessing the World's Largest Industry for the World's Poor*. London: International Institute for Environment and Development.

Rogerson, C.M. (2006) Pro-poor local economic development in South Africa: The role of pro-poor tourism. *Local Environment* 11, 37–60.

Rogerson, C.M. (2012) Tourism–agriculture linkages in rural South Africa: Evidence from the accommodation sector, *Journal of Sustainable Tourism* 20 (3), 477–495.

Rolston, H., III (1981) Values in nature. *Environmental Ethics* 3, 113–128.

Rönnbäck, P., Crona, B. and Ingwall, L. (2007) The return of ecosystem goods and services in replanted mangrove forests: Perspectives from local communities in Kenya. *Environmental Conservation* 34, 313–324.

RSSB (2016) *Rail Sustainable Development Principles.* London: Rail Safety and Standards Board.

Russell, A. and Wallace, G. (2004) Irresponsible ecotourism. *Anthropology Today* 20 (3), 1–2.

Ryan, C. and Saward, J. (2004) The zoo as ecotourism attraction – Visitor reactions, perceptions and management implications: The case of Hamilton Zoo, New Zealand. *Journal of Sustainable Tourism* 12 (3), 245–266.

Ryan, M., Hennessy, T., Buckley, C., Dillon, E.J., Donnellan, T., Hanrahan, K. and Moran, B. (2016) Developing farm-level sustainability indicators for Ireland using the teagasc national farm survey. *Irish Journal of Agricultural and Food Research* 55 (2), 112–125.

Ryanair (2018) *Introducing Europe's Greenest Airline.* Dublin: Ryanair.

Ryanair (2019) *Europe's Lowest Fares, Lowest Emissions Airline,* See https://corporate.ryanair.com/environment/.

Saarinen, J. (2006) Traditions of sustainability in tourism studies. *Annals of Tourism Research* 33 (4), 1121–1140.

Saarinen, J., Rogerson, C. and Manwa, H. (2011) Tourism and millennium development goals: Tourism for global development? *Current Issues in Tourism* 14 (3), 201–203.

Sabaté, J., Harwatt, H. and Soret, S. (2016) Environmental nutrition: A new frontier for public health. *American Journal of Public Health* 106 (5), 815–821.

SAFA Guidelines (2013) *SAFA Sustainability Assessment of Food and Agriculture Systems Guidelines, Version 3.0.* Rome: FAO, See http://www.fao.org/fileadmin/templates/nr/sustainability_pathways/docs/SAFA_Guidelines_Final_122013.pdf.

Saidur, R., Rahim, N.A., Islam, M.R. and Solangi, K.H. (2011) Environmental impact of wind energy. *Renewable and Sustainable Energy Reviews* 15, 2423–2430.

Sakaguchi, L., Pak, N. and Potts, M.D. (2018) Tackling the issue of food waste in restaurants: Options for measurement method, reduction and behavioural change. *Journal of Cleaner Production* 180, 430–436.

Sala, O.E., Chapin, F.S., Armesto, J.J., Berlow, E., Bloomfield, J., Dirzo, R., Huber-Sanwald, E., Huenneke, L.F., Jackson, R.B., Kinzig, A., Leemans, R., Lodge, D.M., Mooney, H.A., Oesterheld, M., Poff, N.L., Sykes, M.T., Walker, B.H. and Hall, D.H. (2000) Global biodiversity scenarios for the year 2100. *Science* 287, 1770–1776.

Samuels, K. (2005) Sustainability and peace building: A key challenge. *Development in Practice* 15 (6), 728–736.

Sanches-Pereira, A., Onguglo, B., Pacini, H., Gomez, M.F., Coelho, S.T. and Muwanga, M.K. (2017) Fostering local sustainable development in Tanzania by enhancing linkages between tourism and small-scale agriculture. *Journal of Cleaner Production* 162, 1567–1581.

Saner, M. and Wilson, J. (2003) *Stewardship, Good Governance and Ethics*. Institute on Governance Policy Brief No. 19, Ottawa: Institute on Governance.

Santas, A. (2014) Aristotelian ethics and Biophilia. *Ethics & the Environment* 19 (1), 95–121.

Sassen, R. and Isenmann, R. (2018) Corporate social responsibility reporting. *Nachhaltigkeits Management Forum* 26 (1–4), 1–2.

Saul, J.R. (2001) *On Equilibrium*. Toronto: Viking.

Savova, N.D. (2009) Heritage kinaesthetic: Local constructivism and UNESCO's Intangible–Tangible Politics at a "Favela" museum. *Anthropological Quarterly* 82 (2), 547–585.

Scarborough (2018) *About Scarborough*, www.scarborough.co.uk/.

Schallaböck, K.O. and Köhn, A. (1997) *Perspektiven Des Luftverkehrs in Nordrhein-Westfalen*. Studie im Auftrag des BUND NRW, Wuppertal: Wuppertal Institut für Klima, Umwelt, Energie.

Schaltegger, S. and Wagner, M. (2011) Sustainable entrepreneurship and sustainability innovation: Categories and interactions. *Business Strategy and the Environment* 20, 222–237.

Schendler, A. (2001) Trouble in paradise: The rough road to sustainability in aspen. *Corporate Environmental Strategy* 8 (4), 293–299.

Scheyvens, R. (1999) Ecotourism and the empowerment of local communities. *Tourism Management* 20, 245–249.

Scheyvens, R. (2002) *Tourism for Development: Empowering Communities*. Harlow: Prentice-Hall.

Scheyvens, R. (2003) *Tourism for Development: Empowering Communities*. Upper Saddle River, NJ: Prentice Hall.

Scheyvens, R., Banks, G. and Hughes, E. (2016) The private sector and the SDGs: The need to move beyond "business as usual". *Sustainable Development* 24 (6), 371–382.

Schianetz, K., Kavanagh, L. and Lockington, D. (2007a) The learning tourism destination: The potential of a learning organisation approach for improving the sustainability of tourism destinations. *Tourism Management* 28, 1485–1496.

Schianetz, K., Kavanagh, L. and Lockington, D. (2007b) Concepts and tools for comprehensive sustainability assessments for tourism destinations: A comparative review. *Journal of Sustainable Tourism* 15 (4), 369–389.

Schilling, B.J., Attavanich, W. and Jin, Y. (2014) Does agritourism enhance farm productivity? *Journal of Agricultural and Resource Economics* 39 (1), 69–87.

Şchiopu, A.F., Pădurean, A.M., Tală, M.L. and Nica, A.M. (2016) The influence of new technologies on tourism consumption behaviour of the millennials. *Amfiteatru Economic* 18 (10), 829–846.

Schoenmaker, D. (2019) Greening Monetary Policy. Research Discussion Paper No. DP13576, London: Centre for Economic Policy.

Schoon, M. and Van Der Leeuw, S. (2015) The shift toward social-ecological systems perspectives: Insights into the human–nature relationship. *Natures Sciences Sociétés* 23, 166–174.

Schumann, U. (1994) On the effect of emissions from aircraft engines on the state of the atmosphere. *Annales Geophysicae* 12, 365–384.

Schwartz, K., Tapper, T. and Font, X. (2008) A sustainable supply chain management framework for tour operators. *Journal of Sustainable Tourism* 16 (3), 298–314.

Schwarzenbach, R.P., Egli, T., Hofstetter, T.B., Von Gunten, U. and Wehrli, B. (2010) Global water pollution and water health. *Annual Review of Environment and Resources* 35, 109–136.

Scott, A. (2000) Price the precautionary principle. *Chemical Week* 162 (1), 48.

Scott, D. (2011) Why sustainable tourism must address climate change. *Journal of Sustainable Tourism* 19 (1), 17–34.

Scott, M. (2015a) Re-theorizing social network analysis and environmental governance: Insights from human geography. *Progress in Human Geography* 39 (4), 449–463.

Scott, N. (2015b) Tourism policy: A strategic review. In C. Cooper (ed.) *Contemporary Tourism Reviews, Volume 1* (pp. 57–90). Oxford: Goodfellow.

Scott, N., Baggio, R. and Cooper, C. (2008) *Network Analysis and Tourism: From Theory to Practice*. Clevedon: Channel View.

Scott, D., Gössling, S. and Hall, C.M. (2012a) International tourism and climate change. *Wires Climate Change* 3, 213–223.

Scott, D., Simpson, M.C. and Sim, R. (2012b) The vulnerability of caribbean coastal tourism to scenarios of climate change related sea level rise. *Journal of Sustainable Tourism* 20 (6), 883–898.

Seaworld (2019) *Corporate Responsibility*, See https://seaworldentertainment.com/about-us/corporate-responsibility/.

Seymour, E., Curtis, A. and Pannell, D. (2010) Understanding the role of assigned values in natural resource management. *Australasian Journal of Environmental Management* 17, 142–153.

Sgroi, F., Di Trapani, A.M., Testa, R. and Tudisca, S. (2014) The rural tourism as development opportunity or farms: The case of direct sales in sicily. *American Journal of Agricultural and Biological Sciences* 9 (3), 407–419.

Shackleford, P. (1985) The world tourism organisation – 30 years of commitment to environmental protection. *International Journal of Environmental Studies* 25, 257–263.

Shannon, G., Larson, C.L., Reed, S.E., Crooks, K.R. and Angeloni, L.M. (2017) Ecological consequences of ecotourism for wildlife populations and communities. In D.T. Blumstein, B. Geffroy, D.S.M. Samia and E. Bessa (eds) *Ecotourism's Promise and Peril: A Biological Evaluation* (pp. 29–46). Cham: Springer.

Sharma, R. and Jha, M. (2017) Values influencing sustainable consumption behaviour: Exploring the contextual relationship. *Journal of Business Research* 76, 77–88.

Shaw, G. and Williams, A. (1992) Tourism, development and the environment: The eternal triangle. In C. Cooper and A. Lockwood (eds) *Progress in Tourism, Recreation and Hospitality Management* (pp. 47–59). London: Belhaven Press.

Sheppard, V. (2017) Resilience and destination governance: Whistler, BC. In R.W. Butler (ed.) *Tourism and Resilience* (pp. 69–80). Wallingford: CABI.

Sheppard, V. and Williams, P.W. (2017) The effects of shocks and stressors on sustainability-focused governance systems. *International Journal of Tourism Policy* 7 (1), 58–80.

Sheu, H.-J., Hu, J.-L. and Shieh, H.-S. (2012a) Going green: Developing a conceptual framework for the green hotel rating system. *Actual Problems of Economics* 134 (8), 521–530.

Sheu, H.-J., Hu, J.-L. and Shieh, H.-S. (2012b) Empirical application of the green hotel rating. *Actual Problems of Economics* 135 (9), 574–583.

Shireman, W. (2003) *A Measurement Guide to Productivity: 50 Powerful Tools to Grow Your Triple Bottom Line.* Tokyo: Asian Productivity Organization.

Shiva, V. (2016) *Earth Democracy: Justice, Sustainability and Peace.* London: Zed Books.

Silva, A. and Stocker, L. (2018) What is a transition? Exploring visual and textual definitions among sustainability transition networks. *Global Environmental Change* 50, 60–74.

Simaika, J.P. and Samways, M.J. (2010) Biophilia as a universal ethic for conserving biodiversity. *Conservation Biology* 24 (3), 903–906.

Simmons, D.G. (2013) Tourism and ecosystem services in New Zealand. In J.R. Dymond (ed.) *Ecosystem Services in New Zealand – Conditions and Trends* (pp. 343–348). Lincoln, NZ: Whenua Press.

Simpson, M.C., Gössling, S., Scott, D., Hall, C.M. and Gladin, E. (2008) *Climate Change Adaptation and Mitigation in the Tourism Sector: Frameworks, Tools and Practices.* Paris: UNEP.

Singer, P. (2009) *Animal Liberation: The Definitive Classic of the Animal Movement.* Toronto: Harper Perennial.

Singh, T. (2015) Pavegen's kinetic walkway in South African mall will power rural villages. *Inhabitat*, 25 March, See https://inhabitat.com/pavegens-kinetic-walkway-in-south-african-mall-will-power-rural-villages/.

Singh Verma, A. (2014) Sustainable supply chain management practices: Selective case studies from Indian hospitality industry. *International Management Review* 10 (2), 13–23.

Singjai, K., Winata, L. and Kummer, T.F. (2017) Green Strategy and its Benefits: An Empirical Study of the Hotel Industry in Thailand in *Proceedings of the 18th Asian Academic Accounting Association Annual Conference, 22–23 November, Bali, Indonesia* (pp. 45–60), Kedah, Malaysia: Asian Academic Accounting Association.

Skål International (2019) *Skål International Sustainable Tourism Awards*, See https://www.skal.org/en/sustainableguidelines.

Skinner, J.A., Lewis, K.A., Bardon, K.S., Tucker, P., Catt, J.A. and Chambers, B.J. (1997) An overview of the environmental impact of agriculture in the UK. *Journal of Environmental Management* 50, 111–128.

Slee, R. (2018) *Defining the Scope of Inclusive Education.* UNESCO Global Education Monitoring Report, See https://unesdoc.unesco.org/ark:/48223/pf0000265773.

Sloan, S., Bertzky, B. and Laurance, W.F. (2017) African development corridors intersect key protected areas. *African Journal of Ecology* 55 (4), 731–737.

Smil, V. (1984) *The Bad Earth: Environmental Degradation in China.* Armonk, NY: M.S. Sharpe.

Smith, A.L. (2017) *Public–Private Partnerships (PPPs) for Sustainable Tourism.* Washington, DC: Inter-American Development Bank.

Smith, M. and Puczkó, L. (2013a) *Health and Wellness Tourism.* New York: Routledge.

Smith, P. and Gregory, P.J. (2013b) Climate change and sustainable food production. *Proceedings of the Nutrition Society* 72, 21–28.

Smith, R.J., Muir, R.D.J., Walpole, M.J., Balmford, A. and Leader-Williams, N. (2003) Governance and the loss of biodiversity. *Nature* 426, 67–70.

Smith, L.C., Alderman, A. and Aduayom, D. (2006) Food Insecurity in Sub-Saharan Africa: New Estimates from Household Expenditure Surveys. Research Report No. 146, Washington, DC: IFPRI.

Smith, A.J., Scherrer, P. and Dowling, R. (2009) Impacts on aboriginal spirituality and culture from tourism in the coastal waterways of the kimberley region, North West Australia. *Journal of Ecotourism* 8 (2), 82–98.

Smith, A.C., Harrison, P.A., Perez Soba, M., Archaux, F., Blicharska, M., Egoh, B.N., Eros, T., Fabrega Domenech, N., Gyorgy, I., Haines-Young, R., Li, S., Lommelen, E., Meiresonne, L., Miguel Ayala, L., Mononen, L., Simpson, G., Stange, E., Turkelboom, F., Uiterwijk, M., Veerkamp, C.J. and Wyllie De Echeverria, V. (2017) How natural capital delivers ecosystem services: A typology derived from a systematic review. *Ecosystem Services* 26, 111–126.

SMTD (Shanghai Maglev Transportation Development Co.) (2018) Website, smtdc.com/en/.

Smulska, G. (1996) (July) 'Świadomość zagrozeń i poczucie bezradności' ['Awareness of Dangers and the Feeling of Helplessness']. *Środowisko Życia* 69, 35–36.

Sonnenburg, S. and Wee, D. (2016) Introduction to touring consumption. *Journal of Consumer Culture* 16 (2), 323–333.

Soper, K. (2008) Alternative hedonism, cultural theory and the role of aesthetic revisioning. *Cultural Studies* 22 (5), 567–587.

Soule, M.E. and Sanjayan, M.A. (1998) Conservation targets: Do they help? *Science* 279, 2060–2061.

SPANA (2018) *Ethical Tourism: The SPANA Holiday Hooves Guide*, See https://spana.org/get-involved/other-ways-to-get-involved/ethical-animal-tourism.

Späth, P. and Rohracher, H. (2012) Local demonstrations for global transitions – Dynamics across governance levels fostering socio-technical regime change towards sustainability. *European Planning Studies* 20 (3), 461–479.

SSI (Sustainable Shipping Initiative) (2018) *Vision 2040: Sustainable Shipping = Success*, See https://www.ssi2040.org.

Stabler, M. (1996) Managing the Leisure Natural Resource Base: Utter Confusion or Evolving Consensus? Paper Presented at the World Leisure and Recreation Association Fourth World Congress, Free Time and the Quality of Life for the 21st Century, Cardiff, July.

Stanciu, P., Hapenciuc, C.V., Moroşan, A.A. and Arionesei, G. (2015) Voluntary integration of ecolabel concept in the entrepreneurial culture of guesthouses in bucovina – Impact study. *International Journal of Economic Practices and Theories* 5 (5), 510–517.

St. Ange, A. (2018) Kenya's First 100% Solar Hotel. *Buzz Travel*, 3 December, See https://www.eturbonews.com/239456/kenyas-first-100-solar-hotel.

Stanojević, A. and Benčina, J. (2019) The construction of an integrated and transparent index of wellbeing. *Social Indicators Research: An International and Interdisciplinary Journal for Quality-of-Life Measurement* 143 (3), 995–1015.

Starbucks (2018) *Global Social Compact.* Seattle: Starbucks Corporation.

Starbucks (2019) Website, www.starbucks.com.

Stark, J.C. (2002) Ethics and ecotourism: Connection and conflicts. *Philosophy & Geography* 5 (1), 101–113.

Statistica (2018) *Number of Tourists in Hotels in Barcelona City from 1990 to 2017(in Millions)*, See https://www.statista.com/statistics/452060/number-of-tourists-in-barcelona-spain/.

Stedman, R.C. and Ingalls, M. (2014) Topophilia, biophilia and greening in the red zone. In K.G. Tidball and M.E. Krasny (eds) *Greening in the Red Zone: Disaster, Resilience and Community Greening* (pp. 129–144). New York: Springer.

Steiner, G. (1972) Social policies for business. *California Management Review* 15 (2), 17–24, in R. Reidenbach and D. Robin (1989) *Ethics and Profits*, Upper Saddle River, NJ: Prentice-Hall.

Stensland, S., Aas, Ø. and Mehmetoglu, M. (2017) Understanding constraints and facilitators to salmon angling participation: Insights from structural equation modeling. *Human Dimensions of Wildlife* 22 (1), 1–17.

Sterling, S. (2001) *Sustainable Education: Re-visioning Learning and Change.* Schumacher Briefings 6, Cambridge: Green Books.

Sterling, S. (2010) Learning for resilience, or the resilient learner? Towards a necessary reconciliation in a paradigm of sustainable education. *Environmental Education Research* 16 (5–6), 511–528.

Steve Gliessman (2016) Transforming food systems withagroecology. *Agroecology and Sustainable Food Systems* 40 (3), 187–189, DOI:10.1080/21683565.2015.1130765

Stigka, E.K., Paravantis, J.A. and Mihalakakou, G.K. (2014) Social acceptance of renewable energy sources: A review of contingent valuation applications. *Renewable and Sustainable Energy Reviews* 32, 100–106.

Stöckli, S., Dorn, M. and Liechti, S. (2018) Normative prompts reduce consumer food waste in restaurants. *Waste Management* 77, 532–536.

Stoll-Kleemann, S., De la Vega-Leinert, A.C. and Schultz, L. (2010) The role of community participation in the effectiveness of UNESCO biosphere reserve management: Evidence and reflections from two parallel global surveys. *Environmental Conservation* 37, 227–238.

Stolton, S. and Dudley, N. (2010) *Vital Sites: The Contribution of Protected Areas to Human Health*. Gland: WWF.

Stolton, S., Maxted, N., Ford-Lloyd, B., Kell, S. and Dudley, N. (2006) *Food Stores: Using Protected Areas to Secure Crop Genetic Diversity*. Gland and Birmingham: WWF and University of Birmingham.

Stonich, S.C. (1998) Political ecology of tourism. *Annals of Tourism Research* 25 (1), 25–54.

Styles, D., Schoenberger, H. and Galvez-Martos, J.L. (2015) Water management in the European hospitality sector: Best practice, performance benchmarks and improvement potential. *Tourism Management* 46, 187–202.

Su, B. and Martens, P. (2017) Public attitudes toward animals and the influential factors in contemporary China. *Animal Welfare* 26, 239–247.

Sumner, J. (2011) Serving social justice: The role of the commons in sustainable food systems. *Studies in Social Justice* 5 (1), 63–75.

Sumner, J. (2012) Dining on the social economy: Local, sustainable food systems and policy development. *Canadian Review of Social Policy* 67, 30–43.

Sun, Y.-Y. (2014) A framework to account for the tourism carbon footprint at Island destinations. *Tourism Management* 45, 16–27.

Svara, J.H., Watt, T.C. and Jang, H.S. (2013) How are U.S. Cities doing sustainability? Who is getting on the sustainability train, and why? *Cityscape: A Journal of Policy Development and Research* 15 (1), 9–44.

Svard, V. (2013) Features of Slow Philosophy in Kajaani: Slow Tourism as a Tool in Destination Development in the Land of Hunger. Thesis, Rovaniemi University of Applied Sciences, Rovaniemi, Finland.

Swarbrooke, J. (1999) *Sustainable Tourism Management.* New York: CABI.

Tabatchnaia-Tamirisa, N., Loke, M.K., Leung, P. and Tucker, K.A. (1997) Energy and tourism in Hawaii. *Annals of Tourism Research* 24 (2), 390–401.

Tang, S.Y. (1991) Institutional arrangements and the management of common pool resources. *Public Administration Review* 51 (1), 42–51.

Tangcharoensathien, V., Mills, A. and Palu, T. (2015) Accelerating health equity: The key role of universal health coverage in the sustainable development goals. *BMC Medicine* 13 (101), 1–5.

Taronga Zoo (2019) *Sustainability*, See https://taronga.org.au/conservation-and-science/sustainability.

Taylor, P. (1986) *Respect for Nature.* Princeton, NJ: Princeton University Press.

Tedmanson, D. and Guerin, P. (2011) Enterprising social wellbeing: Social entrepreneurial and strengths based approaches to mental health and wellbeing in "remote" indigenous community contexts. *Australasian Psychiatry* 19, S30–S33.

TEFI (Tourism Education Futures Initiative) (2019) Website, See http://tourismeducationfutures.org/.

Teilhard de Chardin, P. (1955/2008) *The Phenomenon of Man.* New York: Harper Perennial Modern Classics.

Telfer, D. and Sharpley, R. (2008) *Tourism and Development in the Developing World.* London: Routledge.

Teng, C.-C., Horng, J.-S., Hu, M.-L., Chien, L.-H. and Shen, Y.-C. (2012) Developing energy conservation and carbon reduction indicators for the hotel industry in Taiwan. *International Journal of Hospitality Management* 31, 199–208.

Terry, W.C. (2011) geographic limits to global labor market flexibility: The human resources paradox of the cruise industry. *Geoforum* 42 (6), 660–670.

Thaman, K.H. (2002) Shifting sight: The cultural challenge of sustainability. *International Journal of Sustainability in Higher Education* 3 (3), 233–242.

Thayer, R.L., Jr. (1990) The experience of sustainable landscapes. *Landscape Journal* 8, 101–110.

The Guardian (2017) The latest threat to antarctica: An insect and plant invasion. *The Guardian* 17 June, See https://www.theguardian.com/world/2017/jun/17/antarctica-insect-plant-invasion-house-flies-mosses-warmer-climate.

The Guardian (2019) Prince harry launches sustainable travel initiative travalyst. *The Guardian* 3 September, See https://www.theguardian.com/travel/2019/sep/03/prince-harry-launches-sustainable-travel-initiative-travalyst.

Thenewforest (2019) Website, www.thenewforest.co.uk.

Thériault, M. (2017) Bad taste, aesthetic akrasia, and other "guilty" pleasures. *Journal of Aesthetic Education* 51 (3), 58–71.

Thomas, C.D., Cameron, A., Green, R.E., Bakkenes, M., Beaumont, L.J., Collingham, Y.C., Erasmus, B.F.N., Ferreira de Siqueira, M., Grainger, A., Hannah, L., Hughes, L., Huntley, B., Van Jaarsveld, A.S., Midgley, G.F., Miles, L., Ortega-Huerta, M.A., Townsend Peterson, A., Phillips, O.L. and Williams, S.E. (2004) Extinction risk from climate change. *Nature* 427 (6970), 145–148.

Thomas Cook Group (2018) *A New Approach to Animal Welfare in Tourism.* Peterborough.

Thompson, J., Gannon, M., Curran, R. and Taheri, B. (2017) Activating the Diaspora: Engagement and Satisfaction amongst Philanthropic Transient Volunteer Tourists. Abstract from Academy of Marketing: Tourism Marketing Special Interest Group

(SIG), Edinburgh, https://pureportal.strath.ac.uk/en/publications/activating-the-diaspora-engagement-and-satisfaction-amongst-phila.

Tickner, J. and Raffensberger, C. (1998) The precautionary principle: A framework for sustainable business decision-making. *Environmental Policy* 5 (4), 75–82.

Ting, D.H. and Cheng, F.C. (2014) Experiential Learning and Pro-environmental Behaviour. *Proceedings of the 2014 International Conference on Educational Technologies and Education*, New York: Springer.

Ting, D.H. and Cheng, F.C. (2017) Measuring the marginal effect of pro-environmental behaviour: Guided learning and behavioural enhancement. *Journal of Hospitality, Leisure, Sport & Tourism Education* 20, 16–26.

Tobin, D. (nd) *Whale Watching in the Bay of Fundy*, See http://new-brunswick.net/new-brunswick/whales/ethics.html

Toha, M.A.M. and Ismail, H.N. (2015) A heritage tourism and tourist flow pattern: A perspective on traditional versus modern technologies in tracking the tourists. *International Journal of Built Environment and Sustainability* 2 (2), 85–92.

Tonge, R., Myott, D.E. and Enright, K. (1995) *Why Should Local Government Invest in Tourism?* Melbourne: Country Victoria Tourism Council.

Torrance, A.W. (2010) Patent law, HIPPO, and the biodiversity crisis. *John Marshall Review of Intellectual Property Law* 9, 624–656.

Torres, R. and Momsen, J.H. (2004) Challenges and potential for linking tourism and agriculture to achieve pro-poor tourism objectives. *Progress in Development Studies* 4, 294–318.

Tosun, C. (1999) Towards a typology of community participation in the tourism development process. *International Journal of Tourism and Hospitality* 10, 113–134.

Tosun, C. (2006) Expected nature of community participation in tourism development. *Tourism Management* 27, 493–504.

Tourism Concern (2018) Website, See http://www.tourismconcern.org.uk/.

Tourism Industry Association of Canada (1995) *Code of Ethics and Guidelines for Sustainable Tourism.* Ottawa: Tour Industry Association of Canada.

Towner, N. and Milne, S. (2017) Sustainable surfing tourism development in the Mentawai Islands, Indonesia: Local stakeholder perspectives. *Tourism Planning & Development* 14 (4), 503–526.

Tran, K.C., Euan, J. and Isla, M.L. (2002) Public perception of development issues: Impact of water pollution on a small coastal community. *Ocean & Coastal Management* 45 (6–7), 405–420.

Trathan, P.N., García-Borboroglu, P., Boersma, D., Bost, C.A., Crawford, R.J., Crossin, G.T., Cuthbert, R.J., Dann, P., Davis, L.S., De la Puente, S., Ellenberg, U., Lynch, H.J., Mattern, T., Pütz, K., Seddon, P.J., Trivelpiece, W. and Wienecke, B. (2015) Pollution, habitat loss, fishing, and climate change as critical threats to penguins. *Conservation Biology* 29, 31–41.

Travel Foundation (2013) *Environmental Sustainability for River Cruising: A Guide to Best Practice*. Bristol: Travel Foundation.

Travel Wire News (2019) Skål international and biosphere tourism: Taking sustainable tourism awards to the next level. *Travel Wire News*, 6 April, See https://travelwirenews.com/skal-international-and-biosphere-tourism-taking-sustainable-tourism-awards-to-the-next-level-1326771/.

Travis, C.C. and Hester, S.T. (1991) Global chemical pollution. *Environmental Science and Technology* 25 (5), 814–819.

Treadright Foundation (2019) *Treadright: Make Travel Matter*, See https://www.treadright.org/.

Trivers, R. (1971) The evolution of reciprocal altruism. *Quarterly Review of Biology* 46, 35–57.

Trung, D.N. and Kumar, S. (2005) Resource use and waste management in vietnam hotel industry. *Journal of Cleaner Production* 13, 109–116.

Truong, V.D. and Hall, C.M. (2017) Corporate social marketing in tourism: To sleep or not to sleep with the enemy? *Journal of Sustainable Tourism* 27 (7), 884–902.

Trzyna, T. (2014) *Urban Protected Areas: Profiles and Best Practice Guidelines*. Best Practice Protected Area Guidelines Series No. 22, Gland: IUCN.

Tsai, K.-T., Lin, T.-P., Lin, Y.-H., Tung, C.-H. and Chiu, Y.-T. (2018) The carbon impact of international tourists to an Island country. *Sustainability* 10, 1386.

Tsartas, P. (1992) Socio-economic impacts of tourism on two Greek isles. *Annals of Tourism Research* 19, 516–533.

Tuan, Y.F. (1980) Rootedness versus sense of place. *Landscape* 24, 3–8.

TUI Group (2018) *Animal Welfare and Biodiversity*, See https://www.tuigroup.com/en-en/responsibility/sus_business/destination/animal-welfare.

Tukker, A., Emmert, S., Charter, M., Vezzoli, C., Sto, E., Andersen, M.M., Geerken, T., Tischner, U. and Lahlou, S. (2008) Fostering change to sustainable consumption and production: An evidence based view. *Journal of Cleaner Production* 16, 1218–1225.

Tukker, A., Cohen, M.J., Hubacek, K. and Mont, O. (2010) The impacts of household consumption and options for change. *Journal of Industrial Ecology* 14 (1), 13–30.

Turner, R.K., Pearce, D. and Bateman, I. (1994) *Environmental Economics: An Elementary Introduction*. Hemel Hempstead: Harvester Wheatsheaf.

Twining-Ward, L. (1999) Towards sustainable tourism development: Observations from a distance. *Tourism Management* 20, 187–188.

Tyler, D. and Dangerfield, J.M. (1999) Ecosystem tourism: A resource-based philosophy for ecotourism. *Journal of Sustainable Tourism* 7 (2), 146–158.

Tzschentke, N.A., Kirk, D. and Lynch, P.A. (2008) Going green: Decisional factors in small hospitality operations. *International Journal of Hospitality Management* 27, 126–133.

Uber (2019) *Meet Uber Movement*, See https://www.uber.com/in/en/community/supporting-cities/data/.

Ubrežiová, I. and Moravčiková, K. (2017) How to perceive the corporate social responsibility in the agro-food companies? *Serbian Journal of Management* 12 (2), 201–215.

UC San Diego (2018) *Principles of Sustainability*, See https://sustainability.ucsd.edu/about/principles.html.

Ulrich, R.S. (1979) Visual landscapes and psychological well-being. *Landscape Research* 4, 17–23.

UN (1992) *Convention on Biological Diversity*. New York: United Nations, See https://www.cbd.int/doc/legal/cbd-en.pdf.

UN (2011) *Guiding Principles on Business and Human Rights*, www.ohchr.org/Documents/Publications/GuidingPrinciplesBusinessHR_EN.pdf.

UN (2014) *The Road to Dignity by 2030: Ending Poverty, Transforming All Lives and Protecting the Planet*, A/69/700, See https://digitallibrary.un.org/record/785641.

UN (2015) *Transforming our World: The 2030 Agenda for Sustainable Development*. New York, United Nations, See https://sustainabledevelopment.un.org/post2015/transforming ourworld.

UN (2018a) *Sustainable Development Knowledge Platform: Goals*, See https://sustainabledevelopment.un.org/?menu1300.

UN (2018b) *Food*, See https://www.un.org/en/sections/issues-depth/food/.

UN (2018c) *Promote Sustained, Inclusive and Sustainable Economic Growth, Full and Productive Employment and Decent Work for All*, See https://sustainabledevelopment.un.org/sdg8.

UNDESA (UN Department of Economic and Social Affairs) (1992) *Agenda 21*, www.un.org/esa/sustdev/agenda21.htm.

UNDESA (UN Department of Economic and Social Affairs) (2017) *World Population Prospects: The 2017 Revision*, See https://www.un.org/development/desa/publications/world-population-prospects-the-2017-revision.html.

UNDESA/PD (2012) *World Urbanisation Prospects: The 2011 Revision*. New York: United Nations.

UNDP (2013) *Global Conversation on the Post-2015 Sustainable Development Agenda*. New York: UN Development Program.

UN Economic and Social Council (2003) General Comment No. 15: The Rights to Water. Articles 11 and 12 of the Covenant, UN Committee on Economic, Social and Cultural Rights, Geneva, See https://www.refworld.org/docid/4538838d11.html.

UNEP (1972) *Report of the United Nations Conference on the Human Environment*. Stockholm, 5–16 June, See https://www.un.org/ga/search/view_doc.asp?symbolA/CONF.48/14/REV.1.

UNEP (2014) *Towards a Global Map of Natural Capital: Key Ecosystem Assets*. Nairobi: UN Environment Program.

UNEP (2017) *Indicators of Success: Demonstrating the Shift to Sustainable Consumption and Production: Principles, Process and Methodology*. Nairobi: UN Environment Program.

UNEP (2018a) *Sustainable Development Goals*, See https://www.undp.org/content/undp/en/home/sustainable-development-goals.html.

UNEP (2018b) *Goal No. 1: No Poverty*, See https://www.undp.org/content/undp/en/home/sustainable-development-goals/goal-1-no-poverty.html.

UNEP (2018c) *Goal No. 2: Zero Hunger*, See https://www.undp.org/content/undp/en/home/sustainable-development-goals/goal-2-zero-hunger.html.

UNEP (2018d) *Goal No. 3: Good Health and Well-being*, See https://www.undp.org/content/undp/en/home/sustainable-development-goals/goal-3-good-health-and-well-being.html.

UNEP (2018e) *Goal No. 4: Quality Education*, See https://www.undp.org/content/undp/en/home/sustainable-development-goals/goal-4-quality-education.html.

UNEP (2018f) *Goal No. 5: Gender Equality*, See https://www.undp.org/content/undp/en/home/sustainable-development-goals/goal-5-gender-equality.html.

UNEP (2018g) *Goal No. 8: Decent Work and Economic Growth*, See https://www.undp.org/content/undp/en/home/sustainable-development-goals/goal-8-decent-work-and-economic-growth.html.

UNEP (2018h) *Goal No. 10: Reduced Inequalities*, See https://www.undp.org/content/undp/en/home/sustainable-development-goals/goal-10-reduced-inequalities.html.

UNEP (2018i) *Goal No. 12: Responsible Consumption and Production*, See https://www.undp.org/content/undp/en/home/sustainable-development-goals/goal-12-responsible-consumption-and-production.html.

UNEP (2018j) *Goal No. 16: Peace, Justice and Strong Institutions*, See https://www.undp.org/content/undp/en/home/sustainable-development-goals/goal-16-peace-justice-and-strong-institutions.html.

UNEP/IEO (1989) Environmental Auditing. Report of UNEP/IEO Workshop, Paris (pp. 99–102).

UNEP-WCMC, IUCN and NGS (2018) *Protected Planet Report 2018*. Cambridge, Gland and Washington, DC: UN Environment World Conservation Monitoring Centre, International Union for Conservation of Nature and National Geographic Society.

UNESCO (1992) *The Rio Declaration on Environment and Development*, See http://www.unesco.org/education/pdf/RIO_E.PDF.

UNESCO (2003) *Convention for the Safeguarding of the Intangible Cultural Heritage*. Paris: UNESCO.

UNESCO (2010) *UNESCO Training Module on Sustainable Tourism*. Paris: UNESCO, See https://destinet.eu/resources/...-various-target-groups/individual-puplications/mod16.pdf

UNESCO (2019a) *World Heritage List*, See https://whc.unesco.org/en/list/.

UNESCO (2019b) *The Criteria for Selection*, See https://whc.unesco.org/en/criteria/.

UNESCO (2019c) *UNESCO Global Geoparks*, See http://www.unesco.org/new/en/natural-sciences/environment/earth-sciences/unesco-global-geoparks/.

UNESCO (2019d) *Biosphere Reserves – Learning Sites for Sustainable Development*, See http://www.unesco.org/new/en/natural-sciences/environment/ecological-sciences/biosphere-reserves/.

UNESCO (2019e) *Tsimanampesotse – Nosy Ve Androka Biosphere Reserve, Madagascar*, See https://en.unesco.org/biosphere-reserves/madagascar/tsimanampesotse.

UNESCO (2019f) *Global Action Programme on Education for Sustainable Development*, See https://en.unesco.org/gap.

UNESCO (n.d.) *UNESCO World Heritage Sustainable Tourism Online Toolkit: Guide 4 – Engaging Local Communities and Businesses.* Paris, UNESCO.

UN General Assembly (2014) The Road to Dignity by 2030: Ending Poverty, Transforming All Lives and Protecting the Planet. Synthesis Report of the Secretary-General on the Post-2015 Sustainable Development Agenda, New York: United Nations.

UNGPWM (2018) *Global Partnership on Waste Management*, https://sustainabledevelopment.un.org/partnership/?p7462.

UN HABITAT (2016) *World Cities Report. Urbanization and Development: Emerging Futures.* Nairobi: UN Human Settlements Programme, See https://unhabitat.org/wp-content/uploads/2014/03/WCR-%20Full-Report-2016.pdf.

UNIDO (2011) *Green Industry.* Policies for Supporting Green Industry UNIDO, Vienna.

Universal Studios (2019) *Green is Universal*, See http://www.greenisuniversal.com/learn/about-us/theme-parks/.

UN Report of the Secretary-General (2018a) *The Sustainable Development Goals Report*, See https://unstats.un.org/sdgs/files/report/2018/TheSustainableDevelopmentGoalsReport2018-EN.pdf.

UN Report of the Secretary-General (2018b) *The Sustainable Development Goals Report, SDG 6*, See http://www.undp.org/content/undp/en/home/sustainable-development-goals/goal-6-clean-water-and-sanitation.html.

UN Report of the Secretary-General (2018c) *The Sustainable Development Goals Report, SDG 7*, See http://www.undp.org/content/undp/en/home/sustainable-development-goals/goal-7-affordable-and-clean-energy.html.

UN Report of the Secretary-General (2018d) *The Sustainable Development Goals Report, SDG 11*, See https://sustainabledevelopment.un.org/sdg11.

UN Report of the Secretary-General (2018e) *The Sustainable Development Goals Report, SDG 12*, See https://sustainabledevelopment.un.org/sdg12.

UN Report of the Secretary-General (2018f) *The Sustainable Development Goals Report, SDG 13*, See https://sustainabledevelopment.un.org/sdg13.

UN Report of the Secretary-General (2018g) *The Sustainable Development Goals Report, SDG 14*, See https://sustainabledevelopment.un.org/sdg14.

UN Report of the Secretary-General (2018h) *The Sustainable Development Goals Report, SDG 15*, See https://sustainabledevelopment.un.org/sdg15.

UNWTO (1996) *What Tourism Managers Need to Know: A Practical Guide to the Development and Use of Indicators of Sustainable Tourism.* Madrid: UN World Tourism Organization.

UNWTO (1998) *Guide for Local Authorities on Developing Sustainable Tourism.* ed. Inskeep, E., Madrid, Spain: UN World Tourism Organization.

UNWTO (2004) *Indicators of Sustainable Tourism.* Madrid: UN World Tourism Organization.

UNWTO (2005) *Making Tourism More Sustainable: A Guide for Policy Makers.* Madrid: UN World Tourism Organization.

UNWTO (2007) *A Practical Guide to Destination Management.* Madrid: UN World Tourism Organization.

UNWTO (2008) *Climate Change and Tourism: Responding to Global Challenges.* Madrid: UN World Tourism Organization.

UNWTO (2009) *From Davos to Copenhagen and Beyond: Advancing Tourism's Response to Climate Change.* UNWTO, Madrid.

UNWTO (2010a) *Tourism and Biodiversity.* Achieving Common Goals Towards Sustainability UNWTO, Madrid

UNWTO (2010b) *Demographic Change and Tourism.* Madrid: UN World Tourism Organization.

UNWTO (2011a) *Global Report on Women in Tourism 2010.* Madrid: UN World Tourism Organization.

UNWTO (2011b) *Tourism Towards 2030 – Global Overview.* Madrid: UN World Tourism Organization.

UNWTO (2013) *Recommendations for Accessible Tourism for All.* Madrid: UN World Tourism Organization.

UNWTO (2014) *Towards Measuring the Economic Value of Wildlife Watching Tourism in Africa,* Briefing Paper UNWTO Madrid Delete World Bank.

UNWTO (2015) *Tourism and Gender Portal,* See http://ethics.unwto.org/en/content/tourism-and-gender-portal.

UNWTO (2016) *International Year 2017.* Madrid: UN World Tourism Organization, See http://www.tourism4development2017.org/.

UNWTO (2017) *The Waikato Tourism Monitoring Observatory,* See http://insto.unwto.org/observatories/waikato-region-new-zealand/.

UNWTO (2018a) *Baseline Report on the Integration of Sustainable Consumption and Production Patterns into Tourism Policies.* Madrid: UN World Tourism Organization.

UNWTO (2018b) *Tourism Highlights.* Madrid: UN World Tourism Organization, See https://www.e-unwto.org/doi/pdf/10.18111/9789284419876.

UNWTO (2019a) Website, See http://www2.unwto.org/.

UNWTO (2019b) *Global Code of Ethics for Tourism,* See http://ethics.unwto.org/content/global-code-ethics-tourism.

UNWTO (2019c) *The International Network of Sustainable Tourism Observatories (INSTO): A Network of Tourism Observatories Monitoring Sustainable Tourism Development at Destination Level,* See http://insto.unwto.org/.

UNWTO and European Travel Commission (2018) *Exploring Health Tourism – Executive Summary*. Madrid: UN World Tourism Organization.

US EIA (Energy Information Administration) (2018) *Renewable Energy Explained*, See https://www.eia.gov/energyexplained/renewable-sources/.

Uysal, M., Sirgy, J., Woo, E. and Kim, H. (2016) Quality of life (QOL) and well-being research in tourism. *Tourism Management* 53, 244–261.

Vail, D. and Hultkrantz, L. (2000) Property rights and sustainable nature tourism: Adaptation and Mal-adaptation in Dalarna (Sweden) and Maine (USA). *Ecological Economics* 35, 223–242.

Valdivia, C. and Barbieri, C. (2014) Agritourism as a sustainable adaptation strategy to climate change in the Andean Altiplano. *Tourism Management Perspectives* 11, 18–25.

Van den Belt, M. and Blake, D. (2015) Investing in natural capital and getting returns: An ecosystem service approach. *Business Strategy and the Environment* 24, 667–677.

Van der Duim, R. and Caalders, J. (2002) Biodiversity and tourism: Impacts and interventions. *Annals of Tourism Research* 29 (3), 743–761.

Van Dyke, J.M. (1996) The Rio principles and our responsibilities of ocean stewardship. *Ocean and Coastal Management* 31 (1), 1–23.

VanderZwaag, D. (1994) *CEPA and the Precautionary Principle/Approach*. Hull, Quebec: Minister of Supply and Services.

VanderZwaag, D. (1999) The precautionary principle in environmental law and policy: Elusive Rhetoric and first embraces. *Journal of Environmental Law and Practice* 8, 355–375.

Vargas, F.H. (2009) Penguins on the Equator, hanging on by a thread. In T. De Roy (ed.) *Galápagos: Preserving Darwin's Legacy* (pp. 154–161). Richmond Hill, Ont.: Firefly Books.

Vasconcelos-Vasquez, K., Balbastre-Benavent, F. and Redondo-Cano, A.M. (2011) Is certification for sustainable tourism complementary to IO 9000 certification? The case of the Parque del Lago hotel in Costa Rica. *PASOS* 9 (4), 543–557.

Vernadsky, V.I. (1926) *Biosfera* [*The Biosphere*]. Leningrad: Nauchnoe Khi Miko-Tekhnicheskoe Izdatel'stvo [In Russian].

Vernadsky, V.I. (1929) *La Biosphere*. Paris: Felix Alcan.

Vernadsky, V. (2012) The transition from the biosphere to the Noösphere' (with Introduction by Bill Jones, transl.). *21st Century Science & Technology* 25 (1–2), 10–31.

Vidal González, M. (2008) Intangible heritage tourism and identity. *Tourism Management* 29, 807–810.

Vilà, M. and Pujadas, J. (2001) Land-use and socio-economic correlates of plant invasions in European and North African countries. *Biological Conservation* 100, 397–401.

Virgin Australia (2019) *Sustainability*, See https://www.virginaustralia.com/uk/en/about-us/sustainability/.

VisitCairngorms (2019) Official website, See https://visitcairngorms.com/membership.

VisitCopenhagen (2019) *Sustainable Copenhagen*, See https://www.visitcopenhagen.com/copenhagen/sightseeing/sustainable-copenhagen-0.

VisitEngland (2019) Website, www.visitengland.org.

VisitLošinj (2019) Website, visitlosinj.hr/Default.aspx?langen-GB.

VisitScotland and Future Foundation (2005) *Our Ambition for Scottish Tourism: A Journey to 2025*. Edinburgh: VisitScotland.

Vitasurya, V.R. and Pudianti, A. (2016) Sustainable waste management of traditional craft industry in Lopati Tourism Village, Yogyakarta. *Journal of Agriculture and Built Environment* 43 (2), 123–130.

Voase, R. (2007) Individualism and the "new tourism": A perspective on emulation, personal control and choice. *International Journal of Consumer Studies* 31, 541–547.

Vörösmarty, C.J., McIntyre, P.B., Gessner, M.O., Dudgeon, D., Prusevich, A., Green, P., Glidden, S., Bunn, S.E., Sullivan, C.A., Reidy Liermann, C. and Davies, P.M. (2010) Global threats to human water security and river biodiversity. *Nature* 467, 555–561.

Waayers, D., Newsome, D. and Lee, D. (2006) Observations of non-compliance behaviour by tourists to a voluntary code of conduct: A pilot study of turtle tourism in the exmouth region, Western Australia. *Journal of Ecotourism* 5 (3), 211–221.

Wackernagel, M. and Rees, W.E. (1996) *Our Ecological Footprint: Reducing Human Impact on the Earth*. Gabriola Island, BC: New Society.

Wackernagel, M., Schulz, N.B., Deumling, D., Linares, A.C., Jenkins, M., Kapos, V., Monfreda, C., Loh, J., Myers, N., Norgaard, R. and Randers, J. (2002) Tracking the ecological overshoot of the human economy. *Proceedings of the National Academy of Science USA* 99, 9266–9271.

Wagar, J.A. (1964) *The Carrying Capacity of Wildlands for Recreation*. Forest Service Monograph No. 7, Society of American Foresters, Washington, DC.

Wagle, U. (2002) Rethinking poverty: Definition and measurement. *International Social Science Journal* 54, 155–165.

Waglé, U.R. (2008) *Multidimensional Poverty: An Alternative Measurement Approach for the United States?*, See https://s3.amazonaws.com/academia.edu.documents/3466188/multidimensionalUS.pdf?response-content-disposition=inline%3B%20filename%3DMultidimensional_poverty_An_alternative.pdf&X-Amz-Algorithm=AWS4-HMAC-SHA256&X-Amz-Credential=AKIAIWOWYYGZ2Y53UL3A%2F20191027%2Fus-east-1%2Fs3%2Faws4_request&X-Amz-Date=20191027T201906Z&X-Amz-Expires=3600&X-Amz-SignedHeaders=host&X-Amz-Signature=c1c4b4280bd424992f25ce1b8bff0de2129a1afb6432c2fc75a4af3a4e449b57

Wagner, C.G. (2005) The conscientious tourist. *The Futurist*, September–October, 14–15.

Waitt, G. (1999) Naturalizing the "primitive": A critique of marketing Australia's Indigenous Peoples As "Hunter-gatherers". *Tourism Geographies* 1 (2), 142–163.

Waldrop, M.M. (1992) *Complexity: The Emerging Science at the Edge of Order and Chaos*. New York: Touchstone.

Walker, P.A., Greiner, D., McDonald, D. and Lyne, V. (1998) The tourism futures simulator: A system thinking approach. *Environment Modelling and Software* 14 (1) 59–67.

Wall, G. (1993) International collaboration in the search for sustainable tourism in Bali, Indonesia. *Journal of Sustainable Tourism* 1 (1), 38–47.

Wall, G. (1996) Is ecotourism sustainable? *Environmental Conservation* 2 (3–4), 207–216.

Wall, G. (2018) Beyond sustainable development. *Tourism Recreation Research* 43 (3), 390–399.

Walt Disney Company (2017) *Corporate Social Responsibility Update.* Burbank, CA: Walt Disney Company.

Walt Disney Company (n.d.) *Disney's Environmental Stewardship: Goals and Targets.* Burbank, CA: Walt Disney Company.

Wang, M. and Wang, C.-S. (2018) Tourism, the environment, and energy policies. *Tourism Economics* 24 (7), 821–838.

Wang, S., Hu, Y., He, H. and Wang, G. (2017) Progress and prospects for tourism footprint research. *Sustainability* 9, 1847.

Wang, L., Xue, L., Li, Y., Liu, X., Cheng, S. and Liu, G. (2018) Horeca food waste and its ecological footprint in Lhasa, Tibet, China. *Resources, Conservation & Recycling* 136, 1–8.

Wang, S., Wang, S. and Liu, J. (2019) Life-cycle greenhouse gas emissions of onshore and offshore wind turbines. *Journal of Cleaner Production* 210, 804–810.

Wani, M.A., Dada, Z.A. and Shah, S.A. (2018) Sustainable integrated solid waste management in the trans-himalayan accommodation sector. *African Journal of Hospitality, Tourism and Leisure* 7 (2), 1–17.

Warburton, K. (2003) Deep learning and education for sustainability. *International Journal of Sustainability in Higher Education* 4 (1), 44–56.

Warnken, J., Bradley, M. and Guilding, C. (2004) Exploring methods and practicalities of conducting sector-wide energy consumption accounting in the tourist accommodation industry. *Ecological Economics* 48, 125–141.

WBCSD (2006) *Eco-efficiency Learning Module.* Geneva: World Business Council for Sustainable Development.

WBCSD (World Business Council for Sustainable Development) (2019) Website, See https://www.wbcsd.org/.

Weaver, D.B. (1993) Ecotourism in the small Island Caribbean. *GeoJournal* 31, 457–465.

Weaver, D.B. (2005) Comprehensive and minimalist dimensions of ecotourism. *Annals of Tourism Research* 32 (2), 439–455.

Weaver, D. (2006) *Sustainable Tourism: Theory and Practice.* London: Routledge.

Weaver, D.B. (2010) Indigenous tourism stages and their implications for sustainability. *Journal of Sustainable Tourism* 18 (1), 43–60.

Weaver, D. (2014) The sustainable development of tourism: A state of the art perspective. In A.L. Lew, C.M. Hall and A.M. Williams (eds) *The Wiley Blackwell Companion to Tourism* (pp. 524–533). Chichester: John Wiley.

Weaver, D.B. (2015) *Sustainable Tourism.* London: Routledge.

Weaver, D.B. (2018) creative periphery syndrome? Opportunities for sustainable tourism innovation in timor-leste, an early stage destination. *Tourism Recreation Research* 43 (1), 118–128.

Weaver, D.B. and Fennell, D.A. (1997) The vacation farm industry of saskatchewan: A profile of operators. *Tourism Management* 18 (6), 357–365.

Weaver, D.B. and Jin, X. (2016) *Compassion* as a neglected motivator for sustainable tourism. *Journal of Sustainable Tourism* 24 (5), 657–672.

Webb, G.J.W. (2002) Conservation and sustainable use of wildlife – An evolving concept. *Pacific Conservation Biology* 8 (1), 12–26.

Webster, K. (1999) *Environmental Management in the Hospitality Industry: A Guide for Students and Managers.* Boston, MA: Cengage.

Weeden, C. and Boluk, K. (2014) *Managing Ethical Consumption in Tourism.* London: Routledge.

Weidenfeld, A., Williams, A.M. and Butler, R.W. (2009) Knowledge transfer and innovation among attractions. *Annals of Tourism Research* 37 (3), 604–626.

Weitzman, J. and Bailey, M. (2018) Perceptions of aquaculture ecolabels: A multi-stakeholder approach in Nova Scotia, Canada. *Marine Policy* 87, 12–22.

Welford, R., Ytterhus, B. and Eligh, J. (1999) Tourism and sustainable development: An analysis of policy and guidelines for managing provision and consumption. *Sustainable Development* 7, 165–177.

Wells, N.M. and Lekies, K.S. (2006) Nature and the life course: Pathways from childhood nature experiences to adult environmentalism. *Children, Youth and Environments* 16 (1), 1–24.

Westley, F., Olsson, P., Folke, C., Homer-Dixon, T., Vredenburg, H., Loorbach, D., Thompson, J., Nilsson, M., Lambin, E., Sendzimir, J., Banerjee, B., Galaz, V. and Van Der Leeuw, S. (2011) Tipping toward sustainability: Emerging pathways of transformation. *Ambio* 40, 762–780.

Wheeller, B. (1991) Tourism troubled times: Responsible tourism is not the answer. *Tourism Management* June, 91–96.

Wheeller, B. (1992) Alternative tourism – a deceptive ploy. In C. Cooper and A. Lockwood (eds) *Progress in Tourism, Recreation and Hospitality Management* (pp. 140–145). London: Belhaven Press.

Wheeller, B. (1997) Here we go, here we go, here we go eco. In M.J. Stabler (ed.) *Tourism and Sustainability: From Principles to Practice* (pp. 39–49). Oxford: CABI.

Whitbread (2018) *Sustainability Report 2016/2017.* Dunstable: Whitbread plc.

White, L.T., Jr. (1967) The historical roots of our ecologic crisis. *Science* 155 (3767), 1203–1207.

White, R.L., Eberstein, K. and Scott, D.M. (2018) Birds in the playground: Evaluating the effectiveness of an urban environmental education project in enhancing school children's awareness, knowledge and attitudes towards local wildlife. *PLoS ONE* 13 (3), e0193993.

Whitney-Squire, K. (2016) Sustaining local language relationships through indigenous community-based tourism initiatives. *Journal of Sustainable Tourism* 24 (8–9), 1156–1176.

Wight, P.A. (1994a) The greening of the hospitality industry: Economic and environmental good sense. In A.V. Seaton (ed.) *Tourism: State of the Art* (pp. 665–674). Toronto: John Wiley.

Wight, P.A. (1994b) Limits of acceptable change: A recreational tourism tool for cumulative effects assessment. In A.J. Kennedy (ed.) *Cumulative Effects Assessment in Canada: From Concept to Practice. Papers from the 15th Symposium Held by the Alberta Society of Professional Biologists, Calgary, Alberta* (pp. 159–178). Calgary: Alberta Society of Professional Biologists.

Williams, P.W. and Ponsford, I.F. (2009) confronting tourism's environmental paradox: Transitioning for sustainable tourism. *Futures* 41 (6), 396–404.

Williams, P.W., Penrose, R.W. and Hawkes, S. (1998) Shared decision-making in tourism land use planning. *Annals of Tourism Research* 25 (4), 860–889.

Willis, C. (2015) The contribution of cultural ecosystem services to understanding the tourism nature–wellbeing nexus. *Journal of Outdoor Recreation and Tourism* 10, 38–43.

Wilson, E.O. (1984) *Biophilia*, Cambridge, MA: Harvard University Press.

Wilson, A. (1992) *The Culture of Nature*. Oxford: Blackwell.

Wilson, E.O. (1993) Biophilia and the conservation ethic. In S. Kellert and E.O. Wilson (eds) *The Biophilia Hypothesis* (pp. 31–41). Washington, DC: Island Press.

Wilson, S., Fesenmaier, D.R., Fesenmaier, J. and Van Es, J.C. (2001) Factors for success in rural tourism development. *Journal of Travel Research* 40, 132–138.

Woo, E., Kim, H. and Uysal, M. (2015) Life satisfaction and support for tourism development. *Annals of Tourism Research* 50, 84–97.

Wood, P.J., Greenwood, M.T. and Agnew, M.D. (2003) Pond biodiversity and habitat loss in the UK. *Area* 35 (2), 206–216.

Woodley, A. (1993) Tourism and sustainable development: The community perspective. In J.G. Nelson, R.W. Butler and G. Wall (eds) *Tourism and Sustainable Development: Monitoring, Planning, Managing* (pp. 135–147). Department of Geography Publication No. 37, Waterloo, Ont.: University of Waterloo.

Worboys, G. and De Lacy, T. (2003) *Tourism and the Environment: It's Time! Ecotourism Australia 11th National Conference*. Adelaide: National Wine Centre.

World Bank (2017) *Public–Private Partnerships: Reference Guide, Version 3*. Washington, DC: World Bank.

World Bank (2018) *The Global Invasive Species Program*, See http://documents.worldbank.org/curated/en/659211468164947248/The-global-invasive-species-program.

World Commission on Environment and Development (1987) *Our Common Future*. Oxford University Press, Oxford, England.

World Hunger Education Service (2018) *2018 World Hunger and Poverty Facts and Statistics*, See https://www.worldhunger.org/world-hunger-and-poverty-facts-and-statistics/.

World Meteorological Organization (2017) *Is Set to be in Top Three Hottest Years, with Record-breaking Extreme Weather*, See https://public.wmo.int/en/media/press-release/2017-set-be-top-three-hottest-years-record-breaking-extreme-weather

Worm, B., Barbier, E.B., Beaumont, N., Duffy, E., Folke, C., Halpern, B.S., Jackson, B.C., Lotze, H.K., Micheli, F., Palumbi, S.R., Sala, E., Selkoe, K.A., Stachowicz, J. and Watson, R. (2006) Impacts of biodiversity loss on ocean ecosystem services. *Science* 314, 787–790.

WRAP (2011) *The Composition of Waste Disposed of by the UK Hospitality Industry*. Banbury: WRAP.

Wright, R. (2004) *A Short History of Progress*. Toronto: House of Anansi Press.

WSPA (2012) *The Contribution of Animal Welfare and Sustainable Tourism to Sustainable Development: Heredia, Costa Rica*. London: World Society for the Protection of Animals, See https://www.responsibletravel.org/docs/Sustainable%20Tourism%20&%20Animal%20Welfare.pdf.

WTTC (2015) *Global Talent Trends and Issues for the Travel and Tourism Sector*. London: World Travel and Tourism Council.

WTTC (World Travel and Tourism Council) (2019), Website, See https://www.wttc.org/.

WTTC, UNWTO and the Earth Council (1995) *Agenda 21 for the Travel and Tourism Industry – Towards Environmentally Sustainable Development*. London: World Travel and Tourism Council, UN World Tourism Organization and the Earth Council.

Wu, J.Y. (2014) Everyone vegetarian, world enriching. *Open Journal of Philosophy* 4 (2), 160.

Wu, G., Zhang, Q., Zheng, X., Mu, L. and Dai, L. (2008) Water quality of lugu lake: Changes, causes and measurements. *International Journal of Sustainable Development & World Ecology* 15 (1), 10–17.

WWF (2012) *Shipping and Sustainability*. Gland: World Wildlife Fund.

WWF (World Wildlife Fund) (2017) *The Problems with Current Protected Areas*, See https://wwf.panda.org/our_work/biodiversity/protected_areas/protected_area_problems/.

WWF (2019a) Website, See https://www.wwf.org.uk/.

WWF (World Wildlife Fund) (2019b) *Food Waste: Overview*, www.worldwildlife.org/initiatives/foodwaste.

Wyndham, F.S. (2000) The sphere of the mind: Reviving the Noösphere concept for ecological anthropology. *Journal of Ecological Anthropology* 4, 87–91.

Xiao, W., Mills, J., Guidi, G., Rodríguez-Gonzálvez, P., Barsanti, S.G. and Gonzálvez-Aguilera, D. (2018) Geoinformatics for the conservation and promotion of cultural heritage in support of the UN sustainable development goals. *ISRS Journal of Photogrammetry and Remote Sensing* 142, 389–406.

Xu, F. and Fox, D. (2014) Modelling attitudes to nature, tourism and sustainable development in national parks: A survey of visitors in China and the UK. *Tourism Management* 45, 142–158.

Xu, X. and Gursoy, D. (2015) A conceptual framework of sustainable hospitality supply chain management. *Journal of Hospitality Marketing & Management* 24 (3), 229–259.

Xu, P., Chan, E.H.W. and Qian, Q.K. (2012) Key performance indicators (KPI) for the sustainability of building energy efficiency retrofit (BEER) in hotel buildings in China. *Facilities* 30 (9/10), 432–448.

Yadav, H.P., Lone, S.A., Shah, N., Verma, R., Verma, U.K. and Dewry, R.K. (2018) Effect of heat stress on reproduction in dairy animals. *Journal of Experimental Zoology India* 21 (2), 623–631.

Yang, M., Hens, L., De Wulf, R. and Ou, X. (2011) Measuring tourists' water footprint in a mountain destination of Northwest Yunnan, China. *Journal of Mountain Science* 8, 682–693.

Yedla, S. (2015) Cities and the sustainability dimensions. In S.M. Dev and S. Yedla (eds) *Cities and Sustainability: Issues and Strategic Pathways* (pp. 1–22). New York: Springer.

Yeoman, I. and McMahon-Beattie, U. (2005) Developing a scenario planning process using a blank piece of paper. *Tourism and Hospitality Research* 5 (3), 273–285.

Yeoman, I., Palmomino-Schalscha, M. and McMahon-Beattie, U. (2014) Keeping it pure: Could New Zealand be an Eco Paradise?. *Journal of Tourism Futures* 1 (1), 20–36.

Yi-fong, C. (2012) The indigenous ecotourism and social development in Taroko National Park Area and San-Chan Tribe, Taiwan. *GeoJournal* 77, 805–815.

Yilmaz, H. and Yilmaz, S. (2016) Corporate social responsibility in hotel businesses. *Anadolu University Journal of Social Sciences* 16 (2), 89–100.

Yin, S., Han, Y., Zhang, Y. and Zhang, J. (2016) Depletion control and analysis for groundwater protection and sustainability in the Xingtai Region of China. *Environmental Earth Science* 75, 1246.

Yoon, H., Sauri, D. and Amorós, M.R. (2018) Shifting scarcities? The energy intensity of water supply alternatives in the mass tourism resort of benidorm, Spain. *Sustainability* 10, 824.

Yudina, O. and Fennell, D.A. (2013) Ecofeminism in the tourism context: A discussion of the use of other-than-human animals as food in tourism. *Tourism Recreation Research* 38 (1), 55–69.

Yukon Government (2015) *Tourism Sectors*, See http://www.tc.gov.yk.ca/isu_sectors.html.

Yusta-García, R., Orta-Martínez, M., Mayor, P. and González-Crespo, C. (2017) Water contamination from oil extraction activities in Northern Peruvian Amazonian Rivers. *Environmental Pollution* 225, 370–380.

Zapata, M.J., Hall, C.M., Lindo, P. and Vanderschaeghe, M. (2011) Can community-based tourism contribute to development and poverty alleviation? lessons from Nicaragua. *Current Issues in Tourism* 14 (8), 725–749.

Zein, K., Wazner, M.S. and Meylan, G. (2008) *Best Environmental Practices for the Hotel Industry*. Lausanne: Sustainable Business Associates.

Zeppel, H. and Muloin, S. (2008) Aboriginal interpretation in Australian wildlife tourism. *Journal of Ecotourism* 7 (2&3), 116–136.

Zhang, D., Jiang, Q., Ma, X. and Li, B. (2014a) Drivers for food risk management and corporate social responsibility. *Journal of Cleaner Production* 66, 520–527.

Zhang, J.J., Joglekar, N., Heineke, J. and Verma, R. (2014b) Eco-efficiency of service co-production: Connecting eco-certifications and resource efficiency in U.S. Hotels. *Cornell Hospitality Quarterly* 55 (1), 1–13.

Zhang, W., Goodale, E. and Chen, J. (2014c) How contact with nature affects children's biophilia, biophobia and conservation attitude in China. *Biological Conservation* 177, 109–116.

Zhang, J.-H., Zhang, Y., Zhou, J., Liu, Z.H., Zhang, H.L. and Tian, Q. (2017) Tourism water footprint: An empirical analysis of Mount Huangshan. *Asia Pacific Journal of Tourism Research* 22 (10), 1083–1098.

Zhao, S., Lü, B., Li, R.W., Zhu, A. and Wu, C.W. (2016) A preliminary analysis of fishery resource exhaustion in the context of biodiversity decline. *Science China Earth Sciences* 59 (2), 223–235.

Zimmerman, E. (1951) *World Resources and Industries: A Functional Appraisal of the Availability of Agricultural and Industrial Resources* (2nd edn) New York: Harper.

Zolfani, S.H., Sedaghat, M., Maknoon, R. and Zavadskas, E.K. (2015) Sustainable tourism: A comprehensive literature review on frameworks and applications. *Economic Research – Ekonomska Istraživanja* 28 (1), 1–30.

Zolli, A. (2012) Learning to bounce back. *New York Times*, 2 November, See https://www.nytimes.com/2012/11/03/opinion/forget-sustainability-its-about-resilience.html.

Zorić, J. and Hrovatin, N. (2012) Household willingness to pay for green electricity in Slovenia. *Energy Policy* 47, 180–187.

Zouganeli, S., Trihas, N., Antonaki, M. and Kladou, S. (2012) Aspects of sustainability in the destination branding process: A bottom-up approach. *Journal of Hospitality Marketing & Management* 21 (7), 739–757.

INDEX

Note: References in *italics* are to figures, those in **bold** to tables, 'B' refers to Boxes.